Lecture Notes in Computer Science 2630

Edited by G. Goos, J. Hartmanis, and J. van Leeuwen

Springer
Berlin
Heidelberg
New York
Hong Kong
London
Milan
Paris
Tokyo

Franz Winkler   Ulrich Langer (Eds.)

Symbolic and Numerical Scientific Computation

Second International Conference, SNSC 2001
Hagenberg, Austria, September 12-14, 2001
Revised Papers

Springer

Series Editors

Gerhard Goos, Karlsruhe University, Germany
Juris Hartmanis, Cornell University, NY, USA
Jan van Leeuwen, Utrecht University, The Netherlands

Volume Editors

Franz Winkler
Johannes Kepler University Linz
RISC-Linz
Altenberger Str. 69, 4040 Linz, Austria
E-mail: Franz.Winkler@risc.uni-linz.ac.at

Ulrich Langer
Johannes Kepler University Linz
Institute of Computational Mathematics
Altenberger Str. 69, 4040 Linz, Austria
E-mail: ulanger@numa.uni-linz.ac.at

Cataloging-in-Publication Data applied for

A catalog record for this book is available from the Library of Congress.

Bibliographic information published by Die Deutsche Bibliothek
Die Deutsche Bibliothek lists this publication in the Deutsche Nationalbibliografie;
detailed bibliographic data is available in the Internet at <http://dnb.ddb.de>.

CR Subject Classification (1998): G.1, I.1, J.2, F.2, G.2, I.3.5

ISSN 0302-9743
ISBN 3-540-40554-2 Springer-Verlag Berlin Heidelberg New York

Springer-Verlag Berlin Heidelberg New York
a member of BertelsmannSpringer Science+Business Media GmbH

http://www.springer.de

© Springer-Verlag Berlin Heidelberg 2003
Printed in Germany

Typesetting: Camera-ready by author, data conversion by Steingräber Satztechnik GmbH, Heidelberg
Printed on acid-free paper   SPIN: 10873065   06/3142   5 4 3 2 1 0

Preface

Scientific computation has a long history, dating back to the astronomical tables of the Babylonians, the geometrical achievements of the Egyptians, the calendars of the Mayans, and the number theory of the ancient Chinese. The success of these activities led in many parts of the world to the development of mathematical theories trying to explain the surprising applicability of pure thought to physical phenomena. Probably the most important such development took place in ancient Greece, where mathematics, as we know it today, was created.

Just as practical advances in scientific computation have stimulated the development of theory, so also new theoretical breakthroughs have led to whole new ways of mathematizing science. A prominent example is the development of calculus by Newton and Leibnitz and its application in all areas of science. Mathematical theory, however, became more and more abstract and removed from possible applications.

The twentieth century witnessed the advent of computing devices in the modern sense. Right from the beginning these machines were used for facilitating scientific computations. The early achievements of space exploration in the 1950s and 1960s would have been impossible without computers and scientific computation.

Despite this success, it also became clear that different branches of scientific computation were taking shape and developing apart from each other. Numerical computation saw itself mainly as a computational offspring of calculus, with approximations, convergence, and error analysis forming its core. On the other hand, computer algebra or symbolic computation derived its inspiration from algebra, with its emphasis on solving equations in closed form, i.e., with exact-solution formulae. Both of these branches were very successful in their own right, but barely interacted with each other.

However, this split of scientific computation into numerical and symbolic branches was felt more and more to be a problem. Numerical methods in solving differential equations might make use of an analysis of integrability conditions or symmetries, which can be provided by symbolic methods. Symbolic methods, on the other hand, rely on exact inputs, which are often hard to come by. So they need to be extended to cover certain neighborhoods of their input data. Problems such as these require a well-balanced integration of symbolic and numerical techniques.

In 1998 the Austrian Science Fund (Fonds zur Förderung der wissenschaftlichen Forschung) established the Special Research Program (Spezialforschungsbereich, SFB) "Numerical and Symbolic Scientific Computing" at the Johannes Kepler Universität, Linz, Austria. The speaker of this SFB is Ulrich Langer, and the current co-speaker is Franz Winkler. In this SFB various research groups with backgrounds in numerical analysis and in symbolic computation join forces to bridge the gap between these two approaches to scientific computation. For more detail, please check out the Web page of the SFB: `http://www.sfb013.uni-linz.ac.at`. Besides bringing local researchers together to work on numerical and symbolic problems, this SFB also sees itself as an international forum. Over the years we have organized several conferences and workshops on scientific computation. In particular, in August 1999 the workshop "Symbolic and Numerical Scientific Computation (SNSC '99)" took place in Hagenberg near Linz, the home of the Research Institute for Symbolic Computation (RISC-Linz). SNSC '99 was quite a success. So in September 2001 we held the second such conference, "Symbolic and Numerical Scientific Computation (SNSC '01)", again in Hagenberg.

As for SNSC '99, Franz Winkler served as the organizer and general chair of the conference. SNSC '01 was designed as an open conference of the SFB. The goal was to further the integration of methods in symbolic and numerical scientific computation. Topics of interest for the conference included all aspects of symbolic-numerical scientific computation, in particular: construction of algebraic multigrid methods for solving large-scale finite element equations, computer-aided geometric design, computer-supported derivation of mathematical knowledge, constraint solving, regularization techniques, combinatorics and special functions, symmetry analysis of differential equations, differential geometry, visualization in scientific computation, and symbolic and numerical methods in engineering sciences. The following invited speakers addressed these topics in their presentations:

J. Apel, "Passive Complete Orthonomic Systems of PDEs and Riquier Bases of Polynomial Modules,"
M. Dellnitz, "Set-Oriented Numerical Methods for Dynamical Systems,"
E. Hubert, "Differential Equations from an Algebraic Standpoint,"
F. Schwarz, "Algorithmic Lie Theory for Solving Ordinary Differential Equations,"
H.J. Stetter, "Algebraic Predicates for Empirical Data,"
R. Walentyński, "Solving Symbolic and Numerical Problems in the Theory of Shells with *MATHEMATICA*."

In addition to these invited presentations, 25 contributed talks were given at the conference.

This book contains the proceedings of the conference SNSC '01. The participants were encouraged to submit revised and elaborated papers on their presentations, and many of them did so. These submissions were then carefully refereed by at least two independent referees each, under the supervision of the editors.

The papers in this proceedings volume fall into three categories:

I. Symbolics and Numerics of Differential Equations,
II. Symbolics and Numerics in Algebra and Geometry,
III. Applications in Physics and Engineering.

In the following we briefly describe the various contributions and their interrelations.

I. Symbolics and Numerics of Differential Equations:

Several invited presentations focused on algebraic methods for the investigation of differential systems. The first paper by E. Hubert, "Notes on Triangular Sets and Triangulation-Decomposition Algorithms I: Polynomial Systems," reviews the constructive algebraic theory of the solution of systems of polynomial equations. The core of any such solution method is the elimination of some variables from some of the given equations. In the optimal case we arrive at a triangular system of equations, which can then be solved "one variable at a time." In her second paper, "Notes on Triangular Sets and Triangulation-Decomposition Algorithms II: Differential Systems," E. Hubert describes the extension of the polynomial elimination theory to differential algebra. Differential polynomials are associated with differential equations, and radical differential ideals with differential systems. Hubert investigates questions about the solution sets of differential systems by viewing them in terms of the corresponding radical differential ideals. In his paper "Passive Complete Orthonomic Systems of PDEs and Riquier Bases of Polynomial Modules," J. Apel relates passive complete orthonomic systems of PDEs and Gröbner bases of finitely generated modules, which can be seen as Riquier bases or so-called involutive bases of these differential systems. This correspondence opens the way for an algebraic analysis of differential systems. Lie's symmetry theory allows us to classify differential systems, determine canonical forms of such systems, derive qualitative properties and in certain cases also solve such systems. F. Schwarz investigates "Symmetries of Second- and Third-Order Ordinary Differential Equations" in his contribution.

In the contributed papers in this category K. Schlacher et al. deal with "Symbolic Methods for the Equivalence Problem of Systems of Implicit Ordinary Differential Equations." The goal in this investigation is to identify solutions of an original set of differential equations as solutions of a given normal form. M. Hausdorf and W.M. Seiler are able to generalize ideas from differential algebraic equations to partial differential equations in their paper "On the Numerical Analysis of Overdetermined Linear Partial Differential Systems." Involutive systems provide important normal forms for differential systems. Some of their algebraic and computational aspects are investigated by R. Hemmecke in "Dynamical Aspects of Involutive Bases Computations."

II. Symbolics and Numerics in Algebra and Geometry:

In their invited paper "Congestion and Almost Invariant Sets in Dynamical Systems," M. Dellnitz and R. Preis propose a new combination of numerical and graph theoretic tools for both the identification of the number and the location of almost invariant sets of a dynamical system which are subsets of the state space where typical trajectories stay for a long period of time before entering other parts of the state space. The algorithms derived find interesting applications in the analysis of the dynamics of molecules.

The contributed papers in this category cover a wide range of different symbolic and numeric techniques and also interactions between these techniques. G. Brown proposes certain combinatorial graphs, so-called datagraphs, as a means for classifying and storing information on algebraic varieties. V. Marotta elaborates Stetter's approach of neighborhoods, i.e., inexactly given symbolic objects, in her paper on "Resultants and Neighborhoods of a Polynomial." The issue here is that in practical applications polynomials are given with inexact coefficients. Given such an inexact object, one tries to identify a neighboring object exhibiting a certain structure. "Multi-variate Polynomials and Newton-Puiseux Expansions" are investigated by F. Beringer and F. Richard-Jung. This provides a generalization of local resolution from the classical curve case to the case of hypersurfaces. P. Paule et al. use algebraic methods for constructing wavelets that satisfy alternatives to the vanishing moments conditions in their paper "Wavelets with Scale-Dependent Properties." N. Beaudoin and S.S. Beauchemin propose a numerical Fourier transform in d dimensions that provides an accurate approximation of the corresponding continuous Fourier transform with similar time complexity as the discrete Fourier transform usually realized by Fast Fourier Transform (FFT) algorithms. The paper by G. Bodnár et al. deals with the method of exact real computations that seems to provide a reasonable and effective alternative to the usual symbolic techniques of handling exact computations. In particular, the authors consider some real linear algebra and real polynomial problems such as inversion of nonsingular matrices, computation of pseudoinverses, greatest common divisor computations, and roots computations. Some of these problems are ill-posed and require regularization techniques like the Tikhonov regularization technique. U. Langer et al. develop a new symbolic method for constructing a symmetric, positive definite M-matrix that is as close as possible to some given symmetric, positive definite matrix in the spectral sense. This technique is then directly used for deriving efficient algebraic multigrid preconditioners for large-scale finite element stiffness matrices.

III. Applications in Physics and Engineering:

Applications of both symbolic and numerical techniques in the theory of shells, i.e., thin layers of physical bodies, are exhibited in the invited paper "Solving Symbolic and Numerical Problems in the Theory of Shells with *MATHEMATICA*" by R.A. Walentyński. The author demonstrates the applicability of a typical computer algebra system, in this case *MATHEMATICA*, to mechanical problems.

Several contributed papers describe hybrid symbolic-numerical techniques for practical problems. A.R. Baghai-Wadji makes a conjecture about the diagonalizability of linearized PDEs as models of physically realizable systems and gives a variety of examples of his conjecture in "A Symbolic Procedure for the Diagonalization of Linear PDEs in Accelerated Computational Engineering." The computer algebra system *Maple* is used by V.L. Kalashnikov in his paper "Generation of the Quasi-solitons in the Lasers: Computer Algebra Approach to an Analysis." A. Shermenev investigates the two-dimensional shallow water equation describing the propagation of surface gravity waves in polar coordinates. He presents some solutions to this equation in the form of analytic representations.

The organizers express their thanks to the Austrian Science Fund (FWF), the University of Linz, the Government of Upper Austria, and the City of Linz for moral and financial support. The organizers would like to thank the institute RISC-Linz for its hospitality during the conference. Last, but not least, we thank Rahul Athale for his work in collecting all these contributions and preparing them for printing.

Franz Winkler and Ulrich Langer
Johannes Kepler Universität
Linz, Austria

Table of Contents

Symbolics and Numerics of Differential Equations

Symbolics and Numerics in Algebra and Geometry

Applications in Physics and Engineering

Notes on Triangular Sets
and Triangulation-Decomposition Algorithms
I: Polynomial Systems

Evelyne Hubert

INRIA - Projet CAFE
2004 route des lucioles BP 93
F-06902 Sophia Antipolis
Evelyne.Hubert@inria.fr

Abstract. This is the first in a series of two tutorial articles devoted to triangulation-decomposition algorithms. The value of these notes resides in the uniform presentation of triangulation-decomposition of polynomial and differential radical ideals with detailed proofs of all the presented results. We emphasize the study of the mathematical objects manipulated by the algorithms and show their properties independently of those. We also detail a selection of algorithms, one for each task. We address here polynomial systems and some of the material we develop here will be used in the second part, devoted to differential systems.

1 Introduction

The natural representation of a tower of separable field extensions is a triangular set having some irreducibility properties. Those special triangular sets, called characteristic sets, therefore provide a natural representation for prime ideals in polynomial rings. That idea was in fact first fully developed by J.F.Ritt [Rit32, Rit50] in the context of differential algebra. Ritt gave an algorithm to represent the radical ideal generated by a finite set of polynomials as an intersection of prime ideals defined by their characteristic sets. For want of effectiveness and realistic implementations the irreducibility requirements were to disappear both in the representation of field extensions and of radical ideals.

What we call a triangulation-decomposition algorithm is an algorithm that computes a representation of the radical ideal generated by a finite set of polynomials as an intersection of ideals defined by triangular sets. Many triangulation-decomposition algorithms in fact provide a zero decomposition: the variety defined by a set of polynomials is decomposed into quasi-varieties. To be general, the quasi-varieties are described by a triangular system of equations and a set of inequations. Application or algorithm requirements lead to output systems and decomposition with more or less specific properties, [Wan99] presents quite a few options. Bridges between different approaches have been proposed in [ALMM99, AMM99, Del00a, Del01].

F. Winkler and U. Langer (Eds.): SNSC 2001, LNCS 2630, pp. 1–39, 2003.

Zero decomposition algorithms were investigated and developed for application to geometry theorem proving starting with the work of Wu in the late seventies [Wu78]. Books devoted to the field include [Wu94, Cho88, CGZ94, Wu00]. As for field representation, the D5 system [DDD85, Duv87] was designed to work with algebraic numbers without factorization. Its extension to the dynamical closure system D7 [GD94, Del99, Del00b] is intended to work with towers of extensions. A side product of the D7 system is a zero decomposition algorithm. Last but not least, triangular forms of systems of polynomials are very amenable to resolution. Triangulation-decomposition of polynomial systems are therefore naturally applicable to solving polynomial systems with finite number of solutions [Laz92, MMR95] and parametric systems [GC92, Sch00]. In the case of polynomial system with a variety of positive dimension the decomposition computed is *strongly* unmixed dimensional [Laz91, MM97, Sza98, Aub99]. It therefore gives an excellent description of the variety and can be relatively easily refined into an irreducible decomposition. Solving polynomial system with triangulation-decomposition is particularly adequate when the solution set has relevant components of different dimensions, as is the case of the classification problem solved in [FMM01]. Triangulation-decomposition has also proved adequate for solving positive dimensional systems over the reals [ARSED03]. Also, elementary-algebraic systems [Ric99, BJ01] can be attacked with this technique.

The first focus of this paper is a thorough study of ideals defined by triangular sets. Many of the presented results are either assumed or disseminated in the literature. We will give full proofs to all those results. The second focus will be on Kalkbrener's algorithm. That algorithm was first presented in [Kal93] in a very synthetic and highly recursive way. Our presentation owes to the presentation of [Aub99].

The basic idea of working with triangular sets is to consider multivariate polynomials as univariate polynomial in a distinguished variable, the *leader*. We want to treat systems of polynomials recursively as univariate polynomials. The *reduction* of one polynomial by another will be pseudo-division.

Triangular sets are sets of polynomials with distinct leaders. We look at the ideals they define outside of some singularity sets, given by the *initials* of the polynomials, i.e. the leading coefficients of the polynomials when considered as univariate polynomial in the leader. These ideals have excellent properties: they are unmixed dimensional and the non leading variables give a transcendence basis for all associated primes (see Section 4). Thanks to that structure, we can associate a product of fields to a triangular set in a natural way.

Some specific triangular sets, called *regular chains* (Section 5), are amenable to computations. This is because the cutback of the ideal defined is given by the cutback of the regular chain. Computations can thus be lead recursively in a univariate way. Regular chains give the ideal they define a membership test, as well as a zero-divisor test (Section 8).

Of course, not every ideal can be written as the ideal defined by a regular chain. Our final goal is to give an algorithm that writes the radical ideal generated by some finite set of polynomials as an intersection of (radicals of) *characterizable ideals*. Characterizable ideals can be defined intrinsically and are given by regular chains.

For the zero-divisor test or the decomposition into characterizable ideals, there is an interesting notion coming in, the *pseudo-gcd*. This is a generalization of the gcd of univariate polynomial over a field to the case of univariate polynomials over a product of fields. In our case, we consider univariate polynomials with coefficients taken modulo the ideal defined by a triangular set.

The paper is organized as follow. In Section 2 we shall review some definitions and basic properties on polynomial rings. We in particular give a series of results about zero-divisors. We conclude the section with a simple example of a splitting procedure of a product of fields. This serves as an introduction to one of the two main components of Kalkbrener's algorithm. In Section 3 we review pseudo-division and its property and the definition of pseudo-gcd, as in [Aub99] . After the pseudo-gcd definition, we give an algorithm to compute a pseudo-gcd assuming we are given a splitting procedure. That provides an introduction to the second main component of Kalkbrener's algorithm. In Section 4 triangular sets together with the fundamental properties of the ideals they define are detailed. In Section 5, we define regular chain, review the notion of characteristic set of Ritt and give the equivalence of [ALMM99] between the two approaches. For the characterizable ideals they define we exhibit canonical representatives. Section 6 defines characteristic decompositions and makes the link between irredundant decompositions and decomposition of a product fields. Before the algorithmic part of the article, we assemble in Section 7 a number of results about the radical of an ideal defined by a triangular set. These results are of use to construct the radical of a characterizable ideal and in differential algebra [Hub03]. Section 8 gives the two components of Kalkbrener's algorithm, one to split and one to compute the pseudo-gcd. These components are then applied in Section 9 to provide a characteristic decomposition algorithm. A further application will be given in [Hub03] for the algebraic part of a decomposition algorithm in differential algebra.

2 Preliminaries and Notations in Commutative Algebra

We recall here some basic definitions and results in commutative algebra and take the opportunity to set up some notations. The notions we shall expand on are saturation ideal, zero divisors and product of fields. All of these are central in the techniques of triangular sets and in Kalkbrener's algorithm. Other notions can be found for instance in [Eis94].

2.1 Ideals, Saturation and Equidimensionality

In this section and the following, $\mathcal{R}$ is a Noetherian ring. Let H be a subset of $\mathcal{R}$. We denote by H^∞ the minimal subset of $\mathcal{R}$ that contains 1 and H and is stable by multiplication and division i.e. $a, b \in H^\infty \iff ab \in H^\infty$. When H consists of a unique element h we will write h^∞ instead of $\{h\}^\infty$.

Let I be an ideal of $\mathcal{R}$. We define the saturation of I by a subset H of $\mathcal{R}$ as $I:H^\infty = \{q \in \mathcal{R} \mid \exists h \in H^\infty \text{ s.t. } hq \in I\}$. $I \subset I:H^\infty$ and $I:H^\infty$ is an ideal. Consider $H^{-1}\mathcal{R}$ the localization of $\mathcal{R}$ at H^∞. Let $H^{-1}I$ be the extension of I in $H^{-1}\mathcal{R}$. When $0 \notin H^\infty$, $I:H^\infty = H^{-1}I \cap \mathcal{R}$.

If P is a primary ideal of $\mathcal{R}$, $P:H^\infty$ is either equal to P or $\mathcal{R}$ according to whether $H \cap \sqrt{P}$ is empty or not. This has the following consequence.

PROPOSITION 2.1. *Let J be an ideal of $\mathcal{R}$, and H a subset of $\mathcal{R}$. $J:H^\infty$ is the intersection of those primary components of J the radical of which have an empty intersection with H.*

We shall see that the ideals defined by triangular sets that are central in our approach are *unmixed dimensional*. The following definition is taken from [Vas98].

DEFINITION 2.2. *Let $\mathcal{R}$ be a Noetherian ring. An ideal I in $\mathcal{R}$ is equidimensional if all its minimal primes have the same codimension. An ideal I in $\mathcal{R}$ is unmixed dimensional, if all its associated prime have the same codimension.*

An ideal that is unmixed dimensional is equidimensional and has no embedded primes. Therefore, the set of primary components is uniquely determined. Every zero-dimensional ideal is unmixed dimensional.

We note (F) the ideal generated by a non empty subset F of $\mathcal{R}$. For an ideal I of $\mathcal{R}$, we note $\sqrt{I}$ the radical of I that is $\sqrt{I} = \{r \in \mathcal{R} \mid \exists \alpha \in \mathbb{N} \text{ s.t. } r^\alpha \in I\}$, which is a radical ideal. $\sqrt{(F)}$ will be noted $\langle F \rangle$.

2.2   Zero Divisors

In Kalkbrener's algorithm, deciding if an element is a zero divisor modulo the ideal defined by a triangular set is a central figure. We set up some notations that will enable us to describe the algorithms in a more compact way. We review also a number of results that will be useful, in particular Gauss lemma.

We shall see that to a triangular set is naturally associated a product of fields. Computing modulo the ideal defined by a triangular set amounts to compute over this product of fields. We review some of the features of product of fields and their natural construction with the Chinese remainder theorem.

Given a ring $\mathcal{R}$, $\mathfrak{z}\mathfrak{d}(\mathcal{R})$ will denote the set consisting of 0 and the zero divisors of $\mathcal{R}$. The total quotient ring of $\mathcal{R}$, that is the localization of $\mathcal{R}$ at $\mathcal{R} \setminus \mathfrak{z}\mathfrak{d}(\mathcal{R})$ will be written $\mathfrak{Q}(\mathcal{R})$.

Let I be an ideal in $\mathcal{R}$. An element p of $\mathcal{R}$ is said to be

- invertible modulo I if its canonical image in $\mathcal{R}/I$ is a unit.
- a zero divisor modulo I if its canonical image in $\mathcal{R}/I$ is a zero divisor. By extension we write $p \in \mathfrak{z}\mathfrak{d}(\mathcal{R}/I)$ or $p \in \mathfrak{z}\mathfrak{d}(I)$ for short.

Note that an element of $\mathcal{R}$ is invertible modulo an ideal I iff it is invertible modulo the radical, $\sqrt{I}$. An element p of $\mathcal{R}$ is a zero divisor modulo an ideal I

iff p belongs to an associated prime of I. Thus, when I a unmixed dimensional $3\mathfrak{d}(I) = 3\mathfrak{d}(\sqrt{I})$.

The following theorem is Gauss lemma [Eis94, Exercise 3.4]. It gives a practical test for a univariate polynomial to be a zero divisor. A weak version of this lemma is traditionally used in the literature on triangular sets: the test is made on the leading coefficient only. We will use it in its generality.

THEOREM 2.3. *Consider a polynomial* $f \in \mathcal{R}[x]$ *and note* C_f *the ideal in* $\mathcal{R}$ *generated by the coefficients of* f *(the* content *of* f*). Then* $f \in 3\mathfrak{d}(\mathcal{R}[x])$ $\Leftrightarrow$ $C_f \subset 3\mathfrak{d}(\mathcal{R})$.

2.3 Product of Fields

The Chinese Remainder Theorem [Eis94, Exercise 2.6] is a major tool in constructing product of rings - and thus of fields.

THEOREM 2.4. *Let* $Q_1, \ldots, Q_r$ *be ideals in* $\mathcal{R}$ *such that* $Q_i + Q_j = \mathcal{R}$ *for all* $i \neq j$. *Then*

$$\mathcal{R} / \left(\bigcap_{i=1}^{r} Q_i \right) \cong \prod_{i=1}^{r} \mathcal{R}/Q_i.$$

When the Q_i are maximal ideals in $\mathcal{R}$, $\mathcal{R}/ (\bigcap_{i=1}^{r} Q_i)$ is isomorphic to a product of fields. In particular, the quotient of $\mathcal{R}$ by a zero dimensional radical ideal is isomorphic to a product of fields.

LEMMA 2.5. *Let* $\mathcal{R}$ *be a ring isomorphic to a product of fields. The total quotient ring of* $\mathcal{R}$, $\mathfrak{Q}(\mathcal{R})$, *is equal to* $\mathcal{R}$.

This is easily seen as a nonzero element of $\mathcal{R}$ that is not a zero divisor is a unit.

2.4 A Splitting Procedure

We shall pause here to introduce on a simple case the concept of splitting that comes into Kalkbrener's algorithm. Let $\mathcal{K}$ be a field and c a square-free univariate polynomial over $\mathcal{K}$, say in $\mathcal{K}[\lambda]$. Assume $c = \prod_{i=1}^{r} c_i$ is a decomposition into irreducible factors. By the Chinese remainder theorem $\mathcal{R} = \mathcal{K}[\lambda]/(c)$ is isomorphic to the product of the fields $\mathcal{K}_i = \mathcal{K}[\lambda]/(c_i)$.

To work over $\mathcal{R}$, i.e. modulo the algebraic condition on λ, one can factor c and work over each $\mathcal{K}_i$. Beside factorization, this implies repeating the same computation for several components. Instead, at each step of a computation, we can group the components where the same computation are done. We only need to distinguish the components on which an element is a unit from the components on which it is zero.

Given an element p in $\mathcal{K}[\lambda]$ it is possible to *split* $\mathcal{R}$ into two products of fields, $\mathcal{R}_0$ and $\mathcal{R}_1$, in such a way that the natural projection of p on $\mathcal{R}_0$ is zero while the projection on $\mathcal{R}_1$ is a unit. If $g_0 = \gcd(p, c)$ and $g_1 = \frac{c}{g_0}$ then we can just take $\mathcal{R}_0 = \mathcal{K}[\lambda]/(g_0)$ and $\mathcal{R}_1 = \mathcal{K}[\lambda]/(g_1)$. If $g_0, g_1 \notin \mathcal{K}$ then $\mathcal{R} \cong \mathcal{R}_0 \times \mathcal{R}_1$.

When c is not square-free we consider $\mathcal{R} = \mathcal{K}[\lambda]/\langle c \rangle$. We can take $\mathcal{R}_0 = \mathcal{K}[\lambda]/\langle g_0 \rangle$, where $g_0 = \gcd(p, c)$. But now p and $g_1 = \frac{c}{g_0}$ may have common factors, in which case p is not a unit in $\mathcal{K}[\lambda]/\langle g_1 \rangle$. Consider the sequence starting with g_1 s.t. $g_{k+1} = \frac{g_k}{\gcd(p, g_k)}$. For some k, $g_{k+1} \in \mathcal{K}$. We can then take $\mathcal{R}_1 = \mathcal{K}[\lambda]/\langle g_k \rangle$ so that $\mathcal{R} = \mathcal{R}_0 \times \mathcal{R}_1$.

3 Univariate Polynomials over a Ring

The basic reduction step in the triangular set techniques is pseudo division. We review briefly the properties of such an operation. We then introduce the notion of pseudo-gcd that was defined in [Aub99] to put a mathematical frame on Kalkbrener's algorithm. The pseudo-gcd (or pgcd) of a non empty set of univariate polynomials over a ring is the generalization of gcd in a principal ideal ring, i.e. the generator of the ideal. It is well defined over a product of fields. We shall show how to compute a pseudo-gcd over a product of fields when we have a splitting algorithm. This is a fairly simple extension of Euclid's algorithm. Kalkbrener's algorithm evolves on this basic idea.

3.1 Pseudo Division

Let $\mathcal{R}[x]$ be a ring of univariate polynomials with coefficients in the ring $\mathcal{R}$. For a polynomial $p \in \mathcal{R}[x]$ we will note

- $\deg(p, x)$ the degree of p in x
- $\mathsf{lcoeff}(p, x)$ the coefficient of $x^{\deg(p,x)}$ in p (the leading coefficient).
- $\mathsf{tail}(p, x) = p - \mathsf{lcoeff}(p, x)\, x^{\deg(p,x)}$.

Let $p, q \in \mathcal{R}[x]$, $q \neq 0$, $d = \deg(p, x)$, $e = deg(q, x)$ and $c = \mathsf{lcoeff}(q, x)$. The pseudo-remainder of p w.r.t. q is defined as the unique polynomial $\bar{p}$ such that $\deg(\bar{p}, x) < \deg(q, x)$ and $c^{d-e+1} p \equiv \bar{p} \mod (q)$ when $d \geq e$, p otherwise [GCL92, vzGG99]. A sparse pseudo-remainder is usually defined by taking a power of c as small as possible. One can be slightly more general and define a sparse pseudo-remainder of p with respect to q to be a polynomial r of degree in x strictly lower than that of p for which there exists $h \in c^\infty$ s.t. $h\,p \equiv r \mod (q)$. An equality $h\,p = aq + r$ where $p, q, a, r \in \mathcal{R}[x]$, $\deg(r, x) < \deg(q, x)$ and $h \in \mathsf{lcoeff}(q, x)^\infty$ is called a pseudo-division relationship (of p by q). We will write $r = \mathsf{srem}(p, q, x)$ and $a = \mathsf{squo}(p, q, x)$, though these two quantities depend on the algorithm used to compute them. We nonetheless have a kind of uniqueness property.

PROPOSITION 3.1. Let $f, g \in \mathcal{R}[x]$ such that $\mathsf{lcoeff}(g, x) \notin 3\mathfrak{d}(\mathcal{R})$. Assume $h\,f = q\,g + r$ and $h'\,f = q'\,g + r'$ are two pseudo division relationships. Let $\bar{h}, \bar{h}' \in \mathsf{lcoeff}(g, x)^\infty$ be chosen so that $\bar{h}\,h = \bar{h}'\,h'$. Then $\bar{h}\,q = \bar{h}'\,q'$ and $\bar{h}\,r = \bar{h}'\,r'$

Proof. Indeed $\bar{h}\,h\,f = \bar{h}\,(q\,g + r)$ and $\bar{h}'\,h'\,f = \bar{h}'\,(q'\,g + r')$. Therefore $(\bar{h}\,q - \bar{h}'\,q)\,g = \bar{h}'\,r' - \bar{h}\,r$. The right hand side is of degree strictly less than $\deg(g,x)$. Since $\mathsf{lcoeff}(g,x)$ is not a zero divisor, it follows that $(\bar{h}\,q - \bar{h}'\,q)$ is zero and so is $(\bar{h}'\,r' - \bar{h}\,r)$.

It follows that the pseudo remainder is zero iff all the sparse pseudo remainders are zero. The following property is shown in a similar way.

PROPOSITION 3.2. *Let $f, g \in \mathcal{R}[x]$ be such that $\mathsf{lcoeff}(g,x) \notin \mathfrak{Z}\mathfrak{d}(\mathcal{R})$. The polynomial g divides f in $\mathfrak{Q}(\mathcal{R})[x]$ iff $\mathsf{srem}(f,g,x) = 0$.*

3.2 Pseudo-gcd over a Product of Fields

The notion of pseudo-gcd is present in [Laz91, Kal93, MMR95] We reproduce here the definition given in [Aub99].

DEFINITION 3.3. *Let $\mathcal{R}$ be a ring isomorphic to a product of fields and F a non-empty subset of $\mathcal{R}[x]$. The pseudo-gcd of F over $\mathcal{R}$ is a set of pairs $\{(\mathcal{R}_1, g_1), \ldots, (\mathcal{R}_r, g_r)\}$ such that*

- $(F) = (g_i)$ in $\mathcal{R}_i[x]$, $1 \leq i \leq r$.
- $\mathcal{R}_1, \ldots, \mathcal{R}_r$ are isomorphic to products of fields and $\mathcal{R} \cong \mathcal{R}_1 \times \ldots \times \mathcal{R}_r$.

The existence of a pseudo-gcd is secured by the existence of a gcd over each field that is a component of $\mathcal{R}$.

We shall expand here on the generalization of Euclid's algorithm over a product of fields as it is a good introduction to Kalkbrener's algorithm. Assume that we have for a product of fields $\mathcal{R}$ the procedures

- is-zero that decides if an element of $\mathcal{R}$ is zero
- is-zerodivisor that decides if an element of $\mathcal{R}$ is a zero divisor
- split that splits $\mathcal{R}$ according to a zerodivisor p into $\mathcal{R}_0$ and $\mathcal{R}_1$ s.t. $\mathcal{R} \cong \mathcal{R}_0 \times \mathcal{R}_1$ and $p = 0$ in $\mathcal{R}_0$ while p is a non zero divisor in $\mathcal{R}_1$.

In Section 2.4, we saw for $\mathcal{R} = \mathcal{K}[\lambda]/\langle c \rangle$, c a polynomial in $\mathcal{K}[\lambda]$, how to write these procedures. One can rework Euclid's algorithm to compute the pseudo-gcd of two polynomials in $\mathcal{R}[x]$. It will work with the following property:

PROPOSITION 3.4. *Let $f, g \in \mathcal{R}[x]$ be such that $\mathsf{lcoeff}(g,x) \notin \mathfrak{Z}\mathfrak{d}(\mathcal{R})$. The ideals (f,g) and $(\mathsf{srem}(f,g,x), g)$ are equal when considered as ideals in $\mathfrak{Q}(\mathcal{R})[x]$.*

Recall that $\mathfrak{Q}(\mathcal{R}) = \mathcal{R}$ when $\mathcal{R}$ is a product of fields and so $(f,g) = (\mathsf{srem}(f,g,x), g)$ in $\mathcal{R}[x]$. In the general pgcd algorithm of Section 8.1 we will nonetheless have a slightly stronger result that needs this statement.

The following algorithm computes the pgcd of two univariate polynomials over $\mathcal{R}$. It is described in MAPLE style. As we shall often do in algorithm description, $\mathcal{S}$ is a set of tuples that await more computations while $\mathcal{G}$ contains data for which the computation is over. The command pop returns one element of a set and removes it from that set. Tuples are noted with parenthesis to fit mathematical notation rather than the MAPLE list construction with brackets.

ALGORITHM 3.5. *pgcd*
INPUT:

- $\mathcal{R}$ a ring that is isomorphic to a product of fields.
- p and q polynomials in $\mathcal{R}[x]$.

OUTPUT: A pgcd $\{(\mathcal{R}_1, g_1), \ldots, (\mathcal{R}_r, g_r)\}$ of p and q over $\mathcal{R}$.

   $\mathcal{S} := \{(\mathcal{R}, p, q)\}$;
   $\mathcal{G} := \emptyset$;
   while $\mathcal{S} \neq \emptyset$ do
      $(\mathcal{Q}, a, b) := pop\ (\mathcal{S})$;
      if $b = 0$ then
         $\mathcal{G} := \mathcal{G} \cup \{(\mathcal{Q}, a)\}$;
      else
         $c := \mathsf{lcoeff}(b, x)$;
         if *is-zero*$(\mathcal{Q}, c)$ then
            $\mathcal{S} := \mathcal{S} \cup \{(\mathcal{Q}, a, \mathsf{tail}(b, x))\}$;
         elif *is-zerodivisor*$(\mathcal{Q}, c)$ then
            $\mathcal{Q}_0, \mathcal{Q}_1 := \mathsf{split}\ (\mathcal{Q}, c)$;
            $\mathcal{S} := \mathcal{S} \cup \{(\mathcal{Q}_0, a, \mathsf{tail}(b, x)), (\mathcal{Q}_1, b, \mathsf{srem}(a, b, x))\}$;
         else
            $\mathcal{S} := \mathcal{S} \cup \{(\mathcal{Q}, b, \mathsf{srem}(a, b, x))\}$;
      fi;
   $return(G)$;

Together with what was presented in Section 2.4 we can now compute the pgcd of two polynomials in $(\mathcal{K}[\lambda]/\langle c \rangle)[x]$. The procedure can be made more efficient by using subresultant sequences [AK93, Del00c].

4 Multivariate Polynomials with a Recursive Univariate Viewpoint

We shall see multivariate polynomials as univariate polynomials in a distinguished variable. The coefficients are multivariate polynomials. Triangular sets are the basic objects of our approach. They define ideals with specific structure to which we associate products of fields.

4.1 Vocabulary for Multivariate Polynomials

Consider the polynomial ring $\mathcal{K}[X]$ where X is a set of ordered variables. The order on the indeterminates is called a *ranking* in the line of the terminology introduced by Ritt. A side effect is to avoid possible confusion with the term ordering used in Gröbner bases theory. For $x \in X$ we write $X_{>x}$, and respectively $X_{<x}$, the set of variables greater, and respectively lower, than x in X. We denote also $X_{\leq x} = X_{<x} \cup \{x\}$. When we write $\mathcal{K}[X][y]$ for $\mathcal{K}[X \cup \{y\}]$ it will be understood that $\forall x \in X$, $x < y$.

Let p be a polynomial in $\mathcal{K}[X] \setminus \mathcal{K}$. The *leader* and the *initial* of p are respectively the highest ranking variable appearing in p and the coefficient of its highest power in p. They will be noted $\mathrm{lead}(p)$ and $\mathrm{init}(p)$. If d is the degree of p in its leader, the rank of p is the term $\mathrm{rank}(p) = \mathrm{lead}(p)^d$.

The ranking on X induces a pre-order[1] on the polynomials of $\mathcal{K}[X]$. An element $q \in \mathcal{K}[X] \setminus \mathcal{K}$ is said to have higher rank (or to *rank higher*) than p when its leader, $\mathrm{lead}(q)$, has higher rank than $\mathrm{lead}(p)$ or when $\mathrm{lead}(p) = \mathrm{lead}(q)$ and the degree in this common leader is bigger in q than in p. In that case we write $\mathrm{rank}(q) > \mathrm{rank}(p)$.

A polynomial q is *reduced w.r.t.* p if the degree of q in $\mathrm{lead}(p)$ is strictly less than the degree of p in $\mathrm{lead}(p)$.

4.2 Triangular Sets

From now on we consider a polynomial ring $\mathcal{K}[X]$ endowed with a ranking.

DEFINITION 4.1. *A subset of $\mathcal{K}[X]$ is a triangular set if it has no element in $\mathcal{K}$ and the leaders of its elements are pairwise different.*

A triangular set cannot have more elements than X but can be the empty set. Let $a_1, a_2, \ldots, a_r$ be the elements of a triangular set A such that $\mathrm{lead}(a_1) < \mathrm{lead}(a_2) < \ldots < \mathrm{lead}(a_r)$. We shall write $A = a_1 \triangle a_2 \triangle \ldots \triangle a_r$. We will also use the small triangle $\triangle$ to construct triangular sets by mixing polynomials a and triangular sets A, B in the following ways:

- $a \triangle A$ denotes the triangular set $A \cup \{a\}$ if a is such that $\mathrm{lead}(a)$ ranks lower than the leader of any element of A
- $A \triangle a$ denotes the triangular set $A \cup \{a\}$ if a is such that $\mathrm{lead}(a)$ ranks higher than the leader of any element of A
- $A \triangle B$ denotes the triangular set $A \cup B$ if the leader of any element of A ranks less than the leader of any element of B.

For a triangular set A in $\mathcal{K}[X]$ we note $\mathcal{L}(A)$ and I_A the sets of the leaders and of the initials of the elements of A. We will also note $\mathfrak{T}(A) = X \setminus \mathcal{L}(A)$ the set of the non leading variables of A. For $x \in X$ we define A_x to be the element of A with leader x if there is any, 0 otherwise. $A_{<x} = A \cap \mathcal{K}[X_{<x}]$, $A_{\leq x} = A \cap \mathcal{K}[X_{\leq x}]$ and $A_{>x} = A \setminus (A_{\leq x})$. $A_{<x}$ can be considered as a triangular set of $\mathcal{K}[X]$ or as a triangular set of $\mathcal{K}[X_{<x}]$.

A polynomial p is said to be reduced with respect to a triangular set A if for all x in $\mathcal{L}(A)$ we have $\deg(p, x) < \deg(A_x, x)$. The pseudo division algorithm can be extended to compute the *reduction* of a polynomial in $\mathcal{K}[X]$ with respect to A.

LEMMA 4.2. *Let A be a non empty triangular set and p a polynomial in $\mathcal{K}[X]$. There exists $h \in I_A^\infty$ and $r \in \mathcal{K}[X]$ reduced w.r.t. A such that $hp \equiv r \mod (A)$. We will write $r = \mathrm{red}(p, A)$.*

---

[1] We take the convention that a pre-order is a relation that is reflexive, transitive and connex [BW93]. Therefore the difference with an order is that $a \leq b$ and $b \leq a$ does not imply $a = b$.

The pair (h, r) is not unique but in our use, any such pair will do. If $A = a_1 \triangle \ldots \triangle a_m$ we can take

$$\mathrm{red}(p, A) = \mathrm{srem}(\ldots \mathrm{srem}(\mathrm{srem}(p, a_m, u_m), a_{m-1}, u_{m-1}), \ldots, a_1, u_1) \,,$$

where $u_i = \mathrm{lead}(a_i)$.

Note that if $\mathrm{red}(p, A) = 0$ then $p \in (A) : I_A^\infty$. The best we can expect to represent with a non empty triangular set A of $\mathcal{K}[X]$ is the ideal $(A) : I_A^\infty$. For readability in some equation we note $\mathcal{I}(A) = (A) : I_A^\infty$ and $\mathcal{R}(A) = \sqrt{\mathcal{I}(A)} = \langle A \rangle : I_A^\infty$. For the empty set we take as a convention that $\mathcal{I}(\emptyset) = \mathcal{R}(\emptyset) = (0)$. A triangular set is said to be *consistent* if $1 \notin (A) : I_A^\infty$.

At some points we shall extend the coefficient field to the rational function field in the non leading variables $\mathfrak{T}(A)$ of A. We will write $\mathcal{K}_A = \mathcal{K}(\mathfrak{T}(A))$ and consider the ring of polynomials $\mathcal{K}_A[\mathfrak{L}(A)]$. If $\mathfrak{L}(A)$ is ordered according to the ranking on $\mathcal{K}[X]$ then A is a triangular set in $\mathcal{K}_A[\mathfrak{L}(A)]$. For a consistent triangular set A we furthermore define $\mathfrak{R}(A)$ the ring $\mathcal{K}(\mathfrak{T}(A))[\mathfrak{L}(A)] \,/\, \mathcal{R}(A)$. We shall see that this ring is isomorphic to a product of fields (Proposition 4.7).

4.3   Ideals Defined by Triangular Sets

In this section we examine the properties of the ideal $\mathcal{I}(A) = (A) : I_A^\infty$ defined by a triangular set A. They are unmixed dimensional and all the associated prime share the same transcendence basis, as stated in Theorem 4.4. There are several references for that theorem or its related results [GC92, Kal93, Sza98, Aub99, Hub00]. We also study the cutback properties of ideals defined by triangular sets. Indeed the basic operations on triangular sets is to extend or cut them shorter.

LEMMA 4.3. *Let A be a triangular set in $\mathcal{K}[X]$. Let $a \in \mathcal{K}[X][x]$ be of strictly positive degree in x and $h = \mathrm{init}(a)$. Then*

$$(\mathcal{I}(A) : h^\infty + (a)) : h^\infty = \mathcal{I}(A \triangle a) \quad \text{and} \quad \mathcal{I}(A \triangle a) \cap \mathcal{K}[X] = \mathcal{I}(A) : h^\infty$$

Proof. We prove the first equality. Since $h \in I_{A \triangle a}$, it follows that $\mathcal{I}(A) : h^\infty \subset \mathcal{I}(A \triangle a)$ and thus $((a) + \mathcal{I}(A) : h^\infty) : h^\infty \subset \mathcal{I}(A \triangle a)$.

Let $p \in \mathcal{I}(A \triangle a)$. There exist $\bar{h} \in h^\infty$, $k \in I_A^\infty$ and $\bar{q} \in \mathcal{K}[X][x]$ such that $\bar{h} k p = \bar{q} a \mod (A)$. Since $\bar{h} k \notin \mathfrak{Z}\mathfrak{d}(\mathcal{I}(A) : h^\infty)$, p is divisible by a in $\mathfrak{Q}(\mathcal{K}[X]/\mathcal{I}(A) : h^\infty)[x]$. By Proposition 3.2 there exists $h' \in h^\infty$ and $q' \in \mathcal{K}[X]$ such that $h' p = q' a \mod \mathcal{I}(A) : h^\infty$. Thus $p \in (\mathcal{I}(A) : h^\infty + (a)) : h^\infty$.

For the second equality assume $p \in \mathcal{I}(A \triangle a) = (\mathcal{I}(A) : h^\infty + (a)) : h^\infty$. There exists $h' \in h^\infty, q \in \mathcal{K}[X][x]$ such that $h' p - q a \in \mathcal{I}(A) : h^\infty$, and therefore all the coefficients of the powers of x in $h' p - q a$ belong to $\mathcal{I}(A) : h^\infty$. Since $h = \mathrm{init}(a)$ is not a zero divisor modulo $\mathcal{I}(A) : h^\infty$, if $p \in \mathcal{K}[X]$ then $q = 0$. Thus $\mathcal{I}(A \triangle a) \cap \mathcal{K}[X] \subset \mathcal{I}(A) : h^\infty$ and the other inclusion is immediate from what precedes.

Intuitively this shows that if $\mathcal{I}(A) : h^\infty \neq (1)$, a is not a zero divisor modulo $\mathcal{I}(A) : h^\infty$ and therefore the dimension of $\mathcal{I}(A \triangle a)$ is one less then the dimension of $\mathcal{I}(A)$. Nonetheless an iterative proof of the dimension properties of the ideal defined by a triangular set is not easy. Conversely, the following fundamental property of the ideal $\mathcal{I}(A)$ allows us to understand better the iterative properties of triangular sets.

THEOREM 4.4. *Let A be a consistent triangular set of $\mathcal{K}[X]$. $\mathcal{I}(A)$ is unmixed dimensional of codimension the cardinal of A. Furthermore $\mathfrak{T}(A)$ forms a transcendence basis for each associated prime of $\mathcal{I}(A)$.*

Proof. Let us note $A = a_1 \vartriangle \ldots \vartriangle a_m$ and let u_i denote the leader of a_i. Each a_i introduces the new variable, u_i. Consider the localization $L_A = I_A^{-1} \mathcal{K}[X]$. We note $I_A^{-1}(A)$ the ideal generated by A in L_A. By the principal ideal theorem [Eis94, Theorem 10.2], $I_A^{-1}(A)$ has at most codimension m. Considered as a univariate polynomial in u_i, a_i has one of its coefficient, its initial, invertible in L_A. By Gauss lemma (Lemma 2.3) a_i can not divide zero modulo $I_A^{-1}(a_1, \ldots, a_{i-1})$. It follows that $a_1, \ldots, a_n$ is a L_A-regular sequence. $I_A^{-1}(A)$ has depth greater or equal to m. As L_A is Cohen-Macaulay $I_A^{-1}(A)$ has codimension m and therefore is unmixed dimensional [Eis94, Proposition 18.14 and 18.9]

As $\mathcal{I}(A) = (I_A^{-1}(A)) \cap \mathcal{K}[X]$, an associated prime of $\mathcal{I}(A)$ is the intersection with $\mathcal{K}[X]$ of an associated prime of $I_A^{-1}(A)$. They all have codimension n and there is no embedded prime [Eis94, Exercise 9.4].

Let P be a minimal prime of $\mathcal{I}(A)$. It contains no initial of the elements of A. Therefore all the elements of $\mathfrak{L}(A)$ are algebraic over $\mathfrak{T}(A)$ modulo P. Considering dimensions, $\mathfrak{T}(A)$ provides a transcendence basis for P.

It follows that $\mathfrak{zo}(\mathcal{I}(A)) = \mathfrak{zo}(\mathcal{R}(A))$ and we will simply write $\mathfrak{zo}(A)$.

PROPOSITION 4.5. *Let A be a consistent triangular set of $\mathcal{K}[X]$ and $x \in X$. Assume P is a minimal prime of $\mathcal{I}(A)$. $P \cap \mathcal{K}[X_{\leq x}]$ is a minimal prime of $\mathcal{I}(A_{\leq x})$.*

Proof. Obviously $\mathcal{I}(A_{\leq x}) \subset \mathcal{I}(A)$ in $\mathcal{K}[X]$. Therefore $\mathcal{I}(A_{\leq x}) \subset \mathcal{I}(A) \cap \mathcal{K}[X_{\leq x}]$ in $\mathcal{K}[X_{\leq x}]$. Let P be a minimal prime of $\mathcal{I}(A)$ in $\mathcal{K}[X]$. $P \cap \mathcal{K}[X_{\leq x}]$ is a prime ideal that contains $\mathcal{I}(A_{\leq x})$. It must contain a minimal prime P' of $\mathcal{I}(A_{\leq x})$. Now $P \cap \mathcal{K}[X_{\leq x}]$ admits $\mathfrak{T}(A) \cap X_{\leq x} = \mathfrak{T}(A_{\leq x})$ as a transcendence basis; It has codimension the cardinal of $\mathfrak{L}(A) \cap X_{\leq x} = \mathfrak{L}(A_{\leq x})$. Theorem 4.4 applies to $A_{\leq x}$ in $\mathcal{K}[X_{\leq x}]$. Thus P and P' have the same dimension. They must be equal.

This property admits generally no converse implication. Some minimal prime of $\mathcal{I}(A_{\leq x})$ are not obtained as the intersection with $\mathcal{K}[X_{\leq x}]$ of a minimal prime of $\mathcal{I}(A)$. Lemma 5.8 gives a sufficient condition for this to happen.

EXAMPLE 4.6. *In $\mathbb{Q}[x, y]$ with $x < y$, consider the triangular set $A = x^2 - 1 \vartriangle (x+1) y - 1$. We have $\mathcal{I}(A) = (x - 1, 2y - 1)$ while $\mathcal{I}(A_{\leq x}) = (x - 1) \cap (x + 1)$.*

4.4 Product of Fields Associated to a Triangular Set

The association of a product of fields to a *normalized* triangular set was introduced and used in [Laz91, MM97, Aub99]. This association allows us to define a pseudo-gcd *modulo* a triangular set. Here, to introduce a product of fields we shall use the easy correspondence between positive dimension and dimension zero when dealing with ideals defined by triangular sets.

Let A be a triangular set of $\mathcal{K}[X]$. A can be considered as a triangular set in $\mathcal{K}(\mathfrak{T}(A))[\mathfrak{L}(A)]$. The ideals will be subscripted by $\mathcal{K}$ or by $\mathcal{K}_A = \mathcal{K}(\mathfrak{T}(A))$ to indicate where they are taken when confusion can arise

By Theorem 4.4 $\mathfrak{T}(A)$ is a transcendence basis of each associated prime of $\mathcal{I}(A)$ so that $\mathcal{I}(A)_{\mathcal{K}_A}$ is a zero dimensional ideal of $\mathcal{K}_A[\mathfrak{L}(A)]$ and $\mathcal{I}(A)_{\mathcal{K}_A} \cap \mathcal{K}[X] = \mathcal{I}(A)_{\mathcal{K}}$. Similarly, $\mathcal{R}(A)_{\mathcal{K}_A}$ is a zero dimensional radical ideal of $\mathcal{K}_A[\mathfrak{L}(A)]$ and $\mathcal{R}(A)_{\mathcal{K}_A} \cap \mathcal{K}[X] = \mathcal{R}(A)_{\mathcal{K}}$. The construction of the product of fields associated to A is then an immediate consequence of the Chinese remainder theorem.

PROPOSITION 4.7. *Let A be a consistent triangular set. The ring $\mathfrak{R}(A) = \mathcal{K}_A[\mathfrak{L}(A)] \,/\, \mathcal{R}(A)_{\mathcal{K}_A}$ is isomorphic to a product of fields.*

Proof. $\mathcal{R}(A)_{\mathcal{K}_A}$ is a zero dimensional radical ideal of $\mathcal{K}_A[\mathfrak{L}(A)]$. Let $P_1, \ldots, P_r$ be the associated primes of $\mathcal{R}(A)_{\mathcal{K}_A}$. They are maximal ideals and therefore the $\mathcal{K}_A[\mathfrak{L}(A)] \,/P_i$ are fields. By application of Theorem 2.4, $\mathcal{K}_A[\mathfrak{L}(A)] \,/\, \mathcal{R}(A)_{\mathcal{K}_A} \cong \mathcal{K}_A[\mathfrak{L}(A)] \,/P_1 \times \ldots \times \mathcal{K}_A[\mathfrak{L}(A)] \,/P_r$.

We shall see that when the triangular set is a regular chain, the product of fields defined is computable in the sense that an element can be tested to be a zero divisor and we can write a splitting algorithm. As we have seen in Section 3.2 we can thus write a pseudo-gcd algorithm over $\mathfrak{R}(A) = \mathcal{K}_A[\mathfrak{L}(A)] \,/\, \mathcal{R}(A)$.

Note that $\mathcal{K}(\mathfrak{T}(A))[\mathfrak{L}(A)] \,/\, \mathcal{R}(A)$ is equal to its total field of fraction (Proposition 2.5). Therefore $\mathcal{K}(\mathfrak{T}(A))[\mathfrak{L}(A)] \,/\, \mathcal{R}(A) = \mathfrak{Q}\,(\mathcal{K}[X] \,/\, \mathcal{R}(A))$. For ideals defined by triangular sets, we thus obtain the same product of field as [Aub99]. The benefit of the present presentation is that we control the elements we invert, namely only the elements in $\mathcal{K}[\mathfrak{T}(A)]$.

5   Regular Chains and Characterizable Ideals

Chains are special kinds of triangular set that allow the definition of characteristic set of an ideal. That notion of characteristic set was introduced by J.F. Ritt [Rit30, Rit50]. He defined them as *chains* that are minimal w.r.t. to a certain pre-order. The chains of Ritt are the *autoreduced sets* of Kolchin. The *chains* we define are less restrictive but only slightly.

Characterizable ideals were introduced in [Hub00]. They are ideals that are well defined by their characteristic sets. Ritt and Kolchin made use of the fact that dimension of prime ideals could be read on their characteristic set and that membership to prime ideals could be tested by reduction w.r.t. any of their characteristic set. Characterizable ideals is a wider class of ideals that have those two properties.

Regular chains were introduced by Kalkbrener [Kal93]. It is close to the definition of regular set of [MM97] These definitions were shown to be equivalent in [ALMM99]. The definition we give is equivalent, we only avoid the use of tower of simple extensions.

We show that characterizable ideals are in fact ideals defined by regular chains. Say it otherwise: regular chains are characteristic sets of characterizable ideals. That result appeared in [ALMM99]. Thus, the approach through characteristic sets allows to define characterizable ideals intrinsically while the regular chain approach allow us to construct characterizable ideals.

5.1 Characteristic Sets

The ranking on X induces a pre-order[2] on the set of triangular sets of $\mathcal{K}[X]$. Let $A = a_1 \vartriangle \ldots \vartriangle a_r$ and $B = b_1 \vartriangle \ldots \vartriangle b_s$ be triangular sets. A is said to have lower rank than B when there exists k, $0 \le k \le r, s$, such that $\mathsf{rank}(a_i) = \mathsf{rank}(b_i)$ for all $1 \le i \le k$ and either $k = r < s$ or $\mathsf{rank}(a_k) < \mathsf{rank}(b_k)$. If this is the case, we write $\mathsf{rank}(A) < \mathsf{rank}(B)$.

If $A = B \vartriangle C$ then $\mathsf{rank}(A) < \mathsf{rank}(B)$ reflecting the inclusion of ideals $(B) \subset (A)$. In particular, if B is the empty set and A is not then $\mathsf{rank}(A) < \mathsf{rank}(B)$.

Recall that a relation on a set M is well founded if every non empty subset of M has a minimal element for this relation. This is equivalent to the fact that there is no infinite decreasing sequence in M. See for instance [BW93, Chapter 4].

THEOREM 5.1. *The pre-order on triangular sets is well founded.*

Proof. Assume that the variables are indexed by increasing rank from 1 to n. The rank of a triangular set $A = a_1 \vartriangle \ldots \vartriangle a_r$ can be modeled by an element $(i_1, d_1, \ldots, i_n, d_n)$ of $\mathbb{N}^{2n}$ where, for $j \le r$ i_j and d_j are respectively index of the leader of a_j and the degree of a_j in its leader while for $j > r$ $i_j = n + 1$ and $d_j = 0$. The pre-order on the triangular set is given by the lexicographic order in $\mathbb{N}^{2n}$ and that latter is well founded.

This property allows us to show termination of algorithms and to show that every ideal admits a characteristic set. But for this purpose we need to consider chains and not simply triangular sets. Indeed one can extend a triangular set included in an ideal to a lower rank one still included in the ideal.

EXAMPLE 5.2. *Consider the ideal $I = (x, y)$ in $\mathbb{Q}[x, y, z]$. $x \vartriangle y$ is a triangular set in I as is $x \vartriangle y \vartriangle xz$ which has lower rank. This latter is nonetheless useless in discriminating I.*

DEFINITION 5.3. *A subset of $\mathcal{K}[X]$ is an autoreduced set if any of its element is reduced w.r.t. the others.*
A triangular set A in $\mathcal{K}[X]$ is a chain if for all $x \in \mathcal{L}(A)$, the rank of $\mathsf{red}(A_x, A_{<x})$ is equal to the rank of A_x.

---

[2] A pre-order is a relation that is reflexive, transitive and connex. The difference with an order is that $a \le b$ and $b \le a$ does not imply $a = b$.

An autoreduced set must be a triangular set. The important fact behind the definition of a chain is that to any chain we can associate an autoreduced set B with the same rank. We just take $B = \{\text{red}(A_x, A_{<x}) \mid x \in \mathcal{L}(A)\}$. Unfortunately the definition of chain above depends on the choice of the pseudo-division algorithm used, as illustrated in the example below. It is nonetheless practical to be able to define characteristic sets of ideals that are not necessarily autoreduced. Autoreduced sets have been replaced by *weak ascending chains* [CG90] or *fine* triangular sets [Wan93], i.e. triangular sets such that $\forall x \in \mathcal{L}(A)$ $\text{red}(\text{init}(A_x), A_{<x}) \neq 0$, in Wu-Ritt type of triangulation-decomposition algorithm. Nonetheless, it is not always possible to associate an autoreduced set to a fine triangular set while keeping the same rank. It is indeed possible that $\text{red}(\text{init}(A_x), A_{<x}) \neq 0$ but that $\text{red}(A_x, A_{<x})$ does not have the same rank as A_x. We give an example.

EXAMPLE 5.4. *In $\mathbb{Q}[x, y, z]$ with ranking $x < y < z$ consider*

- $A = x^2 \triangle xy-1 \triangle xz+y$. *Then* $\text{red}(\text{init}(A_z), A_{<z}) = x \neq 0$ *but* $\text{red}(A_z, A_{<z}) = 1$.
- $A = x^2 - x \triangle xy - 1 \triangle (x-1)z + xy$. *If sparse pseudo-division is used then A is a chain as* $\text{red}(A_z, A_{<z}) = (x-1)z + 1$. *If pseudo-division is used then A is not a chain as* $\text{red}(A_z, A_{<z}) = x$.

DEFINITION 5.5. *Let I be a proper ideal in $\mathcal{K}[X]$. A chain A contained in I is a characteristic set of I if one of the following equivalent conditions holds:*

1. *A is of minimal rank among the chains contained in I.*
2. *there is no non zero element of I reduced w.r.t. A.*
3. *$\forall q \in I$, $\text{red}(q, A) = 0$.*

We shall make explicit the equivalences in the definition.

1. $\Rightarrow$ **2.** Assume that there is a nonzero element p in I reduced w.r.t. A. Let $x = \text{lead}(p)$ and consider $B = A_{<x} \triangle p$. B is a chain in I of lower rank than A. This contradicts the hypothesis on A.
2. $\Rightarrow$ **1.** Assume there exists a chain B in I of rank lower than A. We can assume that B is an autoreduced set. Let $A = a_1 \triangle \ldots \triangle a_s$ and $B = b_1 \triangle \ldots \triangle b_r$ and k, $0 \leq k \leq r, s$, such that $\text{rank}(b_i) = \text{rank}(a_i)$ for all $i \leq k$ and $\text{rank}(b_{k+1}) < \text{rank}(a_{k+1})$ or $k = s$. As b_k is reduced w.r.t. $b_1 \triangle \ldots \triangle b_{k-1}$ it is reduced w.r.t. A. That cannot be the case.
3. $\Leftrightarrow$ **3.** is immediate to write down.

From Point 3. in this definition we deduce that if A is a characteristic set of an ideal I of $\mathcal{K}[X]$ then $(A) \subset I \subset (A) : I_A^\infty$. From Point 1. and Theorem 5.1 we deduce that every ideal in $\mathcal{K}[X]$ admits a characteristic set and all the characteristic sets of a given ideal (for a given ranking) have the same rank.

The example below illustrate the fact that a chain A is not obviously a characteristic set of $(A) : I_A^\infty$.

EXAMPLE 5.6. *In $\mathbb{Q}[x, y]$ endowed with a ranking $x < y$ consider, the chain $A = (x-1)y - 1 \triangle x^2 - 1$. Note that $(x+1) \in (A) : I_A^\infty$ though it is reduced w.r.t. A.*

5.2   Regular Chains

The idea behind the definition of regular chains is that a generic zero of $\mathcal{R}(A_{<x})$ can be extended to a generic zero of $\mathcal{R}(A_{\leq x})$. This is expressed in Proposition 5.8 by the fact that any associated prime of $\mathcal{R}(A_{<x})$ does not disapear as in Example 4.6 but is always the cutback of an associated prime of $\mathcal{R}(A)$.

DEFINITION 5.7. *Let A be a triangular set in $\mathcal{K}[X]$. A is a regular chain if for all $x \in \mathfrak{L}(A)$, $\mathrm{init}(A_x)$ is not a zero divisor modulo $\mathcal{I}(A_{<x})$.*

A regular chain is a chain according to Definition 5.3. Indeed, if $\mathrm{rank}(\mathrm{red}(A_x, A_{<x}))$ were different from $\mathrm{rank}(A_x)$ that would imply that $\mathrm{init}(A_x) \in \mathcal{R}(A_{<x})$. If A is a regular chain, the ideal $\mathcal{I}(A)$ is not trivial and the properties of triangular sets apply. Namely

- $\mathcal{I}(A)$ is unmixed dimensional; it has no embedded prime; its set of primary components is unique.
- $\mathfrak{T}(A)$ is a transcendence basis of each associated prime of $\mathcal{I}(A)$.
- $\mathfrak{zo}(\mathcal{I}(A)) = \mathfrak{zo}(\mathcal{R}(A))$ and we write $\mathfrak{zo}(A)$.

Furthermore, by Lemma 4.3, if $A \bigtriangleup a$ is a regular chain in $\mathcal{K}[X]$ then $(\mathcal{I}(A)+(a)):$ $h^\infty = \mathcal{I}(A \bigtriangleup a)$.

PROPOSITION 5.8. *Let C be a regular chain in $\mathcal{K}[X]$ and $x \in X$. $C_{\leq x}$ is a regular chain in $\mathcal{K}[X_{\leq x}]$ and*

- $\mathcal{I}(C) \cap \mathcal{K}[X_{\leq x}] = \mathcal{I}(C_{\leq x})$
- *the associated primes of $\mathcal{I}(C_{\leq x})$ are the intersections with $\mathcal{K}[X_{\leq x}]$ of the associated primes of $\mathcal{I}(C)$.*

Proof. The first point is an immediate consequence of Proposition 4.3 since $\mathrm{init}(C_x) \notin \mathfrak{zo}(\mathcal{I}(C_{\leq x}))$. Together with Theorem 4.5, we can thus say that the set of the minimal primes of $\mathcal{I}(C_{\leq x})$ in $\mathcal{K}[X_{\leq x}]$ is equal to the set of the intersections with $\mathcal{K}[X_{\leq x}]$ of the minimal primes of $\mathcal{I}(C)$.

That implies that an element $p \in \mathcal{K}[X_{\leq x}]$ belongs to $\mathfrak{zo}(C)$ if and only if $p \in \mathfrak{zo}(C_{\leq x})$. We can choose $q \in \mathcal{K}[X_{\leq x}]$ s.t. $pq \in \mathcal{I}(C)$. With this property we can write without confusion, for any regular chain C, $\mathfrak{zo}(C_{\leq x})$ to mean $\mathfrak{zo}(C_{\leq x})$ or $\mathfrak{zo}(C) \cap \mathcal{K}[X_{\leq x}]$ since these two sets are equal in $\mathcal{K}[X_{\leq x}]$ and can be extended to $\mathcal{K}[X]$. Note nonetheless that two different minimal primes of $\mathcal{I}(C)$ can have the same intersection with $\mathcal{K}[X_{\leq x}]$.

EXAMPLE 5.9. *In $\mathbb{Q}[x, y]$ where $x < y$ consider the triangular set $A = x \bigtriangleup y^2 - 1$. We have $\mathcal{R}(A) = \langle x, y - 1\rangle \cap \langle x, y + 1\rangle$. Both the minimal primes contract to $\langle x \rangle$ on $\mathcal{K}[x]$.*

5.3   Characterizable Ideals

Ritt and Kolchin made high use of the fact that if A is a characteristic set of a prime ideal P then $P = (A) : I_A^\infty$ and therefore $p \in P \Leftrightarrow \mathrm{red}(p, A) = 0$. We introduce the very useful wider class of ideals having this property. We give an intrinsic definition and an explicit construction of those ideals.

DEFINITION 5.10. *An ideal I of $\mathcal{K}[X]$ is characterizable if for a characteristic set A of I we have $I = (A) : I_A^\infty$. A is said to characterize I.*

Note that if I is a characterizable ideal characterized by A then $p \in I \Leftrightarrow \mathrm{red}(p, A) = 0$. Prime ideals are characterizable for any ranking but not all primary ideals are characterizable as illustrated in Example 5.11. Characterizable ideals that are not prime do exist but that depends then on the ranking (see Example 5.12). From Theorem 4.4 we see that characterizable ideals have specific dimension properties. A natural question would be to determine if an ideal given by its generators is characterisable. An answer in terms of Gröbner basis is given in [Hub00] for zero dimensional ideals and extended in [CH03] to ideals of positive dimension.

EXAMPLE 5.11. *In $\mathbb{Q}[x, y]$ consider the primary ideal $I = (x^2, x\,y, y^2)$. The generators given here form a reduced Gröbner basis for the lexicographical term ordering $x < y$. Thus a chain of minimal rank in I according to a ranking $x < y$ is given by $A = x^2 \triangle x\,y$. A is thus a characteristic set of I but note that $(A) : I_A^\infty = (1) \neq I$. I is not characterizable.*

EXAMPLE 5.12. *Consider $I = (y^3 - y, 2\,x - y^2 + 2)$ in $\mathbb{Q}[x, y]$. The generators of I given above form a reduced Gröbner basis G for the lexicographical term order where $y < x$. G is also an autoreduced set and therefore it is a characteristic set of I for the ranking $y < x$. We have $I = (G) = (G) : I_G^\infty$ and thus I is characterizable for the ranking $y < x$.*
    The Gröbner basis of I for the lexicographical order $x < y$ is $G' = \{2\,x^2 + 3\,x + 1, 2\,x\,y + y, y^2 - 2\,x - 2\}$. Therefore $A = 2\,x^2 + 3\,x + 1 \triangle 2\,x\,y + y$ is a characteristic set of I for the ranking $x < y$. We can check that $I \neq (A) : I_A^\infty$ so that I is not characterizable for the ranking $x < y$.

The following equivalence was proved in [ALMM99, Theorem 6.1] but a related result appears in [Kol73, Lemma 13, Section 0.14]. It shows that characterizable ideals are in fact ideals defined by regular chains.

THEOREM 5.13. *Let A be a chain in $\mathcal{K}[X]$. A is consistent and A is a characteristic set of $(A) : I_A^\infty$ if and only if A is a regular chain.*

Proof. We first show by induction that if A is a regular chain then $\mathcal{I}(A)$ contains no nonzero element reduced w.r.t. A. This is true for the empty chain which is the only chain of $\mathcal{K}$. Assume this hypothesis is true for all the regular chains of $\mathcal{K}[X]$. Let A be a regular chain of $\mathcal{K}[X][x]$. $A_{<x}$ is a regular chain of $\mathcal{K}[X]$ and $\mathcal{I}(A) \cap \mathcal{K}[X] = \mathcal{I}(A_{<x})$ (Lemma 5.8).

Let $p \in \mathcal{K}[X][x]$ belong to $\mathcal{I}(A)$ and be reduced w.r.t A. We can assume further that $\deg(p, x) > 0$, since otherwise $p \in \mathcal{I}(A) \Leftrightarrow p \in \mathcal{I}(A_{<x})$. If $x \notin \mathfrak{L}(A)$, the coefficients of p being considered as a polynomial in x must belong to $\mathcal{I}(A_{<x})$; they are also reduced w.r.t. $A_{<x}$ and therefore must be zero by induction hypothesis; p is equal to zero. Assume now $x \in \mathfrak{L}(A)$. Since $\mathcal{I}(A) = (\mathcal{I}(A_{<x}), A_x) : \operatorname{init}(A_x)$ there exists $h \in \operatorname{init}(A_x)^\infty$ and $q \in \mathcal{K}[X]$ such that $h\,p \equiv q\,A_x \mod \mathcal{I}(A_{<x})$. If q were non zero, the degree in x of p would be greater or equal to the one of A_x, contradicting thus the hypothesis that p is reduced w.r.t. A. It must be that $q = 0$; therefore p belongs to $\mathcal{I}(A_{<x}) : h^\infty = \mathcal{I}(A_{<x})$ and therefore its coefficients when considered as a polynomials in x belong to $\mathcal{I}(A_{<x})$. The coefficients being reduced w.r.t $A_{<x}$, they must be zero by induction hypothesis. We have thus proved that if A is a regular chain, $1 \notin (A) : I_A^\infty$ and A is a characteristic set of $\mathcal{I}(A)$.

Assume now A is not a regular chain. We shall prove that it is not a characteristic set of $\mathcal{I}(A)$. There exists $x \in X$ such that $\operatorname{init}(A_x) \in \mathfrak{z}\mathfrak{d}(A_{<x})$. Select the smallest such x. $A_{<x}$ is a regular chain so that it is a characteristic set of $\mathcal{I}(A_{<x})$ by the first part of the proof. There exists $q \in \mathcal{K}[X_{<x}]$, $q \notin \mathcal{I}(A_{<x})$ such that $q\operatorname{init}(A_x) \in \mathcal{I}(A_{<x})$. Let $r = \operatorname{red}(q, A_{<x})$; r is nonzero since $q \notin \mathcal{I}(A_{<x})$ and $A_{<x}$ is a characteristic set of this ideal. We have $r\operatorname{init}(A_x) \in \mathcal{I}(A_{<x})$ and therefore $r \in \mathcal{I}(A)$, while r is reduced w.r.t. A. Thus either $1 \in (A) : I_A^\infty$ or A is not a characteristic set of $\mathcal{I}(A)$.

Assume A is a regular chain in $\mathcal{K}[X]$. Playing with definitions and Theorem 5.13, we have:

- $\mathcal{I}(A)$ is a characterizable ideal characterized by A
- $p \in \mathcal{I}(A) \Leftrightarrow \operatorname{red}(p, A) = 0$.

We shall show now that the definition of characterizable differential ideals is independent of the characteristic set chosen in Definition 5.10.

PROPOSITION 5.14. *Let I be a characterizable ideal. Any characteristic set of I characterizes I.*

Proof. Note first that if A is a characteristic set of an ideal I in $\mathcal{K}[X]$ then $A_{<x}$ is a characteristic set of $I \cap \mathcal{K}[X_{<x}]$ for any $x \in X$. We prove the proposition by induction on X.

In $\mathcal{K}$, (0) is the only characterizable ideal and the only chain is the empty set. Assume the proposition is true in $\mathcal{K}[X]$. Let I be a characterizable ideal in $\mathcal{K}[X][x]$: there exists a regular chain C such that $I = \mathcal{I}(C)$. Then $C_{<x}$ is a regular chain and a characteristic set of $\mathcal{I}(C_{<x})$ in $\mathcal{K}[X]$. Let A be another characteristic set of I; A and C have the same rank. $A_{<x}$ is a characteristic set of $I \cap \mathcal{K}[X] = \mathcal{I}(C_{<x})$. By induction hypothesis, $A_{<x}$ is a regular chain characterizing $I \cap \mathcal{K}[X]$. Thus $I \cap \mathcal{K}[X] = \mathcal{I}(A_{<x}) = \mathcal{I}(C_{<x})$.

We shall show that A is a regular chain. If $x \notin \mathfrak{L}(A)$ then $I = \mathcal{I}(C_{<x})$ and the result comes from the induction hypothesis. Assume now $x \in \mathfrak{L}(A)$. Since $A_x \in I = (\mathcal{I}(C_{<x}), C_x) : \operatorname{init}(C_x)^\infty$, there exists $h \in \operatorname{init}(C_x)^\infty$ such that

$h\,A_x \equiv q\,C_x \mod \mathcal{I}(C_{<x})$, where $q \in \mathcal{K}$ since the degrees in x of A_x and C_x are equal. If $q = 0$ then $A_x \in \mathcal{I}(C_{<x}) : h^\infty = \mathcal{I}(C_{<x}) = I \cap \mathcal{K}[X]$: the coefficients of A_x all belong to $I \cap \mathcal{K}[X]$ and therefore are reduced to zero by $A_{<x}$ which is a characteristic set of $I \cap \mathcal{K}[X]$. This contradicts the fact that A is a chain. Thus $q \in \mathcal{K} \setminus \{0\}$. Equating the leading coefficients on both side of the pseudo-division relationship, we have that $h\,\mathrm{init}(A_x) \equiv q\,\mathrm{init}(C_x) \mod \mathcal{I}(C_{<x})$. Thus $\mathrm{init}(A_x)$ cannot divide zero modulo $\mathcal{I}(C_{<x})$. Since $\mathcal{I}(A_{<x}) = \mathcal{I}(C_{<x})$, A is a regular chain.

Furthermore, A being a characteristic set of I we have $A \subset \mathcal{I}(C) \subset \mathcal{I}(A)$ and thus $\mathcal{I}(A) = \mathcal{I}(C) : I_A^\infty = \mathcal{I}(C)$ since $I_{A_{<x}}$ and $\mathrm{init}(A_x)$ are not zero divisor of $\mathcal{I}(C)$.

5.4   Canonical Representatives of Characterizable Ideals

We shall exhibit canonical representatives of characterizable ideals. These canonical representatives are taken in what we call Gröbner chains. They appear as *p-chain* in [GC92] for their relevance in deciding for which parameters there is a (regular) zero. In [BL00], where they are called *strongly normalized triangular sets*, they serve the purpose of computing normal forms of (differential) polynomials modulo a (differential) characterizable ideal. The name owes to Proposition 5.16 that makes the link between the characteristic set approach to represent ideals and the Gröbner bases approach.

DEFINITION 5.15. *A triangular set A such that $\forall x \in \mathfrak{L}(A)$, $\mathrm{init}(A_x) \in \mathcal{K}[\mathfrak{T}(A)]$ is a Gröbner chain. A Gröbner chain is reduced if it is an autoreduced set such that none of its element has factors in $\mathcal{K}[\mathfrak{T}(A)]$.*

Theorem 4.4 shows that, for a triangular set A in $\mathcal{K}[X]$, $\mathcal{K}[\mathfrak{T}(A)]$ contains no zero divisor modulo $\mathcal{I}(A)$. Therefore a Gröbner chain is a regular chain.

PROPOSITION 5.16. *A (reduced) Gröbner chain A in $\mathcal{K}[X]$ is a (reduced) Gröbner basis in $\mathcal{K}(\mathfrak{T}(A))[\mathfrak{L}(A)]$ according to the lexicographical term order on $\mathfrak{L}(A)$ induced by the ranking on X.*

Proof. The leading terms have no common divisors. According to the first Buchberger criterion the S-polynomials are reduced to zero.

PROPOSITION 5.17. *Every characterizable ideal in $\mathcal{K}[X]$ admits a (unique) characteristic set that is a (reduced) Gröbner chain.*

Proof. Let I be a characterizable ideal. There exists A, a regular chain, such that $I = \mathcal{I}(A)$. For all $y \in \mathfrak{L}(A)$ we define q_y and r_y as follow. If $\mathrm{init}(A_y) \in \mathcal{K}[\mathfrak{T}(A)]$ then $q_y = 1$ and $r_y = 0$. Otherwise since $\mathcal{I}(A_{<y})_{\mathcal{K}_{A_{<y}}}$ is a zero dimensional ideal in $\mathcal{K}(A_{<y})[\mathfrak{L}(A_{<y})]$ and $\mathrm{init}(A_y) \notin \mathfrak{Z}\mathfrak{o}(A_{<y})$ there exists $q_y \in \mathcal{K}[X_{<y}]$ and $u_y \in \mathcal{K}[\mathfrak{T}(A_{<y})]$ such that $q_y\,\mathrm{init}(A_y) = u_y + r_y$ where $r_y \in \mathcal{I}(A_{<y})$. Let $C = \bigtriangleup_{y \in \mathfrak{L}(A)}(q_y\,A_y - r_y\,\mathrm{init}(A_y)\,\mathrm{rank}(A_y))$. C is a Groebner chain. It characterizes I as $C \subset I$ and $\mathrm{rank}(C) = \mathrm{rank}(A)$.

The uniqueness of the second point owes to the canonical properties of reduced and minimal Gröbner bases.

Given a regular chain A it is possible to compute the canonical Gröbner chain of the characterizable ideal defined. First, one can think of making the proof constructive by a generalization of the extended Euclidean algorithm [GC92, MM97, BL00]. The algorithm of [MM97, BL00] might require nonetheless to split the ideal[3]. Alternatively, we can use Buchberger algorithm. Indeed the desired Gröbner chain is the Gröbner basis of (A) in $\mathcal{K}(\mathfrak{T}(A))[\mathfrak{L}(A)]$ w.r.t. the lexicographical term order induced by the ranking on $\mathfrak{L}(A)$. This works thanks to the following proposition derived from [Aub99].

PROPOSITION 5.18. *If A is a regular chain then the ideal $(A) : I_A^\infty$ is equal to the ideal (A) in $\mathcal{K}(\mathfrak{T}(A))[\mathfrak{L}(A)]$.*

Proof. Assume $A = a_1 \triangle \ldots \triangle a_r$ and note $y_1, \ldots, y_r$ the leaders of $a_1, \ldots, a_r$. The proof works by induction. $\mathrm{init}(a_1) \in \mathcal{K}_A$ so that $(a_1) : \mathrm{init}(a_1)^\infty = (a_1)$. Assume that $\mathcal{I}(a_1 \triangle \ldots \triangle a_k) = (a_1, \ldots, a_k)$.

Since $\mathrm{init}(a_k) \notin \mathfrak{Z}\mathfrak{d}(a_1 \triangle \ldots \triangle a_k)$ and $\mathcal{I}(a_1 \triangle \ldots \triangle a_k)$ is a zero dimensional ideal in $\mathcal{K}_A[y_1, \ldots, y_k]$ there exists $q \in \mathcal{K}_A[y_1, \ldots, y_k]$ such that $q \, \mathrm{init}(a_{k+1}) \equiv 1$ mod $\mathcal{I}(a_1 \triangle \ldots \triangle a_k)$. If p belongs to $\mathcal{I}(a_1 \triangle \ldots \triangle a_{k+1})$ that is equal to $(\mathcal{I}(a_1 \triangle \ldots \triangle a_k) + (a_{k+1})) : \mathrm{init}(a_{k+1})^\infty$ then there exists $e \in \mathbb{N}$ s.t. $\mathrm{init}(a_{k+1})^e p \in (\mathcal{I}(a_1 \triangle \ldots \triangle a_k) + (a_{k+1}))$. Premultiplying by q^e we obtain that $p \in (\mathcal{I}(a_1 \triangle \ldots \triangle a_k) + (a_{k+1}))$ so that by induction hypothesis $p \in (a_1, \ldots, a_{k+1})$.

The latter proposition entails that if a polynomial p does not belong to $\mathfrak{Z}\mathfrak{d}(A)$, where A is a regular chain, then p is a unit modulo (A) in $\mathcal{K}(\mathfrak{T}(A))[\mathfrak{L}(A)]$ and therefore (A, p) contains an element in $\mathcal{K}[\mathfrak{T}(A)]$. Those properties are in fact definitions for a polynomial to be *invertible w.r.t.* A in [MKRS98, BKRM01].

6 Decomposition into Characterizable Ideals

In Section 9 we shall see that we can compute for any finite set of polynomials a decomposition of the radical of the ideal generated by these polynomials into characterizable ideals. This provides a membership test to this radical ideal. It also provides a *strongly* unmixed dimensional decomposition. We define here characteristic decomposition and make the link to decomposition of product of fields.

Let $J, J_1, \ldots, J_r$ be radical ideals such that the equality $J = J_1 \cap \ldots \cap J_r$ holds. This equality defines a decomposition of J and the J_i are called the component of J (in this decomposition). The decomposition is *irredundant* if the sets of prime components of the J_i gives a partition of the set of prime components of J. In other words, if P is a minimal prime of J, there exists a unique i such that $J_i \subset P$.

The following proposition is then trivial but useful.

PROPOSITION 6.1. *Let J be a radical ideal in $\mathcal{K}[X]$. If $J = J_1 \cap \ldots \cap J_r$ is an irredundant decomposition then $\mathfrak{Z}\mathfrak{d}(J) = \mathfrak{Z}\mathfrak{d}(J_1) \cup \ldots \cup \mathfrak{Z}\mathfrak{d}(J_r)$.*

---

[3] Personal communication of F. Lemaire

DEFINITION 6.2. *Let J be a non trivial radical ideal in $\mathcal{K}[X]$. A set of regular chains $\mathcal{C} = \{C_1, \ldots, C_r\}$ defines a characteristic decomposition of J if*

$$\sqrt{I} = \mathcal{R}(C_1) \cap \ldots \cap \mathcal{R}(C_r) \text{ i.e. } \sqrt{I} = \langle C_1 \rangle : I_{\bar{C}_1}^\infty \cap \ldots \cap \langle C_r \rangle : I_{\bar{C}_r}^\infty.$$

We take as convention that the empty set is a characteristic decomposition of $J = \mathcal{K}[X]$.

PROPOSITION 6.3. *Let A be a consistent triangular set of $\mathcal{K}[X]$. If $\mathcal{R}(A) = \mathcal{R}(C_1) \cap \ldots \cap \mathcal{R}(C_r)$ is an irredundant characteristic decomposition then $\mathfrak{L}(A) = \mathfrak{L}(C_1) = \ldots = \mathfrak{L}(C_r)$.*

Proof. By Theorem 4.4, $\mathfrak{T}(A)$ is a basis of transcendence of all the associated prime of $\mathcal{R}(A)$ and thus of all the associated primes of $(C_i) : I_{\bar{C}_i}^\infty$ while, for $1 \le i \le r$, $\mathfrak{T}(C_i)$ is a basis of transcendence of the associated primes of $(C_i) : I_{\bar{C}_i}^\infty$. Thus $\mathfrak{T}(A)$ and $\mathfrak{T}(C_i)$ have the same cardinal and it is sufficient to prove that $\mathfrak{L}(A) \subset \mathfrak{L}(C)$. For $1 \le i \le r$, no element of I_A is a zero divisor of $(C_1) : I_{\bar{C}_1}^\infty$. Assume there exists $x \in \mathfrak{L}(A)$ such that $x \notin \mathfrak{L}(C_i)$. Then $\{A_x\} \cup C_i$ is a regular chain lower then C_i in $(C_i) : I_{\bar{C}_i}^\infty$. This contradicts the fact that C_i is a characteristic set of $(C_i) : I_{\bar{C}_i}^\infty$. The conclusion follows: $\mathfrak{L}(A) \subset \mathfrak{L}(C)$.

The same way we can consider A as a triangular set in $\mathcal{K}_A[\mathfrak{L}(A)]$ we can also consider the C_i as regular chains in $\mathcal{K}_A[\mathfrak{L}(A)]$. An irredundant decomposition of $\mathcal{R}(A)$ provides in fact a decomposition of the product of fields associated to A. Recall we defined $\mathfrak{R}(A) = \mathcal{K}_A[\mathfrak{L}(A)]/\mathcal{R}(A)$ which is equal to $\mathfrak{Q}(\mathcal{K}[X]/\mathcal{R}(A))$.

PROPOSITION 6.4. *Let A be a consistent triangular set of $\mathcal{K}[X]$. Assume $\mathcal{R}(A) = \mathcal{R}(C_1) \cap \ldots \cap \mathcal{R}(C_r)$ is an irredundant characteristic decomposition in $\mathcal{K}[X]$. Then $\mathcal{R}(A)_{\mathcal{K}_A} = \mathcal{R}(C_1)_{\mathcal{K}_A} \cap \ldots \cap \mathcal{R}(C_r)_{\mathcal{K}_A}$ is an irredundant characteristic decomposition in $\mathcal{K}_A[\mathfrak{L}(A)]$ and $\mathfrak{R}(A) \cong \mathfrak{R}(C_1) \times \ldots \times \mathfrak{R}(C_r)$.*

Proof. We have $\mathfrak{L}(A) = \mathfrak{L}(C_1) = \ldots = \mathfrak{L}(C_r)$. Thus $A, C_1, \ldots, C_r$ can be considered as triangular sets in $\mathcal{K}(\mathfrak{T}(A))[\mathfrak{L}(A)]$ and we overline them when we consider them as such. Let $\bar{P}$ be an associated prime of $\mathcal{R}(\bar{A})_{\mathcal{K}_A}$. If $\bar{P}$ contained $\mathcal{R}(\bar{C}_i)$ and $\mathcal{R}(\bar{C}_j)$ for $i \ne j$ then $\bar{P} \cap \mathcal{K}[X]$ would contain $\mathcal{R}(\bar{C}_i) \cap \mathcal{K}[X] = \mathcal{R}(C_i)$ and $\mathcal{R}(\bar{C}_j) \cap \mathcal{K}[X] = \mathcal{R}(C_j)$. Since $\bar{P} \cap \mathcal{K}[X]$ is a prime component of $\mathcal{R}(A)$ and the decomposition $\mathcal{R}(A) = \mathcal{R}(C_1) \cap \ldots \cap \mathcal{R}(C_r)$ is irredundant this is not possible. Therefore $\mathcal{R}(\bar{A}) = \mathcal{R}(\bar{C}_1) \cap \ldots \cap \mathcal{R}(\bar{C}_r)$ is irredundant.

The radical ideals $\mathcal{R}(\bar{C}_i)_{\mathcal{K}_A}$ are zero dimensional and have no prime divisor in common. They are thus comaximal: $\mathcal{R}(\bar{C}_i) + \mathcal{R}(\bar{C}_j) = \mathcal{K}_A[\mathfrak{L}(A)]$. By the Chinese remainder theorem (Theorem 2.4) we have that

$$\mathcal{K}_A[\mathfrak{L}(A)]/\mathcal{R}(\bar{A}) \cong \mathcal{K}_A[\mathfrak{L}(A)]/\mathcal{R}(\bar{C}_1) \times \ldots \times \mathcal{K}_A[\mathfrak{L}(A)]/\mathcal{R}(\bar{C}_r)$$

The converse property is in fact also true. If $\mathcal{R}(\bar{A}) = \mathcal{R}(\bar{C}_1) \cap \ldots \cap \mathcal{R}(\bar{C}_r)$ is an irredundant characteristic decomposition in $\mathcal{K}_A[\mathfrak{L}(A)]$ then it can be *lifted* to the irredundant characteristic decomposition $\mathcal{R}(A) = \mathcal{R}(C_1) \cap \ldots \cap \mathcal{R}(C_r)$ in $\mathcal{K}[X]$ where a C_i is obtained from $\bar{C}_i$ by cleaning the denominators and the factors in $\mathcal{K}[\mathfrak{T}(A)]$ (see [Hub00]).

7   Radical Ideals Defined by Triangular Sets

In this section, we generalize to triangular sets the following properties for $p \in \mathcal{K}[x]$:

- $q = \frac{p}{\gcd(p, \frac{\partial p}{\partial x})}$ is squarefree and $\sqrt{(p)} = (q)$ (Theorem 7.1).
- p is squarefree iff $\gcd(p, \frac{\partial p}{\partial x}) = 1$ (Corollary 7.3).
- $(p) : (\frac{\partial p}{\partial x})^\infty$ is radical (Theorem 7.5).

Some of these results can be found in [Laz91, BLOP95, BLOP97, SL98, Mor99, Del99, Aub99, Hub00]. They can be proved by applying the Jacobian criterion for regularity (see [Eis94, Chapter 16] and [Vas98, Chapter 5]).

Let p be an element of $\mathcal{K}[X]$. The *separant* of p is the derivative of p w.r.t. its leader: $\mathsf{sep}(p) = \frac{\partial p}{\partial \mathsf{lead}(p)}$. Thus if $x = \mathsf{lead}(p)$ and we can write $p = a_d x^d + \ldots + a_1 x + a_0$ then $\mathsf{sep}(p) = d a_d x^{d-1} + \ldots + a_1$. For a triangular set A in $\mathcal{K}[X]$ we denote S_A the set formed up by the separants of the elements of A and H_A the set of the initials and separants of the elements of A.

THEOREM 7.1. *Let A be a triangular set in $\mathcal{K}[X]$. Let s be the product of the separants of the elements of A. $\mathcal{R}(A) = \mathcal{I}(A) : s$. In other words $\sqrt{(A) : I_A^\infty} = \{ p \in \mathcal{K}[X] \mid s\, p \in (A) : I_A^\infty \}$.*

Proof. As seen in Theorem 4.4, $\mathcal{I}(A)$ and $\mathcal{R}(A)$ have no zero divisor in $\mathcal{K}[\mathfrak{T}(A)]$. It follows that $\mathcal{I}(A)_\mathcal{K} = \mathcal{I}(A)_{\mathcal{K}_A} \cap \mathcal{K}[X]$ and $\mathcal{R}(A)_\mathcal{K} = \mathcal{R}(A)_{\mathcal{K}_A} \cap \mathcal{K}[X]$.

Consider A as a triangular set of $\mathcal{K}_A[\mathfrak{L}(A)]$. The Jacobian ideal of (A) is generated by s. By [Vas98, Theorem 5.4.2.], $(A)_{\mathcal{K}_A} : s = \sqrt{(A)}_{\mathcal{K}_A}$. Thus $\mathcal{I}(A)_{\mathcal{K}_A} : s = \mathcal{R}(A)_{\mathcal{K}_A}$. Taking the intersection with $\mathcal{K}[X]$ we have the equality announced.

This gives us a way to construct $\mathcal{R}(A)$ that will be used in Section 9. It also gives a criterion for $\mathcal{I}(A)$ to be radical: $\mathcal{I}(A)$ is radical if and only if no separant of the elements of A is a zero divisor modulo $\mathcal{I}(A)$. This criterion motivates the following definition and gives the corollary after it.

DEFINITION 7.2. *A regular chain C in $\mathcal{K}[X]$ is squarefree if $S_C \cap \mathfrak{z} \mathfrak{d}(C) = \emptyset$ i.e. no separant of C is a zero divisor modulo $\mathcal{I}(C)$.*

COROLLARY 7.3. *If C is a regular chain of $\mathcal{K}[X]$ then $(C) : I_C^\infty$ is radical if and only if C is squarefree.*

Note nonetheless that the radical of a characterizable ideal is not always characterizable. The example below is taken from [Aub99, Section 4.5].

EXAMPLE 7.4. *In $\mathcal{K}[x, y]$ consider $A = x^2 - x \wedge y^2 - x$. A is a regular chain of $\mathcal{K}[x, y]$. We have $\mathcal{I}(A) = (A)$. A Gröbner basis of $\mathcal{R}(A) = \sqrt{(A)}$ for a lexicographical term oder satisfying $x < y$ is given by $G = \{ x^2 - x, (x - 1) y, y^2 - x \}$. We can extract from it the characteristic set $B = x^2 - x \wedge (x - 1) y$ of $\mathcal{R}(A)$. It is not a regular chain. Thus $\mathcal{R}(A)$ is not characterizable.*

The following theorem is central in constructive differential algebra. It was first enunciated in [BLOP95] and named there after one of the coauthors, D. Lazard. The second part of the statement appeared already in [Kol73]. It shows that we have similar dimension properties whether we saturate by the initials or by the separants.

THEOREM 7.5. *Let A be a triangular set of $\mathcal{K}[X]$. $(A):S_A^\infty$ is a radical ideal. If it is non trivial, $(A):S_A^\infty$ is unmixed dimensional and $\mathcal{L}(A)$ is the set of leaders of the characteristic set of any associated prime of $(A):S_A^\infty$.*

Proof. The principal ideal theorem [Eis94, Theorem 10.2] implies that no associated prime of (A) has codimension bigger than card(A). The product s of the separants of A is a maximal minor of the Jacobian matrix $\frac{\partial A}{\partial X}$. Let P be a prime ideal that does not contain s. The rank of the Jacobian matrix modulo P is maximal and equal to card(A). By [Eis94, Theorem 16.9] the localization of $\mathcal{K}[X]/(A)$ at P is an integral ring and the codimension of $(A)_P$ is card(A). Thus P contains a single primary component of (A) and that component is prime and of codimension card(A).

$(A):S_A^\infty$ is the intersection of the primary components of (A) the radical of which does not contain s. From what precedes, these components have to be prime and of codimension card(A). Therefore $(A):S_A^\infty$ is the intersection of prime ideals and so is radical.

We are left to show that $\mathcal{L}(A) \subset \mathcal{L}(C)$ for any characteristic set C of P. For any $x \in \mathcal{L}(A)$, $\mathrm{sep}(A_x) \notin P$. Thus one of the coefficients of a positive power of x in A_x does not belong to P. As any characteristic set of P must reduce A_x to zero, it must contain an element with leader x.

The theorem and its corollary is obviously true if we replace S_A by any set H that contains S_A. The case $H = H_A$ is of special use in differential algebra (see [Hub03]). The relationship between $(A):I_A^\infty$, $(A):S_A^\infty$ and $(A):H_A^\infty$ for regular chains is made explicit below.

PROPOSITION 7.6. *If A is a regular chain in $\mathcal{K}[X]$ then $(A):H_A^\infty = (A):S_A^\infty$. If A is a squarefree regular chain then $(A):I_A^\infty = (A):S_A^\infty$.*

Proof. Assume A is a regular chain and $a \in A$. In $\mathcal{K}(\mathfrak{T}(A))[\mathcal{L}(A)]$, $(A) = (A):I_A^\infty$ (Proposition 5.18). Any prime ideal of $\mathcal{K}[X]$ that contains A and $\mathrm{init}(a)$, for some $a \in A$, must contain an element in $\mathcal{K}[\mathfrak{T}(A)]$. This is the case of no associated prime of $(A):S_A^\infty$. Thus $\mathrm{init}(a)$ is not a zero divisor of $(A):S_A^\infty$ and the equality of ideals $(A) : H_A^\infty = (A) : S_A^\infty$ follows. If A is a squarefree regular chain we furthermore have that $(A):I_A^\infty = (A):H_A^\infty$.

8   Kalkbrener's Pseudo-gcd and Splitting Algorithm

In Section 3.2 we saw that computing a pgcd over $\mathcal{K}[\lambda]/\langle c \rangle$ relied on a splitting algorithm for $\mathcal{K}[\lambda]/\langle c \rangle$. In Section 2.4 we saw how to write the splitting algorithm for $\mathcal{K}[\lambda]/\langle c \rangle$ in terms of a gcd over $\mathcal{K}$. We shall show how to generalize this process to compute a pgcd over the product of fields $\mathfrak{R}(C) = \mathfrak{Q}\,(\mathcal{K}[X]/\mathcal{R}(C)) = \mathcal{K}_C[\mathfrak{L}(C)]/\mathcal{R}(C)$ associated to a regular chain C.

The algorithms are presented in [Kal93, Kal98] in a very synthetic and highly recursive way. Improvements in view of an efficient implementation are presented in [Aub99]. The present presentation owes to the presentation of Aubry: the recursive calls are limited so that one sees better where the work is done. We shall give complete proofs that lead to a sharper description of the outputs.

The two procedures we shall describe, pgcd and split, work in interaction and recursively. Basically, pgcd($\mathcal{K}[X][x], *, *$) calls split($\mathcal{K}[X], *, *$) which in turn calls pgcd($\mathcal{K}[X], *, *$).

The input of split consist of a regular chain C and a polynomial f. It then returns a pair of set $(\mathcal{Z}, \mathcal{U})$ s.t. $\mathcal{R}(C) = \bigcap_{A \in \mathcal{Z} \cup \mathcal{U}} \mathcal{R}(A)$ is an irredundant characteristic decomposition and $f \in \mathcal{I}(A), \forall A \in \mathcal{Z}$ while $f \notin \mathfrak{z}\mathfrak{0}(A), \forall A \in \mathcal{U}$. We thus have a splitting $\mathfrak{R}(C) \cong \mathfrak{R}_{\mathcal{Z}} \times \mathfrak{R}_{\mathcal{U}}$ where $\mathfrak{R}_{\mathcal{Z}} = \prod_{A \in \mathcal{Z}} \mathfrak{R}(A)$ and $\mathfrak{R}_{\mathcal{U}} = \prod_{A \in \mathcal{U}} \mathfrak{R}(A)$ so that the projection of f on $\mathfrak{R}_{\mathcal{Z}}$ is zero while the projection of f on $\mathfrak{R}_{\mathcal{U}}$ is a unit.

As for pgcd, it computes a pseudo-gcd over some product of field $\mathfrak{R}(C)$ defined by a regular chain. The pgcd computed has additional properties. In fact the output description of these algorithms are sharper in this paper than in [Kal93]. This nicely avoids extraneous complications in the proofs and the application to characteristic decomposition. We shall also make explicit that no redundancy nor multiplicities are introduced in the computations. If we start with a squarefree regular chain, the output regular chains will all be squarefree. Also, it may happen that the output consists of squarefree regular chains even if the input regular chain is not squarefree.

Termination and Correctness: One will easily check that if $X = \emptyset$, split($\mathcal{K}, \emptyset, f$) decides if $f = 0$ or not and pgcd($\mathcal{K}[x], \emptyset, F$) computes the gcd of (F) over $\mathcal{K}$. The proof of pgcd and split is inductive. Assuming that pgcd($\mathcal{K}[X_{\leq y}], *, *$) is correct and terminates, for all $y \in X$, we shall prove that split($\mathcal{K}[X], *, *$) is correct. Assuming that split($\mathcal{K}[X], *, *$) is correct, we shall prove that pgcd($\mathcal{K}[X][x], *, *$) is correct.

An intermediate algorithm relatively-prime, that is a direct application of pgcd, generalizes the recursion presented at the end of Section 2.4 in the case c is not squarefree. It is used in the splitting algorithm split.

Conventions: We shall describe accurately the output of the algorithms and their properties. Correctness of the algorithms and their outputs will be proved by exhibiting invariants for the loops of the algorithms. These invariants are also useful to understand what goes on in the algorithm. We shall need to name precisely one property of the output of an algorithm. For that purpose we number the output properties. We will refer to the ith property of the output of an algorithm, say pgcd, by pgcd.i.

The algorithms are described in Maple style but we use parentheses to denote ordered tuples. We make use of sets $\mathcal{S}$ to stack the data awaiting more computations. The command pop chooses one of those, removes it from the set and returns it. The set $\mathcal{C}$, $\mathcal{U}$, $\mathcal{Z}$ contains the data for which computation is completed.

8.1 Pseudo-gcd Algorithm

This algorithm is no different than the one given in Section 3.2 only more specific. The product of fields are replaced by regular chains defining them and we use a more specific output of the splitting algorithm.

ALGORITHM 8.1. **pgcd**
INPUT:

- $\mathcal{K}[X][x]$ a ring of polynomials.
- C a regular chain in $\mathcal{K}[X]$
- F a subset of $\mathcal{K}[X][x]$ s.t. $F \not\subset \{0\}$.

OUTPUT: A set of pairs $\{(A_1, g_1), \ldots, (A_r, g_r)\}$ such that

1. $\mathcal{R}(C) = \mathcal{R}(A_1) \cap \ldots \cap \mathcal{R}(A_r)$ is an irredundant characteristic decomposition.
2. $\mathcal{I}(C) \subset \mathcal{I}(A_i)$, $1 \leq i \leq r$.
3. $(F) = (g_i)$ in $\mathfrak{Q}(\mathcal{K}[X]/\mathcal{I}(A_i))[x]$.
4. $g_i \in (F) + \mathcal{I}(A_i)$.
5. $g_i = 0$ or $\mathsf{lcoeff}(g_i, x)$ does not belong to $3\mathfrak{d}(A_i)$.

$\mathcal{C} := \emptyset;$
$\mathcal{S} := \{ (C, F) \};$
while $\mathcal{S} \neq \emptyset$ do
    $(B, G) := \mathsf{pop}(\mathcal{S});$
    $g :=$ an element of G of lowest degree in $x;$
    $G := G \setminus \{g\};$
    $(\mathcal{Z}, \mathcal{U}) := \mathsf{split}(\mathcal{K}[X], B, \mathsf{lcoeff}(g, x));$
    if $\mathcal{Z} \neq \emptyset$ then
        $G' := G \cup \{\mathsf{tail}(g, x)\};$
        if $G' \subset \{0\}$ then
            $\mathcal{C} := \mathcal{C} \cup \{(A, 0) \mid A \in \mathcal{Z}\};$
        else
            $\mathcal{S} := \mathcal{S} \cup \{(A, G' \setminus \{0\}) \mid A \in \mathcal{Z}\};$
        fi;
    fi
    if $\mathcal{U} \neq \emptyset$ then
        $G' := \{\mathsf{srem}(f, g, x) \mid f \in G\};$
        if $G' \subset \{0\}$ then
            $\mathcal{C} := \mathcal{C} \cup \{(A, g) \mid A \in \mathcal{U}\};$
        else
            $\mathcal{S} := \mathcal{S} \cup \{(A, G' \cup \{g\}) \mid A \in \mathcal{U}\};$
        fi;

fi;
od;
return (C);

It is also possible (recommended) to reduce the sets G' by A, or only some element of A, before inserting the pairs $[A, G']$ in $\mathcal{S}$.

Termination: The algorithm can be seen as constructing a tree with root (C, F). Each node has a finite number of sons. The sons of a node (B, G) are some (A, G') where either $G' \subset \{0\}$, in which case it is a leaf, or the sum of the degrees in x of the elements of G' is lower than for G. A path in the tree thus gives a sequence of strictly decreasing positive integers. Any path must be finite and therefore the tree is finite. If split $(\mathcal{K}[X], *, *)$ terminates, pgcd $(\mathcal{K}[X][x], *, *)$ terminates.

Correctness. Assuming the correctness of split $(\mathcal{K}[X], *, *)$ we shall show that the while loop has the following invariants.

I1 $\mathcal{R}(C) = \bigcap_{(A,g) \in \mathcal{C}} \mathcal{R}(A) \cap \bigcap_{(A,G) \in \mathcal{S}} \mathcal{R}(A)$ is an irredundant characteristic decomposition.

I2 $\mathcal{I}(C) \subset \mathcal{I}(B)$ for all $(B, G) \in \mathcal{S}$,

I2' $\mathcal{I}(C) \subset \mathcal{I}(A)$ for all $(A, g) \in \mathcal{C}$,

I3 $(\Gamma) = (G)$ in $\mathfrak{Q}(\mathcal{K}[X]/\mathcal{I}(A))[x]$ for all $(A, G) \in \mathcal{S}$

I3' $(F) = (g)$ in $\mathfrak{Q}(\mathcal{K}[X]/\mathcal{I}(A))[x]$ for all $(A, g) \in \mathcal{C}$

I4 $G \subset (F) + \mathcal{I}(A)$, for all $(A, G) \in \mathcal{S}$

I4' $g \in (F) + \mathcal{I}(A)$, for all $(A, g) \in \mathcal{C}$

The invariants are easily checked to be true before the loop. That I1 and I2 are preserved comes from Output 1 and 2 of split. I2' is a direct consequence of I2.

I3 is preserved because, if the selected $g \in G$ is such that

- lcoeff$(g, x) = 0$ in $\mathcal{K}[X]/\mathcal{I}(B)$ then $g = \text{tail}(g, x)$ in $\mathcal{K}[X]/\mathcal{I}(B)$.
- init$(g) \notin \mathfrak{z}\mathfrak{d}(B)$ so that $(f, g) = (\text{srem}(f, g), g)$ in $\mathfrak{Q}(\mathcal{K}[X]/\mathcal{I}(B))[x]$, by Proposition 3.4.

I3' is a direct consequence of I3.

The set G in the algorithm starts from F and evolves by pseudo-division by an element in G or by setting to zero an element that belongs to $\mathcal{I}(A)$. I4 is thus preserved and I4' is a consequence of it.

8.2   Useful Properties for Splitting

This section contains the ingredients for the proof of the splitting algorithm. We first give the simple splits that can always be done on radical ideals and two properties to play with characteristic decompositions. We then give two additional properties of the outputs of the pgcd algorithm.

PROPOSITION 8.2. *Let F be a non empty subset of $\mathcal{K}[X]$. Then*

- $\langle (F) + (a\,b) \rangle = \langle (F) + (a) \rangle \cap \langle (F) + (b) \rangle$ *for any a, $b \in \mathcal{K}[X]$.*
- $\langle F \rangle = \langle F \rangle : H^\infty \cap \bigcap_{h \in H} \langle (F) + (h) \rangle$ *for any finite subset H of $\mathcal{K}[X]$.*

PROPOSITION 8.3. *Let $C \bigtriangleup c$ be a regular chain in $\mathcal{K}[X][x]$ s.t. lead$(c) = x$. Consider $b \in \mathcal{K}[X][x]$ s.t. $\deg(b, x) = \deg(c, x)$ and $h \in \mathcal{K}[X]$, $h \notin 3\mathfrak{d}(C)$ s.t. $h\,c \equiv b \mod \mathcal{I}(C)$. Then $\mathcal{I}(C \bigtriangleup c) = \mathcal{I}(C \bigtriangleup b)$.*

Proof. Let us note $h_c = \text{init}(c)$ and $h_b = \text{init}(b)$. Note that $h_b \equiv h\,h_c \mod \mathcal{I}(C)$. Therefore $h_b \notin 3\mathfrak{d}(C)$ and $C \bigtriangleup b$ is a regular chain. $\mathcal{I}(C \bigtriangleup b) = (\mathcal{I}(C) + (b)) : h_b^\infty = (\mathcal{I}(C) + (h\,c)) : (h h_c)^\infty$. We have $c \in (\mathcal{I}(C) + (h\,c)) : (h h_c)^\infty$ and $(\mathcal{I}(C) + (h\,c)) : (h h_c)^\infty \subset \mathcal{I}(C \bigtriangleup c)$ since $h \notin 3\mathfrak{d}(C \bigtriangleup c)$. The equality follows.

PROPOSITION 8.4. *Let $C \bigtriangleup c$ be a regular chain. Assume $\mathcal{R}(C) = \mathcal{R}(C_1) \cap \ldots \cap \mathcal{R}(C_r)$ is an irredundant characteristic decomposition. Then $\mathcal{R}(C \bigtriangleup c) = \mathcal{R}(C_1 \bigtriangleup c) \cap \ldots \cap \mathcal{R}(C_r \bigtriangleup c)$ is an irredundant characteristic decomposition.*

Proof. Since the decomposition is irredundant, $3\mathfrak{d}(C) = 3\mathfrak{d}(C_1) \cup \ldots \cup 3\mathfrak{d}(C_r)$. Thus init$(c) \notin 3\mathfrak{d}(C_i)$, $1 \leq i \leq r$, so that $C_1 \bigtriangleup c, \ldots, C_r \bigtriangleup c$ are regular chains.

Obviously $\mathcal{R}(C \bigtriangleup c) \subset \mathcal{R}(C_i \bigtriangleup c)$, for $1 \leq i \leq r$. Let P be a minimal prime of $\mathcal{R}(C \bigtriangleup c)$. $P \cap \mathcal{K}[X]$ is a minimal prime of $\mathcal{R}(C)$ (Proposition 4.5). By the irredundancy of the decomposition of $\mathcal{R}(C)$, there exists a unique i, $1 \leq i \leq r$, such that $P \cap \mathcal{K}[X]$ contains $\mathcal{R}(C_i)$. P contains thus a unique $\mathcal{R}(C_i \bigtriangleup c) = (\mathcal{R}(C_i), c) : \text{init}(c)^\infty$. Given the dimension of the minimal primes of $\mathcal{R}(C_i \bigtriangleup c)$ and $\mathcal{R}(C \bigtriangleup c)$ (Proposition 4.4), P is a minimal prime of $\mathcal{R}(C_i \bigtriangleup c)$. We have

$$\mathcal{R}(C \bigtriangleup c) \subset \mathcal{R}(C_1 \bigtriangleup c) \cap \ldots \cap \mathcal{R}(C_r \bigtriangleup c) \subset \bigcap_{P \text{ a minimal prime of } \mathcal{R}(C \bigtriangleup c)} P$$

We thus have proved that $\mathcal{R}(C \bigtriangleup c) = \mathcal{R}(C_1 \bigtriangleup c) \cap \ldots \cap \mathcal{R}(C_r \bigtriangleup c)$ is an irredundant characteristic decomposition.

Iterating the process we obtain the following result.

COROLLARY 8.5. *Let $A \bigtriangleup B$ be a regular chain and assume $\mathcal{R}(A) = \mathcal{R}(A_1) \cap \ldots \cap \mathcal{R}(A_r)$ is an irredundant characteristic decomposition. Then $\mathcal{R}(A \bigtriangleup B) = \mathcal{R}(A_1 \bigtriangleup B) \cap \ldots \cap \mathcal{R}(A_r \bigtriangleup B)$ is an irredundant characteristic decomposition.*

PROPOSITION 8.6. *Assume the pair (B, g) belongs to the output of pgcd $(\mathcal{K}[X][x], C, F)$. If $\deg(g, x) > 0$ then $B \bigtriangleup g$ is a regular chain and $((F) + \mathcal{I}(B)) : h^\infty = \mathcal{I}(B \bigtriangleup g)$, where $h = \text{init}(g)$.*

Proof. By pgcd.5, init$(g) = \text{lcoeff}(g, x) \notin 3\mathfrak{d}(B)$. Therefore $B \bigtriangleup g$ is a regular chain. From pgcd.3, $(F) = (g)$ in $\mathfrak{Q}(\mathcal{K}[X]/\mathcal{I}(B))[x]$. From Proposition 3.2 we have that for all $f \in F$ there exists $k \in h^\infty$ and $q \in \mathcal{K}[X][x]$ such that $k\,f = q\,g \mod \mathcal{I}(B)$. Therefore $F \subset (\mathcal{I}(B) + (g)) : h^\infty$. From pgcd.4, $(g) \subset (F) + \mathcal{I}(B)$. As $\mathcal{I}(B \bigtriangleup g) = (\mathcal{I}(B) + (g)) : h^\infty$, the equality follows.

PROPOSITION 8.7. *Let $C \vartriangle c$ be a regular chain in $\mathcal{K}[X][x]$ such that $\mathsf{lead}(c) = x$ and $f \in \mathcal{K}[X][x]$ such that $\deg(f,x) > 0$. Assume the pair (B,g) belongs to the output of $\mathsf{pgcd}(\mathcal{K}[X][x], C, \{f,c\})$. Then $B \vartriangle c$ is a regular chain, $g \neq 0$ and*

- *if $\deg(g,x) = 0$ then $f \notin 3\eth(B \vartriangle c)$ so that $\mathcal{I}(B \vartriangle c) : f^\infty = \mathcal{I}(B \vartriangle c)$.*
- *if $\deg(g,x) > 0$ then*
  1. *$\mathcal{R}(B \vartriangle c) : f^\infty = \langle \mathcal{R}(B) + (q) \rangle : (f\, h_q)^\infty$*
  2. *$\mathcal{I}(B \vartriangle c) \subset (\mathcal{I}(B) + (q)) : h_q^\infty$*

  where $q = \mathsf{squo}(c,g,x)$ and $h_q = \mathsf{lcoeff}(q,x) \notin 3\eth(B)$.

Proof. $B \vartriangle c$ is a regular chain by $\mathsf{kpgcd.1}$ and Proposition 8.4. By $\mathsf{pgcd.3}$, $g \neq 0$ since c can not be zero modulo $\mathcal{I}(B)$.

From $\mathsf{pgcd.4}$, there exists $a \in \mathcal{K}[X][x]$ such that $g = a\, f \mod \mathcal{I}(B \vartriangle c)$. If f belongs to $3\eth(B \vartriangle c)$ so does g. If $\deg(g,x) = 0$ then $g = \mathsf{lcoeff}(g,x) \notin 3\eth(B \vartriangle c)$ by $\mathsf{pgcd.5}$. This proves the first point.

If $\deg(g,x) > 0$, $B \vartriangle g$ is a regular chain and $c \in \mathcal{I}(B \vartriangle g)$, as seen in Proposition 8.6. Let us write $h_g = \mathsf{init}(g)$. There exists $h \in h_g^\infty$ such that $h\,c \equiv q\,g \mod \mathcal{I}(B)$ where $q = \mathsf{squo}(c,g,x)$. We have $h\,\mathsf{init}(c) \equiv h_q\, h_g \mod \mathcal{I}(B)$ and therefore $h_q \notin 3\eth(B)$.

By Proposition 8.3 and Lemma 4.3, $\mathcal{I}(B \vartriangle c) = \mathcal{I}(B \vartriangle q\,g) = (\mathcal{I}(B) + (q\,g)) : (h_q\, h_g)^\infty$. We first have the inclusion property $\mathcal{I}(B \vartriangle c) \subset (\mathcal{I}(B) + (q)) : h_q^\infty$, since the latter ideal is equal to $\mathcal{K}[X][x]$ when $\deg(q,x) = 0$ and equal to $\mathcal{I}(B \vartriangle q)$ when $\deg(q,x) > 0$. Then, by Proposition 8.2 $\langle \mathcal{I}(B) + (q\,g) \rangle = \langle \mathcal{I}(B) + (q) \rangle \cap \langle \mathcal{I}(B) + (g) \rangle$ so that $\mathcal{R}(B \vartriangle c) = \langle \mathcal{I}(B) + (q) \rangle : (h_q\, h_g)^\infty \cap \mathcal{R}(B \vartriangle g)$ and consequently $\mathcal{R}(B \vartriangle c) : f^\infty = \langle \mathcal{I}(B) + (q) \rangle : (f\, h_q\, h_g)^\infty$. Reasoning on the degree of q again we conclude $\langle \mathcal{I}(B) + (q) \rangle : (f\, h_q\, h_g)^\infty = \langle \mathcal{I}(B) + (q) \rangle : (f\, h_q)^\infty$.

8.3   A Sub-algorithm of the Split

The relatively-prime algorithm presented is a sub-algorithm of split. It generalizes the recursion at the end of Section 2.2 when dealing with non squarefree polynomials. Its role is to compute an irredundant decomposition of the saturation of a characterisable ideal by a polynomial. The call of split to relatively-prime requires in fact to compute saturations of ideals of $\mathcal{K}[X][x]$ of the type $\langle \mathcal{I}(C) + (c) \rangle : \mathsf{lcoeff}(c,x)^\infty$ as it is possible that $\deg(c,x) = 0$.

ALGORITHM 8.8. **Krelatively-prime**
INPUT:

- $\mathcal{K}[X][x]$ *a ring of polynomials.*
- C *a regular chain of $\mathcal{K}[X]$*
- $c \in \mathcal{K}[X][x]$ *such that $\mathsf{lcoeff}(c,x) \notin 3\eth(C)$*
- $f \in \mathcal{K}[X][x]$, $\deg(f,x) > 0$

OUTPUT: *A set $\bar{\mathcal{C}}$ of regular chains in $\mathcal{K}[X][x]$ such that $\bar{\mathcal{C}}$ is empty if $(\mathcal{I}(C) + (c)) : \mathsf{lcoeff}(c,x)^\infty : f^\infty = \mathcal{K}[X][x]$ and otherwise*

1. *$\mathcal{R}(C \vartriangle c) : f^\infty = \bigcap_{A \in \bar{\mathcal{C}}} \mathcal{R}(A)$ is an irredundant characteristic decomposition*

2. $\mathcal{I}(C \bigtriangleup c) \subset \mathcal{I}(A)$, $\forall A \in \bar{\mathcal{C}}$

$\mathcal{C} := \emptyset;$
$\mathcal{S} := \{(C,c)\};$
while $\mathcal{S} \neq \emptyset$ do
    $(A, a) := pop(\mathcal{S});$
    if $\deg(a, x) \neq 0$ then
        $G := pgcd\ (\mathcal{K}[X][x],\ A,\ \{a, f\});$
        $\mathcal{C} := \mathcal{C} \cup \{(B, a) \mid (B, b) \in G\ \text{and}\ \deg(b, x) = 0\};$
        $\mathcal{S} := \mathcal{S} \cup \{(B, squo(a, b, x)) \mid (B, b) \in G\ \text{and}\ \deg(b, x) > 0\};$
    fi;
od;
return ($\{A \bigtriangleup a \mid (A, a) \in \mathcal{C}\}$);

Note that if $\deg(c, x) = 0$ then $(\mathcal{R}(C) + (c)) : \mathsf{lcoeff}(c, x)^\infty = \mathcal{K}[X][x]$ and Krelatively-prime returns an empty set.

Termination: The degree in x is lower in $\mathsf{squo}(a, b, x)$ than in a, $\forall (B, b) \in G_0$. Termination is proved thanks to the same analogy to a tree as for pgcd.

Correctness. We shall check that the following properties are invariants of the while loop. The correctness then follows.

I0 $\bigcap_{(A,a) \in \mathcal{C} \cup \mathcal{S}} \mathcal{R}(A)$ is an irredundant characteristic decomposition.
I1 $(\mathcal{I}(C) + (c)) : \mathsf{lcoeff}(c, x)^\infty \subset (\mathcal{I}(A) + (a)) : \mathsf{lcoeff}(a, x)^\infty$, for all $(A, a) \in \mathcal{S}$
I1' $\mathcal{I}(C \bigtriangleup c) \subset \mathcal{I}(A \bigtriangleup a)$, for all $(A, a) \in \mathcal{C}$
I2 $\mathsf{lcoeff}(a, x) \notin \mathfrak{z0}(A)$, for all $(A, a) \in \mathcal{S}$
I2' $A \bigtriangleup a$ is a regular chain for all $(A, a) \in \mathcal{C}$
I3 $f \notin \mathfrak{z0}(A \bigtriangleup a)$, for all $(A, a) \in \mathcal{C}$.
I4 $\langle \mathcal{I}(C) + (c) \rangle : (f\,\mathsf{lcoeff}(c, x))^\infty =$
$$\bigcap_{(A,a) \in \mathcal{S}} \langle \mathcal{I}(A) + (a) \rangle : (f\,\mathsf{lcoeff}(a, x))^\infty \cap \bigcap_{(A,a) \in \mathcal{C}} \mathcal{R}(A \bigtriangleup a)$$

The invariants are satisfied before the while loop. Assume they are now true at the beginning of an iteration. If $\deg(a, x) = 0$ then $a = \mathsf{lcoeff}(a, x)$ and thus $1 \in \langle \mathcal{I}(A) + (a) \rangle : \mathsf{lcoeff}(a, x)^\infty$. Dropping this component does not affect I4. Nor does it affect I0, I1, I2, I2' and I3.

Let us consider the case $\deg(a, x) > 0$. This case can only happen when the input is such that $\deg(c, x) > 0$. By induction hypothesis on I2, $A \bigtriangleup a$ is a regular chain.

By pgcd.1 and Proposition 8.4, $\mathcal{R}(A) = \bigcap_{(B,b) \in G} \mathcal{R}(B)$ and $\mathcal{R}(A \bigtriangleup a) = \bigcap_{(B,b) \in G} \mathcal{R}(B \bigtriangleup a)$ are irredundant characteristic decompositions. I0 will be preserved. From pgcd.2, $\mathcal{I}(A) \subset \mathcal{I}(B)$.

Let (B, b) be an element of G. Note first that $b \neq 0$ since $a \notin \mathfrak{z0}(A)$. If $\deg(b, x) = 0$, then $f \notin \mathfrak{z0}(B \bigtriangleup a)$ by Proposition 8.7. The pair (B, a) is put in C so that I2' and I3 are preserved. From pgcd.2 we can say that $\mathcal{I}(A \vartriangle a) \subset \mathcal{I}(B \vartriangle a)$ and thus I1' is preserved by induction hypothesis on I1.

If $\deg(b, x) > 0$, by Proposition 8.7, $\mathcal{I}(B \vartriangle a) \subset (\mathcal{I}(B) + (q)) : h_{ab}^{\infty}$ and $\mathcal{R}(B \vartriangle a) : f^{\infty} = \langle \mathcal{R}(B) + (q_{ab}) \rangle : (f\,h_{ab})^{\infty}$, where $q_{ab} = \mathsf{squo}(a, b, x)$ and $h_{ab} = \mathsf{lcoeff}(q_{ab}, x) \notin \mathfrak{Z}\mathfrak{d}(B)$. On the one hand I1 and I2 are preserved and on the other hand we can write

$$\mathcal{R}(A \vartriangle a) : f^{\infty} = \bigcap_{(B,b) \in G} \mathcal{R}(B \vartriangle a) : f^{\infty}$$
$$= \bigcap_{\substack{(B,b) \in G \\ \deg(b,x) = 0}} \mathcal{R}(B \vartriangle a) \quad \cap \bigcap_{\substack{(B,b) \in G \\ \deg(b,x) > 0}} (\mathcal{R}(B) + (q_{ab})) : (f\,h_{ab})^{\infty} .$$

That insures that I4 is preserved.

8.4   Splitting Algorithm

In Section 2.4 we saw a way of splitting a product of field $\mathcal{K}[x] / \langle c \rangle$ according to a polynomial in $\mathcal{K}[x]$. The process relied on gcd computations over $\mathcal{K}$. The split algorithm we describe here is an extension to the product of field $\mathcal{K}(\mathfrak{T}(C))[\mathfrak{L}(C)]$ where C is a regular chain in $\mathcal{K}[X]$. It relies essentially on a pgcd algorithm and Gauss lemma Theorem 2.3 .

The algorithm split has the side effect of decreasing multiplicities. Take the simple case where we examine the splitting of $\mathcal{K}[x] / \langle c \rangle$, where $c = x(x + 1)^r$ according to the polynomial $f - (x+1)^e$ where $c < r$. We will obtain $\mathcal{K}[x] / \langle c \rangle = \mathcal{K}[x] / \langle g_0 \rangle \times \mathcal{K}[x] / \langle x \rangle$ where $g_0 = (x + 1)^e$ if $e < r$ and $g_1 = x$.

ALGORITHM 8.9. **split**
INPUT:

- $\mathcal{K}[X]$ a polynomial ring
- C a regular chain of $\mathcal{K}[X]$
- f a polynomial in $\mathcal{K}[X]$

OUTPUT: A pair $(\mathcal{Z}, \mathcal{U})$ of sets of regular chains in $\mathcal{K}[X]$ such that

1. $\mathcal{R}(C) = \bigcap_{A \in \mathcal{Z} \cup \mathcal{U}} \mathcal{R}(A)$ is an irredundant characteristic decomposition.
2. $\mathcal{I}(C) \subset \mathcal{I}(A), \forall A \in \mathcal{Z} \cup \mathcal{U},$
3. $f \in \mathcal{I}(A), \forall A \in \mathcal{Z}.$
4. $f \notin \mathfrak{Z}\mathfrak{d}(A), \forall A \in \mathcal{U}.$

   if $f = 0$ then
     return$(\ (\{C\}, \emptyset)\ )$;
   elif $f \in \mathcal{K}$ then
     return$((\emptyset, \{C\})\ )$;
   fi;
   $x := \mathsf{lead}(f)$;

if $x \notin \mathfrak{L}(C)$ then
   $F :=$ the set of coefficients of f, seen as a polynomial in x;
   $\mathcal{U} := \emptyset; \quad \mathcal{Z} = \emptyset;$
   $S := \{ (C_{<x}, F) \};$
   while $S \neq \emptyset$ do
      $(B, E) :=$ pop $(S);$
      $g :=$ pop$(E);$
      $(Z, U) :=$ split $(\mathcal{K}[X_{<x}], B, g);$
      $\mathcal{U} := \mathcal{U} \cup U;$
      if $E = \emptyset$ then
         $\mathcal{Z} := \mathcal{Z} \cup Z;$
      else
         $S := S \cup \{(A, E) \mid A \in Z\};$
      fi;
   od;
else
   $c :=$ the element of C with leader x;
   $G :=$ pgcd$(\mathcal{K}[X_{<x}][x], C_{<x}, \{f, c\});$
   $\mathcal{Z} := \{A \triangle a \mid (A, a) \in G, \deg(a, x) > 0\};$
   $\mathcal{U} := \{A \triangle c \mid \exists a, (A, a) \in G, \deg(a, x) = 0\};$
   $\mathcal{U} := \mathcal{U} \cup \displaystyle\bigcup_{\substack{(A,a) \in G \\ \deg(a,x) > 0}}$ relatively-prime$(\mathcal{K}[X][x], A, $squo$(c, a, x), f);$

fi;
$\mathcal{Z} := \{A \triangle C_{>x} \mid A \in \mathcal{Z}\};$
$\mathcal{U} := \{A \triangle C_{>x} \mid A \in \mathcal{U}\};$
return$((\mathcal{Z}, \mathcal{U}));$

In the case $C_x = 0$, we use Gauss lemma. Consider for instance $C = x\,(x+1)\,(x+2)$ as a regular chain in $\mathcal{K}[x, y]$ and $f = x\,y^2 + (x+1)\,y + 1$. To decide if f is not a zero divisor one needs to decide if one its coefficients is not a zero divisor. A simple inspection leads us to the fact $f \notin 3\mathfrak{d}(C)$ since one of its coefficient belongs to $\mathcal{K}$. Therefore $(\emptyset, C)$ is a valid output. [Kal93] and [Aub99] use a weakened version of Gauss lemma. The successive initials of f are inspected. For the example presented here, split would be recursively called for x and then for $x+1$. The output would therefore be $(\emptyset, \{C_1, C_2, C_3\})$ leading to more components and therefore redundancies in the computations.

Termination. Termination follows simply from termination of split$(\mathcal{K}[X_{<y}], *, *)$ and of pgcd$(\mathcal{K}[X_{<y}][y], *, *)$ for any $y \in X$.

Correctness. We shall assume that split$(\mathcal{K}[X_{<y}], *, *])$ and pgcd$(\mathcal{K}[X_{<y}][y], *, *])$ is correct for any $y \in X$ and prove correctness for split$(\mathcal{K}[X], *, *])$. For that we need to prove that after the conditional branching if $x \notin \mathfrak{L}(C)$ then [...] else [...] we have:

- $\mathcal{R}(C_{\leq x}) = \bigcap_{A \in \mathcal{U} \cup \mathcal{Z}} \mathcal{R}(A)$ is an irredundant characteristic decomposition.
- $f \equiv 0 \mod \mathcal{I}(A), \forall A \in \mathcal{Z}$
- $f \notin 3\eth(A), \forall A \in \mathcal{U}$.

Indeed Proposition 5.8 and 8.4 allow then to conclude. There are two different cases according to whether or not $x = \mathsf{lead}(f)$ appears as a leader of an element of C.

The case $x \notin \mathfrak{L}(C)$

Then $\mathcal{K}[X][x]/\mathcal{I}(C_{\leq x}) = (\mathcal{K}[X]/\mathcal{I}(C_{<x}))[x]$. Thus, according to Gauss lemma (Theorem 2.3), $f \in 3\eth(C_{\leq x})$ if and only if F, the set of coefficients of f considered as a polynomial in x, is included in $3\eth(C_{<x})$. The *while* loop inspects each coefficient in turn (in any order). It has the following invariants:

I0 $\mathcal{R}(C_{<x}) = \bigcap_{(A,E) \in \mathcal{S}} \mathcal{R}(A) \cap \bigcap_{A \in \mathcal{U} \cup \mathcal{Z}} \mathcal{R}(A)$ is an irredundant characteristic decomposition

I1 $\mathcal{I}(C_{<x}) \subset \mathcal{I}(A)$ for all $A \in \mathcal{Z} \cup \mathcal{U}$ and all $(A, E) \in \mathcal{S}$

I2 $F \setminus E \subset \mathcal{I}(A)$, for all $(A, E) \in \mathcal{S}$

I2' $F \not\subset 3\eth(A)$, for all $A \in \mathcal{U}$

I2" $F \subset \mathcal{I}(A)$, for $A \in \mathcal{Z}$

These invariants are obviously satisfied before the *while* loop. At each iteration I0 and I1 are kept true because of split.1 and split.2 on $\mathcal{K}[X_{<x}]$. I2 is easily seen to be kept true since $\mathcal{S}$ is augmented with the components modulo which the element of F treated is 0. I2' and I2" are consequences of I2 given how are augmented $\mathcal{Z}$ and $\mathcal{U}$.

The case $x \in \mathfrak{L}(C)$

Thanks to pgcd.1, pgcd.2 and Proposition 8.4 we have that $\mathcal{R}(C_{\leq x}) = \bigcap_{(A,a) \in G} \mathcal{R}(A \triangle c)$ is an irredundant characteristic decomposition and $\mathcal{I}(C_{\leq x}) \subset \mathcal{I}(A \triangle c), \forall (A, a) \in G$. For any pair $(A, a) \in G$, $a \neq 0$ because $c \neq 0 \mod \mathcal{I}(A)$.

For $(A, a) \in G$ such that $\deg(a, x) = 0$, $f \notin 3\eth(A \triangle c)$ by Proposition 8.7. That justifies the initialization of $\mathcal{U}$.

For (A, a) with $\deg(a, x) > 0$, $h_a = \mathsf{init}(a) = \mathsf{lcoeff}(a, x)$ does not belong to $3\eth(A)$, by pgcd.5. By pgcd.3, $f \in \mathcal{I}(A \triangle a)$ which justifies the value given to $\mathcal{Z}$. Let $h_c = \mathsf{init}(c)$. We have $\mathcal{R}(A \triangle c) = \mathcal{R}(A \triangle c) : h_a^\infty = \langle \mathcal{I}(A) + (c, f) \rangle : (h_c h_a)^\infty \cap \mathcal{R}(A \triangle c) : f^\infty$ by Lemma 4.3 and Proposition 8.2. This decomposition is irredundant if $1 \notin \mathcal{R}(A \triangle c) : f^\infty$. Now $\langle \mathcal{I}(A) + (c, f) \rangle : (h_a)^\infty = \mathcal{R}(A \triangle a)$ by Proposition 8.6. Since $h_c \notin 3\eth(A \triangle a)$ the previous decomposition can be written $\mathcal{R}(A \triangle c) = \mathcal{R}(A \triangle a) \cap \mathcal{R}(A \triangle c) : f^\infty$. By Proposition 8.7, $\mathcal{R}(A \triangle c) : f^\infty = \langle \mathcal{I}(A) + (q) \rangle : (f h_q)^\infty$ and $\mathcal{I}(A \triangle c) \subset (\mathcal{I}(A) + (q)) : h_q^\infty$ where $q = \mathsf{squo}(c, a, x)$ and $h_q = \mathsf{lcoeff}(q, x)$. The output properties of relatively-prime allow to conclude.

9   Characteristic Decomposition Algorithm

A pseudo-gcd algorithm with respect to a characterizable ideal given by a regular chain can be applied to compute a characteristic decomposition of the radical ideal generated by a finite family of polynomials. After describing and proving the algorithm, we shall discuss the membership test it gives to the radical ideal as well as how to refine the characteristic decomposition obtained.

In the description of this algorithm again, S is a set containing the data awaiting more computations, while C is a set of data for which the computation is completed. An element $(\hat{F}, \check{F}, C)$ of S, where $\hat{F}, \check{F}$ are subsets of $\mathcal{K}[X]$ and C is a regular chain, represents the radical ideal $\left\langle (\hat{F}) + \mathcal{I}(C) \right\rangle$. $\check{F}$ correspond to the already considered polynomials of F, in the sense that $\check{F} \subset \mathcal{I}(C)$.

ALGORITHM 9.1. **decompose**
INPUT:

 - $\mathcal{K}[X]$ a polynomial ring
 - F a nonempty set of polynomials in $\mathcal{K}[X]$

OUTPUT: A set C of regular chains such that

 - C is empty if $\langle F \rangle = \mathcal{K}[X]$.
 - $\langle F \rangle = \displaystyle\bigcap_{C \in \mathcal{C}} \mathcal{R}(C)$ otherwise.

$C := \emptyset$;
$S := \{(F, \emptyset, \emptyset)\}$;
while $S \neq \emptyset$ do
    $(\hat{F}, \check{F}, C) := pop \ (S)$;
    if $\hat{F} \subset \{0\}$ then
        $C := C \cup \{C\}$
    elif $\hat{F} \cap \mathcal{K} \neq \emptyset$
        $x := \min\{y \in X \mid \hat{F} \cap \mathcal{K}[X_{\leq y}] \neq \emptyset\}$;
        $F_x := \hat{F} \cap \mathcal{K}[X_{\leq x}]$;
        $G := pgcd \ (\mathcal{K}[X_{<x}][x], C, F_x)$;
        $S_0 := \{(\hat{F} \setminus F_x, \check{F} \cup F_x, B) \mid (B, 0) \in G\}$;
        $S_1 := \{(\hat{F} \cup \check{F} \cup B \cup \{g\}, \emptyset, \emptyset) \mid (B, g) \in G, \ \deg(g, x) = 0, \ g \neq 0\}$;
        $S_2 := \{(\hat{F} \setminus F_x, \check{F} \cup F_x, B \bigtriangleup g) \mid (B, g) \in G, \ \deg(g, x) > 0\}$;
        $S_2' := \{(\hat{F} \cup \check{F} \cup B \cup \{init(g)\}, \emptyset, \emptyset) \mid (B, g) \in G, \ \deg(g, x) > 0\}$;
        $S := S \cup S_0 \cup S_1 \cup S_2 \cup S_2'$
    fi;
od;
return(C);

Note the following difference with the versions of [Kal93, Aub99]: the regular chains computed up to a point are reintroduced in the components of S_1 and S_2' where the computation basically starts over. Computations are then easier if there is already a triangular set to start with.

Termination. We can visualize the algorithm as constructing a tree with root $(F, \emptyset, \emptyset)$. A node is given by a 3-tuple $(\hat{F}, \check{F}, C)$. A son of a node $(\hat{F}, \check{F}, C)$ is an element of the constructed sets S_0, S_1, S_2 or S_2'. A leaf is an element $(\emptyset, *, *)$.

For convenience, we shall introduce a dummy variable x_0 that we assume to be lower than all the variables of X. We write $\bar{X} = X \cup \{x_0\}$. We extend each 3-tuple $(\hat{F}, \check{F}, C)$ to a 4-tuple $(\hat{F}, \check{F}, C, y)$ where y is such that $C \subset \mathcal{K}[\bar{X}_{\le y}]$ and $\hat{E} \cap \mathcal{K}[\bar{X}_{\le y}] = \emptyset$. The root is now $(F, \emptyset, \emptyset, x_0)$.

A son $(\hat{F}, \check{F}, C, y)$ of a node $(\hat{E}, \check{E}, B, x)$ falls into one of the two following categories

Type 1: It is such that $\hat{F} \cup \check{F} = \hat{E} \cup \check{E}$ and $y > x$. This is the case of the 4-tuples in S_0 and S_2.

Type 2: $\check{F} = \emptyset$, $C = \emptyset$ and $y = x_0$. This is the case of the 4-tuples in S_1 and S_2'. Their main property is that the ideal generated by the set $(\hat{E} \cup \check{E}) \cap \mathcal{K}[\bar{X}_{<y}] = \check{E} \cap \mathcal{K}[\bar{X}_{<y}]$ is strictly included in the ideal generated by the set $(\hat{F} \cup \check{F}) \cap \mathcal{K}[\bar{X}_{<y}] = \hat{F} \cap \mathcal{K}[\bar{X}_{<y}]$. Indeed we introduce in $\hat{F}$ g or init(g) and we shall see in the correctness part of the proof that they do not belong to $(\check{E}) \subset \mathcal{I}(B)$. We also have $\hat{E} \cup \check{E} \subset \hat{F} \cup \check{F}$.

Assume that there is an infinite path in the tree. Since the set $\bar{X}$ is finite, there will be on this path an infinite sequence of nodes $(\hat{F}_i, \emptyset, \emptyset, x_0)$ of type 2 with a father having the same y as 4th component. This sequence defines a strictly increasing sequence of ideals in $\mathcal{K}[\bar{X}_{<y}]$, namely the ideals $(\hat{F}_i \cap \mathcal{K}[\bar{X}_{<y}])$. This contradicts the fact that $\mathcal{K}[\bar{X}_{<y}]$ is Noetherian.

Correctness. We shall show that the while loop has the following invariants.

I0 $F \subset \hat{F} \cup \check{F}$ and $\check{F} \subset \mathcal{I}(C)$, for all $(\hat{F}, \check{F}, C) \in \mathcal{S}$

I1 $\langle F \rangle = \bigcap_{(\hat{F}, \check{F}, C) \in \mathcal{S}} \left\langle (\hat{F}) + \mathcal{I}(C) \right\rangle \cap \bigcap_{C \in \mathcal{C}} \mathcal{R}(C)$

We first give a couple of easy properties that are used implicitly in the proof.

PROPOSITION 9.2. *Let I and J be ideals in a ring $\mathcal{K}[X]$. $\sqrt{I + J} = \sqrt{I + \sqrt{J}}$.*

PROPOSITION 9.3. *Let $I_0, I_1, \ldots, I_r$ be ideals in a ring $\mathcal{K}[X]$. Then $\sqrt{I_0 + \bigcap_{j=1}^{r} I_j} = \bigcap_{k=1}^{r} \sqrt{I_0 + I_k}$.*

I0 and I1 are obviously true before the while loop. Assume that they are true at the beginning of a new iteration treating the tuple $(\hat{F}, \check{F}, C)$. If $\hat{F} \subset \{0\}$ then $\left\langle (\hat{F}) + \mathcal{I}(C) \right\rangle = \mathcal{R}(C)$. If $\hat{F} \cap \mathcal{K} \setminus \{0\}$ then $\left\langle (\hat{F}) + \mathcal{I}(C) \right\rangle = \mathcal{K}[X]$ and the component can be dropped. We assume from now on that $\hat{F} \cap \mathcal{K} = \emptyset$.

By pgcd.1 and pgcd.2, $\check{F} \subset \mathcal{I}(C) \subset \mathcal{I}(B)$ and

$$\left\langle (\hat{F}) + \mathcal{I}(C) \right\rangle = \bigcap_{(B,g) \in G} \left\langle (\hat{F}) + \mathcal{I}(B) \right\rangle. \tag{1}$$

Take $(B, g) \in G$. There are three cases according to whether $g = 0$, $\deg(g, x) = 0$ or $\deg(g, x) > 0$. We examine these three cases separately.

If $g = 0$, we know that $F_x \subset \mathcal{I}(B)$ by pgcd.3. Thus

$$\left\langle (\hat{F}) + \mathcal{I}(B) \right\rangle = \left\langle (\hat{F} \setminus F_x) + \mathcal{I}(B) \right\rangle \tag{2}$$

If $\deg(g, x) = 0$ but $g \neq 0$, from pgcd.4, $g \in (F_x) + \mathcal{I}(B)$. By induction hypothesis on I0 we thus obtain

$$\langle F \rangle \subset \left\langle (\hat{F} \cup \check{F}) + (g) + (B) \right\rangle \subset \left\langle (\hat{F}) + \mathcal{I}(B) \right\rangle \tag{3}$$

If $\deg(g, x) > 0$, let $h_g = \operatorname{init}(g) = \operatorname{lcoeff}(g, x)$. Thanks to Proposition 8.2

$$\left\langle (\hat{F}) + \mathcal{I}(B) \right\rangle = \left\langle (\hat{F} \setminus F_x) + (F_x) + \mathcal{I}(B) \right\rangle$$
$$= \left\langle (\hat{F} \setminus F_x) + \langle (F_x) + \mathcal{I}(B) \rangle : h_g^\infty \right\rangle \cap \left\langle (\hat{F} \setminus F_x) + \langle (F_x) + \mathcal{I}(B) + (h_g) \rangle \right\rangle. \tag{4}$$

By Proposition 8.6, $B \triangle g$ is a regular chain and $((F_x) + \mathcal{I}(B)) : h_g^\infty = \mathcal{I}(B \triangle g)$. Equation (4) becomes

$$\left\langle (\hat{F}) + \mathcal{I}(B) \right\rangle = \left\langle (\hat{F} \setminus F_x) + \mathcal{I}(B \triangle g) \right\rangle \cap \left\langle (\hat{F}) + \mathcal{I}(B) + (h_g) \right\rangle \tag{5}$$

By induction hypothesis on I0, $F \subset \hat{F} \cup \check{F}$ and with pgcd.2, $\check{F} \subset \mathcal{I}(C) \subset \mathcal{I}(B)$. Thus

$$\langle F \rangle \subset \left\langle (\hat{F} \cup \check{F}) + (B) + (h_g) \right\rangle \subset \left\langle (\hat{F}) + \mathcal{I}(B) + (h_g) \right\rangle. \tag{6}$$

$\mathcal{S}$ is the set deprived from the tuple $\{(\hat{F}, \check{F}, C)\}$. With the induction hypothesis on I1 we can write

$$\langle F \rangle = \bigcap_{(\hat{E}, \check{E}, D) \in \mathcal{S}} \left\langle (\hat{E}) + \mathcal{I}(D) \right\rangle \cap \left\langle (\hat{F}) + \mathcal{I}(C) \right\rangle$$

With (1), (2) and (5) we can rewrite this equation as

$$\langle F \rangle = \bigcap_{(\hat{E}, \check{E}, D) \in \mathcal{S}} \left\langle (\hat{E}) + \mathcal{I}(D) \right\rangle \cap \bigcap_{(B, 0) \in G} \left\langle (\hat{F} \setminus F_x) + \mathcal{I}(B) \right\rangle \cap \bigcap_{\substack{(B, g) \in G \\ \deg(g, x) = 0}} \left\langle (\hat{F}) + \mathcal{I}(B) \right\rangle$$
$$\cap \bigcap_{\substack{(B, g) \in G \\ \deg(g, x) > 0}} \left(\left\langle (\hat{F} \setminus F_x) + (B \triangle g) : I_{B \triangle g}^\infty \right\rangle \cap \left\langle (\hat{F}) + \mathcal{I}(B) + (h_g) \right\rangle \right)$$

Intersecting both sides of this latter equation by $\left\langle (\hat{F} \cup \check{F}) + (g) + (B) \right\rangle$, for all $(B, g) \in G$ with $\deg(g, x) = 0, g \neq 0$, and $\left\langle (\hat{F} \cup \check{F}) + (B) + (h_g) \right\rangle$ for all $(B, g) \in G$ with $\deg(g, x) > 0$, we obtain, thanks to (3) and (6),

$$\langle F \rangle = \bigcap_{(\hat{E}, \check{E}, D) \in \mathcal{S}} \left\langle (\hat{E}) + \mathcal{I}(D) \right\rangle \cap \bigcap_{(B, 0) \in G} \left\langle (\hat{F} \setminus F_x) + \mathcal{I}(B) \right\rangle$$
$$\cap \bigcap_{\substack{(B, g) \in G \\ \deg(g, x) = 0}} \left\langle (\hat{F} \cup \check{F}) + (g) + (B) \right\rangle$$
$$\cap \bigcap_{\substack{(B, g) \in G \\ \deg(g, x) > 0}} \left(\left\langle (\hat{F} \setminus F_x) + \mathcal{I}(B \triangle g) \right\rangle \cap \left\langle (\hat{F} \cup \check{F}) + (B) + (h_g) \right\rangle \right)$$

This justifies that the elements pushed on the stack preserve the invariants I0 and I1.

Membership Test. With a characteristic decomposition, as computed by decompose, it is possible to test membership to the radical ideal generated by a finite set of polynomials. This is explained below. Nonetheless, since the decomposition is not always irredundant, it is not possible to give sufficient or necessary conditions for f to be invertible or a zero divisor modulo $\langle F \rangle$.

Consider a finite set of polynomials F in $\mathcal{K}[X]$ and compute its characteristic decomposition

$$\langle F \rangle = \mathcal{R}(C_1) \cap \ldots \cap \mathcal{R}(C_r).$$

For an element f of $\mathcal{K}[X]$, let $(\mathcal{Z}_i, \mathcal{U}_i)$ be the output of $\mathsf{split}(\mathcal{K}[X], C_i, f)$. For f to belong to $\langle F \rangle$ it is necessary and sufficient that all $\mathcal{U}_i$ be empty.

If in the characteristic decomposition all the regular chains C_i are squarefree, the test is simpler. In this case a necessary and sufficient condition for f to belong to $\langle F \rangle$ is that $\mathsf{srem}(f, C_i) = 0$ for all $1 \leq i \leq r$. In the next paragraph we show how to obtain a squarefree decomposition.

Refinement to Squarefree Regular Chains. It is possible to refine the output of decompose so that all the components are squarefree regular chains. In which case we have a decomposition that can be written

$$\langle F \rangle = \mathcal{I}(C_1) \cap \ldots \cap \mathcal{I}(C_r).$$

One way to proceed is to apply the following algorithm to each component of the output of decompose.

ALGORITHM 9.4. **sqrfree-decomposition**
INPUT:

- $\mathcal{K}[X]$ a ring of polynomials.
- C a regular chain of $\mathcal{K}[X]$

OUTPUT: A non empty set $\mathcal{C}$ of squarefree regular chains in $\mathcal{K}[X][x]$ such that $\mathcal{R}(C) = \bigcap_{A \in \mathcal{C}} \mathcal{I}(A)$ is an irredundant characteristic decomposition

$$
\begin{aligned}
&\mathcal{C} := \emptyset; \\
&\mathcal{S} := \{(C, \emptyset)\}; \\
&\text{while } \mathcal{S} \neq \emptyset \text{ do} \\
&\quad (\hat{B}, \check{B}) := pop(\mathcal{S}); \\
&\quad \text{if } \hat{B} = \emptyset \text{ then} \\
&\quad\quad \mathcal{C} := \mathcal{C} \cup \{C\}; \\
&\quad \text{else} \\
&\quad\quad x := \text{the lowest variable of } \mathfrak{L}(\hat{B}); \\
&\quad\quad b := \text{the element of } \hat{B} \text{ with leader } x;
\end{aligned}
$$

$$G := pgcd\ (\mathcal{K}[X][x], \check{B}, \{b, sep(b)\});$$
$$\mathcal{S} := \mathcal{S} \cup \{(\hat{B}_{>x}, A \bigtriangleup squo(c, a, x)) \mid (A, a) \in G\};$$
    fi;

od;

return (C);

The main ingredient to prove that the following properties are invariants of the while loop is Theorem 7.1.

I1 $\check{B}$ is a squarefree regular chain, for all $(\hat{B}, \check{B}) \in \mathcal{S}$

I1' B is a squarefree regular chain, for all $B \in \mathcal{C}$

I2 $\mathcal{R}(C) = \displaystyle\bigcap_{(\hat{B},\check{B})\in\mathcal{S}} \mathcal{R}(\check{B} \bigtriangleup \hat{B}) \cap \bigcap_{B\in\mathcal{C}} \mathcal{I}(B)$ is an irredundant characteristic decomposition.

References

[AK93]   S. A. Abramov and K. Y. Kvashenko. The greatest common divisor of polynomials that depend on a parameter. *Vestnik Moskovskogo Universiteta. Seriya XV. Vychislitel' naya Matematika i Kibernetika*, 2:65–71, 1993. translation in Moscow Univ. Comput. Math. Cybernet. 1993, no. 2, 59–64.

[ALMM99] P. Aubry, D. Lazard, and M. Moreno-Maza. On the theories of triangular sets. *Journal of Symbolic Computation*, 28(1-2), 1999.

[AMM99] P. Aubry and M. Moreno-Maza. Triangular sets for solving polynomial systems: a comparative implementation of four methods. *Journal of Symbolic Computation*, 28(1-2):125–154, 1999.

[ARSED03] P. Aubry, F. Rouillier, and M. Safey El Din. Real solving for positive dimensional systems. *Journal of Symbolic Computation*, 33(6):543–560, 2003.

[Aub99]   P. Aubry. *Ensembles triangulaires de polynomes et résolution de systèmes algébriques. Implantation en Axiom.* PhD thesis, Université de Paris 6, 1999.

[BJ01]   A. M. Bellido and V. Jalby. Spécifications des algorithmes de triangulation de systèmes algébro-élémentaires. *C. R. Acad. Sci. Paris*, Ser. I(334):155–159, 2001.

[BKRM01] D. Bouziane, A. Kandri Rody, and H. Maârouf. Unmixed-dimensional decomposition of a finitely generated perfect differential ideal. *Journal of Symbolic Computation*, 31(6):631–649, 2001.

[BL00]   F. Boulier and F. Lemaire. Computing canonical representatives of regular differential ideals. In C. Traverso, editor, *ISSAC*. ACM-SIGSAM, ACM, 2000.

[BLOP95] F. Boulier, D. Lazard, F. Ollivier, and M. Petitot. Representation for the radical of a finitely generated differential ideal. In A. H. M. Levelt, editor, *ISSAC'95*. ACM Press, New York, 1995.

[BLOP97] F. Boulier, D. Lazard, F. Ollivier, and M. Petitot. Computing representations for radicals of finitely generated differential ideals. Technical Report IT-306, LIFL, 1997.

[BW93]   T. Becker and V. Weispfenning. *Gröbner Bases - A Computational Approach to Commutative Algebra*. Springer-Verlag, New York, 1993.

[CG90]   S. C. Chou and X. S. Gao. Ritt-wu's decomposition algorithm and geometry theorem proving. In M. E. Stickel, editor, *10th International Conference on Automated Deduction*, number 449 in Lecture Notes in Computer Sicences, pages 207–220. Springer-Verlag, 1990.

[CGZ94]   S. C. Chou, X. S. Gao, and J. Z. Zhang. *Machine proofs in geometry*. World Scientific Publishing Co. Inc., River Edge, NJ, 1994. Automated production of readable proofs for geometry theorems, With a foreword by Robert S. Boyer.

[CH03]   T. Cluzeau and E. Hubert. Resolvent representation for regular differential ideals. *Applicable Algebra in Engineering, Communication and Computing*, 13(5):395–425, 2003.

[Cho88]   S. C. Chou. *Mechanical geometry theorem proving*. D. Reidel Publishing Co., Dordrecht, 1988. With a foreword by Larry Wos.

[DDD85]   J. Della Dora, C. Dicrescenzo, and D. Duval. About a new method for computing in algebraic number fields. In *Proceedings of EUROCAL 85*, number 204 in Lecture Notes in Computer Science, pages 289–290. Springer-Verlag, 1985.

[Del99]   S. Dellière. *Triangularisation de systèmes constructibles. Application à l'évaluation dynamique*. PhD thesis, Université de Limoges, 1999.

[Del00a] S. Dellière. D. M. Wang simple systems and dynamic constructible closure. Technical Report 16, Laboratoire d'Arithmétique, de Calcul Formel et d'Optimisation, Limoges, http://www.unilim.fr/laco, 2000.

[Del00b] S. Dellière. A first course to D7 with examples. www-lmc.imag.fr/lmc-cf/Claire.Dicrescenzo/D7, 2000.

[Del00c] S. Dellière. Pgcd de deux polynômes à paramètres : approche par la clôture constructible dynamique et généralisation de la méthode de S. A. Abramov, K. Yu. Kvashenko. Technical Report RR-3882, INRIA, Sophia Antipolis, 2000. http://www.inria.fr/RRRT/RR-3882.html.

[Del01]   S. Dellière. On the links between triangular sets and dynamic constructible closure. *J. Pure Appl. Algebra*, 163(1):49–68, 2001.

[Duv87]   D. Duval. *Diverses questions relatives au calcul formel avec des nombres algébriques*. PhD thesis, Institut Fourier, Grenoble, 1987.

[Eis94]   D. Eisenbud. *Commutative Algebra with a View toward Algebraic Geometry*. Graduate Texts in Mathematics. Springer-Verlag New York, 1994.

[FMM01]   M. V. Foursov and M. Moreno Maza. On computer-assisted classification of coupled integrable equations. In *Proceedings of the 2001 international symposium on Symbolic and algebraic computation*, pages 129–136. ACM Press, 2001.

[GC92]   X. S. Gao and S. C. Chou. Solving parametric algebraic sytems. In *ISSAC 92*, pages 335–341. ACM Press, New York, NY, USA, 1992.

[GCL92]   K. O. Geddes, S. R. Czapor, and G. Labahn. *Algorithms for computer algebra*. Kluwer Academic Publishers, Boston, MA, 1992.

[GD94]   T. Gómez-Dîaz. *Quelques applications de l'évaluation dynamique*. PhD thesis, Université de Limoges, 1994.

[Hub00]   E. Hubert. Factorisation free decomposition algorithms in differential algebra. *Journal of Symbolic Computation*, 29(4-5):641–662, 2000.

[Hub03]   E. Hubert. Notes on triangular sets and triangulation-decomposition algorithms. II Differential systems. In this volume, 2003.

[Kal93]   M. Kalkbrener. A generalized Euclidean algorithm for computing triangular representations of algebraic varieties. *Journal of Symbolic Computation*, 15(2):143–167, 1993.

[Kal98]   M. Kalkbrener. Algorithmic properties of polynomial rings. *Journal of Symbolic Computation*, 26(5):525–582, November 1998.

[Kol73]   E. R. Kolchin. *Differential Algebra and Algebraic Groups*, volume 54 of *Pure and Applied Mathematics*. Academic Press, New York-London, 1973.

[Laz91]   D. Lazard. A new method for solving algebraic systems of positive dimension. *Discrete and Applied Mathematics*, 33:147–160, 1991.

[Laz92]   D. Lazard. Solving zero dimensional algebraic systems. *Journal of Symbolic Computation*, 15:117–132, 1992.

[MKRS98]   H. Maârouf, A. Kandri Rody, and M. Ssafini. Triviality and dimension of a system of algebraic differential equations. *Journal of Automated Reasoning*, 20(3):365–385, 1998.

[MM97]   M. Moreno-Maza. *Calculs de pgcd au-dessus des tours d'extensions simples et résolution des systèmes d'équations algébriques*. PhD thesis, Université Paris 6, 1997.

[MMR95]   M. Moreno Maza and R. Rioboo. Polynomial gcd computations over towers of algebraic extensions. In *Applied algebra, algebraic algorithms and error-correcting codes (Paris, 1995)*, pages 365–382. Springer, Berlin, 1995.

[Mor99]   S. Morrison. The differential ideal $[P]:M^\infty$. *Journal of Symbolic Computation*, 28(4-5):631–656, 1999.

[Ric99]   D. Richardson. Weak Wu stratification in $\mathbf{R}^n$. *Journal of Symbolic Computation*, 28(1-2):213–223, 1999. Polynomial elimination—algorithms and applications.

[Rit30]   J. F. Ritt. Manifolds of functions defined by systems of algebraic differential equations. *Transaction of the American Mathematical Society*, 32:569–598, 1930.

[Rit32]   J. F. Ritt. *Differential Equations from the Algebraic Standpoint*. Amer. Math. Soc. Colloq. Publ., 1932.

[Rit50]   J. F. Ritt. *Differential Algebra*, volume XXXIII of *Colloquium publications*. American Mathematical Society, 1950. Reprinted by Dover Publications, Inc (1966).

[Sch00]   E. Schost. *Sur la résolution des systèmes polynomiaux à paramètres*. PhD thesis, École polytechnique, 2000.

[SL98]   J. Schicho and Z. Li. A construction of radical ideals in polynomial algebra. Technical Report 98-17, RISC-Linz, 1998. ftp://ftp.risc.uni-linz.ac.at/pub/techreports/1998/98-17.ps.gz.

[Sza98]   A. Szanto. *Computation with polynomial systems*. PhD thesis, Cornell University, 1998.

[Vas98]   W. V. Vasconcelos. *Computational Methods in Commutative Algebra and Algebraic Geometry*, volume 2 of *Algorithms and Computation in Mathematics*. Springer, Berlin, 1998.

[vzGG99]   J. von zur Gathen and J. Gerhard. *Modern computer algebra*. Cambridge University Press, New York, 1999.

[Wan93]   D. Wang. An elimination method for polynomial systems. *Journal of Symbolic Computation*, 16(2):83–114, August 1993.

[Wan98]   D. Wang. Decomposing polynomial systems into simple systems. *Journal of Symbolic Computation*, 25(3):295–314, 1998.

[Wan99]   D. Wang. *Elimination methods*. Texts and Monographs in Symbolic Computation. Springer-Verlag Wien, 1999.

[Wan00]   D. Wang. Computing triangular systems and regular systems. *Journal of Symbolic Computation*, 30(2):221–236, 2000.

[Wu78]   W. T. Wu. On the decision problem and the mechanization of theorem-proving in elementary geometry. *Sci. Sinica*, 21(2):159–172, 1978.

[Wu94]   W. T. Wu. *Mechanical theorem proving in geometries*. Springer-Verlag, Vienna, 1994. Basic principles, Translated from the 1984 Chinese original by Xiao Fan Jin and Dong Ming Wang.

[Wu00]   W. T. Wu. *Mathematics mechanization*. Kluwer Academic Publishers Group, Dordrecht, 2000. Mechanical geometry theorem-proving, mechanical geometry problem-solving and polynomial equations-solving.

Notes on Triangular Sets
and Triangulation-Decomposition Algorithms
II: Differential Systems

Evelyne Hubert

INRIA - Projet CAFE
2004 route des lucioles BP 93
F-06902 Sophia Antipolis
Evelyne.Hubert@inria.fr

Abstract. This is the second in a series of two tutorial articles devoted to triangulation-decomposition algorithms. The value of these notes resides in the uniform presentation of triangulation-decomposition of polynomial and differential radical ideals with detailed proofs of all the presented results. We emphasize the study of the mathematical objects manipulated by the algorithms and show their properties independently of those. We also detail a selection of algorithms, one for each task. The present article deals with differential systems. It uses results presented in the first article on polynomial systems but can be read independently.

1 Introduction

Given a system of partial differential equations we wish to compute a representation that is equivalent but for which we can analyze the solution set. The first essential problem arising is whether the system admits any solution, analytic or in terms of formal power series. The second central problem is to describe the arbitrariness coming into the solution set i.e. how many initial conditions can be chosen to ensure the existence and uniqueness of a solution. Those problems were addressed in the late nineteenth century by Cartan and Riquier with different viewpoints [18, 57]. We will not attempt here any historical review but wish to report on recent algorithmic development in *differential algebra* in the line of the work of Ritt and Kolchin [60, 40], that is partly based on the work of Riquier. Few bridges have been established between the different algebraic approaches and their algorithmic developments. W. Seiler gives a review of the different algebraic and algorithmic approaches in [67] and B. Malgrange advocates some mutual interaction in [45]. The article by A. Buium and P. J. Cassidy [15] gives an excellent account of the emergence and development of differential algebra and the collections of surveys and tutorials [39, 30] show the implications of differential algebra and the work of J. F. Ritt and E. Kolchin with diverse branches of mathematics. In particular the reader shall find the tutorial [68], that was communicated to the author late in the preparation of these notes, complementary to the present one.

F. Winkler and U. Langer (Eds.): SNSC 2001, LNCS 2630, pp. 40–87, 2003.

Differential algebra is an extension of polynomial algebra aimed at the analysis of systems of ordinary or partial differential equations that are polynomially nonlinear. To a differential equation we associate a differential polynomial and to a differential system we associate a radical differential ideal. Questions about the solution set of the differential system are best expressed in terms of that radical differential ideal. There are differential analogues to the Nullstellensatz, the Hilbert basis theorem and the decomposition into prime ideals. The latter point gives light to the old problem of singular solution of a single differential equation: the radical differential ideal of a single differential polynomial may split into several prime differential ideals. One of those describes the general solution and the others the singular solutions.

Loosely speaking, a triangulation-decomposition algorithm takes as input a system of differential equations and outputs a finite set of differential systems with specific shape and properties. The set of solutions of the original system is equal to the union of the nonsingular solutions of the output systems. The output systems are given by *coherent differential triangular sets* of differential polynomials. The notion of *differential triangular sets* relies on the definition of a ranking that is a total order on the derivatives compatible with derivation. For any ranking, existence and uniqueness of formal power series solutions are secured for coherent differential triangular systems [63] but convergence of these power series depends on the ranking in use [42]. Also, for so called orderly ranking, we can compute the *differential dimension polynomial* [40] of a coherent differential triangular set. It has some resemblance with the Hilbert polynomial and measures the arbitrariness coming into the nonsingular solutions i.e. the number of arbitrary functions. Triangulation-decomposition algorithms also belong to the class of differential elimination algorithm. For a system of differential equations $\Sigma = 0$ and an appropriate choice of ranking we can answer the following typical questions:

- is a differential equation (not apparent in $\Sigma = 0$) satisfied by all the solutions of the system $\Sigma = 0$?
- what are the differential equations satisfied by the solutions of $\Sigma = 0$ in a subset of the dependent variables? If Σ is a differential system in the unknown functions $y_1, \ldots, y_n$, one might be interested in knowing the equations governing the behavior of the component y_1 independently of the others.
- what are the lower order differential equations satisfied by the solutions of $\Sigma = 0$? In particular, one might inquire if the solutions of the system are constrained by purely algebraic equations i.e. differential equations of order zero.
- what are the ordinary differential equations in one of the independent variables satisfied by the solution set of $\Sigma = 0$? If $\Sigma = 0$ is a differential system where $t_1, \ldots, t_m$ are the independent variables, finding the ordinary differential equations in t_1 satisfied by the solutions might be used to solve explicitly the system.

In Section 8 we give examples on how those questions arise in some applications. The reader might want to look at that section for motivation.

All those questions require in fact, a way or another, a membership test to the radical differential ideal generated by the set of differential polynomials Σ. We consider thus that we have a good representation for the radical differential ideal generated by a finite set of differential polynomials if it provides a membership test. We shall seek a representation of that radical differential ideal as an intersection of *characterizable differential ideals*. In practice, characterizable differential ideals are defined by regular differential chains, i.e. a special kind of coherent differential triangular sets, that allow testing membership through *differential reduction*. That final decomposition is called a characteristic decomposition. To compute the characteristic decomposition of the radical differential ideal generated by a finite set of differential polynomials we shall proceed as follow in this paper: First compute a decomposition into differential ideals defined by a regular differential systems, i.e. a system consisting of a coherent differential triangular set of equations and some inequations. Then, a characteristic decomposition of each of the found *regular differential ideals* can be computed by purely algebraic means, i.e. without performing any differentiation [33]. For the decomposition into regular differential ideals we expose the Rosenfeld-Gröbner algorithm by Boulier *et al.* taken from [10]. Other alternatives are the factorization free versions of the algorithm of Ritt and Kolchin [40, IV.9] proposed in [33] and [13]. Rosenfeld-Gröbner is essentially different from those as the splitting scheme owes to the elimination theory of Seidenberg [66]. For the decomposition of regular differential ideals into characterizable differential ideals we apply the pseudo-gcd algorithm of Kalkbrener [37, 34]. There are also known alternatives [33, 11].

Ritt's approach was essentially constructive and already in [58] he gave a triangulation-decomposition algorithm for finitely generated radical differential ideals in the case of ordinary differential equations. The components of the output there were prime differential ideals instead of the wider class of characterizable ideals. Later Ritt gave a triangulation-decomposition algorithm for partial differential equations based on Riquier's notion of passivity [57, 36, 60] and Kolchin gave a triangulation-decomposition algorithm based on Rosenfeld's notion of coherence [61, 40]. Algorithmic aspects of differential algebra have drawn attention again around 1990 in view of applications to theorem proving, control theory, analysis of over determined differential systems and computation of Lie symmetries [74, 46, 27, 55, 56]. For want of effectiveness, the primality of the decomposition that requires factorization in towers of extension was to be removed. A key point for factorization free triangulation-decomposition algorithms was revealed in [9]. It is a new application of a lemma by Rosenfeld [61]. Rosenfeld's lemma establishes a bridge between differential algebra and polynomial algebra through *coherent autoreduced sets* and by generalization through *regular differential systems*, defined in [9]. Gröbner basis approaches to the problem of representation of differential ideals have been studied in [16, 52, 46]. Those approaches have practical interests and theoretical difficulties as *differential Gröbner bases* can be infinite.

This paper is organized as follows.

In Section 2 the fundamental definitions and results in differential algebra are reviewed. We introduce radical differential ideals and their decomposition into prime differential ideals. We state the differential basis theorem and discuss the differential Nullstellensatz. This section is all based on the reference books [60, 40].

In Section 3 we define differential rankings and show that they are well orders. We then define the central objects that are the differential triangular sets and give the algorithm of reduction by them. We also introduce differential characteristic sets.

Section 4 introduces coherence for differential triangular sets and proves Rosenfeld's lemma. Rosenfeld's lemma allows us to lift results about ideals defined by triangular sets in polynomial algebra to results about differential ideals defined by regular differential systems. These are the structure theorems on which effective algorithms rely. They first appeared in [9, 10].

Section 5 introduces characterizable differential ideals and *differential regular chains* as a way of constructing all such differential ideals. We show the fundamental principle that allows to compute a characteristic decomposition of a regular differential ideals by purely algebraic means. This result appeared in [33].

Section 6 presents the Rosenfeld-Gröbner algorithm of [10] to compute a decomposition into differential ideals defined by regular differential systems. The presentation is basically the original one. We nonetheless left out the delicate implementation of the analogue of the second Buchberger criterion.

Section 7 then gives an algorithm based on Kalkbrener's pseudo-gcd to refine the decomposition into a characteristic decomposition.

We give in Section 8 a couple of by now classical examples of applications of differential elimination algorithms. These include solving ordinary differential systems, index reduction of differential-algebraic equations, observability in control theory and symmetry analysis. Small examples computed with the MAPLE *diffalg* package [8] are presented. Many of these applications appeared and were given successful achievements with other algorithms and related software that are cited in the text.

For this paper we take the risk of introducing a new notation for the radical differential ideal generated by a set Σ. The classical notation $\{\Sigma\}$ can indeed induce confusion between sets and radical differential ideals. We suggest the notation $[\![\Sigma]\!]$, given that the differential ideal generated by Σ is $[\Sigma]$.

The algorithms presented are described by pseudo-code in the MAPLE style.

2   Differential Algebra

Consider the differential system

$$y_{ss}^2 - 2y_t y_{st} = y_t^2 - 1, \quad y_s^2 + y_t^2 = 1$$

where y is a function of s and t and subscripts indicate derivation w.r.t. these variables. The differential system is defined by the *differential polynomials*

$p = y_{ss}^2 - 2y_ty_{st} - y_t^2 + 1$ and $q = y_s^2 + y_t^2 - 1$. An analytic solution of the system is a common *zero*[1] A common zero of p and q must be a zero of any derivative of p or q and of any linear combination of those. For instance it is a zero of $r = p + q + \delta_s(q) = y_{ss}^2 + 2y_sy_{ss} + y_s^2$, where δ_s indicates the derivation according to s. Note that $r = (y_{ss} + y_s)^2$. Therefore a common zero of p and q is a zero of $y_{ss} + y_s$. The set of all the *differential and algebraic consequences* of p and q is the *radical differential ideal* generated by p and q. It has the same zero set as p and q and is in fact the biggest set having this property.

We start by giving the formal definitions and properties of differential rings and differential ideals. We then introduce the ring of differential polynomials together with the fundamental results.

2.1   Differential Rings and Ideals

All rings $\mathcal{R}$ that we consider are commutative and contain the field $\mathbb{Q}$ of rational numbers. A derivation δ on $\mathcal{R}$ is a map from $\mathcal{R}$ to $\mathcal{R}$ such that for all $a, b \in \mathcal{R}$

$$\delta(a + b) = \delta(a) + \delta(b) \qquad \delta(a\,b) = a\,\delta(b) + \delta(a)\,b$$

EXAMPLE 2.1. *On the polynomial ring* $\mathbb{Q}[s, t]$ *we can define the usual derivations* $\delta_s = \frac{\partial}{\partial s}$ *and* $\delta_t = \frac{\partial}{\partial t}$ *so that* $\delta_s(s) = 1$ *and* $\delta_s(t) = 0$ *and similarly for* δ_t. *These two derivations commute:* $\delta_s\,\delta_t = \delta_t\,\delta_s$. *We could also consider the Euler derivations* $\bar{\delta}_s = s\frac{\partial}{\partial s}$ *and* $\bar{\delta}_t = t\frac{\partial}{\partial t}$ *so that* $\bar{\delta}_s(s) = s$, $\bar{\delta}_s(t) = 0$ *and similarly for* $\bar{\delta}_t$. *For all examples in this paper we consider the usual derivations* δ_s *and* δ_t. *They extend uniquely to* $\mathbb{Q}(s, t)$.

Let $\Delta = \{\delta_1, \ldots, \delta_m\}$ be a set of pairwise commuting derivations on a ring $\mathcal{R}$ that are linearly independent over $\mathcal{R}$. We note Θ the free commutative monoid with a unit generated by Δ. An element θ of Θ is a *derivation operator*. It can be written $\theta = \delta_1^{e_1} \ldots \delta_m^{e_m}$ for some $(e_1, \ldots, e_m) \in \mathbb{N}^m$. Its *order* is then $\mathrm{ord}(\theta) = e_1 + \ldots + e_m$. The only derivation operator of order zero is the identity. We note Θ_r, $r \in \mathbb{N}$, the set of derivation operators of order r or lower and Θ^+ the set of derivation operators of positive order.

Endowed with a set Δ of commuting derivations, a ring (or a field) $\mathcal{R}$ becomes a *differential ring* (or a *differential field*). An element c of $\mathcal{R}$ is a *constant* if $\delta c = 0$ for all $\delta \in \Delta$. When Δ consists of a single derivation δ we speak of an ordinary differential ring. If Δ has more than one element we have a partial differential ring. From now on we shall speak of differential ring assuming that the set $\Delta = \{\delta_1, \ldots, \delta_m\}$ of derivations has been fixed.

In a differential ring $\mathcal{R}$, an ideal I of $\mathcal{R}$ is a *differential ideal* if it is stable under derivation, that is $\delta(a) \in I$, for all $a \in I$ and $\delta \in \Delta$. A differential ideal I is radical if $a^e \in I$, for $e \in \mathbb{N} \setminus \{0\}$, implies $a \in I$. It is prime if whenever a product ab belongs to I at least one of the factor, a or b, belongs to I.

---

[1] A precise definition of zero will appear later. It is nonetheless appropriate to think analytic or meromorphic solution of the associated system when one reads zero of a set of polynomial.

Given a set of elements Σ in a differential ring $\mathcal{R}$, we note $[\Sigma]$ the differential ideal generated by Σ i.e. the intersection of all the differential ideals containing Σ. $[\Sigma]$ is in fact the ideal generated by the elements of Σ together with all their derivatives: $[\Sigma] = (\Theta\Sigma)$. $[\![\Sigma]\!]$ denotes the radical differential ideal generated by Σ, i.e. the intersection of all radical differential ideals containing Σ. Because we assumed that $\mathbb{Q} \subset \mathcal{R}$, $[\![\Sigma]\!]$ is the radical of $[\Sigma]$ i.e. the set $\{q \in \mathcal{R} \mid \exists e \in \mathbb{N}, q^e \in [\Sigma]\}$.

Let H be a subset of $\mathcal{R}$. We denote by H^∞ the minimal subset of $\mathcal{R}$ that contains 1 and H and is stable by multiplication and division i.e. $a, b \in H^\infty \Leftrightarrow ab \in H^\infty$. When H consists of a unique element h we write h^∞ instead of $\{h\}^\infty$. For a differential ideal I we define the *saturation* of I by a subset H of $\mathcal{R}$ as $I : H^\infty = \{q \in \mathcal{R} \mid \exists h \in H^\infty \text{ s.t. } h\,q \in I\}$. $I \subset I : H^\infty$ and $I : H^\infty$ is easily seen to be a differential ideal with Proposition 2.2 below. If I is a prime differential ideal $I : H^\infty$ is either equal to I or $\mathcal{F}[\![Y]\!]$ according to whether $H \cap I$ is empty or not.

PROPOSITION 2.2. *Let a, b be elements of the differential ring $\mathcal{R}$. For any $\theta \in \Theta_e$, $a^{e+1} \theta b$ belongs to the ideal generated by the set $\{\psi(a\,b) \mid \psi \in \Theta, \psi \text{ divides } \theta\}$.*

Thus $a^{e+1} \theta(b) \in [a\,b]$, for all $\theta \in \Theta_e$ and $\phi(a)\psi(b) \in [\![ab]\!]$, for all $\phi, \psi \in \Theta$.

Proof. This is trivially true if $\mathrm{ord}(\theta) = 0$. Assume the property is true for all θ of order e or less. Let θ be a derivation operator of order $e+1$. We can write $\theta = \delta\bar{\theta}$ for some $\delta \in \Delta$ and $\bar{\theta} \in \Theta_e$. By induction hypothesis $a^{e+1}\bar{\theta}(b) = \sum_{\psi|\bar{\theta}} \alpha_\psi \psi(a\,b)$ for some $\alpha_\psi \in \mathcal{R}$. Thus

$$a^{e+2}\delta\bar{\theta}(b) = a\left(\delta\left(a^{e+1}\bar{\theta}(b)\right) - (e+1)a^e\delta(a)\bar{\theta}(b)\right)$$

$$= a\sum_{\psi|\bar{\theta}}\left(\delta(\alpha_\psi)\psi(a\,b) + \alpha_\psi\delta\psi(ab)\right) - (e+1)\delta(a)\sum_{\psi|\bar{\theta}}\alpha_\psi\psi(a\,b).$$

The conclusion follows.

2.2   Differential Polynomial Rings

Let $\mathcal{F}$ be a differential field for the derivations $\Delta = \{\delta_1, \ldots, \delta_m\}$. It is assumed that $\mathcal{F}$ contains $\mathbb{Q}$. Given a set of *differential indeterminates* $Y = \{y_1, \ldots, y_n\}$, we construct the ring of *differential polynomials*, $\mathcal{F}[\![Y]\!]$ that is a ring in the infinitely many variables $\Theta Y = \{\theta y, \ y \in Y, \ \theta \in \Theta\}$ called the *derivatives*. This is naturally a differential ring for Δ. A differential polynomial p corresponds to a differential equation $p = 0$.

EXAMPLE 2.3. *The ordinary differential equation $y'^2 - t\,y' + y = 0$ is modeled by the differential polynomial $q = (\delta_t y)^2 - t\,\delta_t y + y$ in the ordinary differential polynomial ring $\mathbb{Q}(t)[\![y]\!]$ where the derivation on $\mathbb{Q}(t)$ is the usual one, i.e. $\delta_t = \frac{d}{dt}$.*

Let us now look at a partial differential ring. Consider $\mathcal{F} = \mathbb{Q}(s, t)$ endowed with the derivations δ_s and δ_t of the previous example and $Y = \{y\}$. The differential polynomial $p = \delta_s y\,\delta_t y + s\,\delta_s y + t\,\delta_t y - y$ represent the differential equation

$y_s\, y_t + s\, y_s + t\, y_s - y = 0$ where we took the standard notation $y_s = \frac{\partial y}{\partial s}$, $y_t = \frac{\partial y}{\partial t}$. In the examples we shall in fact use this notation for differential polynomials so that p shall be written $y_s\, y_t + s\, y_s + t\, y_s - y$.

The analogue of the Hilbert basis theorem for polynomial rings is given by the basis or Ritt-Raudenbush theorem. The most general proof is given in [40]. The proof for ordinary differential ring is given in [60, 38].

THEOREM 2.4. *If J is any radical differential ideal in $\mathcal{F}\llbracket Y \rrbracket$ there exists a finite subset Σ of $\mathcal{F}\llbracket Y \rrbracket$ s.t. $J = \llbracket \Sigma \rrbracket$.*

This is equivalent to the fact that any increasing sequence of radical differential ideals is stationary. The result does not hold for differential ideals.

THEOREM 2.5. *Any radical differential ideal J in $\mathcal{F}\llbracket Y \rrbracket$ is the intersection of a finite number of prime differential ideals. When minimal, this decomposition is unique.*

The decomposition is minimal if none of the components contains another one. The prime differential ideals coming into the minimal decomposition are called the *essential prime components* of J.

A n-tuple $\xi = (\xi_1, \ldots, \xi_n)$ with components in a differential field extension $\mathcal{F}'$ of $\mathcal{F}$ is a zero of a differential polynomial $p \in \mathcal{F}\llbracket y_1, \ldots, y_n \rrbracket$ if p vanishes when we replace the y_i by the ξ_i.

EXAMPLE 2.6. *We considered in Example 2.3 the differential polynomial $p = y_t{}^2 - t\, y_t + y$ in $\mathbb{Q}(t)\llbracket y \rrbracket$. The elements $\frac{1}{4}t^2$ and $a(t - a)$ of $\mathbb{Q}(a,t)$, where a is any constant (it could be transcendental), are zeros of p. In a matching way, the minimal prime decomposition of $\llbracket p \rrbracket$ is given by*

$$\llbracket p \rrbracket = \llbracket p, y_{tt} \rrbracket \cap \llbracket 4\, y - t^2 \rrbracket \,.$$

The zero of the second component has to be $\frac{1}{4}\, t^2$. The other zero, $a\, (t - a)$, is a zero of the first component.

A zero of a differential polynomial is also a zero of any of its derivatives. If ξ is a common zero to a set of differential polynomials Σ, it is a zero of any linear combination over $\mathcal{F}[Y]$ of the elements of Σ and their derivatives. This amounts to say that ξ is a zero of $[\Sigma]$. One can see it is also a zero of $\llbracket \Sigma \rrbracket$ since any element q of $\llbracket \Sigma \rrbracket$ is such that $q^e \in [\Sigma]$ for some $e \in \mathbb{N} \setminus \{0\}$. One can inquire if there is a bigger set in $\mathcal{F}[Y]$ that admits the same zeros as Σ. The answer is given by the differential Nullstellensatz.

To any prime differential ideal P we can associate the differential field that is the quotient field of the integral differential ring $\mathcal{F}\llbracket Y \rrbracket / P$. The canonical image of $(y_1, \ldots, y_n)$ on that field is a zero of P. It is in fact a *generic zero*, i.e. every differential polynomial that vanishes on that zero belongs to P. If P contains $\llbracket \Sigma \rrbracket$, that image is also a zero of $\llbracket \Sigma \rrbracket$. That abstract construction of zeros allows us to state the differential Nullstellensatz, known also as the theorem of zeros.

THEOREM 2.7. *Let Σ be a subset of $\mathcal{F}[[Y]]$.*

- Σ *admits a zero iff* $1 \notin [[\Sigma]]$
- *a differential polynomial of $\mathcal{F}[[Y]]$ vanishes on all the zeros of Σ iff it belongs to $[[\Sigma]]$.*

Indeed, if $g \in \mathcal{F}[[Y]]$ vanishes on all zeros of Σ it must vanish in particular on the generic zeros of the essential prime components of $[[\Sigma]]$ and therefore it belongs to $[[\Sigma]]$. In addition to this abstract setting, Ritt details the *analytic case*, i.e. the case where $\mathcal{F}$ is a field of meromorphic functions in a region. If g does not belong to a prime differential ideal P, an analytic zero of P can be constructed so that it does not make g vanish. In the partial differential case, this relies on the Riquier's existence theorem [57].

Instead of trying to study directly the solutions of the differential system $\Sigma = 0$ we shall inspect $[[\Sigma]]$. Our goal is to give an adequate representation of $[[\Sigma]]$. This representation shall allow us to test membership to $[[\Sigma]]$ and to answer question about the zero set of Σ.

We shall discuss now systems of equations and inequations like $\Sigma = 0, H \neq 0$ where Σ and H are two finite but non empty set of differential polynomials. Its best differential ideal representation is $[[\Sigma]] : H^\infty$. Let us note $\mathcal{Z}(\Sigma/H)$ the subset of the zeros of Σ that are not zeros of any element of H. A differential polynomial p vanishes on $\mathcal{Z}(\Sigma/H)$ iff $p \in [[\Sigma]] : H^\infty$. Indeed, if h is the product the element of H then $h p$ vanishes on all the zeros of Σ so that $h p \in [[\Sigma]]$. Note that if J is a radical differential ideal of $\mathcal{F}[[Y]]$, $J : H^\infty$ is the intersection of the essential prime components of J that have an empty intersection with H. The generic zeros of the essential components of $[[\Sigma]] : H^\infty$ are elements of $\mathcal{Z}(\Sigma/H)$. Note nonetheless that a zero of $[[\Sigma]] : H^\infty$ can make a zero of an element of H. In the ordinary and analytic case, Ritt proves that those zeros are *adherent* to $\mathcal{Z}(\Sigma/H)$ [60].

EXAMPLE 2.8. *In the ordinary differential polynomial ring $\mathbb{Q}(t)[[y]]$, consider the differential polynomial $p = y_t^2 - 4 y^3$. The common zero to p and $h = y_t$ is $\xi = 0$. You can nonetheless show [59] that $[[p]] : h^\infty = [[p]]$ so that $\xi = 0$ is a zero of $[[p]] : h^\infty$. A general zero of p is given by $\frac{c^2}{(ct-1)^2}$ so that 0 is analytically embedded in a family of zeros of p for which h does not vanish.*

3   The Tools for Algorithms

This section defines the objects we manipulate for algorithms. We introduce differential rankings that are orders on the derivatives that are compatible with derivations. They provide well orders. Then it is possible to define consistently leaders and differential triangular sets in an analogous way to what is done in the polynomial case [34]. We give a reduction algorithm by differential triangular sets. A differential ranking induces a pre-order on differential triangular set and this allows to define differential characteristic set and prove their existence. We introduce all these concepts to set up algorithms. These tools are nonetheless fundamental in showing the Ritt-Raudenbush theorem of previous section.

In polynomial algebra, ideals defined by triangular sets have good structural properties. In differential algebra this happens if we add a condition, namely *coherence*. This is studied in next section.

In this section, and the following, we have a fixed differential polynomial ring $\mathcal{F}[Y] = \mathcal{F}[y_1, \ldots, y_n]$ with derivations $\Delta = \{\delta_1, \ldots, \delta_m\}$. The set of derivation operators is denoted Θ.

3.1 Differential Ranking

DEFINITION 3.1. *A differential ranking, or d-ranking, on a set of differential indeterminates Y is a total order on the set of derivatives ΘY that satisfies for all $u, v \in \Theta Y$ and all $\delta \in \Delta$ the two conditions*

$$u \leq \delta u, \qquad u \leq v \Rightarrow \delta u \leq \delta v$$

This of course implies that for any $\theta \in \Theta$ we have $u \leq \theta u$ and $u \leq v \Rightarrow \theta u \leq \theta v$. If $Y = \{y_1, \ldots, y_n\}$ and the derivations are $\Delta = \{\delta_1, \ldots, \delta_m\}$, any derivative of ΘY can be written as $\delta_1^{e_1} \ldots \delta_m^{e_m} y_i$ and is thus determined by a $(m+1)$-tuple $(i, e_1, \ldots, e_m)$ in $\mathbb{N}_n \times \mathbb{N}^m$, where $\mathbb{N}_n = \{1, \ldots, n\}$. Consequently, a d-ranking is a total order on $\mathbb{N}_n \times \mathbb{N}^m$ that satisfies for any $i, j \in \mathbb{N}_n$ and $e, f, g \in \mathbb{N}^m$ $(i, e) \leq (i, e+g)$ and $(i, e) \leq (j, f) \Rightarrow (i, e+g) \leq (j, f+g)$. The restriction of this order to $\{i\} \times \mathbb{N}^m$, for some $i \in \mathbb{N}_n$, is an extension of the direct product order on $\mathbb{N}^m$. Conversely, if there is a single differential indeterminate y, any order on $\mathbb{N}^m$ that is an extension of the direct product order of $\mathbb{N}^m$, noted $\preceq_{\mathbb{N}^m}$, provides a d-ranking in a natural way. Recall that a relation on a set M is well founded if every non empty subset of M has a minimal element for this relation. This is equivalent to the fact that there is no infinite decreasing sequence in M. See for instance [5, Chapter 4]. A well founded order is a well-order.

PROPOSITION 3.2. *A d-ranking is a well order.*

The essential ingredients for that is the first defining property of a d-ranking and Dickson's lemma.

Proof. Assume we have an infinite decreasing sequence of derivatives in ΘY. Since Y is finite, we can extract from it an infinite subsequence of derivatives $\{\theta_k y\}_{k \in \mathbb{N}}$ of a single differential indeterminate $y \in Y$ that is strictly decreasing. The infinite sequence $\{\theta_k y\}_{k \in \mathbb{N}}$ can be viewed as an infinite sequence of $\mathbb{N}^m$. By Dickson's lemma [5, Corollary 4.48], there exists a subsequence $\{\theta_{k_i} y\}_{i \in \mathbb{N}}$ such that $\theta_{k_i} y \preceq_{\mathbb{N}^m} \theta_{k_j} y$ for all $i < j$. This implies that $\theta_{k_i} y \leq \theta_{k_j} y$ for all $i < j$. This contradicts our assumption that the original sequence is strictly decreasing.

Study and classification of d-rankings are examined in [17, 62]. We introduce here only the most commonly used d-rankings. In Section 8 we shall see that a d-ranking is chosen according to the kind of properties on the solution set of a differential system we want to exhibit. A d-ranking is *orderly* if it satisfies $\psi y < \phi x$, for $x, y \in Y$ and $\phi, \psi \in \Theta$, as soon as the order of ϕ is greater that the order of ψ. Let Z be a subset of Y. A d-ranking on Y *eliminates* Z if $\phi y < \psi z$ for any $y \in Y \setminus Z$, $z \in Z$, $\psi, \phi \in \Theta$. We write $Y \setminus Z \ll Z$. A pure elimination d-ranking is such that $y_{\sigma(1)} \ll y_{\sigma(2)} \ll \cdots \ll y_{\sigma(n)}$ for some permutation σ of $\mathbb{N}_n$.

EXAMPLE 3.3. *Consider an ordinary differential polynomial ring* $\mathbb{Q}(t)\,[\![Y]\!]$, *with derivation* δ_t. *If there is a single differential indeterminate,* $Y = \{y\}$, *there is only one possible d-ranking:* $y < y_t < y_{tt} < \ldots$. *If* $Y = \{y, z\}$, *an orderly d-ranking is given by* $y < z < y_t < z_t < y_{tt} < \ldots$. *An elimination d-ranking is given by* $y < y_t < y_{tt} < \cdots < z < z_t < z_{tt} < \ldots$.

EXAMPLE 3.4. *Assume we have two derivations,* δ_s *and* δ_t, *and a single differential indeterminate* $Y = \{y\}$. *The lexicographic order on* $\mathbb{N}^2$ *induces the d-ranking* $y < y_s < y_{ss} < \cdots < y_t < y_{st} < y_{sst} < \ldots$. *An example of orderly d-ranking is:* $y < y_s < y_t < y_{ss} < y_{st} < y_{tt} < \ldots$.

From now on, when we speak of the differential polynomial ring $\mathcal{F}\,[\![Y]\!]$ we assume that we are given a differential ranking on Y. For $u \in \Theta Y$, we note $\Theta Y_{<u}$ the set of derivatives that rank lower than u. Similarly $\Theta Y_{\leq u}$ denotes the set of derivatives that are equal to or lower than u. These sets are finite when we have an orderly ranking but this needs not be the case for other d-rankings.

Let p be a differential polynomial of $\mathcal{F}\,[\![Y]\!]$ that is not in $\mathcal{F}$. The *leader* and the *initial* of p are respectively the highest ranking derivative appearing in p and the coefficient of its highest power in p. We shall denote them as $\mathsf{lead}(p)$ and $\mathsf{init}(p)$ respectively. The *separant* of p is the formal derivative of p w.r.t. its leader, i.e. $\dfrac{\partial p}{\partial \mathsf{lead}(p)}$. It is denoted by $\mathsf{sep}(p)$.

If we take here $u = \mathsf{lead}(p)$ we can write

$$p = a_d\, u^d + a_{d-1}\, u^{d-1} + \cdots + a_0,$$

where $a_i \in \mathcal{F}[\Theta Y_{<u}]$, $a_d \neq 0$. Then $\mathsf{init}(p) = a_d$ and $\mathsf{sep}(p) = d\, a_d\, u^{d-1} + (d - 1)\, a_{d-1} u^{d-2} + \cdots + a_1$. We define furthermore the *rank* of p, $\mathsf{rank}(p)$, to be the term u^d and $\mathsf{tail}(p) = p - \mathsf{init}(p)\mathsf{rank}(p)$ that is $a_{d-1}\, u^{d-1} + \cdots + a_0$ in the above write up.

For $\delta \in \Delta$, the derivatives appearing in $\delta(p)$ consists of derivatives v appearing in p and of their derivatives δv. The second defining property of d-ranking allows us thus to state that $\delta \mathsf{lead}(p) = \mathsf{lead}(\delta p)$ for any $\delta \in \Delta$. Furthermore, from the definition of the separant, we can write

$$\delta(p) = \mathsf{sep}(p)\, \delta(\mathsf{lead}(p)) + \sum_{v \in \Theta Y} \frac{\partial p}{\partial v}\, \delta v.$$

Thus $\delta(p)$ has degree one in its leader, $\delta(\mathsf{lead}(p))$ and $\mathsf{sep}(p)$ is the initial of $\delta(p)$. By induction, for any $\theta \in \Theta^+$, $\theta\mathsf{lead}(p)$ and $\mathsf{sep}(p)$ are respectively the leader and the initial of θp and θp has degree one in $\theta\mathsf{lead}(p)$.

For two elements $p, q \in \mathcal{F}\,[\![Y]\!]$, we say that p has higher rank (or ranks higher) than q, and write $\mathsf{rank}(q) < \mathsf{rank}(p)$, if either

- $q \in \mathcal{F}$ and $p \notin \mathcal{F}$
- $\mathsf{lead}(q)$ ranks lower than $\mathsf{lead}(p)$
- $\mathsf{lead}(q) = \mathsf{lead}(p) = u$ and $\deg(q, u) < \deg(p, u)$.

The d-ranking on the differential indeterminates thus induces a pre-order[2] on $\mathcal{F}[\![Y]\!]$.

A differential polynomial q is *partially reduced* w.r.t. p if no proper derivative of $\mathsf{lead}(p)$ appears in q; q is *reduced* w.r.t. p if q is partially reduced w.r.t. to p and the degree of q in $\mathsf{lead}(p)$ is strictly less than the degree of p in $\mathsf{lead}(p)$.

EXAMPLE 3.5. *In* $\mathbb{Q}(s,t)\{y\}$ *endowed with a d-ranking such that* $y < y_s < y_t < y_{ss} < y_{st} < y_{tt}$,

- $q = y_{ss} - y_t$ *is partially reduced with respect to* $p = y_{tt} + y_s$ *but not w.r.t.* $p_2 = y_s + y$
- $q = (y_t)^3 + y_s$ *is partially reduced w.r.t.* $p = y_t^2 + y_s^2$ *but not reduced.*

Let F be a finite subset of $\mathcal{F}[\![Y]\!]$ and u a derivative of ΘY. We note ΘF the set of differential polynomials consisting of the element of F together with all their derivatives. $\Theta F_{<u}$ (respectively $\Theta F_{\leq u}$) denotes the set consisting of all elements of F together with all their derivatives the leaders of which are of lower (respectively lower or equal) rank than u. In other words $\Theta F_{<u} = \Theta F \cap \mathcal{F}[\Theta Y_{<u}]$ and $\Theta F_{\leq u} = \Theta F \cap \mathcal{F}[\Theta Y_{\leq u}]$. $\Theta F_{<u}$ and $\Theta F_{\leq u}$ need not be finite. The following property follows directly from Proposition 2.2.

PROPOSITION 3.6. *Let* F *and* H *be finite subsets of* $\mathcal{F}[\![Y]\!]$ *and* $u \in \Theta Y$. *If* $p \in (\Theta F_{<u}) : H^\infty$ *then* $\theta p \in (\Theta F_{<\theta u}) : H^\infty$ *for any* $\theta \in \Theta$.

3.2 Differential Triangular Sets and Differential Reduction

In the ordinary differential case and w.r.t to an elimination ranking, the *chains* of [60] coincide with the *autoreduced set* of [40]. The *differential chains* defined here are slightly less restrictive but still allow to give the Ritt-Kolchin definition and properties of *characteristic set* of differential ideals. An intermediate step in defining differential chains is the notion of *differential triangular sets* from [9] and reduction by these. In the algorithm of Section 6 the additional notion of *weak differential triangular set* arises.

DEFINITION 3.7. *A subset* A *of* $\mathcal{F}[\![Y]\!]$ *is a* weak differential triangular set, *if*

- *no element of* A *belongs to* $\mathcal{F}$
- *for any two distinct elements* a, b *of* A $\mathsf{lead}(a)$ *is not a derivative of* $\mathsf{lead}(b)$.

It is a *differential triangular set*, or *d-triangular set* for short, if furthermore any element of A is partially reduced w.r.t. the other ones.

It is an *autoreduced set* if no element of A belongs to $\mathcal{F}$ and each element of A is reduced w.r.t. all the others.

---

[2] We take the convention that a pre-order is a relation that is reflexive, transitive and connex. Therefore the difference with an order is that $a \leq b$ and $b \leq a$ does not imply $a = b$.

An autoreduced set is a (weak) d-triangular set.

EXAMPLE 3.8. *With the d-ranking $y < y_s < y_t < y_{ss} < y_{st} < y_{tt} < \ldots$ the subset $\{y_{tt} - y_{ss}^2, y_s - y\}$ is a weak d-triangular, $\{y_{tt} - y_s^2, y_s - y\}$ is a d-triangular set and $\{y_{tt}, y_s\}$ is an autoreduced set.*

PROPOSITION 3.9. *A weak d-triangular set is finite.*

Proof. This is of course another application of Dickson's lemma. Assume for contradiction that a weak d-triangular set is infinite. Take $\{u_l\}_{l \in \mathbb{N}}$ the sequence of the leaders of this infinite weak d-triangular set. Since Y is finite, there must be an infinite subsequence $\{u_{l_k}\}_{k \in \mathbb{N}}$ such that $u_{l_k} = \theta_k y$ for a $y \in Y$. $\{\theta_k y\}_{k \in \mathbb{N}}$ can be viewed as an infinite sequence of $\mathbb{N}^m$. By Dickson's lemma, there exists a subsequence $\{\theta_{k_i} y\}_{i \in \mathbb{N}}$ such that $\theta_{k_i} y \preceq_{\mathbb{N}^m} \theta_{k_j} y$ for all $i < j$. This means that $\theta_{k_j} y$ is a derivative of $\theta_{k_i} y$. This contradicts the definition.

Let $a_1, a_2, \ldots, a_r$ be the elements of a weak d-triangular set A indexed such that $\mathsf{lead}(a_1) < \mathsf{lead}(a_2) < \ldots < \mathsf{lead}(a_r)$. We write $A = a_1 \triangle a_2 \triangle \ldots \triangle a_r$. We can also use the $\triangle$ notation to construct weak d-triangular sets by mixing differential polynomials a and weak d-triangular sets A, B in the following ways: $a \triangle A$ (respectively $A \triangle a$) denotes the weak d-triangular set $A \cup \{a\}$ if the leader of a ranks lower (respectively higher) than the leader of any element of A; $A \triangle B$ denotes the weak d-triangular set $A \cup B$ if the leader of any element of A ranks lower than the leader of any element of B and $A \cup B$ forms a weak d-triangular set.

We denote $\mathfrak{L}(A) = \mathsf{lead}(A)$, $I_A = \mathsf{init}(A)$ and $S_A = \mathsf{sep}(A)$ the sets of leaders, initials and separants of the elements of A. Also $H_A = I_A \cup S_A$ is the set of the initials and the separants of A.

For $u \in \mathfrak{L}(A)$, A_u denotes the element of A having u as leader. For $u \in \Theta Y$, $A_{<u}$ (respectively $A_{\leq u}$) denotes the elements of A with leader ranking lower than u (respectively lower or equal to u). $A_{>u}$ denotes the elements of A with leader ranking higher than u. $A_{<u}$ and $A_{\leq u}$ do not include any derivative of the elements of A contrary to $\Theta A_{<u} = \Theta A \cap \mathcal{F}[\Theta Y_{<u}]$ and $\Theta A_{\leq u} = \Theta A \cap \mathcal{F}[\Theta Y_{\leq u}]$. Note that $\Theta A_{<u}$ and $\Theta A_{\leq u}$ need not be finite nor triangular.

EXAMPLE 3.10. *In $\mathbb{Q}(s,t) [\![y]\!]$ endowed with the d-ranking s.t. $y < y_s < y_t < y_{ss} < y_{tt} < \ldots$ consider the d-triangular set $A = y_s - sy \triangle y_t + y$. Then $\Theta A_{\leq y_{tt}} = \{y_s - sy, y_t + y, y_{ss} - sy_s - y, y_{st} - sy_t, y_{tt} + y_s, y_{tt} + y_t\}$.*

When A is the empty set we take the convention that $[A] = [\![A]\!] = [0]$.

A differential polynomial is said to be (partially) reduced w.r.t. a weak d-triangular set A when it is (partially) reduced w.r.t. each element of A. Given an element $q \in \mathcal{F} [\![Y]\!]$ we can compute $s \in S_A^\infty$ and $\mathsf{pd\text{-}red}(q, A)$ that is partially reduced w.r.t. A such that $s\,q \equiv \mathsf{pd\text{-}red}(q, A) \mod [A]$. Similarly, we can compute $h \in H_A^\infty$ and $\mathsf{d\text{-}red}(q, A)$ that is reduced w.r.t. A such that $h\,q \equiv \mathsf{d\text{-}red}(q, A) \mod [A]$. We give examples of algorithms for differential reduction and partial differential reduction. In the partial differential case, it is neither enough to reduce successively by the elements of A starting with the highest ranking one nor starting with the lowest ranking one. This is illustrated in the next example.

EXAMPLE 3.11. In $\mathbb{Q}(s,t)\{y\}$ endowed with the ranking $y < y_s < y_t < y_{ss} < y_{st} < y_{tt} < \dots$ consider the differential polynomials $a_1 = y_{ss} - y_t$, $a_2 = y_t^2 - y_s$, $p = y_{st}$ and $q = y_{sss}$. $A = a_1 \triangle a_2$ is an autoreduced set. On the one hand $\mathsf{d\text{-}red}(p, a_2) = p$ and $\mathsf{d\text{-}red}(p, a_1) = y_{ss}$ that is not reduced w.r.t. A. On the other hand $\mathsf{d\text{-}red}(q, a_1) = q$ and $\mathsf{d\text{-}red}(q, a_2) = y_{st}$ that is not reduced w.r.t. A.

ALGORITHM 3.12. **pd-red**.
  INPUT: $q \in \mathcal{F}[Y]$, A a weak d-triangular set of $\mathcal{F}[Y]$.
  OUTPUT: $\bar{q} \in \mathcal{F}[Y]$ s.t.
  $-$ $\bar{q}$ is partially reduced w.r.t. A
  $-$ $\exists s \in S_A^\infty$ s.t. $s\,q \equiv \bar{q} \mod (\Theta A_{\leq u})$ where $u = \mathsf{lead}(q)$
  $-$ $\mathsf{rank}(\bar{q}) \leq \mathsf{rank}(q)$
  $\bar{q} := q;$
  while $\bar{q}$ is not partially reduced w.r.t. A do
    $v :=$ the highest ranking element of $\Theta^+ \mathcal{L}(A)$ that appears in $\bar{q}$.
    $a :=$ any element of $\{b \in A \mid v$ is a derivative of $\mathsf{lead}(b)\};$
    $\theta :=$ the element of Θ s.t. $\theta\,\mathsf{lead}(a) = v;$
    $\tilde{q} := \mathsf{srem}(\bar{q}, \theta a, v);$
    $\bar{q} := \tilde{q};$
  od;
  $return(\bar{q});$

Contrary to the algorithm proposed in [40, I.9], this algorithm taken from [10] is not completely deterministic in the sense that at each step there may be a choice among the elements of A by which one can reduce. This freedom is an advantage in the sense we can make criteria to choose elements with lower degree or smaller coefficients in order to limit expression swell.

By definition of weak d-triangular sets, $\Theta^+ \mathcal{L}(A) \cap \mathcal{L}(A) = \emptyset$. Therefore θ needs to be in Θ^+. Consequently θa is of degree one in v with initial $\mathsf{sep}(a)$. The pseudo-division relationship obtained is $\tilde{s}\bar{q} \equiv \tilde{q} \mod (\theta a)$ for some $\tilde{s} \in \mathsf{sep}(a)^\infty$. Furthermore $\tilde{q}$ is free of v, and the derivatives appearing in $\tilde{q}$ consist of derivatives that were in $\bar{q}$ or derivatives lower than v. The highest ranking element of $\Theta^+ \mathsf{lead}(A)$ that appears in $\tilde{q}$ is of lower rank than v. The sequence of the v's appearing in the while loop is thus strictly decreasing. In virtue of Proposition 3.2, the algorithm terminates. Note that the first v to be selected has to rank lower than $u = \mathsf{lead}(p)$. Thus the differential polynomials used for pseudo-division belong to $\Theta A_{\leq u}$. The final relationship $s\,q \equiv \bar{q} \mod (\Theta A_{\leq u})$, for some $s \in S_A^\infty$, follows from the intermediate pseudo-division relationships.

The algorithm to compute a full reduction is no different. For a derivative $u \in \Theta Y$ occurring in $\bar{q}$, we shall say that the differential polynomial $\bar{q} \in \mathcal{F}[Y]$ is reduced w.r.t. A at u if u is not a proper derivative of an element of $\mathcal{L}(A)$ or, if u belongs to $\mathcal{L}(A)$, the degree of q in u is lower than the degree of A_u in u.

ALGORITHM 3.13. **d-red**.
  INPUT: $q \in \mathcal{F}[Y]$, A a weak d-triangular set of $\mathcal{F}[Y]$.
  OUTPUT: $\bar{q} \in \mathcal{F}[Y]$ s.t.

- $\bar{q}$ reduced w.r.t. A
- $\exists h \in H_A^\infty$ s.t. $h\,q \equiv \bar{q} \mod (\Theta A_{\leq u})$ where $u = lead(q)$
- $rank(\bar{q}) \leq rank(q)$

$\bar{q} := q;$

while $\bar{q}$ is not reduced w.r.t. A do

    $v :=$ the highest ranking derivative $u \in \Theta\mathcal{L}(A)$ in $\bar{q}$ s.t. $\bar{q}$ is not reduced

    w.r.t. A at u;

    $a :=$ any element of $\{b \in A \mid v$ is a derivative of $lead(b)\};$

    $\theta :=$ the element of Θ s.t. $\theta\,lead(a) = v;$

    $\tilde{q} := srem(\bar{q}, \theta a, v);$

    $\bar{q} := \tilde{q};$

od;

return$(\bar{q});$

The full reduction algorithm of [40, I.9] by an autoreduced set first proceed by a partial reduction then an algebraic reduction. That strategy is not possible when we reduce by a weak d-triangular set.

Note that $\tilde{q}$ is reduced w.r.t. to A at v and that the other derivatives appearing in $\tilde{q}$ consist of derivatives that were in $\bar{q}$ or derivatives lower than v. The sequence of the v's appearing in the while loop is therefore strictly decreasing and thus the algorithm terminates (Proposition 3.2). According to whether $v \in lead(A)$ or $v \in \Theta^+lead(A)$, we have a pseudo-division relationship $\tilde{h}\bar{q} \equiv \tilde{q}$ mod (θa), where $h \in sep(a)^\infty$ or $h \in init(a)^\infty$. The final relationship $h\,q \equiv \bar{q}$ mod $[A]$, for some $h \in H_A^\infty$ follows from these intermediate pseudo-division relationships.

3.3 Differential Characteristic Sets

In order to introduce and show the existence of differential characteristic sets we define a pre-order on d-triangular sets. This pre-order is also of use in proving the termination of algorithms.

DEFINITION 3.14. Let $A = a_0 \triangle \ldots \triangle a_r$ and $B = b_0 \triangle \ldots \triangle b_s$ be two weak d-triangular sets. We say that A has lower rank than B and write $rank(A) < rank(B)$ if there exists k, $0 \leq k \leq r,s$, such that $rank(a_i) = rank(b_i)$ for all $1 \leq i \leq k$ and either $k = s < r$ or $rank(a_k) < rank(b_k)$.

If $A = B \triangle C$ then $rank(A) < rank(B)$ reflecting the inclusion of differential ideals $[B] \subset [A]$. In particular, if B is the empty set and A is not then $rank(A) < rank(B)$.

PROPOSITION 3.15. The pre-order on weak d-triangular set is well-founded.

Proof. Assume we have an infinite sequence of weak d-triangular sets $\{A_j\}_{j \in \mathbb{N}}$ that is strictly decreasing. We shall construct from it an infinite autoreduced set of derivatives. Since this is not possible by Proposition 3.9, the result is proved by contradiction.

Let $a_j^{(1)}$ be the element of lowest rank in A_j. The sequence $\{\text{rank}(a_j^{(1)})\}_{j \in \mathbb{N}}$ is decreasing and thus must be stationary after a certain $j_1 \in \mathbb{N}$. Let $u_1 = \text{lead}(a_j^{(1)}) = u_1$ for $j \geq j_1$. For $j > j_1$, A_j must have at least two elements. Let $a_j^{(2)}$ be the second lowest ranked element in A_j, for $j > j_1$. The sequence $\{\text{rank}(a_j^{(2)})\}_{j > j_1}$ is decreasing and therefore must become stationary after a $j_2 > j_1$. Let $u_2 = \text{lead}(a_j^{(2)})$ for $j \geq j_2$. Because u_1 and u_2 are the leaders of the two lowest elements of A_{j_2} they form an autoreduced set. Continuing this way, we construct an infinite autoreduced set of derivatives $\{u_n\}_{n \in \mathbb{N}}$.

For the same reason as in the polynomial case we need to introduce specific d-triangular sets in order to define differential characteristic sets.

DEFINITION 3.16. *A differential triangular set A is a differential chain if for all $x \in \mathfrak{L}(A)$ red$(A_x, A_{<x})$ has the same rank as A_x.*

Note that owing to the definition of d-triangular sets, A_x is partially reduced w.r.t. A and therefore only algebraic reduction is needed to reduce A_x by $A_{<x}$. Basically this definition says that we can associate an autoreduced set to a differential chain and the same discussion than in the polynomial case applies.

The definition of a differential characteristic set of a differential ideal is no different than the definition of a characteristic set of an ideal in the polynomial case and the equivalence between the different definitions is shown similarly.

DEFINITION 3.17. *Let I be a differential ideal in $\mathcal{F}[\![Y]\!]$. A differential chain A contained in I is a differential characteristic set of I if one of the following equivalent conditions holds:*

1. A is of minimal rank among the differential chains contained in I.

2. there is no non zero element of I reduced w.r.t. A.

3. $\forall q \in I$, d-red$(q, A) = 0$.

The existence of a differential characteristic set for any differential ideal is secured by the first definition and by Proposition 3.15. The first definition also secures that any two characteristic sets of a differential ideal have the same rank.

Note that if d-red$(p, A) = 0$, for $p \in \mathcal{F}[\![Y]\!]$ and A a weak d-triangular set, then $p \in [A] : H_A^\infty$. If A is a differential characteristic set of a differential ideal I we thus have $A \subset I \subset [A] : H_A^\infty$.

PROPOSITION 3.18. *If C is a differential characteristic set of a prime differential ideal P then $P = [C] : H_C^\infty = [C] : S_C^\infty$.*

To prove that $P = [C] : H_C^\infty$ it is sufficient to note that no initial and no separant of C can belong to P and therefore if d-red$(p, C) = 0$ then $p \in P$. The fact that $P = [C] : S_C^\infty$ will be proved more generally in Section 5. Indeed, recent algorithmic improvement in differential algebra owes to the idea of using a wider class of differential ideals than prime differential ideals. We define in Section 5 a less restrictive class of differential ideals that have those properties of prime differential ideals exhibited in the proposition above.

4   Regular Differential Systems

A necessary and sufficient condition for the existence of a solution to a system of differential equations such as

$$\frac{\partial y}{\partial s} = f(s, t), \quad \frac{\partial y}{\partial t} = g(s, t)$$

is that $\dfrac{\partial f}{\partial t} = \dfrac{\partial g}{\partial s}$. Coherence is a property that generalizes this condition for systems given by d-triangular sets. Coherence implies *formal integrability* and is a cousin concept of Riquier's *passivity* [57, 36]. Riquier-Janet approach is discussed in [2]. Informally speaking, the approach through coherence is to the approach through passivity what Gröbner bases are to *involutive* or *Riquier bases*.

The theorem that F. Boulier named Rosenfeld's lemma [6], appeared in [61]. A generalization of use for specialization in positive characteristic appears in [40, III.8]. As for algorithms, in characteristic zero, Kolchin applied this theorem only in the special case of prime differential ideals [40, IV-9] to compute prime characteristic decompositions.

F. Boulier and coauthors [9] showed the fundamental relevance of this theorem for effective algorithms to compute representation of radical differential ideals that allow a membership test. Since then, several generalization were shown. The original theorem applies to ideals $[A] : H_A^\infty$ defined by a coherent autoreduced set A. We give the version of [10] that applies to differential ideals $[A] : H^\infty$ defined by a *regular differential system* (A, H). The proof is nonetheless no different than the original one. The approach of E. Mansfield [46] does not go through d-triangular sets explicitly. [51] gives a generalization of Rosenfeld's lemma that applies in this context.

As a general principle Rosenfeld's lemma makes the bridge between differential algebras and polynomial algebras through ideals that are defined by coherent sets of differential polynomials. In virtue of this principle, many a property of ideals defined by d-triangular sets can be *lifted* to differential ideals defined by coherent d-triangular sets.

In this section we first define the coherence of a d-triangular set and give a finite test for it. We then prove Rosenfeld lemma and give the structure theorem for differential ideals defined by regular differential systems.

4.1   Coherence

We define *coherence of a d-triangular set A away from a set H*. Classically coherence of A is away from H_A. In Section 6, H will be the set of differential polynomials by which it has been allowed to premultiply for reduction and therefore are intrinsically assumed not to vanish identically. H can thus involve separants or initials of elements that are no longer in A.

Testing coherence from its definition is an impossible task since it imposes infinitely many conditions. We give the usual finite test that is analogous to the construction of S-polynomials in the Gröbner bases theory. These analogues to the S-polynomial we call the Δ-polynomials[3]. Inspired by the Gröbner bases theory, an analogue to the second Buchberger criterion was given in [10]. We present this criterion but the real difficulty of it is in its implementation.

The ideals $(A):H^\infty$, where A is a weak d-triangular set and H is a finite set, we shall encounter can be understood as living in the polynomial ring $\mathcal{F}[\Theta Y]$ but also as living in $\mathcal{F}[X]$ where X is a subset of derivatives that include all the ones present in the differential polynomials of A and H. If V is a finite set of derivatives such that $A, H \subset \mathcal{F}[V]$ and I is the ideal $(A):H^\infty$ considered in $\mathcal{F}[V]$ then, since $\mathcal{F}[V]$ is Noehterian,

– I is finitely generated: there exists a finite set $G \subset \mathcal{F}[V]$ such that $I = (G)$
– I admits a primary decomposition in $\mathcal{F}[V]$ say $I = Q_1 \cap \ldots \cap Q_r$.

If X is any set of derivatives such that $V \subset X$ and $I^e, Q_1^e, \ldots, Q_2^e$ are the extension of $I, Q_1, \ldots Q_r$ to $\mathcal{F}[X]$, then $I^e \cap \mathcal{F}[V] = I$ and $Q_1^e \cap \mathcal{F}[V] = Q_1, \ldots, Q_r^e \cap \mathcal{F}[V] = Q_r$. The point is that the polynomial ring in which $(A):H^\infty$ is considered is not very relevant. Since A and H are finite sets, $(A):H^\infty$ is finitely generated, admits a primary decomposition and any computations on $(A):H^\infty$ can be made in a polynomial ring with finitely many variables.

Two derivatives u and v have a *common derivative* if there exist $\psi, \phi \in \Theta$ such that $\psi u = \phi v$. This happens when u and v are the derivative of the same differential indeterminate. Assume $u = \delta_1^{e_1} \ldots \delta_m^{e_m} y$ and $v = \delta_1^{f_1} \ldots \delta_m^{f_m} y$. Then any $\delta_1^{g_1} \ldots \delta_m^{g_m} y$ with $g_i \geq \max(e_i, f_i)$ is a common derivative of u and v. If we take $g_i = \max(e_i, f_i)$ we obtain the *lowest common derivative* of u and v and that is noted $\mathsf{lcd}(u, v)$.

DEFINITION 4.1. *Let A be a weak d-triangular set in $\mathcal{F}[\![Y]\!]$ and H a subset of $\mathcal{F}[\![Y]\!]$. A is said to be coherent away from H (or H-coherent for short) if:*

– $S_A \subset H^\infty$
– *whenever $a, b \in A$ are such that $\mathsf{lead}(a)$ and $\mathsf{lead}(b)$ have a common derivative, say $v = \psi(\mathsf{lead}(a)) = \phi(\mathsf{lead}(b))$, $\phi, \psi \in \Theta$, then*

$$\mathsf{sep}(b)\,\psi(a) - \mathsf{sep}(a)\,\phi(b) \in (\Theta A_{<v}):H^\infty.$$

Note that the leader of $\mathsf{sep}(b)\,\psi(a) - \mathsf{sep}(a)\,\phi(b)$ is lower than v and obviously $\mathsf{sep}(b)\,\psi(a) - \mathsf{sep}(a)\,\phi(b) \in [A]:H^\infty$. What is required here is that $\mathsf{sep}(b)\,\psi(a) - \mathsf{sep}(a)\,\phi(b)$ can be obtained algebraically over the derivatives of the element of A with leader lower than v.

Testing coherence can be done in fact with finitely many tests. For each pair of differential polynomial a and b in the weak d-triangular set it is sufficient to look at the differential polynomial corresponding to the lowest common derivative between $\mathsf{lead}(a)$ and $\mathsf{lead}(b)$.

---

[3] Though the analogy goes this way today, it is worth recalling that this construction of Δ-polynomial dates from 1959's paper of Rosenfeld.

DEFINITION 4.2. *Let a and b be differential polynomials in $\mathcal{F}[\![Y]\!]$. We note $\Delta(a,b)$ the Δ-polynomial of a and b that is defined as follow. If lead(a) and lead(b) have no common derivative, then $\Delta(a,b) = 0$. Otherwise let $\psi, \phi \in \Theta$ be s.t. lcd(lead(a), lead(b)) $= \psi(lead(a)) = \phi(lead(b))$. Then*

$$\Delta(a,b) = \mathsf{sep}(b)\,\psi(a) - \mathsf{sep}(a)\,\phi(b)$$

We could in fact replace sep(a) and sep(b) by their respective quotients with gcd(sep(a), sep(b)) to limit expression swell when we compute the Δ-polynomials.

PROPOSITION 4.3. *Let A be a weak d-triangular set and H a subset of $\mathcal{F}[\![Y]\!]$, $S_A \subset H$. If for all $a, b \in A$ we have $\Delta(a,b) \in (\Theta A_{<v}) : H^\infty$, where $v =$ lcd(lead(a), lead(b)), then A is H-coherent.*

Proof. Note first that for any $h, p \in \mathcal{F}[\![Y]\!]$ and $\theta \in \Theta$, $\theta(hp) \equiv h\theta(p) \mod (\gamma(p)\ |\ \gamma$ divides θ, $\gamma \neq \theta$). Assume that $v = $ lcd(lead(a), lead(b)) $= \psi(lead(a)) = \phi(lead(b))$ for some $\psi, \phi \in \Theta$. Then $\Delta(a,b) = \mathsf{sep}(b)\,\psi(a) - \mathsf{sep}(a)\,\phi(b)$. Any other common derivative of lead(a) and lead(b) can be written θv for some $\theta \in \Theta$. By the first remark $\theta(\Delta(a,b)) \equiv \mathsf{sep}(b)\,\theta\psi(a) - \mathsf{sep}(a)\,\theta\phi(b) \mod (\gamma\psi(a), \gamma\phi(b)\ |\ \gamma$ divides θ, $\gamma \neq \theta$). According to the hypotheses and Proposition 3.6, $\theta(\Delta(a,b)) \in (\Theta A_{<\theta v}) : H^\infty$. It follows that $\mathsf{sep}(b)\,\theta\psi(a) - \mathsf{sep}(a)\,\theta\phi(b) \in (\Theta A_{<\theta v}) : H^\infty$. $\square$

The simplest test for coherence is thus the following. It gives only a sufficient condition as shows the example.

PROPOSITION 4.4. *A weak d-triangular set A in $\mathcal{F}[\![Y]\!]$ is H_A-coherent if d-red($\Delta(a,b), A$) $= 0$, for all $a, b \in A$.*

EXAMPLE 4.5. *With the orderly d-ranking $y < y_t < y_s < \ldots$ consider the d-triangular set $A = a \bigtriangleup b$ where $a = y_t^2 - t^2$ and $b = (y_t - t)\,y_s - 2t + ts$. Then $\Delta(b,a) = 2\,y_t\delta_t(b) - (y_t - t)\delta_s(a) = 2\,y_t(y_s\,y_{tt} - y_s + s - 2)$ is such that d-red($\Delta(a,b), A$) $= (2 - s)(y_t + t)$. Nonetheless, $\Delta(a,b) \in (\Theta A_{<y_{st}}) : H_A^\infty = (\delta_t b, a, b) : (y_t(y_t - t))^\infty$ since $y_t + t \in (A) : H_A^\infty$.*

Note that a differential characteristic set C of a differential ideal I must be H_C-coherent. Indeed $\Delta(a,b) \in [C] \subset I$ for all $a, b \in C$. Thus $\Delta(a,b)$ is reduced to zero by C as prescribed by Definition 3.17.

We proceed to give the analogue of the second Buchberger criterion introduced in [10]. This allows to check coherence by forming less Δ-polynomials. To that end, let us introduce a notation for lowest common derivation operators. Assume $\phi, \psi \in \Theta$ can be written $\phi = \delta_1^{e_1} \ldots \delta_m^{e_m}$ and $\psi = \delta_1^{f_1} \ldots \delta_m^{f_m}$. We define the lowest common derivation operator of ϕ and ψ to be $\phi \diamond \psi = \delta_1^{g_1} \ldots \delta_m^{g_m}$ where $g_i = \max(e_i, f_i)$. In order to get familiar with that notation in use for the rest of the section, note that if lead(a) $= \phi y$ and lead(b) $= \psi y$ for some $y \in Y$ and $\phi, \psi \in \Theta$ then

$$\Delta(a,b) = \mathsf{sep}(b)\frac{\phi \diamond \psi}{\phi}(a) - \mathsf{sep}(a)\frac{\phi \diamond \psi}{\psi}(b)$$

and the coherence condition is that

$$\Delta(a,b) \in (\Theta A_{<\phi \diamond \psi y}):H^\infty.$$

THEOREM 4.6. *Let A be a weak d-triangular set and H a subset of $\mathcal{F}[\![Y]\!]$ that contains the separants of A. Let $p, q, r \in A$ be s.t. lead$(p) = \phi y$, lead$(q) = \psi y$ and lead$(r) = \theta y$ for some $y \in Y$ and $\phi, \psi, \theta \in \Theta$. If the three following conditions are satisfied*

1. *θ divides $\phi \diamond \psi$*
2. *$\Delta(r, p) \in (\Theta A_{<\theta \diamond \phi y}):H^\infty$*
3. *$\Delta(r, q) \in (\Theta A_{<\theta \diamond \psi y}):H^\infty$*

then $\Delta(p, q) \in (\Theta A_{<\phi \diamond \psi y}):H^\infty$

Proof. Note first the easy equivalence $\phi \diamond \theta$ divides $\phi \diamond \psi$ $\Leftrightarrow$ θ divides $\phi \diamond \psi$ $\Leftrightarrow$ $\psi \diamond \theta$ divides $\phi \diamond \psi$. By application of Proposition 3.6 to the hypotheses we obtain:

$$\frac{\phi \diamond \psi}{\phi \diamond \theta}(\Delta(r,p)) \in (\Theta A_{<\phi \diamond \psi\, y}):H^\infty, \qquad \frac{\phi \diamond \psi}{\psi \diamond \theta}(\Delta(r,p)) \in (\Theta A_{<\phi \diamond \psi\, y}):H^\infty.$$

Since for all $\gamma \in \Theta$ we have $\gamma(hp) \equiv h\gamma(p) \mod (\beta(p) \mid \beta \text{ divides } \gamma,\ \beta \neq \gamma)$ we can write from the above memberships

$$s_p \frac{\phi \diamond \psi}{\theta}(r) - s_r \frac{\phi \diamond \psi}{\phi}(p) \in (\Theta A_{<\phi \diamond \psi\, y}):H^\infty,$$

$$s_p \frac{\phi \diamond \psi}{\theta}(r) - s_r \frac{\phi \diamond \psi}{\psi}(q) \in (\Theta A_{<\phi \diamond \psi\, y}):H^\infty.$$

Multiplying the differential polynomials by s_q and s_p respectively and subtracting one to the other, we get $s_r \Delta(p, q) \in (\Theta A_{<\phi \diamond \psi\, y}):H^\infty$. As $s_r \in H$, the conclusion follows.

4.2   Rosenfeld's Lemma

To apply Rosenfeld lemma to H-coherent d-triangular sets we need some additional reduction hypothesis on the pair (A, H). These properties were encapsulated in the name *regular differential systems* in [9, 10].

DEFINITION 4.7. *A pair (A, H) is a regular differential system if*

- *A is a d-triangular set*
- *H is a set of nonzero differential polynomials partially reduced w.r.t. A*
- *$S_A \subset H^\infty$*
- *for all $a, b \in A$, $\Delta(a, b) \in (\Theta A_{<v}):H^\infty$ where $v = \text{lcd}(\text{lead}(a), \text{lead}(b))$.*

Note that the fact that A is a d-triangular set implies that the separants of A are partially reduced w.r.t. A.

It is practical to define *regular differential ideals* as differential ideals of the type $[A]:H^\infty$ where (A, H) is a regular differential system.

THEOREM 4.8. *Let (A, H) be a regular differential system in $\mathcal{F}[\![Y]\!]$. A differential polynomial that is partially reduced w.r.t. A belongs to $[A]:H^\infty$ if and only if it belongs to $(A):H^\infty$.*

Proof. For $a \in A$ we note u_a and s_a respectively the leader and the separant of a. Let us consider $p \in [A]:H^\infty$. There thus exists a finite subset D of $\Theta^+ \times A$ s.t. for some $h \in H^\infty$ we can write

$$h\,p = \sum_{(\theta,a)\in D\subset\Theta^+\times A} \alpha_{\theta,a}\,\theta(a) + \sum_{a\in A} \alpha_a\, a \tag{1}$$

for some $\alpha_a, \alpha_{\theta,a} \in \mathcal{F}[\![Y]\!]$. For each equation of type (1) we consider v to be the highest ranking derivative of $\Theta^+\mathcal{L}(A)$ that appears effectively in the right hand side.

Assume that p is partially reduced w.r.t. A. If the set D is empty then $p \in (A):H^\infty$. Assume, for contradiction, that there is no relation of type (1) with an empty D for p. Among all the possible relationships (1) that can be written, we consider one for which v is minimal.

Consider $E = \{(\theta, a) \in D \mid \theta(u_a) = v\}$ and single out any $(\bar{\theta}, \bar{a})$ of E. As A is H-coherent, for all (θ, a) of E we have $s_{\bar{a}}\,\theta(a) \equiv s_a\,\bar{\theta}(\bar{a}) \mod (\Theta A_{<v}):H^\infty$. Thus

$$s_{\bar{a}}\,h\,p \equiv \left(\sum_{(\theta,a)\in E} s_a\,\alpha_{\theta,a}\right)\bar{\theta}(\bar{a}) + \sum_{(\theta,a)\in \mathcal{D}\backslash E} s_{\bar{a}}\alpha_{\theta,a}\,\theta a + \sum_{a\in A} s_{\bar{a}}\alpha_a\, a \mod (\Theta A_{<v}):H^\infty \tag{2}$$

so that we can find $k \in H^\infty$ s.t.

$$s_{\bar{a}}k\,p = \beta_{\bar{\theta},\bar{a}}\,\bar{\theta}(\bar{a}) + \sum_{\substack{(\theta,\, a)\, \in\, \Theta\, \times\, A, \\ \theta(u_a)\, <\, v}} \beta_{\theta,a}\,\theta a + \sum_{a\in A} \beta_a\, a \tag{3}$$

for some $\beta_a, \beta_{\theta,a} \in \mathcal{F}[\![Y]\!]$.

We proceed now to eliminate v from the coefficients $\beta_{\bar{\theta},\bar{a}}$. We make use of the fact that $s_{\bar{a}}\,v = \bar{\theta}(\bar{a}) - \mathsf{tail}(\bar{\theta}(\bar{a}))$. Recall that $\mathsf{tail}(\bar{\theta}(\bar{a}))$ contains only derivatives lower than v. Multiplying both sides of (3) by $s_{\bar{a}}^d$, where d is the degree of v in the right hand side, and replacing $s_{\bar{a}}\,v$ by $\bar{\theta}(\bar{a}) - \mathsf{tail}(\bar{\theta}(\bar{a}))$ we can rewrite the relationship obtained as

$$s_{\bar{a}}^{\,d+1}k\,p = \gamma_d\,\bar{\theta}(\bar{a})^d + \ldots + \gamma_1\,\bar{\theta}(\bar{a}) + \gamma_0 \tag{4}$$

where $\gamma_0, \gamma_1, \ldots, \gamma_d$ no longer contain v and $\gamma_0 \in (\Theta A_{<v})$. The only occurrences of v in that right hand side is through $\bar{\theta}(\bar{a})$. Because p and the elements of H are partially reduced w.r.t. A, v does not appear in the left hand side. The coefficients γ_i, for $1 \le i \le d$ must be zero. We have thus exhibited a relationship like (1) with a v lower than what we started from. This contradicts our hypotheses.

In Section 6.4 we will see that for a weak d-triangular set B that is K-coherent, we can compute a regular differential system (A, H) such that $\mathcal{L}(A) =$

$\mathcal{L}(B)$ and $[B] : K^{\infty} = [A] : H^{\infty}$. Some of the consequences of Rosenfeld lemma we shall see can therefore be applied to coherent weak d-triangular sets. But Rosenfeld lemma does not, as shows this example.

EXAMPLE 4.9. *In* $\mathbb{Q}(s,t)\, [\![y]\!]$ *endowed with the d-ranking* $y < y_s < y_t < y_{ss} < y_{st} < y_{tt} < \ldots$ *consider the weak d-triangular set* $A = y_s \bigtriangleup y_{tt} - y_{ss}$ *that is 1-coherent.* y_{tt} *is partially reduced w.r.t.* A *and belongs to* $[A]$. *Nonetheless* y_{tt} *does not belong to* (A).

There is a number of corollaries that immediately derives from Rosenfeld's lemma. The first one provides a membership test to regular differential ideals. The second one conveys the fact that a regular differential system is formally integrable (outside the union of the zeros of the elements of H). These corollaries are in fact simple consequences of the property: *A differential polynomial that is partially reduced w.r.t.* A *belongs to* $[A] : H^{\infty}$ *if and only if it belongs to* $(A) : H^{\infty}$. We saw that coherence, i.e. the fact that (A, H) is a regular differential system, is a sufficient condition to get that property. It is not a necessary condition as discussed in [68].

COROLLARY 4.10. *Let* (A, H) *be a regular differential system. Then* $p \in [A] : H^{\infty}$ *iff* **pd-red**$(p, A) \in (A) : H^{\infty}$.

COROLLARY 4.11. *Let* (A, H) *be a regular differential system and* $u \in \Theta Y$ *ranking higher than all the leaders of the elements of* A. *Then* $[A] : H^{\infty} \cap \mathcal{F}[\Theta Y_{\leq u}] = (\Theta A_{\leq u}) : H^{\infty}$.

This trivially implies that if $v > u$, $(\Theta A_{\leq v}) : H^{\infty} \cap \mathcal{F}[\Theta Y_{\leq u}] = (\Theta A_{\leq u}) : H^{\infty}$. If we have an orderly d-ranking and $v(n)$ is the highest derivative of order n, then $\Theta A_{\leq v(n)}$ is the *prolongation* of A at order n. The property implies that *projection* of the prolongation does not introduce new conditions: $(\Theta A_{\leq v(n+r)}) : H^{\infty} \cap \mathcal{F}[\Theta_n Y] = (\Theta A_{\leq v(n)}) : H^{\infty}$. A procedure to compute a power series solution according to any d-ranking is described in [63, 42].

4.3   Regular Differential Ideals

Rosenfeld lemma allows us to lift the properties of the polynomial ideal $(A) : H^{\infty}$ to the differential ideal $[A] : H^{\infty}$ when (A, H) is a regular differential system. The first result in this line, Theorem 4.12, states that $[A] : H^{\infty}$ is radical. This is the key to effective algorithm. Theorem 4.13 shows that we can inspect on the d-triangular set some properties of the essential prime components of the regular differential ideal defined. Theorem 4.12 is the direct lift of Lazard lemma and was first given in [9]. Theorem 4.13 appeared in [10] and we give the proof of [33].

THEOREM 4.12. *Let* (A, H) *be a regular differential system of* $\mathcal{F} [\![Y]\!]$. *Then* $[A] : H^{\infty}$ *is a radical differential ideal.*

Proof. Assume p^e belongs to $[A]:H^\infty$ and let $\bar{p} = \mathsf{pd\text{-}red}(p, A)$. $\bar{p}^e$ belongs to $[A]:H^\infty$ and is partially reduced w.r.t. A so that it belongs to $(A):H^\infty$. By [34, Theorem 7.5] $(A):S_A^\infty$ is radical. So is $(A):H^\infty$ since $S_A \subset H^\infty$. Thus $\bar{p}$ belongs to $(A):H^\infty$ and therefore p belongs to $[A]:H^\infty$. It follows that $[A]:H^\infty$ is radical.

THEOREM 4.13. *Let (A, H) be a regular differential system of $\mathcal{F}[\![Y]\!]$ such that $1 \notin (A):H^\infty$. C is a characteristic set of an associated prime of $(A):H^\infty$ iff C is a differential characteristic set of an essential prime component of $[A]:H^\infty$. Then $\mathfrak{L}(C) = \mathfrak{L}(A)$.*

There is therefore a one-to-one correspondence between the minimal primes of $(A):H^\infty$ and the essential prime components of $[A]:H^\infty$.

Proof. Since we assume $1 \notin (A):H^\infty$, $1 \notin [A]:H^\infty$ by Theorem 4.8 and $[A]:H^\infty$ is radical by Theorem 4.12.

Let X to be the set of the derivatives that are partially reduced w.r.t. A. Then $A, H \subset \mathcal{F}[X]$. Assume the minimal prime decomposition of $[A]:H^\infty$ is $[A]:H^\infty = \bigcap_{i=1}^r P_i$. By Theorem 4.8, $[A]:H^\infty \cap \mathcal{F}[X] = \bigcap_{i=1}^r (P_i \cap \mathcal{F}[X]) = (A):H^\infty$. All the $\tilde{P}_i = P_i \cap \mathcal{F}[X]$ are prime ideals in $\mathcal{F}[X]$ and therefore the minimal primes of $(A):H^\infty$ are to be taken among these $\tilde{P}_i$. We shall show first that all the $\tilde{P}_i$ are minimal primes of $(A):H^\infty$.

$\tilde{P}_i$ must contain at least one minimal prime $\bar{P}_i$ of $(A):H^\infty$. Let $\bar{p}$ be an element of $\mathcal{F}[X]$ that belongs to $\tilde{P}_i$ but does not belong to $(A):H^\infty$ and therefore does not belong to $[A]:H^\infty$ by Theorem 4.8. There exists $q \subset \mathcal{F}[\![Y]\!]$, $q \notin P_i$ such that $q\bar{p} \in [A]:H^\infty$. Let $\bar{q} = \mathsf{pd\text{-}red}(q, A)$ so that there exists $s \in S_A^\infty$ such that $sq \equiv \bar{q} \mod [A]$. We have that $\bar{q} \notin (A):H^\infty$ otherwise q would belong to $[A]:H^\infty$ and therefore to P_i. Nonetheless, $\bar{q}\bar{p}$ belongs to $[A]:H^\infty$ and thus to $(A):H^\infty$ since it is partially reduced w.r.t. A. This says that $\bar{p}$ belongs to a minimal prime of $(A):H^\infty$. Thus $\tilde{P}_i$ is contained in the union of minimal primes of $(A):H^\infty$. By the prime avoidance theorem [28, Lemma 3.3] $\tilde{P}_i$ must be contained in one of the minimal primes, say $\bar{P}_i'$, of $(A):H^\infty$. Thus $\bar{P}_i \subset \tilde{P}_i \subset \bar{P}_i'$. We must have $\bar{P}_i' = \bar{P}_i$ and therefore $\tilde{P}_i$ is a minimal prime of $(A):H^\infty$.

If $\tilde{C}_i$ is the characteristic set of a minimal prime $\tilde{P}_i = P_i \cap \mathcal{F}[X]$ of $(A):H^\infty$, we saw in [34, Theorem 7.5] that $\mathfrak{L}(C_i) = \mathfrak{L}(A)$. Thus $\tilde{C}_i$ is a d-triangular set. Let p be an element of P_i and $\bar{p} = \mathsf{d\text{-}red}(p, \tilde{C}_i)$. Then $\bar{p} \in P_i \cap \mathcal{F}[X] = \tilde{P}_i$. C_i being a characteristic set of $\tilde{P}_i$, $\bar{p}$ must be zero. Therefore $\tilde{C}_i$ is a differential characteristic set of P_i.

As all differential characteristic sets of P_i have the same rank, if C_i is another differential characteristic set of P_i, it is included in $\mathcal{F}[X]$. P_i does not contain any non-zero element reduced w.r.t. C_i and so neither does $\tilde{P}_i = P_i \cap \mathcal{F}[X]$. C_i is a characteristic set of $\tilde{P}_i$.

5   Characterizable Differential Ideals

If A is a d-triangular set and p is a differential polynomial s.t. $\mathsf{d\text{-}red}(p, A) = 0$ then $p \in [A]:H_A^\infty$. What Ritt and Kolchin used in their developments is the fact that if

A is the characteristic of a prime differential ideal P then $P = [A]:H_A^\infty = [A]:S_A^\infty$ and $\mathsf{d\text{-}red}(p, A) = 0 \Leftrightarrow p \in [A]:H_A^\infty$. Recent algorithmic improvement in differential algebra owes to the idea of using a wider class of differential ideals and d-triangular set. We shall introduce *differential regular chains* [42] and show that they are the d-triangular sets A for which $p \in [A]:H_A^\infty \Leftrightarrow \mathsf{d\text{-}red}(p, A) = 0$ and then $[A]:H_A^\infty = [A]:S_A^\infty$. They define *characterizable differential ideals* that are particular cases of regular differential ideals. Contrary to regular differential ideals though, characterizable differential ideals can be defined intrinsically. They are the best thing after prime differential ideals.

We proceed to define the *characteristic decomposition* of a radical differential ideal as the representation of this ideal into an intersection of characterizable differential ideals.

We finally show that the characteristic decomposition of a regular differential ideal can be trivially lifted from a characteristic decomposition in polynomial algebra [33]. The remarkable thing is that it is independent of the algorithm used to compute the algebraic characteristic decomposition. That is the last main theorem on which is based the complete algorithm to compute characteristic decompositions for finitely generated radical differential ideals.

5.1   Differential Regular Chains

DEFINITION 5.1. *A differential ideal I is characterizable if for a differential characteristic set C of I we have $I = [C]:H_C^\infty$. C is said to characterize I.*

Membership to a characterizable differential ideal can thus be tested by a simple differential reduction: if I is characterized by C then $p \in I \Leftrightarrow \mathsf{d\text{-}red}(p, C) = 0$. Prime differential ideals are characterizable for any d-ranking. Characterizable differential ideals that are not prime do exist, but that depends on the d-ranking.

We need nonetheless to determine when a d-triangular set A characterizes $[A]:H_A^\infty$. The criterion was provided in [33].

THEOREM 5.2. *Let A be a d-triangular set of $\mathcal{F}[\![Y]\!]$. A is a differential characteristic set of $[A]:H_A^\infty$ if and only if A is H_A-coherent and A is a characteristic set of $(A):H_A^\infty$.*

Proof. We claim that if A is a differential characteristic set of $[A]:H_A^\infty$ then A must be H_A-coherent. For any $a, b \in A$, $\Delta(a, b) \in [A] \subset [A]:H_A^\infty$. Therefore $\mathsf{d\text{-}red}(\Delta(a, b), A) = 0$ so that $\Delta(a, b) \in (\Theta A_{<v}):H_A^\infty$, where $v = \mathsf{lcd}(\mathsf{lead}(a), \mathsf{lead}(b))$. Now if A is a differential characteristic set of $[A]:H_A^\infty$, $[A]:H_A^\infty$ has no non-zero element reduced w.r.t. A. It is then obviously also the case for $(A):H_A^\infty$.

Conversely, assume A is H_A-coherent and a characteristic set of $(A):H_A^\infty$. If there would exist a non-zero differential polynomial p in $[A]:H_A^\infty$ reduced w.r.t. A then, by Rosenfeld's lemma (Theorem 4.8) it would belong to $(A):H_A^\infty$. It cannot be so since A is a characteristic set of $(A):H_A^\infty$. Thus A is a differential characteristic set of $[A]:H_A^\infty$.

By application of Theorem 4.12 and Theorem 4.13, a characterizable differential ideal $[C]:H_C^\infty$ is radical and the characteristic sets of its essential prime components have the same set of leaders as C.

Together with the results seen in [34, Section 5 and 7], we can give now a characterization of practical use in terms of regular chains.

COROLLARY 5.3. *Let A be a d-triangular set of $\mathcal{F}\llbracket Y\rrbracket$. A is a differential characteristic set of $[A]:H_A^\infty$ iff the two following conditions are satisfied:*

- *A is H_A-coherent*
- *A is a squarefree regular chain.*

Proof. If A is a squarefree regular chain then A is a characteristic set of $(A):I_A^\infty$ and $(A):H_A^\infty = (A):I_A^\infty$ [34, Theorem 5.13 and Theorem 7.1]. Conversely, if A is a characteristic set of $(A):H_A^\infty$ then any element of $p \in (A):H_A^\infty$ is such that $\mathrm{red}(p,A) = 0$ and therefore $p \in (A):I_A^\infty$. Thus $(A):H_A^\infty = (A):I_A^\infty$ so that A is a squarefree regular chain. We can conclude thanks to Theorem 5.2.

It is reasonable to call a d-triangular set satisfying the two conditions of the preceding corollary a *differential regular chain* so that we can state the previous corollary in an analogous way to the polynomial counterpart seen in [34, Theorem 5.13]: A differential triangular set A is a differential characteristic set of $[A]:H_A^\infty$ iff it is a differential regular chain.

We saw in [34] that when A is a squarefree regular chain then $(A):I_A^\infty = (A):H_A^\infty = (A):S_A^\infty$. This has the following implication which is used in [13].

COROLLARY 5.4. *If A is a differential regular chain then $[A]:H_A^\infty = [A]:S_A^\infty$.*

Proof. Let $p \in [A]:H_A^\infty$. Then $\mathsf{pd\text{-}red}(p,A)$ belongs to $(A):H_A^\infty$. Since $(A):H_A^\infty = (A):S_A^\infty$, $\mathsf{pd\text{-}red}(p,A)$ belongs to $(A):S_A^\infty$ and therefore $p \in [A]:S_A^\infty$.

5.2 Characteristic Decomposition

Let J be a radical differential ideal of $\mathcal{F}\llbracket Y\rrbracket$. We call a *characteristic decomposition* of J a representation of J as an intersection of characterizable differential ideals, called *components* of J. Such a characteristic decomposition of J exists: J is the intersection of prime differential ideals and prime differential ideals are characterizable.

Given a finite set Σ of differential polynomials of $\mathcal{F}\llbracket Y\rrbracket$, computing a characteristic decomposition of $\llbracket\Sigma\rrbracket$ means finding the characteristic sets of its components. In other words, given Σ, we want to compute differential regular chains $C_1, \ldots, C_r$ such that $\llbracket\Sigma\rrbracket = \cap_{i=1}^r [C_i]:S_{C_i}^\infty$. This provides naturally a membership test to $\llbracket\Sigma\rrbracket$:

$$p \in \llbracket\Sigma\rrbracket \iff \mathsf{d\text{-}red}(p,C_i) = 0 \; \forall i, \; 1 \leq i \leq r.$$

According to the d-ranking chosen this allows also to exhibit diverse properties of $\llbracket\Sigma\rrbracket$ or equivalently of the zeros of Σ. As an example, assume we want to know if the zeros of Σ are zeros of differential polynomials in only a subset

Z of the differential indeterminates Y. Compute a characteristic decomposition according to an elimination ranking $Z \ll Y \setminus Z$. If there exists such a differential polynomial in Z, it must reduce to zero by the differential regular chains in the decomposition. That is possible only if there is a differential polynomial in all the differential regular chains that has a leader in ΘZ. In view of the d-ranking, all the derivatives of this differential polynomial must be in ΘZ. We can thus read on the differential regular chains the answer.

A characteristic decomposition of a radical differential ideal J is *irredundant* if associating each component in the decomposition with the set of its essential prime components yields a partition of the set of the essential prime components of J. In other words, consider a characteristic decomposition of J, $J = \cap_{i=1}^{r} [C_i] : S_{C_i}^{\infty}$. This decomposition is irredundant if any prime differential ideal that contains two distinct components $[C_i] : S_{C_i}^{\infty}$ is not an essential prime component of J.

5.3 Characteristic Decomposition of Regular Differential Ideals

We present now the theorem on which is based the last step of the characteristic decomposition algorithm. It shows that the characteristic decomposition of a regular differential ideal can be achieved algebraically, i.e. without performing any derivations. The result was presented in [33].

THEOREM 5.5. *If (A, H) is a regular differential system and $(A) : H^{\infty} = \cap_{i=1}^{r} (C_i) : I_{C_i}^{\infty}$ is an irredundant characteristic decomposition then each C_i is a differential regular chain and $[A] : H^{\infty} = \cap_{i=1}^{r} [C_i] : S_{C_i}^{\infty}$ is an irredundant characteristic decomposition.*

Proof. From [34, Theorem 7.5], $(A) : H^{\infty}$ is radical as $S_A \subset H$ and therefore all the components $(C_i) : I_{C_i}^{\infty}$ in the irredundant decomposition are radical ideals. It follows that each C_i is a squarefree regular chains and $(C_i) : I_{C_i}^{\infty} = (C_i) : H_{C_i}^{\infty}$.

Let $B_{i,j}$, $1 \le j \le r_i$, be characteristic sets for the minimal primes of $(C_i) : I_{C_i}^{\infty}$ so that $(C_i) : I_{C_i}^{\infty} = \cap_{j=1}^{r_i} (B_{ij}) : I_{B_{ij}}^{\infty}$ is an irredundant prime characteristic decomposition. Thus $(A) : H^{\infty} = \cap_{i,j} (B_{ij}) : I_{B_{ij}}^{\infty}$ is an irredundant prime characteristic decomposition. By Theorem 4.13, $[A] : H^{\infty} = \cap_{i,j} [B_{ij}] : H_{B_{ij}}^{\infty}$ is an irredundant prime characteristic decomposition and $\mathcal{L}(B_{ij}) = \mathcal{L}(A) = \mathcal{L}(C_i)$. In particular C_i is a d-triangular set.

We prove now that C_i is H_{C_i}-coherent. Let $a, b \in C_i$. Since a and b belong to $\cap_{j=1}^{r_i} [B_{ij}] : H_{B_{ij}}^{\infty}$, so does $\Delta(a, b)$ and $\mathsf{d\text{-}red}(\Delta(a, b), C_i)$. Since $\mathsf{d\text{-}red}(\Delta(a, b), C_i)$ is partially reduced w.r.t. all the B_{ij} it must belong to $\cap_{j=1}^{r_i} (B_{ij}) : H_{B_{ij}}^{\infty} = (C_i) : H_{C_i}^{\infty}$ in virtue of Theorem 4.8. Now, C_i is a characteristic set of $(C_i) : H_{C_i}^{\infty}$ so that $\mathsf{d\text{-}red}(\Delta(a, b), C_i)$, that is reduced w.r.t. C_i, must be zero. C_i is thus H_{C_i}-coherent.

We thus have proved that C_i is a differential regular chain. By Theorem 4.13 and Corollary 5.4 $[C_i] : S_{C_i}^{\infty} = [C_i] : H_{C_i}^{\infty} = \cap_{j=1}^{r_i} [B_{ij}] : H_{B_{ij}}^{\infty}$ is an irredundant prime decomposition. We can conclude that $[A] : H^{\infty} = \cap_{i=1}^{r} [C_i] : S_{C_i}^{\infty}$ is an irredundant characteristic decomposition.

6 The Rosenfeld Gröbner Algorithm

We saw that Rosenfeld's lemma was a link between differential algebra and polynomial algebra and is therefore the key to effective algorithms in differential algebra. Boulier *et al.* gave a first algorithm called Rosenfeld-Gröbner in [9]. For a finite set of differential polynomials, the algorithm computes in a factorization free way a representation of the radical differential ideal generated that allows to test membership. The radical differential ideal is represented as an intersection of regular differential ideals. Each regular differential ideal was given by a *regular differential system* (A, H) together with a Gröbner basis of the polynomial ideal $(A) : H^\infty$, hence the Gröbner part of the name. The membership test was thus into two parts: first a differential reduction by A and a membership test to $(A) : H^\infty$ by mean of the computed Gröbner basis.

The algorithm for the decomposition into regular differential ideals was greatly improved in [10]. That is the algorithm we reproduce here less the application of the differential analogue of Buchberger second criterion. The full algorithm was implemented by F. Boulier in the *diffalg* package of MAPLE [8].

The splitting strategy of the Rosenfeld-Gröbner algorithm owes to the differential elimination scheme of Seidenberg [66]. Alternative effective algorithms evolving from the Ritt and Kolchin algorithm [40, IV.9] were presented in [33] and [13]. The algorithm of [33] was the first to implement the present strategy that is, first compute a decomposition into regular differential ideals then refine it into a characteristic decomposition by purely algebraic means. The algorithm of [13] is closer to Ritt and Kolchin's algorithm. It intertwines purely algebraic and differential computations recursively to compute a characteristic decomposition. Doing so one is able to compute a *zero decomposition* and that is finer than the radical differential ideal decomposition under study here.

6.1 Philosophy and Data Representation

The input to the algorithm are two finite sets F and S of differential polynomials. The output is a finite set of regular differential systems, $\{(A_1, H_1), \ldots, (A_r, H_r)\}$, such that

$$\llbracket F \rrbracket : S^\infty = [A_1] : H_1^\infty \cap \ldots \cap [A_r] : H_r^\infty$$

Before we obtain regular differential systems we work with *RG-quadruples* that we process towards being regular differential systems.

DEFINITION 6.1. *A RG-quadruple T is defined by a sequence of four finite subsets G, D, A, H of $\mathcal{F}\llbracket Y \rrbracket$ such that*

- *A is a weak d-triangular set*
- *$H_A \subset H$.*
- *D is a set of Δ-polynomials*
- *for all $a, b \in A$ either $\Delta(a, b) = 0$ or $\Delta(a, b) \in D$ or $\Delta(a, b) \in \left(\Theta(A \cup G)_{<u} \right) :$ H_u^∞, where $u = lcd(lead(a), lead(b))$ and $H_u = H_{A_{<u}} \cup (H \setminus H_A) \cap \mathcal{F}[\Theta Y_{<u}]$.*

To a RG-quadruple noted $T = (G, D, A, H)$ we associate the radical differential ideal $\mathcal{J}(T) = [\![G \cup D \cup A]\!] : H^\infty$.

The weak d-triangular set A consists of the *processed* differential polynomials, while G is the set of differential polynomials to be processed. Rosenfeld-Gröbner starts with only one RG-quadruple, namely $(F, \emptyset, \emptyset, S)$. Each step of the Rosenfeld-Gröbner algorithm consists in processing one step further a RG-quadruple (G, D, A, H). An element of $G \cup D$ is taken out and reduced w.r.t. A. If the element does not reduce to an element of $\mathcal{F}$, we *insert* it into A. This is the role of the algorithm update presented in Section 6.5. New Δ-polynomials are added to D and the separant and the initial of the new element are added to H. To keep the decomposition correct, we need to introduce new RG-quadruples containing the separant and the initial of the new element.

A RG-quadruple is nearly processed when it is of the type $(\emptyset, \emptyset, A, H)$. It is the role of auto-partial-reduce in Section 6.4. to make the final step of taking (A, H) to a regular system (B, K) so that $[\![A]\!] : H^\infty = [B] : K^\infty$. A regular differential system (A, H) can be considered as a RG-quadruple $(\emptyset, \emptyset, A, H)$. At all time the intersection of the differential ideals defined by the present RG-quadruples is equal to $[\![F]\!] : S^\infty$.

The *coherence condition* in the definition of a RG-quadruple is seemingly difficult and the reason for that precision is dictated by the proof of the last part of the algorithm auto-partial-reduce. Basically, the Δ-polynomial between two elements of A is either awaiting treatment, and it belongs to D, or it has been reduced to zero by A. It could also have been discarded by some criterion.

First are presented the lemmas that allow to show the termination of the algorithm, then the lemmas that allow the splittings. The algorithm is afterwards divided in three parts.

6.2   Finiteness Lemmas

We shall introduce a partial order on RG-quadruples that is a refinement of the pre-order on d-triangular sets. This order is used to prove the termination of the algorithm.

DEFINITION 6.2. *Let E and F be two finite subsets of $\mathcal{F}[\![Y]\!]$. We say that F precedes E w.r.t. $\prec_s$ if F is obtained from E by replacing only one of its element by a finite number of elements of strictly lower rank. In other words*

$$F \prec_s E \iff \exists p \in E \text{ s.t. } F = \begin{cases} E \setminus \{p\} \cup \{p_1, \ldots p_k\}, & \text{where } \mathsf{rank}(p_i) < \mathsf{rank}(p) \\ \text{or } E \setminus \{p\} \end{cases}$$

PROPOSITION 6.3. *The partial order $\prec_s$ on finite sets of $\mathcal{F}[\![Y]\!]$ is well founded.*

Proof. Let $\{E_i\}_{i \in \mathbb{N}}$ be a strictly decreasing sequence of subsets of $\mathcal{F}[\![Y]\!]$. It can be represented by a finite number of trees. The roots of these trees are the elements of E_1. At each i, we transform one leaf, i.e. one element p of E_i, into a node. The descendants of this node are the differential polynomials of lower ranks $p_1, \ldots, p_k$, $k \in \mathbb{N}$, by which p is replaced. A final leaf correspond to the sheer elimination of a polynomial. Therefore, the paths in the trees consist of strictly decreasing sequences of differential polynomials. The trees must have finite depth and thus the sequence $\{E_i\}_{i \in \mathbb{N}}$ must be finite.

DEFINITION 6.4. *Let* $T = (G, D, A, H)$ *and* $\bar{T} = (\bar{G}, \bar{D}, \bar{A}, \bar{H})$ *be quadruples. We say that* $T \prec_t \bar{T}$ *if either*

- *rank*$(A) <$ *rank*$(\bar{A})$
- *rank*$(A) =$ *rank*$(\bar{A})$ *and* $G \cup D \prec_s \bar{G} \cup \bar{D}$

PROPOSITION 6.5. *The partial order* $\prec_t$ *on RG-quadruple is well founded.*

Proof. It follows immediately from Proposition 3.15 and Proposition 6.3.

6.3   Splitting

At each step of the algorithm we want to replace a differential polynomial by its reduction by a weak d-triangular set. This implies premultiplying by initials and separants and intrinsically assuming that we are seeking zeros where those do not vanish. We need to look for the zeros where the separants and the initials vanish, by introducing a branch split.

Let Σ and H be finite subset of differential polynomials. The first type of split expresses that the zero set of Σ can be decomposed into the union of the set of zeros of Σ that make no element of H vanish with the set of zeros common to Σ and an element of H. This split is used in Ritt-Kolchin style of algorithm [40, 33, 13]. The second kind of split introduced in [66] and used in Rosenfeld-Gröbner algorithm [9, 10] is somewhat finer.

PROPOSITION 6.6. *Let* Σ *be a non-empty subset of a differential ring* $\mathcal{R}$.

- $[\![\Sigma \cup \{ab\}]\!] = [\![\Sigma \cup \{a\}]\!] \cap [\![\Sigma \cup \{b\}]\!]$ *for all* $a, b \in \mathcal{R}$.
- $[\![\Sigma]\!] = [\![\Sigma]\!] : H^\infty \cap [\![\Sigma \cup \{\prod_{h \in H} h\}]\!]$ *for all finite subset* H *of* $\mathcal{R}$.

Proof. As $[\![\Sigma \cup \{ab\}]\!] \subset [\![\Sigma \cup \{a\}]\!], [\![\Sigma \cup \{b\}]\!]$, the first inclusion is trivial. Let $p \in [\![\Sigma \cup \{a\}]\!] \cap [\![\Sigma \cup \{b\}]\!]$. There exist $e, f \in \mathbb{N}^*$ s.t. $p^e = q + \alpha$ and $p^f = r + \beta$ where $q, r \in [\Sigma]$ while $\alpha \in [a]$ and $\beta \in [b]$. Thus $p^{e+f} = (q + \alpha)r + \beta q + \alpha \beta$. By Proposition 2.2, $\alpha \beta \in [\![ab]\!]$ so that $p \in [\![\Sigma \cup \{ab\}]\!]$.

Consider $\bar{h}$ the product of the elements of H. We have $[\![\Sigma]\!] : H^\infty = [\![\Sigma]\!] : \bar{h}^\infty$. Let $p \in [\![\Sigma]\!] : \bar{h}^\infty \cap [\![\Sigma \cup \{\bar{h}\}]\!]$. On the one hand there exists $e \in \mathbb{N}^*$ s.t. $p^e = q + \sum_{\theta \in \Theta} \alpha_\theta \theta(\bar{h})$ where $q \in [\Sigma]$ and $\{\alpha_\theta\}_\theta$ is a family of elements of $\mathcal{F}[Y]$ with finite support. Then $p^{e+1} = qp + \sum_{\theta \in \Theta} \alpha_\theta p\, \theta(\bar{h})$ so that $p^{e+1} - qp \in [\![\bar{h}p]\!]$ by Proposition 2.2. On the other hand $[\![\bar{h}p]\!] \subset [\![\Sigma]\!]$. It follows that p^{e+1} and therefore p belongs to $[\![\Sigma]\!]$.

If $H = \{h_1, h_2\}$ we can thus write $[\![\Sigma]\!] = [\![\Sigma]\!] : H^\infty \cap [\![\Sigma \cup \{h_1\}]\!] \cap [\![\Sigma \cup \{h_2\}]\!]$. Informally speaking, if we note $\mathcal{Z}(\Sigma)$ the set of zeros of Σ and $\mathcal{Z}(\Sigma/H)$ the subset of zeros of Σ that do not make any element of H vanish, we can write $\mathcal{Z}(\Sigma) = \mathcal{Z}(\Sigma/\{h_1, h_2\}) \cup \mathcal{Z}(\Sigma \cup \{h_1\}) \cup \mathcal{Z}(\Sigma \cup \{h_2\})$. We could also make the following decomposition that introduces less redundancies: $\mathcal{Z}(\Sigma) = \mathcal{Z}(\Sigma/\{h_1, h_2\}) \cup \mathcal{Z}(\Sigma \cup \{h_1\}/\{h_2\}) \cup \mathcal{Z}(\Sigma \cup \{h_2\})$. In terms of radical differential ideals it translates into $[\![\Sigma]\!] = [\![\Sigma]\!] : \{h_1, h_2\}^\infty \cap [\![\Sigma \cup \{h_1\}]\!] : h_2^\infty \cap [\![\Sigma \cup \{h_2\}]\!]$. We give and prove the formal and general property that is used in Rosenfeld-Gröbner algorithm.

PROPOSITION 6.7. *Let Σ be a non-empty subset of $\mathcal{R}$.*

- $[\![\Sigma \cup \{\prod_{i=1}^r a_i\}]\!] = \bigcap_{i=1}^{r-1} [\![\Sigma \cup \{a_i\}]\!] : \{a_{i+1}, \ldots, a_r\}^\infty \cap [\![\Sigma \cup \{a_r\}]\!]$ *for any* $a_1, \ldots, a_r \in \mathcal{R}$.
- $[\![\Sigma]\!] = [\![\Sigma]\!] : H^\infty \cap \bigcap_{1 \le i < r} [\![\Sigma \cup \{h_i\}]\!] : \{h_{i+1}, \ldots, h_r\}^\infty \cap [\![\Sigma \cup \{h_1, \ldots, h_r\}]\!]$ *for any* $H = \{h_1, \ldots, h_r\} \subset \mathcal{F}[Y]$.

Proof. The result is trivial for $r = 1$. Assume this is true up to $r - 1$, for $r \ge 2$. We have

$$[\![\Sigma \cup \{a_1 \ldots a_r\}]\!] = [\![\Sigma \cup \{a_1\}]\!] \cap [\![\Sigma \cup \{a_2 \ldots a_r\}]\!]$$
$$= [\![\Sigma \cup \{a_1\}]\!] : \{a_2, \ldots, a_r\}^\infty \cap [\![\Sigma \cup \{a_1\} \cup \{a_2 \ldots a_r\}]\!] \cap [\![\Sigma \cup \{a_2 \ldots a_r\}]\!].$$

by application of Proposition 6.6. Noting that the last component is included in the middle component, we can remove this latter so that $[\![\Sigma \cup \{a_1 \ldots a_r\}]\!] = [\![\Sigma \cup \{a_1\}]\!] : \{a_2, \ldots, a_r\}^\infty \cap [\![\Sigma \cup \{a_2 \ldots a_r\}]\!]$. We obtain the result by applying the induction hypothesis on $[\![\Sigma \cup \{a_2 \ldots a_r\}]\!]$. The second point follows immediately since $[\![\Sigma]\!] = [\![\Sigma]\!] : H^\infty \cap [\![\Sigma \cup \{h_1 \ldots h_r\}]\!]$ by Proposition 6.6.

6.4   Computing Regular Systems Equivalent to Weak Coherent d-Triangular Sets

We are going to give the last bit of the algorithm. For a pair (A, H) s.t. $(\emptyset, \emptyset, A, H)$ is a RG-quadruple we compute a pair (B, K) that is a regular differential system with the property that $[A] : H^\infty = [B] : K^\infty$. The algorithm consists simply in making partial reductions of the elements of A and H, but the proof is quite involved.

ALGORITHM 6.8. **auto-partial-reduce**
  INPUT: *Two finite subsets A and H s.t. $(\emptyset, \emptyset, A, H)$ is a RG-quadruple*
  OUTPUT:
    – *the empty set if it is detected that $1 \in [A] : H^\infty$.*
    – *a set with a single regular differential system (B, K) with $\mathfrak{L}(A) = \mathfrak{L}(B)$, $H_B \subset K$ and $[A] : H^\infty = [B] : K^\infty$*
  $B := \emptyset$;
  for $u \in \mathfrak{L}(A)$ increasingly do
    $b := \text{pd-red}(A_u, B)$;
    if $\text{rank}(b) = \text{rank}(A_u)$ then

$B := B \triangle b;$
    else
        $return(\emptyset);$
    fi;
od;
$K := H_B \cup \{\text{pd-red}(h, B) \mid h \in H \setminus H_A\};$
if $0 \in K$ then
    $return(\emptyset);$
else
    $return(\{(B, K)\});$
fi;

Correctness

In the proof we use implicitly the following lemma that is immediate to establish.

LEMMA 6.9. *Let H and K be two subsets and I a (differential) ideal in a (differential) ring $\mathcal{R}$. If $H^\infty K \subset H^\infty + I$ then $I : H^\infty = I : (H \cup K)^\infty$*

The condition $H^\infty K \subset H^\infty + I$ means that for all $k \in K$ there exists h and $\bar{h}$ in H^∞ s.t. $h\,k = \bar{h} + p$ for some $p \in I$.

We shall prove first that after the for loop, either $1 \in [A] : S_A^\infty$ or for any $u \in \mathfrak{L}(A)$ we have:

I0 $\text{rank}(A_{\leq u}) = \text{rank}(B_{\leq u})$ and B_u is partially reduced w.r.t. $B_{<u}$.
I1 $S_{B_{<u}}^\infty H_{A_{\leq u}} \subset H_{B_{\leq u}}^\infty + (\Theta B_{<u})$
I2 $H_{B_{\leq u}} \subset H_{A_{\leq u}}^\infty + \overline{(\Theta B_{<u})}$
I3 $(\Theta A_{\leq u}) : H_{A_{\leq u}}^\infty = (\Theta B_{\leq u}) : H_{B_{\leq u}}^\infty$

The properties are true for the lowest $u \in \mathfrak{L}(A)$ since then $A_u = B_u$. Let us assume that the properties I0, I1, I2, I3 hold for all $v \in \mathfrak{L}(A)$ with $v < u$, that is as we start a new iteration for u.

Note that induction hypothesis I3 implies that $(\Theta A_{<u}) : H_{A_{<u}}^\infty = (\Theta B_{<u}) : H_{B_{<u}}^\infty$ in virtue of Proposition 3.6 and induction hypothesis I1 and I2.

By induction hypothesis I0, $\mathfrak{L}(B_{<u}) = \mathfrak{L}(A_{<u})$ and thus $u \notin \Theta \mathfrak{L}(B_{<u})$ as A is a weak d-triangular set. The reduction of A_u by $B_{<u}$ is in fact a reduction on the coefficients of A_u seen as a polynomial in u. If $\text{rank}(b) \neq \text{rank}(A_u)$ then $\text{init}(A_u)$ belongs to $(\Theta B_{<u}) : S_{B_{<u}}^\infty$ and thus to $(\Theta A_{\leq u}) : H_{A_{\leq u}}^\infty$ by I3 so that $1 \in [A] : H_A^\infty$. Otherwise $\text{rank}(b) = \text{rank}(A_u)$, so that I0 is preserved, and there exists $s \in S_{B_{<u}}^\infty$ such that

$$s\,A_u \equiv B_u \quad \text{mod}\ (\Theta B_{<u})$$
$$s\,\text{sep}(A_u) \equiv \text{sep}(B_u) \quad \text{mod}\ (\Theta B_{<u})$$
$$s\,\text{init}(A_u) \equiv \text{init}(B_u) \quad \text{mod}\ (\Theta B_{<u})$$

From the latter equations, we see immediately that I1 and I2 are kept true.

From the pseudo-division relationship $s\,A_u \equiv B_u \mod (\Theta B_{<u})$ and the induction hypothesis I3 we see that $(\Theta A_{<u}):H^\infty_{A_{<u}} + (A_u) \subset ((B_u) + (\Theta B_{<u}):$ $H^\infty_{B_{<u}}):H^\infty_{B_{<u}}$ so that, by Lemma 6.9 and what we just have seen (property I1) $(\Theta A_{\leq u}):H^\infty_{A_{\leq u}} \subset (\Theta B_{\leq u}):H^\infty_{B_{\leq u}}$. Similarly the pseudo-division relationship, induction hypothesis I3 and property I2 just shown give the converse inclusion. Thus $(\Theta A_{\leq u}):H^\infty_{A_{\leq u}} = (\Theta B_{\leq u}):H^\infty_{B_{\leq u}}$ so that I3 is preserved.

I0, I1, I2 and I3 are thus proved for all $u \in \mathfrak{L}(A)$ unless we found that $1 \in [A]:H^\infty$. They imply that B is a d-triangular set and $(\Theta A_{<v}):H^\infty_{A_{<v}} = (\Theta B_{<v}):H^\infty_{B_{<v}}$ for all $v \in \Theta Y$ and $[A]:H^\infty_A = [B]:H^\infty_B$.

Consider $\bar{H} = H \setminus H_A$ and for $v \in \Theta Y$, write $\bar{H}_v = \bar{H} \cap \mathcal{F}[\Theta Y_{<v}]$. We shall show now first that $(\Theta A_{<v}):(H_{A_{<v}} \cup \bar{H}_v)^\infty \subset (\Theta B_{<v}):K^\infty$ for all $v \in \Theta Y$ and $[A]:H^\infty = [B]:K^\infty$. We shall prove then that (B, K) is a regular system.

By construction $S^\infty_{B_{<v}} \bar{H}_v \subset K^\infty + (\Theta B_{<v})$ thus $(\Theta A_{<v}):(H_{A_{<v}} \cup \bar{H}_v)^\infty = (\Theta B_{<v}):(H_{B_{<v}} \cup \bar{H}_v)^\infty \subset (\Theta B_{<v}):K^\infty$.

The latter inclusion implies that $[A]:H^\infty \subset [B]:K^\infty$. For the converse inclusion, we saw that $H_B \subset H^\infty_A + [B]$ and thus $K \subset H^\infty + [A]:H^\infty_A$ so that $[B]:K^\infty = [A]:(H_A \cup K \setminus H_B)^\infty \subset [A]:H^\infty$. We thus have the desired equality $[A]:H^\infty = [B]:K^\infty$.

Let $u, v \in \mathfrak{L}(B)$ have common derivatives. We want to show that $\Delta(B_u, B_v) \in (\Theta B_{<w}):K^\infty$, where $w = \mathsf{lcd}(u, v)$. By construction there exists $s_u \in S^\infty_{B_{<u}}$ s.t. $B_u \equiv s_u\,A_u \mod (\Theta B_{<u})$ and $\mathsf{sep}(B_u) \equiv s_u\,\mathsf{sep}(A_u) \mod (\Theta B_{<u})$. Thus $B_u \equiv s_u\,A_u \mod (\Theta A_{<u}):H^\infty_{A_{<u}}$ and $\mathsf{sep}(B_u) \equiv s_u\,\mathsf{sep}(A_u) \mod (\Theta A_{<u}):H^\infty_{A_{<u}}$. Similarly there exists $s_v \in S^\infty_{B_{<v}}$ s.t. $B_v \equiv s_v\,A_v \mod (\Theta A_{<v}):H^\infty_{A_{<v}}$ and $\mathsf{sep}(B_v) \equiv s_v\,\mathsf{sep}(A_v) \mod (\Theta A_{<v}):H^\infty_{A_{<v}}$.

Take $\psi, \phi \in \Theta$ s.t. $w = \mathsf{lcd}(u, v) = \phi u = \psi v$. Since $\phi(s_u A_u) \equiv s_u \phi(A_u) \mod (\Theta A_{<\phi u})$ and similarly for $\psi(s_v A_v)$ we can write:

$$\begin{aligned}
\Delta(B_u, B_v) &= \mathsf{sep}(B_v)\phi(B_u) - \mathsf{sep}(B_u)\psi(B_v) \\
&\equiv s_v\mathsf{sep}(A_v)\phi(s_u A_u) - s_u\mathsf{sep}(A_u)\psi(s_v A_v) \mod (\Theta A_{<w}):H^\infty_{A_{<w}} \\
&\equiv s_u s_v \Delta(A_u, A_v) \mod (\Theta A_{<w}):H^\infty_{A_{<w}}
\end{aligned}$$

From the hypothesis on (A, H), $\Delta(A_u, A_v) \in (\Theta A_{<w}):(H_{A_{<w}} \cup \bar{H}_w)^\infty$. Thus $\Delta(B_u, B_v) \in (\Theta B_{<w}):K^\infty$ so that (B, K) is a regular differential system.

Let us note here the difficulty encountered with a lighter coherence condition for RG-quadruples. Assume that the hypothesis on (A, H) had been $\Delta(A_u, A_v) \in (\Theta A_{<w}):H^\infty$ for all $u, v \in \mathfrak{L}(A)$. We could then not conclude that $\Delta(B_u, B_v) \in (\Theta B_{<w}):K^\infty$ as the element of H could have required reduction by some element of ΘB with leader bigger than w.

6.5   A Simple Updating of RG-Quadruple

At each step we need to augment the weak d-triangular set of a RG-quadruple with a new differential polynomial. We need to do it so as to preserve a RG-quadruple structure. The sub-algorithm update we present in this section is meant for that. It is in this procedure that the analogue of second Buchberger criterion (Theorem 4.6) should be applied to prune the set of Δ-polynomials. As in the polynomial case, this is a rather delicate matter [5, 10, 7] and only a simple version of the update algorithm is given.

ALGORITHM 6.10. **update**
  INPUT:
    - a 4-tuple (G, D, A, H) of finite sets of $\mathcal{F}[\![Y]\!]$
    - $p \in \mathcal{F}[\![Y]\!]$ reduced w.r.t. A s.t. $(G \cup \{p\}, D, A, H)$ is a RG-quadruple
  OUTPUT: $\bar{T} = (\bar{G}, \bar{D}, \bar{A}, \bar{H})$ a RG-quadruple such that
    - $\bar{A} \prec A$
    - $\mathcal{J}(\bar{G}, \bar{D}, \bar{A}, \bar{H}) = \mathcal{J}(G \cup \{p\}, D, A, H) : \{sep(p), init(p)\}^{\infty}$
  $u := lead(p)$;
  $G_A := \{a \in A \mid lead(a) \in \Theta u\}$;
  $\bar{A} := A \setminus G_A$;
  $\bar{G} := G \cup G_A$;
  $\bar{D} := D \cup \{\Delta(p, a) \mid a \in \bar{A}\} \setminus \{0\}$;
  $\bar{H} = H \cup \{sep(p), init(p)\}$;
  return($(\bar{G}, \bar{D}, \bar{A}_{<u} \Delta p \Delta \bar{A}_{>u}, \bar{H})$);

$(\bar{G}, \bar{D}, \bar{A}_{<u} \Delta p \Delta \bar{A}_{>u}, \bar{H})$ is a RG-quadruple since:
- p is reduced w.r.t. A and therefore w.r.t. $\bar{A}$ so that $\bar{A}_{<u} \Delta p \Delta \bar{A}_{>u}$ is a weak d-triangular set.
- $\bar{H}$ contains all the initials and separants of $\bar{A}_{<u} \Delta p \Delta \bar{A}_{>u}$
- $\bar{A} \cup \bar{G} = A \cup G$. Thus if a, b are elements of $\bar{A}$, they are elements of A so that, by hypothesis on the input, either $\Delta(a, b) = 0$ or $\Delta(a, b) \in D$ or $\Delta(a, b) \in \left(\Theta(A \cup G \cup \{p\})_{<v}\right) : H_v^{\infty}$ where $v = \text{lcd}(lead(a), lead(b))$ and $H_v = H_{A_{<v}} \cup (H \setminus H_A) \cap \mathcal{F}[\Theta Y_{<v}]$. Now $D \subset \bar{D}$ and $H_v \subset H_{\bar{A}_{<v}} \cup (H \setminus H_{\bar{A}} \cap \mathcal{F}[\Theta Y_{<v}])$.
- the Δ-polynomials of elements of $\bar{A}$ with p are added to $\bar{D}$.

Because p is reduced w.r.t. A, $\text{rank}(\bar{A}_{<u} \Delta p) < \text{rank}(A_{\leq u})$ so that $\text{rank}(A_{<u} \Delta p \Delta \bar{A}_{>u}) < \text{rank}(A)$.

Furthermore the equality of differential ideals requested is immediate from the facts that $G \cup A = \bar{G} \cup \bar{A}$ and all the $\Delta(p, a)$ introduced in $\bar{D}$ belong to $[\bar{A}]$.

As it stands in this presentation, we could remove from $\bar{D}$ the Δ-polynomials involving elements of G_A.

6.6   Core of the Algorithm

After giving the algorithm we shall write down the invariants of the while loop that help understanding the algorithm. In the algorithm description, the set S contains the RG-quadruples that are still to be processed while $\mathcal{A}$ contains the regular differential systems already obtained.

ALGORITHM 6.11. **Rosenfeld-Gröbner**
  INPUT: F, K finite subsets of $\mathcal{F}[\![Y]\!]$
  OUTPUT: A set $\mathcal{A}$ of regular differential systems s.t.
    - $\mathcal{A}$ is empty if it has been detected that $1 \in [\![F]\!] : K^{\infty}$

$$- \ [\![F]\!] : K^\infty = \bigcap_{(A,H) \in \mathcal{A}} [A] : H^\infty \ otherwise$$

$- \ H_A \subset H \ for \ all \ (A, H) \in \mathcal{A}$

$\mathcal{S} := \{(F, \emptyset, \emptyset, K)\}$;

$\mathcal{A} := \emptyset$;

while $\mathcal{S} \neq \emptyset$ do

    $(G, D, A, H) :=$ an element of $\mathcal{S}$;

    $\bar{\mathcal{S}} := \mathcal{S} \setminus \{(G, D, A, K)\}$

    if $G \cup D = \emptyset$ then

        $\bar{A} := \bar{A} \cup$ auto-partial-reduce (A, H);

    else

        $p :=$ an element of $G \cup D$;

        $\bar{G}, \ \bar{D} := G \setminus \{p\}, \ D \setminus \{p\}$;

        $\bar{p} :=$ d-red(p, A) ;

        if $\bar{p} = 0$ then

            $\bar{\mathcal{S}} := \bar{\mathcal{S}} \cup \{(\bar{G}, \bar{D}, A, K)\}$;

        elif $\bar{p} \notin \mathcal{F}$ then

            $\bar{p}_i := \bar{p} -$ init$(\bar{p})$ rank$(\bar{p})$;

            $\bar{p}_s := \deg(\bar{p}, \text{lead}(\bar{p})) \, \bar{p} - \text{lead}(\bar{p}) \, \text{sep}(\bar{p})$;

            $\bar{\mathcal{S}} := \bar{\mathcal{S}} \cup \{$ update $(\bar{G}, \bar{D}, A, H, \bar{p})$,

                        $(G \cup \{\bar{p}_s, \text{sep}(\bar{p})\}, \ \bar{D}, \ A, \ H \cup \{\text{init}(\bar{p})\})$,

                        $(\bar{G} \cup \{\bar{p}_i, \text{init}(\bar{p})\}, \ \bar{D}, \ A, \ H)\}$;

        fi;

    fi;

    $\mathcal{S} := \bar{\mathcal{S}}$

od;

return$(\mathcal{A})$;

Correctness

We shall prove that the following properties are invariants of the while loop.

I0 All elements of $\mathcal{S}$ are RG-quadruple

I1 All elements of $\mathcal{A}$ are regular differential systems

I2 $[\![F]\!] : K^\infty = \bigcap_{T \in \mathcal{S}} \mathcal{J}(T) \cap \bigcap_{(A,H) \in \mathcal{A}} [A] : H^\infty$

These properties are trivially satisfied before the while loop. Assume they are true at the beginning of a new iteration. A RG-quadruple $T = (G, D, A, H)$ is selected.

    When $G \cup D = \emptyset$

I0 is trivially preserved. The output of auto-partial-reduce is the empty set only if $1 \in [\![A]\!] : H^\infty$. Otherwise, the output of auto-partial-reduce is the set $\{(B, K)\}$ where (B, K) is a regular differential system s.t. $[A] : H^\infty = [B] : K^\infty$. By Theorem 4.12, $[B] : K^\infty$ is radical and therefore $\mathcal{J}(T) = [\![A]\!] : H^\infty = [B] : K^\infty$. I1 and I2 are thus preserved.

If $G \cup D$ is not empty

I1 is trivially preserved. There exists $h \in H^{\infty}_{A_{\le u}}$ s.t. $hp \equiv \bar{p} \mod (\Theta A_{\le u})$ where $u = \text{lead}(p)$. Consequently $\mathcal{J}(T) = \llbracket \bar{G} \cup \bar{D} \cup \{\bar{p}\} \cup A \rrbracket : H^{\infty}$ since $H_A \subset H$.

We check that for all $a, b \in A$ we have that either $\Delta(a, b) = 0$ or $\Delta(a, b) \in \bar{D}$ or $\Delta(a, b) \in \left(\Theta(A \cup \bar{G} \cup \{\bar{p}\})_{<v} \right) : H^{\infty}_v$, where $v = \text{lcd}(\text{lead}(a), \text{lead}(b))$ and $H_v = H_{A_{<v}} \cup (H \setminus H_A) \cap \mathcal{F}[\Theta Y_{<v}]$. If p was taken in D and $p = \Delta(a, b)$ then $u = \text{lead}(p)$ is lower than $v = \text{lcd}(\text{lead}(a), \text{lead}(b))$ and thus $\Delta(a, b) \in \left(\Theta A_{<v} \cup \{\bar{p}\} \right) : H^{\infty}_{A_{<v}}$. Nothing changes for other elements of D. If p was taken in G, we have to note that $\theta p \in \left(\Theta(A \cup \{\bar{p}\})_{\le \theta u} \right) : H^{\infty}_{A_{\le u}}$ so that $\left(\Theta(A \cup G)_{<v} \right) : H^{\infty}_v \subset \left(\Theta(A \cup \bar{G} \cup \{\bar{p}\})_{<v} \right) : H^{\infty}_v$ for all $v \in \Theta Y$.

If $\bar{p} = 0$, $(\bar{G}, \bar{D}, A, H)$ is thus a RG-quadruple by induction hypothesis on I0 and we have $\mathcal{J}(T) = \mathcal{J}\left(\bar{G}, \bar{D}, A, H \right)$. Therefore I0 and I3 are preserved. If $\bar{p} \in \mathcal{F}$, $\bar{p} \neq 0$, then $1 \in \mathcal{J}(T)$ so that this component can be dropped. Otherwise, let us write $\bar{s} = \text{scp}(\bar{p})$ and $\bar{\imath} - \text{init}(\bar{p})$. By Proposition 6.7

$$\mathcal{J}(T) = \llbracket \bar{G} \cup \bar{D} \cup \{\bar{p}\} \cup A \rrbracket : H^{\infty}$$
$$= \llbracket \bar{G} \cup \bar{D} \cup \{\bar{p}\} \cup A \rrbracket : (H \cup \{\bar{\imath}, \bar{s}\})^{\infty}$$
$$\cap \llbracket \bar{G} \cup \bar{D} \cup \{\bar{p}, \bar{s}\} \cup A \rrbracket : (H \cup \{\bar{\imath}\})^{\infty} \cap \llbracket \bar{G} \cup \bar{D} \cup \{\bar{p}, \bar{\imath}\} \cup A \rrbracket : H^{\infty}$$

$(\bar{G}, \bar{D}, A, H)$ and $\bar{p}$ satisfy the input specifications of update (Algorithm 6.10). Then, by the output properties of that latter, $\llbracket \bar{G} \cup \bar{D} \cup \{\bar{p}\} \cup A \rrbracket : (H \cup \{\bar{\imath}, \bar{s}\})^{\infty} = \mathcal{J}\left(\text{update}(\bar{G}, \bar{D}, A, H, \bar{p}) \right)$ and $\text{update}(\bar{G}, \bar{D}, A, H, \bar{p})$ is a RG-quadruple. Since $[\bar{p}, \bar{s}] = [\bar{p}_s, \bar{s}]$ and $[\bar{p}, \bar{\imath}] = [\bar{p}_i, \bar{\imath}]$ we can replace $\bar{p}$ by respectively $\bar{p}_s$ and p_i in the last two components of the above decomposition. We do that to ensure that the corresponding RG-quadruples, $(\bar{G} \cup \{\bar{p}_i, \bar{\imath}\}, \bar{D}, A, H)$ and $(G \cup \{\bar{p}_s, \bar{s}\}, \bar{D}, A, H \cup \{\text{init}(\bar{p})\})$, are lower than (G, D, A, H) for the partial order of Definition 6.4. I0 is preserved and so is I3 since we proved that

$$\mathcal{J}(G, D, A, H) = \mathcal{J}\left(\text{update}(\bar{G}, \bar{D}, A, H, \bar{p}) \right)$$
$$\cap \mathcal{J}\left(\bar{G} \cup \{\bar{p}_s, \bar{s}\}, \bar{D}, A, H \cup \{\bar{\imath}\} \right) \cap \mathcal{J}\left(\bar{G} \cup \{\bar{p}_i, \bar{\imath}\}, \bar{D}, A, H \right)$$

Termination

The algorithm proceeds by constructing a tree where the nodes are RG-quadruples, the root being $(F, \emptyset, \emptyset, K)$. The final leafs are RG-quadruples of the type $(\emptyset, \emptyset, A, H)$ or RG-quadruples that define differential ideals containing 1. At each iteration, we give to a RG-quadruple $T = (G, D, A, H)$ in $\mathcal{S}$ that is not a final leaf one or three descendants according to whether $\bar{p} = 0$ or not. These descendant are RG-quadruples that precede T w.r.t. $\prec_q$ (Definition 6.4) so that each path in the tree describes a decreasing sequence of RG-quadruples. By Proposition 6.5 it must be finite.

Improvements

In order to detect earlier inconsistencies, it is recommended to partially reduce the sets H in a RG-quadruple by the weak d-triangular set A. Also, it is worth noting that there is no theoretical need for the initials to appear in the output regular differential systems. One can consider adding them to the sets H and splitting according to them only when they are used for premultiplication in a reduction.

If the differential polynomial $\bar{p}$ has multiple factors, it has common factors with its separant and then inconsistencies and redundancies will appear in the computations. In the implementation it is worth making a partial factorization of the differential polynomial $\bar{p}$ and split according to its *regular* factors. A differential polynomial is said *regular* if it has no common factor with its separant. We can simplify a squarefree factorization scheme to produce the regular differential polynomials $p_1, p_2 \ldots, p_r$ s.t. $\bar{p} = p_1^{e_1} p_2^{e_2} \ldots p_r^{e_r}$ for some $e_i \in \mathbb{N}$. Then

$$
\begin{aligned}
[\![\bar{G} \cup \bar{D} \cup \{\bar{p}\}]\!] : H^\infty &= [\![\bar{G} \cup \bar{D} \cup \{p_1\}]\!] : (H \cup \{p_2, \ldots, p_r\})^\infty \\
&\cap [\![\bar{G} \cup \bar{D} \cup \{p_2\}]\!] : (H \cup \{p_3, \ldots, p_r\})^\infty \cap \ldots \cap [\![\bar{G} \cup \bar{D} \cup \{p_r\}]\!] : H^\infty
\end{aligned}
$$

in virtue of Proposition 6.7. The related RG-quadruples can then be added to $\mathcal{S}$.

7   Characteristic Decomposition Algorithm

In this section we give first an algorithm to compute the irredundant characteristic decomposition of a regular differential ideal. We showed in Theorem 5.5 that an irredundant characteristic decomposition of a regular differential ideal $[A] : H^\infty$ could be trivially lifted from an irredundant characteristic decomposition of $(A) : H^\infty$ taken in $\mathcal{F}[X]$ where X is the set of derivatives appearing in A and H. Together with the Rosenfeld-Gröbner algorithm that provides a full algorithm to compute a characteristic decomposition of any finitely generated radical differential ideal. The whole decomposition might nonetheless be redundant. If one furthermore wants a prime characteristic decomposition, it is just a matter of computing prime characteristic decomposition of some zero dimensional radical ideals in some polynomial algebra. It is an open problem to make that prime characteristic decomposition minimal.

7.1   Characteristic Decomposition for Regular Differential Ideals

This part is purely algebraic. We describe an algorithm based on Kalkbrener's pseudo-gcd algorithm[4] to compute the irredundant characteristic decomposition of the an ideal defined by a triangular set saturated by at least its initial and separants. Alternative algorithms are given in [33] and in [11]. The algorithm in [33] is based on the computation of Gröbner bases. The algorithm in [11] evolves on the ideas of [41, 50]. The outputs of both these algorithms consist of Gröbner chains. They work by first reducing the problem to dimension 0 by extending the coefficient field with the non leading variables and compute an irredundant

---

[4] Late in the preparation of these notes the author was pointed out to [4] where the same idea has been developed independently

decomposition according to the ranking induced on the leading variables. In those approaches it needs to be proved additionally that the decomposition obtained in dimension zero can be lifted back to positive dimension. A crucial point is to prove that the triangular sets obtained by considering the non leading variables in the coefficient field are triangular sets for the original ranking on all the variables. This was shown to be true independently of the algorithm used in [33, Theorem 3.10]. The three algorithms have been implemented by their respective authors.

ALGORITHM 7.1. **Irredundant-Characteristic-Decomposition**
INPUT:
- $\mathcal{F}[X]$ *a ring of polynomials.*
- *A a triangular set of* $\mathcal{F}[X]$
- *H a finite subset of* $\mathcal{F}[X]$ *such that* $S_A, I_A \subset H^\infty$

OUTPUT: *A set* $\mathcal{C}$ *of squarefree regular chains such that*
- $\mathcal{C}$ *is empty iff* $(A):H^\infty = (1)$.
- *Otherwise* $(A):H^\infty = \bigcap_{C \in \mathcal{C}} (C):I_C^\infty$ *is an irredundant decomposition.*

$\mathcal{C} := \{\emptyset\}$
for x in X increasingly do
    $H_x := \{h \mid h \in H,\ \text{lead}(h) = x\};$
    if $x \notin \mathfrak{L}(A)$ then
        for h in H_x do
            $\mathcal{C} := \bigcup_{C \in \mathcal{C}} \mathcal{U}_C$ where $(\mathcal{Z}_C, \mathcal{U}_C)$ is the output of $\text{Ksplit}(\mathcal{F}[X_{\leq x}], C, h);$
        od;
    else
        for h in H_x do
            $\mathcal{C} := \bigcup_{C \in \mathcal{C}} \text{Krelatively-prime}\ (\mathcal{F}[X_{\leq x}][x], C, A_x, h)$
        od;
    fi;
od;
return($\mathcal{C}$);

Termination is not an issue here since Ksplit and Krelatively-prime are called a finite number of time.

Correctness

We show that the outer for loop has the following invariants.

I1 C is a squarefree regular chain, for all $C \in \mathcal{C}$.
I2 $(A_{\leq x}):H_{\leq x}^\infty = \bigcap_{C \in \mathcal{C}} (C):I_C^\infty$ is an irredundant characteristic decomposition.

The invariants are satisfied before the outer for loop when $\mathcal{C} = \{\emptyset\}$ and $A_{<x} = \emptyset$. Let $x \in X$ and assume the invariants are satisfied for all $y \in X_{<x}$.

If $x \notin \mathfrak{L}(A)$

By induction hypothesis on I1, any C in $\mathcal{C}$ is a squarefree regular chain so that the output of $\mathsf{Ksplit}(\mathcal{F}[X_{\leq x}], C, *)$ consists of squarefree regular chains. I1 is preserved.

If $(\mathcal{Z}_C, \mathcal{U}_C)$ is the output of $\mathsf{Ksplit}(\mathcal{F}[X_{\leq x}], C, h)$, where C is a squarefree regular chain, then $(C) : I_C^\infty : h^\infty = \cap_{B \in \mathcal{U}_C}(B) : I_B^\infty$. The inner for loop thus computes an irredundant characteristic decomposition of $(C) : I_C^\infty : H_x^\infty$. I2 is preserved.

If $x \in \mathfrak{L}(A)$

By induction hypothesis on I2 and because $H_A \subset H$, $\mathrm{init}(A_x) \in H_{<x}$ is not a zero divisor modulo any $(C) : I_C^\infty$ for any $C \in \mathcal{C}$. Thus $C \bigtriangleup A_x$ is a regular chain, for any $C \in \mathcal{C}$. The inputs of $\mathsf{Krelatively\text{-}prime}$ are therefore correct. The output of $\mathsf{Krelatively\text{-}prime}(\mathcal{F}[X_{\leq x}][x], C, A_x, h)$ is an irredundant characteristic decomposition $\langle C \bigtriangleup A_x \rangle : I_{C \bigtriangleup A_x}^\infty : h^\infty = \langle B_1 \rangle : I_{B_1}^\infty \cap \ldots \cap \langle B_r \rangle : I_{B_r}^\infty$ so that $(C \bigtriangleup A_x) : I_{C \bigtriangleup A_x}^\infty : h^\infty \subset (B_i) : I_{B_i}^\infty$. The inner loop thus computes an irredundant characteristic decompositions of $\langle C \bigtriangleup A_x \rangle : I_{C \bigtriangleup A_x}^\infty : H_x^\infty = \langle C_1 \rangle : I_{C_1}^\infty \cap \ldots \cap \langle C_r \rangle : I_{C_r}^\infty$ so that $(C \bigtriangleup A_x) : I_{C \bigtriangleup A_x}^\infty : H_x^\infty \subset (C_i) : I_{C_i}^\infty$. If $\deg(A_x, x) = 1$ then $C \bigtriangleup A_x$ is a squarefree regular chain and so are the C_i. If $\deg(A_x, x) > 1$ then $\mathsf{sep}(A_x)$ belongs to H_x so that $(C \bigtriangleup A_x) : I_{C \bigtriangleup A_x}^\infty : H_x^\infty$ is radical. It follows that all the $(C_i) : I_{C_i}^\infty$ are radical and therefore the C_i are squarefree regular chains. I1 and I2 are preserved.

7.2   The Complete Algorithm

We now have all the material to present an algorithm to compute a characteristic decomposition of the radical differential ideal $[\![F]\!] : K^\infty$ given by two finite set F and K of differential polynomials in $\mathcal{F}[\![Y]\!]$.

For a regular differential system (A, H) we define $X_{A,H}$ to be the finite set of derivatives appearing in A or H. They are ordered according to the underlying d-ranking on $\mathcal{F}[\![Y]\!]$.

ALGORITHM 7.2. **Differential-Characteristic-Decomposition**
  INPUT: F, K finite subsets of $\mathcal{F}[\![Y]\!]$
  OUTPUT: A set $\mathcal{C}$ of regular differential chains s.t.
  $-$ $\mathcal{C}$ is empty iff $1 \in [\![F]\!] : K^\infty$
  $-$ $[\![F]\!] : K^\infty = \bigcup_{C \in \mathcal{C}} [C] : S_C^\infty$ otherwise

$\mathcal{A} := Rosenfeld\text{-}Gröbner\ (F, K);$
$\mathcal{C} := \bigcap_{(A,H) \in \mathcal{A}} Irredundant\text{-}Characteristic\text{-}Decomposition\ (\mathcal{F}[X_{A,H}], A, H)$
return($\mathcal{C}$);

If $\mathcal{A}$ is not the empty set, the output properties of $\mathsf{Rosenfeld\text{-}Gröbner}$ imply that $[\![F]\!] : K^\infty = \bigcap_{(A,H) \in \mathcal{A}}[A] : H^\infty$ where all $(A, H) \in \mathcal{A}$ are regular differential systems. If $\mathcal{C}_{A,H}$ is the output of $\mathsf{Irredundant\text{-}Characteristic\text{-}Decomposition}$

$(\mathcal{F}[X_{A,H}], A, H)$, $(A) : H^\infty = \bigcap_{C \in \mathcal{C}_{A,H}} (C) : I_C^\infty$ is an irredundant characteristic decomposition. By Theorem 5.5 $[A] : H^\infty = \bigcap_{C \in \mathcal{C}_{A,H}} [C] : S_C^\infty$ is an irredundant characteristic decomposition. Thus $[\![F]\!] : K^\infty = \bigcap_{C \in \mathcal{C}} [C] : S_C^\infty$ is a characteristic decomposition.

In virtue of Theorem 4.13 or Theorem 5.5, the decomposition could be refined to a prime characteristic decomposition by purely algebraic means. By [34, Proposition 5.18] it is in fact a matter of decomposing the radical zero dimensional ideals (C) in $\mathcal{F}(\mathfrak{T}(C))[\mathfrak{L}(C)]$ into prime ideals. That can be achieved by factorizations in towers of extensions as described in [60]. One can alternatively use Gröbner bases techniques and factorization of univariate polynomials [5].

Note that in general the characteristic decomposition computed is not irredundant. In the case we have computed a prime characteristic decomposition, it is possible that one component contains another one.

The *generalized Ritt problem* is the following: given B and C differential characteristic sets of prime differential ideals, can we algorithmically decide whether $[B] : S_B^\infty \subset [C] : S_C^\infty$. This problem is equivalent to be able to compute a basis for a prime differential ideal defined by its differential characteristic set. Already in the case where B and C consist each of a single differential polynomial the problem has remained unsolved.

If one can solve the Ritt problem one can for sure eliminate the redundancy in the characteristic decomposition obtained by the presented algorithm. But this might not be necessary. Indeed, in the case of the radical differential ideal generated by a single polynomial p we can remove the redundancy in a characteristic decomposition. The process is described in [40, Chapter IV] and a factorization free algorithm starting from a decomposition into regular differential ideals is given in [32]. It is based on the *component theorem* and the *low power theorem*, which are among Ritt's deeper results in differential algebra [59]. The link between this algebraic theorem and the analysis of singular solution is extremely interesting and still bears many open problems. For lack of space we had to discard the discussion from the present paper.

8   Examples of Applications

In this section we give examples of typical problems where the questions mentioned in the introduction arise. They are answered by computing a characteristic decomposition. We treat here small examples with *diffalg*, a MAPLE package. The *diffalg* package was developed by F. Boulier for the main library of MAPLE V.5. The author has contributed extensions and improvements to the package for MAPLE V.5.1, 6 and 7. We provide a MAPLE worksheet online to demonstrate how to use the package on the presented examples and more space demanding examples [8].

The different questions in the introduction require different d-rankings. Assume we want to know if the solution of a differential system $F = 0$ satisfy some algebraic constraints. That translates into determining if the radical differential ideal $[\![F]\!]$ contains a differential polynomial of order 0. Take an orderly ranking and compute a characteristic decomposition of $[\![F]\!]$. If there exists a differential polynomial of order 0 in $[\![F]\!]$ it must reduce to zero by the differential regular chains defining the decomposition. This is possible only if there are differential polynomials of order 0 in all differential regular chains. We can thus read on the output differential regular chains the answer.

Similarly, assume we want to know if there is an ordinary differential polynomial in one of the independent variables in $[\![F]\!]$. That translates into determining if there is a differential polynomials where all the derivatives are given by a power of a single derivation, say δ_1. We introduce a d-ranking s.t. $e_1 < f_1$ implies $\delta_1^{e_1} \ldots \delta_m^{e_m} y < \delta_1^{f_1} \ldots \delta_m^{e_m} z$ for any $y, z \in Y$. If the radical differential ideal $[\![F]\!]$ contains a differential polynomial with only δ_1 in the derivatives, it must reduce to zero by the regular differential chains defining the characteristic decomposition. These regular differential chains must thus contain such a differential polynomial.

Unfortunately, no example of application to mechanical theorem proving in differential geometry is given. The reader can refer to [74, 23, 43, 4].

8.1   Solving Systems of Ordinary Differential Equations

The abilities of the non linear differential system solver, *dsolve*, were enhanced in MAPLE 6 by using the *diffalg* package. For a single differential equation, a characteristic decomposition exhibits the differential equations for the *singular solution*. If they are solved, the complete set of solutions is now returned by *dsolve*.

The simple idea for solving ordinary differential systems is to take advantage of the solver for single non linear differential equation [21, 22, 20]. The input system is thus decomposed with respect to an elimination ranking. We attempt then to solve the system in close form by solving iteratively single differential equations in a unique indeterminate. These applications were developed by E. Cheb Terrab in collaboration with the author. We give an example for each problem.

EXAMPLE 8.1. *Consider the differential equation* $(y'' + y^3 y')^2 = (yy')^2(4y' + y^4)$. *That differential equation came up in Chazy's work to extend the Painlevé analysis to 3rd order differential equations [19, 35]. The particularity of this equation is that the* general *solution has no movable singularity whereas one of the* singular *solution has.*

Let p be the differential polynomial $(y_{tt} + y^3 y_t)^2 - (yy_t)^2(4y_t + y^4)$ *in* $\mathbb{Q}(t)[\![y]\!]$. *The characteristic decomposition obtained for* $[\![p]\!]$ *is* $[\![p]\!] = [p] : s_p^\infty \cap [q] : s_q^\infty \cap [y]$ *where* $q = y_t(4y_t + y^4)$ *and s_p and s_q are the respective separants of p and q. The complete set of solutions is now found by* MAPLE 6 *or 7. Given semi-explicitely they are*

$$y(t) = 0, \quad y(t) = a, \quad y(t)^3 = \frac{4}{3(x+a)}, \quad y(t) = a\tan(a^3 x + b).$$

EXAMPLE 8.2. *Let us consider the differential system*

$$x' = y - x^2, \qquad y' = 4\,yx - 4\,x^3, \qquad z' = z^2 - 2\,x^2 + y$$

to which we associate the set of differential polynomials $F = \{x_t - y + x^2, y_t - 4\,yx + 4\,x^3, z_t - z^2 + 2\,x^2 - y\}$ *in the differential ring* $\mathbb{Q}(t)\,[\![x, y, z]\!]$. *We use the elimination ranking such that* $x \ll y \ll z$. *The characteristic decomposition computed is* $[\![F]\!] = [C] : H_C^\infty$ *where* $C = x_{tt} - 2\,xx_t \,\triangle\, y - x_t - x^2 \,\triangle\, z_t - z^2 + x^2 - x_t$. *The solution of the original system is thus given by the solutions of the differential system below that is amenable, by a chain resolution, for a solver of differential equations.*

$$x'' = 2\,xx', \qquad y = x' + x^2, \qquad z' = z^2 - x^2 + x'.$$

The solutions are $(x(t), y(t), z(t)) = \big(a\tan(at + b), a^2 + 2\,a^2\tan(at + b)^2,$ $a\tan(at + c)\big)$ *where* a, b, c *are arbitrary constants.*

Another way of solving regular differential systems is to use the results of [26]. It is shown there that any regular differential system is equivalent to a single differential equation. A method to compute that differential equation is given.

8.2 New Classes of Ordinary Differential Equations

Another application of characteristic decomposition algorithms in the line of solving ODE was explored by E. Cheb Terrab. The idea is to create new classes of ordinary differential equations, say of first order, that can be solved together with a way of recognizing whether a given ordinary differential equation is in the class.

EXAMPLE 8.3. *We look for the condition on the function* f *for the ordinary first order differential equation* $\frac{dy}{dx} = f(x, y)$ *to admit a group of symmetry the infinitesimal generator of which,* $\xi(x, y)\frac{\partial}{\partial x} + \phi(x, y)\frac{\partial}{\partial y}$, *have its coefficient satisfying*

$$\xi_y + 1 = 0, \ x\xi_x - y - \xi = 0, \ \phi_x - 1 = 0, \ y\phi_y - \phi + x = 0, \ \phi_y - \xi_x = 0. \quad (5)$$

Those equations say that $\xi(x, y) = \alpha\,x - y, \ \phi = x + \alpha\,y$, *where* α *is an arbitrary constant. If* f *is as desired,* $\frac{dy}{dx} = f(x, y)$ *can be transformed to a quadrature* $\frac{dv}{du} = g(u)$ *with the change of variables*[5] $u = \frac{1}{2}\ln(x^2 + y^2) - \alpha\arctan\left(\frac{y}{x}\right), \ v = \arctan\left(\frac{y}{x}\right).$

The equations $\frac{dy}{dx} = f(x, y)$ *that admit a group of symmetry with infinitesimal generator* $\xi(x, y)\frac{\partial}{\partial x} + \phi(x, y)\frac{\partial}{\partial y}$ *are the ones that satisfy*

$$\phi_x + \phi_y\,f - \xi_x\,f - \xi_y\,f^2 - \xi\,f_x - \phi\,f_y = 0 \qquad (6)$$

---

[5] [53, Chapter 2.5] details how to find this change of variables from the knowledge of the infinitesimal generator.

We therefore consider $\mathbb{Q}(x,y)\,[\![f,\phi,\xi]\!]$, endowed with derivations according to x and y, and the set F of differential polynomials that correspond to equations (5) and (6). To find the conditions on f we must assign a d-ranking that eliminates ξ and ϕ. We choose $f < f_y < f_x < f_{yy} < f_{xy} < f_{yy} < \cdots < \xi < \xi_y < \xi_x < \cdots < \phi < \phi_y < \phi_x < \cdots$. The characteristic decomposition computed is $\{\Sigma\} = [C_1]\!:\!H_{C_1}^\infty \cap [C_2]\!:\!H_{C_2}^\infty$ where C_1 is

$$
\begin{aligned}
&\left((y^2+x^2)f_y - x(f^2+1)\right)\underline{f_{xy}} - \left((y^2+x^2)f_x + y(f^2+1)\right)f_{yy} \\
&+2\,f_y f(f_y y + x f_x) - f_y(f^2+1) + x(f_x{}^2 + f_y{}^2)
\end{aligned}
$$

$$
\begin{aligned}
\Delta\ &\left((y^2+x^2)f_y - x(f^2+1)\right)^2 \underline{f_{xx}} - \left((y^2+x^2)f_x + y(f^2+1)\right)^2 f_{yy} \\
&+ \left((y^2+x^2)(4\,xf-y)f_y - 2\,x(f^2+1)(xf-y)\right)f_x^2 \\
&+ \left((y^2+x^2)(x+4\,yf)f_x + 2\,y(f^2+1)(yf+x)\right)f_y^2 \\
&+ (y^2+x^2)(xf_x{}^3 - yf_y{}^3) \\
&- (f^2+1)\left(2\,(y^2+x^2)f_x f_y + (f^2+1)(yf_y - xf_x)\right)
\end{aligned}
$$

$$
\Delta\ (f_y y + x f_x)\,\underline{\phi} - (y^2+x^2)\,f_x - y(f^2+1)
$$

$$
\Delta\ (f_y y + x f_x)\,\underline{\xi} + (y^2+x^2)\,f_y - x\left(f^2+1\right).
$$

and C_2 is

$$
(x^2+y^2)\,\underline{f_y} - x(1+f^2) \qquad \Delta \qquad (x^2+y^2)\,\underline{f_y} + y(1+f^2)
$$

$$
\Delta\ y\underline{\phi_y} - \phi + x, \qquad \Delta \qquad \underline{\phi_x} - 1, \qquad \Delta \qquad y\underline{\xi} - x\phi + x^2 + y^2.
$$

The leaders have been underlined.

Let us be given f. If the two first differential polynomials of either C_1 or C_2 vanish on f, we are able to reduce the differential equation $\frac{dy}{dx} = f(x,y)$ to a quadrature with the change of variables given above.

In the case determined by C_2, the zero of the two first differential polynomials is given by a formula of the type $\frac{x+\alpha y}{\alpha x - y}$. The solutions are given implicitly as $\frac{1}{2}\ln(x^2+y^2) + \alpha \arctan\left(\frac{y}{x}\right) = c$, where c is an arbitrary constant.

In the case determined by C_1, the two last differential polynomials allow us to determine the right pair ξ, ϕ. For instance, one can check that the function $f(x,y) = \frac{y + x\,H(x^2+y^2)}{x - y\,H(x^2+y^2)}$, where H is an arbitrary function of one variable, makes the two first differential polynomials of C_1 vanish and entails $\xi(x,y) = -y$, $\phi(x,y) = x$. The change of variables $u = \frac{1}{2}\ln(x^2+y^2)$, $v = \arctan\left(\frac{y}{x}\right)$ then transforms $\frac{dy}{dx} = f(x,y)$ to the quadrature $\frac{dv}{du} = H(e^{2u})$.

8.3   Differential Algebraic Systems

Consider implicit differential systems $F(t,Y,Y') = 0$, where $Y = (y_1,\ldots,y_n)$ and $F = (f_1,\ldots,f_n)$ is a polynomial map. When the Jacobian w.r.t. Y', is identically zero, it indicates that there are algebraic constraints on the components of Y and classical schemes of numerical integration fail. The difficulty of numerical integration has been measured by the differential index and specific numerical schemes can handle low index cases [14, 31, 3].

Algorithms specifically designed to reduce a quasi-linear differential-algebraic system to a system a priori amenable to numerical computations were given in

[69, 71]. The difficulty is that the expressions can become really big and one is faced with their numerical evaluation. A successful study of a sample of high order index systems was carried out in [71] and the related software and test set are available [70]. We give here only one common example. Interestingly, the computation could not be completed before the results of [33] were implemented.

EXAMPLE 8.4. *We consider the set of differential polynomials in*
$\mathbb{Q}(t) [\![\lambda, x, y, \nu, u, v]\!]$

$$F = \{\, x_{tt} - 2\,\lambda x + 2\,\nu u,\ y_{tt} - 2\,\lambda y + 2\,\nu v,\quad x^2 + y^2 - 1,$$
$$u_{tt} + x_{tt} - 2\,\nu u,\ v_{tt} + y_{tt} - 2\,\nu v + 1,\ u^2 + v^2 - 1 \,\}.$$

The defined system of ordinary differential equations describes the motion of a double pendulum in Cartesian coordinates. The indeterminates λ, ν are the Lagrangian multipliers used to obtain the equations. It is of course possible to transform it into a set of differential polynomials of order one or less by introducing new differential indeterminates, but in the present approach it is better not to. If we are interested in knowing the trajectory of the two masses only, we choose a d-ranking s.t. $\{x, y, u, v\} \ll \{\lambda, \nu\}$ and the d-ranking on $\{x, y, u, v\}$ is orderly. The characteristic decomposition obtained is $[\![F]\!] = [C_1] : S_{C_1}^\infty \cap [C_2] : S_{C_2}^\infty$ where

$$
\begin{aligned}
C_1 = \ & \underline{x^2 + y^2 - 1} \ \triangle \ \underline{u^2 + v^2 - 1} \\
& \triangle \ x^2 u^2 (2\,yv(yv - xu) - y^2 - v^2 - 1)\,\underline{y_{tt}} \\
& \quad -(yu - xv)(x^3 v_t^2 + u^3 y_t^2 - x^3 u^2 v) + yu^2 y_t^2 + x^4 u^2 \\
& \triangle \ x^3 u^2 \,\underline{v_{tt}} - x^2 u^3 (xu + yv) y_{tt} + x^3 v v_t^2 - u^3 v y_t^2 - x^3 u^4 \\
& \triangle \ 2u^2 x^3 y\,\underline{\lambda} - x^2 u^2 (v(xv - yu) + x) y_{tt} + x^3 v v_t^2 + u^3 v y_t^2 - x^3 u^2 (v^2 + 1) \\
& \triangle \ 2v\,\underline{\nu} - y_{tt} - v_{tt} - 1
\end{aligned}
$$

and

$$C_2 = \underline{y^2} - 1 \ \triangle \ \underline{x} \ \triangle \ \underline{v^2} - 1 \ \triangle \ \underline{u} \ \triangle \ \underline{\lambda} - y \ \triangle \ 2\underline{\nu} - v.$$

The leaders have been underlined. C_2 represent the equilibrium, while C_1 represent the motion. It is described by two second order differential polynomials together with two constraints. The solutions thus depends on only four arbitrary constants (initial conditions) only.

8.4   Observability of Control Systems

The work of M. Fliess, T. Glad and coworkers showed the relevance of constructive differential algebra in control theory [29]. We consider systems of the type

$$X' = F(X, U, \Lambda), \quad Y = G(X, U, \Lambda),$$

where $X = (x_1, \ldots, x_n)$ are the state variables, $U = (u_1, \ldots, u_k)$ are the control variables, $\Lambda = (\lambda_1, \ldots, \lambda_p)$ are parameters and $Y = (y_1, \ldots, y_d)$ are the output variables. The questions that one can answer with a characteristic decomposition, when F and G are rational maps, are

- what are the differential equations giving Y in terms of U (the input-output representation).
- can the sate variables be deduced from the knowledge of the output Y and the controls U. In which case the system is observable.
- can the parameters be deduced uniquely from the knowledge of the output Y and the controls U. In which case the system is identifiable.

The first step is to consider the set of differential polynomials C obtained from the above equations together with the additional differential polynomials $\lambda_{1t}, \ldots, \lambda_{pt}$ and the set S of numerators coming into F and G. The answer to the previous questions can be read from a characteristic decomposition of $[\![F]\!] : S^\infty$ w.r.t. a d-ranking s.t. $U \ll Y \ll X \cup \Lambda$.

EXAMPLE 8.5. *Consider the following system with one control and one parameter:*

$$x_1' = x_1 - \lambda u, \quad x_2' = x_2(1 - x_1), \quad y = \lambda x_1.$$

We compute the characteristic decomposition of $F = \{x_{1t} - x_1 + \lambda u, x_{2t} - x_2(1 - x_1), y - \lambda x_1, \lambda_t\}$ in $\mathbb{Q}[\![x_1, x_2, y, \lambda]\!]$ endowed with a d-ranking s.t. $u \ll y \ll \lambda \ll \{x_1, x_2\}$. We obtain $[\![F]\!] = [C_1] : S_{C_1}^\infty \cap [C_2] : H_{C_2}^\infty \cap [C_3] : H_{C_3}^\infty$ where

$$C_1 = u\underline{y_{tt}} - uy_t - u_t y_t + u_t y \vartriangle u\underline{\lambda}^2 - y_t + y \vartriangle (y_t - y)\underline{x_1} - \lambda u y$$
$$\vartriangle (y_t - y)\underline{x_{2t}} - x_2(y_t + y + \lambda y u)$$

$$C_2 = \underline{u} \vartriangle \underline{y_t} - y \vartriangle \lambda\underline{x_1} - y \vartriangle \lambda\underline{x_{2t}} - x_2(\lambda + y)$$

$$C_3 = \underline{y} \vartriangle \underline{\lambda} \vartriangle \underline{x_{2t}} - x_2(1 - x_1) \vartriangle \underline{x_{1t}} - x_1$$

We read from C_1 that generically, the parameter λ is algebraically identifiable and x_1 is algebraically observable: for a given input and output there are only a finite number of possibilities for their values. On the contrary x_2 is not observable. From C_2 we see that when $u = 0$ the parameter λ is no longer identifiable.

A number of biological models arising in the literature are examined under variants of the characteristic decomposition algorithms in [49]. Another approach is to note that C is a characteristic set for a d-ranking $U \cup \Lambda \ll X \cup Y$ with an orderly ranking on $X \cup Y$ and that $[C] : S^\infty$ is a prime differential ideal. We can take advantage of this fact to avoid splitting when computing a characteristic decomposition for the d-ranking $U \cup Y \ll X \cup \Lambda$. This approach is brought to a very efficient algorithm in [12]. Also, A. Sedoglavic developed an algorithm specific to the question of observability and identifiability [65]. The probabilistic aspect of it brings it to a polynomial time algorithm.

8.5   Symmetry Analysis of Partial Differential Equations

A source of over-determined systems of partial differential equations comes from symmetry analysis. The determining equations of the infinitesimal generators of the Lie symmetry group of a differential equation form a set of linear partial differential equations. *Reducing* this set allows to determine if there is a symmetry, to analyze the structure of the group and apply that knowledge to classification problems or to find a reduction of the original equation [53]. Because the system is linear, there is a number of alternative algorithms that can be applied. For a bibliography on the subject, the reader is invited to check [1, 64]. We just give an example that will illustrate a way to solve systems of partial differential systems.

EXAMPLE 8.6. *The determining equations for the infinitesimal generators of the Lie symmetry group of the Burgers' equation* $u_t = u_{ss} - uu_s$ *are given by the set F of differential polynomials below.*

$$F = \{ -\tau_u - \xi_{s,u}, \xi_{s,s} + 2\tau_s - \xi_t, -u\phi_s + \phi_{s,s} - \phi_t, 2\tau_{s,u} + 2u\tau_u - \phi_{u,u},$$
$$\phi - \tau_t + u\tau_s - 2\phi_{s,u} + \tau_{s,s}, -\xi_{u,u}, -\xi_s, -\xi_u, -\tau_{u,u} \}$$

We shall use a d-ranking that is induced by a lexicographical order on the derivation variables s, t, u, i.e. a d-ranking such that

$$\xi < \tau < \phi, \xi_u < \tau_u < \phi_u < \xi_{uu} < \tau_{uu} < \phi_{uu} < \cdots$$
$$< \xi_t < \tau_t < \phi_t < \xi_{tu} < \tau_{tu} < \phi_{tu} < \xi_{tuu} < \cdots$$

in order to exhibit the ordinary differential equations in u satisfied by the zeros of F. The characteristic decomposition for $[\![F]\!]$ has a single component as the differential polynomials of F are linear. It corresponds to the differential system:

$$\xi_u = 0, \ \tau_u = 0, \ \phi_{uu} = 0,$$
$$\xi_t = -2\phi_u, \ \tau_t = \phi - u\phi_u, \ \phi_{tu} = \frac{\phi_t}{u}, \ \phi_{tt} = 0,$$
$$\xi_s = 0, \ \tau_s = -\phi_u, \ \phi_s = \frac{\phi_t}{u}.$$

Form this differential characteristic set we can determine that the symmetry group is of dimension 5. We can even solve the determining equations, starting with solving the ordinary differential polynomials in the above set. We find the solution $(\phi(t, s, u), \tau(t, s, u), \xi(t, s, u)) = (a + bs - 1/2\, cu - btu, d + at + 1/2\, cs + bst, e + ct + bt^2)$ *where a, b, c, d, e are arbitrary constants.*

When we look for classes of differential equations [54] or when we search non-classical symmetry reduction [24], the determining equation are non-linear. The work of E.L. Mansfield and coauthors lead, after some monster computations with the competitor MAPLE packages *rif* [56, 73] and *diffgrob2* [46, 47] to new closed form solutions to physically relevant partial differential equations [25, 48].

References

[1] R. L. Anderson, V. A. Baikov, R. K. Gazizov, W. Hereman, N. H. Ibragimov, F. M. Mahomed, S. V. Meleshko, M. C. Nucci, P. J. Olver, M. B. Sheftel, A. V. Turbiner, and E. M. Vorob'ev. *CRC handbook of Lie group analysis of differential equations. Vol. 3.* CRC Press, Boca Raton, FL, 1996. New trends in theoretical developments and computational methods.

[2] J. Apel. *Passive Complete Orthonormal Systems of PDEs and Riquier Bases of Polynomial Modules*. 2003. In this volume.

[3] U. M. Ascher and L. R. Petzold. *Computer methods for ordinary differential equations and differential-algebraic equations*. Society for Industrial and Applied Mathematics (SIAM), Philadelphia, PA, 1998.

[4] P. Aubry and D. Wang. Reasonning about surfaces using differential zero and ideal decomposition. In J. Richet-Gebert and D. Wang, editors, *ADG 2000*, number 2061 in LNAI, pages 154–174, 2001.

[5] T. Becker and V. Weispfenning. *Gröbner Bases - A Computational Approach to Commutative Algebra*. Springer-Verlag, New York, 1993.

[6] F. Boulier. *Étude et Implantation de Quelques Algorithmes en Algèbre Différentielle*. PhD thesis, Université de Lille, 1994.

[7] F. Boulier. A new criterion to avoid useless critical pairs in buchberger's algorithm. Technical Report 2001-07, LIFL, Université de Lille, ftp://ftp.lifl.fr/pub/reports/internal/2001-07.ps, 2001.

[8] F. Boulier and E. Hubert. *DIFFALG: description, help pages and examples of use*. Symbolic Computation Group, University of Waterloo, Ontario, Canada, 1998. Now available at http://www.inria.fr/cafe/Evelyne.Hubert/webdiffalg.

[9] F. Boulier, D. Lazard, F. Ollivier, and M. Petitot. Representation for the radical of a finitely generated differential ideal. In A. H. M. Levelt, editor, *ISSAC'95*. ACM Press, New York, 1995.

[10] F. Boulier, D. Lazard, F. Ollivier, and M. Petitot. Computing representations for radicals of finitely generated differential ideals. Technical Report IT-306, LIFL, 1997.

[11] F. Boulier and F. Lemaire. Computing canonical representatives of regular differential ideals. In C. Traverso, editor, *ISSAC*. ACM-SIGSAM, ACM, 2000.

[12] F. Boulier, F. Lemaire, and M. Moreno-Maza. Pardi! In *ISSAC 2001*. pp 38–47, ACM, 2001.

[13] D. Bouziane, A. Kandri Rody, and H. Maârouf. Unmixed-dimensional decomposition of a finitely generated perfect differential ideal. *Journal of Symbolic Computation*, 31(6):631–649, 2001.

[14] K. E. Brenan, S. L. Campbell, and L. R. Petzold. *Numerical solution of initial value problems in differential-algebraic equations*. North-Holland, 1989.

[15] A. Buium and P. J. Cassidy. Differential algebraic geometry and differential algebraic groups: from algebraic differential equations to diophantine geometry. In Bass et al. [39].

[16] G. Carra Ferro. Gröbner bases and differential algebra. In *AAECC*, volume 356 of *Lecture Notes in Computer Science*. Springer-Verlag Berlin, 1987.

[17] G. Carrà Ferro and W. Y. Sit. On term-orderings and rankings. In *Computational algebra (Fairfax, VA, 1993)*, number 151 in Lecture Notes in Pure and Applied Mathematics, pages 31–77. Dekker, New York, 1994.

[18] E. Cartan. *Les systèmes Différentiels Extérieurs et leurs Applications Géométrique*. Hermann, 1945.

[19] J. Chazy. Sur les équations différentielles du troisième ordre et d'ordre supérieur dont l'intégrale générale a ses points critiques fixes. *Acta Mathematica*, 34:317–385, 1911.

[20] E. S. Cheb-Terrab. Odetools: A maple package for studying and solving ordinary differential equations. http://lie.uwaterloo.ca/description/odetools, 1998. presented for the incoming update of "Handbook of Computer Algebra".

[21] E. S. Cheb-Terrab, L. G. S. Duarte, and L. A. C. P. da Mota. Computer algebra solving of first order ODEs using symmetry methods. *Computer Physics Communications. An International Journal and Program Library for Computational Physics and Physical Chemistry*, 101(3):254–268, 1997.

[22] E. S. Cheb-Terrab, L. G. S. Duarte, and L. A. C. P. da Mota. Computer algebra solving of second order ODEs using symmetry methods. *Computer Physics Communications. An International Journal and Program Library for Computational Physics and Physical Chemistry*, 108(1):90–114, 1998.

[23] S-C. Chou and X-S. Gao. Automated reasonning in differential geometry and mechanics using the characteristic set method. part II. mechanical theorem proving. *Journal of Automated Reasonning*, 10:173–189, 1993.

[24] P. A. Clarkson, D. K. Ludlow, and T. J. Priestley. The classical, direct, and non-classical methods for symmetry reductions of nonlinear partial differential equations. *Methods and Applications of Analysis*, 4(2):173–195, 1997. Dedicated to Martin David Kruskal.

[25] P. A. Clarkson and E. L. Mansfield. Symmetry reductions and exact solutions of a class of non-linear heat equations. *Physica*, D70:250–288, 1994.

[26] T. Cluzeau and E. Hubert. Resolvent representation for regular differential ideals. *Applicable Algebra in Engineering, Communication and Computing*, 13(5):395–425, 2003.

[27] S. Diop. Differential-algebraic decision methods and some applications to system theory. *Theoretical Computer Science*, 98(1):137–161, 1992. Second Workshop on Algebraic and Computer-theoretic Aspects of Formal Power Series (Paris, 1990).

[28] D. Eisenbud. *Commutative Algebra with a View toward Algebraic Geometry*. Graduate Texts in Mathematics. Springer-Verlag New York, 1994.

[29] M. Fliess and S.T. Glad. An algebraic approach to linear and nonlinear control. In H.L. Trentelman and J.C. Willems, editors, *Essays on control: Perspectives in the theory and its applications*, volume 14 of *PCST*, pages 223–265. Birkhäuser, Boston, 1993.

[30] L. Guo, W. F. Keigher, P. J. Cassidy, and W. Y. Sit, editors. *Differential Algebra and Related Topics*. World Scientific Publishing Co., 2002.

[31] E. Hairer and G. Wanner. *Solving Ordinary Differential Equations II - Stiff and Differential- Algebraic Problems - Second Revised Edition*. Springer, 1996.

[32] E. Hubert. Essential components of an algebraic differential equation. *Journal of Symbolic Computation*, 28(4-5):657–680, 1999.

[33] E. Hubert. Factorisation free decomposition algorithms in differential algebra. *Journal of Symbolic Computation*, 29(4-5):641–662, 2000.

[34] E. Hubert. Notes on triangular sets and triangulation-decomposition algorithms. I Polynomial systems. In this volume, 2003.

[35] E. L. Ince. *Ordinary Differential Equations*. Dover Publications, Inc., 1956.

[36] M. Janet. *Sur les systèmes d'équations aux dérivées paritelles*. Gauthier-Villars, 1929.

[37] M. Kalkbrener. A generalized Euclidean algorithm for computing triangular representations of algebraic varieties. *Journal of Symbolic Computation*, 15(2):143–167, 1993.

[38] I. Kaplansky. *An Introduction to Differential Algebra*. Hermann, Paris, 1970.

[39] E. Kolchin. *Selected works of Ellis Kolchin with commentary*. Commentaries by Armand Borel, Michael F. Singer, Bruno Poizat, Alexandru Buium and Phyllis J. Cassidy, Edited and with a preface by Hyman Bass, Buium and Cassidy. American Mathematical Society, Providence, RI, 1999.

[40] E. R. Kolchin. *Differential Algebra and Algebraic Groups*, volume 54 of *Pure and Applied Mathematics*. Academic Press, New York-London, 1973.

[41] D. Lazard. Solving zero dimensional algebraic systems. *Journal of Symbolic Computation*, 15:117–132, 1992.

[42] F. Lemaire. *Contribution à l'algorithmique en algèbre différentielle*. PhD thesis, Université des Sciences et Technoligies de Lille, http://www.lifl.fr/~lemaire/pub, 2002.

[43] Z. Li. Mechanical theorem proving in the local theory of surfaces. *Annals of Mathematics and Artificial Intelligence*, 13(1-2):25–46, 1995.

[44] Z. Li and D. Wang. Coherent, regular and simple system in zero decompositions of partial differential systems. *Systems Sciences and Mathematical Sciences*, 12 suppl, 1999.

[45] B. Malgrange. Differential algebra and differential geometry. In *Differential Geometry - Kyoto Nara 2000*, 2000. to appear.

[46] E. L. Mansfield. *Differential Gröbner Bases*. PhD thesis, University of Sydney, 1991.

[47] E. L. Mansfield. *DIFFGROB2: a symbolic algebra package for analysing systems of PDE using MAPLE*. University of Exeter, 1994. Preprint M/94/4.

[48] E. L. Mansfield, G. J. Reid, and P. A. Clarkson. Nonclassical reductions of a 3+1-cubic nonlinear Schrödinger system. *Computer Physics Communications*, 115:460–488, 1998.

[49] G. Margaria, E. Riccomagno, M. J. Chappell, and H. P. Wynn. Differential algebra methods for the study of the structural identifiability of rational function state-space models in the biosciences. *Mathematical Biosciences*, 174(1):1–26, 2001.

[50] M. Moreno Maza and R. Rioboo. Polynomial gcd computations over towers of algebraic extensions. In *Applied algebra, algebraic algorithms and error-correcting codes (Paris, 1995)*, pages 365–382. Springer, Berlin, 1995.

[51] S. Morrison. The differential ideal $[P] : M^\infty$. *Journal of Symbolic Computation*, 28(4-5):631–656, 1999.

[52] F. Ollivier. Canonical bases : Relations with standard bases, finiteness conditions and application to tame automorphisms. In Costiglioncello, editor, *MEGA'90*. Birkhauser, August 1990.

[53] P. Olver. *Applications of Lie Groups to Differential Equations*. Number 107 in Graduate texts in Mathematics. Springer-Verlag, New York, 1986.

[54] G. J. Reid. Algorithms for reducing a system of pde to standard form, determining the dimension of its solution space and calculating its taylor series solution. *European Journal Of Applied Mathematics*, 2:293–318, 1991.

[55] G. J. Reid. Finding abstract Lie symmetry algebras of differential equations without integrating determining equations. *European Journal of Applied Mathematics*, 2(4):319–340, 1991.

[56] G. J. Reid, A. D. Wittkopf, and A. Boulton. Reduction of systems of nonlinear partial differential equations to simplified involutive forms. *Eur. J. of Appl. Math.*, 7:604 – 635, 1996.

[57] C. Riquier. *Les systèmes d'équations aux dérivés partielles*. Gauthier-Villars, Paris, 1910.

[58] J. F. Ritt. *Differential Equations from the Algebraic Standpoint*. Amer. Math. Soc. Colloq. Publ., 1932.

[59] J. F. Ritt. On the singular solutions of algebraic differential equations. *Annals of Mathematics*, 37(3):552–617, 1936.

[60] J. F. Ritt. *Differential Algebra*, volume XXXIII of *Colloquium publications*. American Mathematical Society, 1950. Reprinted by Dover Publications, Inc (1966).

[61] A. Rosenfeld. Specializations in differential algebra. *Transaction of the American Mathematical Society*, 90:394–407, 1959.

[62] C. J. Rust and G. J. Reid. Rankings of partial derivatives. In *Proceedings of the 1997 International Symposium on Symbolic and Algebraic Computation (Kihei, HI)*, pages 9–16, New York, 1997. ACM.

[63] C. J. Rust, G. J. Reid, and A. D. Wittkopf. Existence and uniqueness theorems for formal power series solutions of analytic differential systems. In *Proceedings of the 1999 International Symposium on Symbolic and Algebraic Computation (Vancouver, BC)*, pages 105–112, New York, 1999. ACM.

[64] F. Schwarz. Algorithmic Lie theory for solving ordinary differential equations. 2003. In this volume.

[65] A. Sedoglavic. A probabilistic algorithm to test local algebraic observability in polynomial time. In B. Mourrain, editor, *Proceedings of the 2001 International Symposium on Symbolic and Algebraic Computation (London, Canada, July 22–25 2001)*, pages 309–316, 2001.

[66] A. Seidenberg. An elimination theory for differential algebra. *University of California Publications in Mathematics*, 3(2):31–66, 1956.

[67] W. Seiler. Computer algebra and differential equations - an overview. *mathPAD* 7, 7:34–49, 1997.

[68] W. Sit. The Ritt-Kolchin theory for differential polynomials. In Guo et al. [30].

[69] G. Thomas. *Contributions Théoriques et Algorithmiques à l'Étude des Équations Différentielles-Algébriques. Approche par le Calcul Formel.* PhD thesis, Institut National Polytechnique de Grenoble, Juillet 1997.

[70] J. Visconti. HIDAES (Higher Index Differential Algebraic Equations Solver). Software and test set. Technical report, LMC-IMAG, http://www-lmc.imag.fr/CF/LOGICIELS/page_dae.html, 1999.

[71] J. Visconti. *Résolution numérique des équations algébro-différentielles, Estimation de l'erreur globale et réduction formel de l'indice.* PhD thesis, Institut National Polytechnique de Grenoble, 1999. http://www-lmc.imag.fr/CF/publi/theses/visconti.ps.gz.

[72] D. Wang. An elimination method for differential polynomial systems. I. *Systems Science and Mathematical Sciences*, 9(3):216–228, 1996.

[73] A. Witkopf and G. Reid. *The RIF package*. CECM - Simon Fraser University - Vancouver, http://www.cecm.sfu.ca/ wittkopf/rif.html.

[74] W. T. Wu. Mechanical theorem proving of differential geometries and some of its applications in mechanics. *Journal of Automated Reasoning*, 7(2):171–191, 1991.

Passive Complete Orthonomic Systems of PDEs and Involutive Bases of Polynomial Modules

Joachim Apel

Universität Leipzig, Mathematisches Institut
Augustusplatz 10–11,
04109 Leipzig, Germany
apel@mathematik.uni-leipzig.de

Abstract. The objective of this article is to enlighten the relationship between the two classical theories of passive complete orthonomic systems of PDEs on the one hand and Gröbner bases of finitely generated modules over polynomial rings on the other hand. The link between both types of canonical forms are the involutive bases which are both, a particular type of Gröbner bases which carry some additional structure and a natural translation of the notion of passive complete orthonomic systems of linear PDEs with constant coefficients into the language of polynomial modules.

We will point out some desirable applications which a "good" notion of involutive bases could provide. Unfortunately, these desires turn out to collide and we will discuss the problem of finding a reasonable compromise.

1 Introduction

Our objective is to discuss a general framework of involutive bases of finitely generated modules over polynomial rings which on the one hand still provides a uniform algorithmic treatment but on the other hand includes useful instances for some natural applications.

First of all, we will shortly introduce the two major pillars of involutive bases which consist in the passive complete orthonomic form of systems of partial differential equations (PDEs) and the Gröbner bases of polynomial modules. Passive complete orthonomic systems of PDEs are the main idea behind Riquier's existence theorems (c.f. Riquier (1910); Thomas (1929)) and the corresponding algorithms for solving systems of PDEs due to Janet (1929). Gröbner bases and Buchberger's algorithm (c.f. (Buchberger , 1965, 1985)) for their computation, nowadays, have countless applications in commutative and non-commutative algebra as well as in algebraic geometry. Moreover, both methods are fundamental for lots of applications in- and outside mathematics due to their impact on respective solution methods for systems of differential or algebraic equations. Riquier's and Buchberger's theories are related to each other by the fact that the translation of the passive complete orthonomic form of systems of linear PDEs with constant coefficients into the language of polynomial modules provides a

F. Winkler and U. Langer (Eds.): SNSC 2001, LNCS 2630, pp. 88–107, 2003.

particular type of Gröbner bases carrying an additional structure which can be roughly described by saying that in addition to the direct access to the initial module $in_\prec(M)$, which is characteristic for all Gröbner bases of a module M, an involutive basis of M provides a decomposition of $in_\prec(M)$ into a certain direct sum of finitely many free modules. Indeed, the overlapping area of both theories is larger than only linear PDEs with constant coefficients on the one hand and polynomial modules on the other hand side. In Section 4 we will discuss the connection a little more detailed.

The notion *involutive bases* for the Gröbner bases with the additional initial ideal decomposition property was introduced by Gerdt and Blinkov, who were inspired by the modern terminology used in the theory of PDEs. We follow this notion since it is well-established in the community of people working in this field. But a warning might be helpful at this place. In the theory of PDEs the adjective 'involutive' describes intrinsic geometric properties of the system and its meaning is coordinate free. In contrast, Riquier's normal forms as well as Buchberger's Gröbner bases are defined within a given coordinate system. Therefore, 'involutive bases' is a coordinate free notion assigned to a coordinate dependent object. This fact can cause heavy confusion unless one forgets about the intrinsic meaning of the adjective 'involutive' when working in ideal theory. The classical meaning of 'involutive' suggests the distinction of a property of ideals rather than a property of particular generating sets.

Riquier's theory leaves some freedom in choosing the multipliers for each of the equations in order to complete an orthonomic system of PDEs (see Section 2). Depending on what application is concerned different ways of defining the sets of multipliers have been introduced in the past. Janet suggested a very sophisticated assignment rule leading to an efficient algorithm (see (Janet , 1929)). In Section 6 we will explain the advantage of Janet's assignments for algorithmic purposes but we will also show how to further improve them. The main objective of Thomas was to give a most elegant and easy understandable presentation of Riquier's existence theorems. This aim is fulfilled best by concentrating on only the principal algorithmic aspects and disregarding any efficiency issues (see (Thomas , 1929, 1937)). Yet another objective can be found in Pommaret's coordinate free studies of systems of PDEs, where a multiplier notion optimal for systems of PDEs in generic coordinates is applied (see (Pommaret , 1978)).

The use of different multiplier notions for different applications was motivation for Gerdt and Blinkov (1998) to think about a general framework for fixing the multipliers. In this article we will discuss and motivate the ideas which stand behind the alternative concept introduced in (Apel , 1998).

Sometimes it might seem that the presented examples are not characteristic since many of them consider only the most simple case of monomial ideals. However, the situation is not much different for general modules M because after fixing an admissible term order $\prec$ the main algorithmic part takes place only in the initial module $in_\prec(M)$, which indeed is monomial. Also the second restriction to ideals is not essential. If M, and consequently also $in_\prec(M)$, are submodules of a module F of higher rank k then we can decompose $in_\prec(M)$ in a direct sum of k monomial submodules of modules of rank 1, i.e. ideals, and treat each of them separately.

2 Orthonomic Systems of Partial Differential Equations

Let $x_1, \ldots, x_n$ denote the independent variables, $u_1, \ldots, u_k$ the unknown functions and $\mathcal{D}$ the set of all differential operators of the form

$$\frac{\partial^{i_1 + \cdots + i_n}}{\partial x_1^{i_1} \cdots \partial x_n^{i_n}} . \tag{1}$$

Furthermore, let the set $\mathcal{D}\mathcal{U} = \{Du_j \mid D \in \mathcal{D}, 1 \le j \le k\}$ of all formal derivatives of the unknown functions by differential operators from $\mathcal{D}$ be ordered by a well-order $\prec$ which is compatible with differentiation in the sense $\frac{\partial}{\partial x_i} D_1 u_{j_1} \prec \frac{\partial}{\partial x_i} D_2 u_{j_2}$ for all $i \in \{1, \ldots, n\}$ and $D_1 u_{j_1}, D_2 u_{j_2} \in \mathcal{D}\mathcal{U}$ such that $D_1 u_{j_1} \prec D_2 u_{j_2}$. Usually, the additional assumption that $\prec$ has to be compatible with the order of derivation is made, i.e. $D_1 u_{j_1} \prec D_2 u_{j_2}$ holds for all $j_1, j_2 \in \{1, \ldots, k\}$ and $D_1 = \frac{\partial^{i_1 + \cdots + i_n}}{\partial x_1^{i_1} \cdots \partial x_n^{i_n}}, D_2 = \frac{\partial^{i'_1 + \cdots + i'_n}}{\partial x_1^{i'_1} \cdots \partial x_n^{i'_n}} \in \mathcal{D}$ such that $i_1 + \cdots + i_n < i'_1 + \cdots + i'_n$.

Consider a system of finitely many partial differential equations which are all of the form

$$D_i u_{j_i} = G_i , \tag{2}$$

where $D_i u_{j_i} \in \mathcal{D}\mathcal{U}$ and G_i is a function of the independent variables and the derivatives which are less than $D_i u_{j_i}$ with respect to $\prec$, and the left hand sides of the equations are pairwise distinct. Then we call $Du_j \in \mathcal{D}\mathcal{U}$ a *principal* derivative of the system if it can be obtained by differentiation from the left hand side of one of the equations and we call it a *parametric* derivative of the system, otherwise. A system of PDEs is called *orthonomic* if it satisfies the above conditions, no right hand side contains any principal derivative, and each right hand side can be developed in a Taylor series in the region under consideration. Let us illustrate the notions introduced so far by an example.

Example 1 (Part 1). Consider the system

$$u_{xx} = yv_x \tag{3}$$

$$u_{yy} = -x^2 u \tag{4}$$

of two partial differential equations in the unknown functions $u = u(x, y)$ and $v = v(x, y)$ in two independent variables x and y. As usual, subscripts at the unknown functions indicate partial derivatives, e.g. $u_{xy} = \frac{\partial^2}{\partial x \partial y} u$. This system is in orthonomic form with respect to any order $\prec$ which is compatible with the order of derivation. Obviously, the right hand sides considered as functions in x, y and the occurring derivatives of u and v have Taylor series expansions everywhere. Since the derivatives on the right hand side of the equations are of lower order than the derivatives on the left hand side, the derivatives on the right hand side are smaller than the left hand sides with respect to $\prec$, in particular, no derivative on the right hand side can be obtained by differentiation of a left hand side.

All derivatives of v are parametric. The parametric derivatives of u are u, u_x, u_y and u_{xy}. The remaining derivatives of u are principal.

Let $D_1 u_{j_1} = G_1$ and $D_2 u_{j_2} = G_2$ be two equations of the system such that differentiation of the left hand sides by $\Delta_1, \Delta_2 \in \mathcal{D}$, respectively, leads to the same principal derivative $\Delta_1(D_1 u_{j_1}) = \Delta_2(D_2 u_{j_2})$. In particular we must have $j_1 = j_2$ and any solution of our system of PDEs has to satisfy also the equation $\Delta_1 G_1 - \Delta_2 G_2 = 0$ which is called an *integrability condition* for the system. The system is called *passive* if the left hand side of any of its integrability conditions becomes identical zero after replacement of all principal derivatives in an obvious way. In what follows by calling a system passive we will subsume that it is also orthonomic.

In general, replacing all principal derivatives in the left hand side of an integrability condition yields a function depending on the independent variables and the parametric derivatives. Obviously, the system of PDEs is incompatible if for some integrability condition this function is non-zero and depends only on the independent variables. If neither such an integrability condition exists nor the system is passive we can solve the non-zero integrability conditions for their highest derivative, write them in form (2) and add them to the system of PDEs. Repeating this process eventually we will produce a passive system or will encounter that the system is incompatible.

Example 1 (Part 2). Differentiating Equation (3) twice by y and Equation (4) twice by x yields the integrability condition

$$yv_{xyy} + 2v_{xy} + x^2 u_{xx} + 4xu_x + 2u = 0$$

Replacing the principal derivative u_{xx} by the right hand side of Equation (3) yields

$$yv_{xyy} + 2v_{xy} + x^2 yv_x + 4xu_x + 2u = 0 \ .$$

Hence, the system is neither passive nor *proved* to be incompatible. We solve the equation for the highest occurring derivative v_{xyy} and add the new equation

$$v_{xyy} = -2\frac{v_{xy} + 2xu_x + u}{y} - x^2 v_x \tag{5}$$

to our system. The system formed by Equations (3), (4) and (5) is passive since there are no further (essential) integrability conditions. For the new system we have

$$\mathcal{PR} = \left\{ \frac{\partial^{i+j}}{\partial x^i \partial y^j} u \ \middle| \ i \geq 2 \text{ or } j \geq 2 \right\} \cup \left\{ \frac{\partial^{i+j}}{\partial x^i \partial y^j} v \ \middle| \ i \geq 1 \text{ and } j \geq 2 \right\}$$

$$\mathcal{PA} = \left\{ u, \frac{\partial}{\partial x} u, \frac{\partial}{\partial y} u, \frac{\partial^2}{\partial x \partial y} u, v \right\} \cup \left\{ \frac{\partial^i}{\partial y^i} v, \frac{\partial^i}{\partial x^i} v, \frac{\partial^{i+1}}{\partial x^i \partial y} v \ \middle| \ i = 1, 2, \ldots \right\}$$

The observation that any compatible system of PDEs can be transformed in passive orthonomic form is due to Tresse (1894). Riquier surpassed this result by showing that any system of this particular form has a unique holomorphic solution for arbitrary given initial determinations of a certain type prescribed by the system (see (Riquier , 1910)). Riquier's result makes use of decompositions of each of the sets $\mathcal{PR}$ of principal and $\mathcal{PA}$ of parametric derivatives, and hence also of $\mathcal{DU}$, in finitely many pairwise disjoint cones. Such a cone is given by its vertex which is a derivative $Du_j \in \mathcal{DU}$ and a subset $Y \subseteq \{x_1, \ldots, x_n\}$, where $\Delta(Du_j)$ belongs to the cone if and only if $\Delta \in \mathcal{D}$ is a differential operator involving only variables from Y. We will use the denotation $[Du_j, Y]$ for this cone. The elements of Y are called the *multipliers* and the remaining independent variables the *non-multipliers* for Du_j. Such a decomposition of the set $\mathcal{DU}$ allows to prove the existence of a convergent power series solution for the unknown functions u_j, $j \in \{1, \ldots, k\}$, at an arbitrary given point $(\widehat{x}_1, \ldots, \widehat{x}_n) \in \mathbb{C}^n$ belonging to a region where the system is orthonomic. The particular importance of the cones forming $\mathcal{PA}$ is the description of the initial determinations, more precisely, each cone $[Du_j, Y]$ belonging to $\mathcal{PA}$ provides an initial condition of the form $Du_j|_{\forall x_i \notin Y : x_i = \widehat{x}_i} = \varphi(y_1, \ldots, y_r)$, where $Y = \{y_1, \ldots, y_r\}$ and φ can be chosen as an arbitrary holomorphic function. We will call an orthonomic system of PDEs *complete* if there exists a disjoint decomposition of $\mathcal{PR}$ into cones such that each vertex is the left hand side of an equation of the system (see (Thomas , 1929)).

Example 1 (Part 3). We have the disjoint decompositions

$$\mathcal{PR} = [u_{xx}, \{x\}] \cup [u_{xxy}, \{x\}] \cup [u_{yy}, \{x, y\}] \cup [v_{xyy}, \{x, y\}]$$
$$\mathcal{PA} = [u, \emptyset] \cup [u_x, \emptyset] \cup [u_y, \emptyset] \cup [u_{xy}, \emptyset] \cup [v, \{y\}] \cup [v_x, \{x\}] \cup [v_{xy}, \{x\}]$$

Hence, we can extend our system of PDEs to the passive complete form

$$
\begin{aligned}
u_{xx} &= yv_x \\
u_{xxy} &= yv_{xy} + v_x \\
u_{yy} &= -x^2 u \\
v_{xyy} &= -2\frac{v_{xy} + 2xu_x + u}{y} - x^2 v_x \ .
\end{aligned}
$$

The system has a uniquely determined holomorphic solution for an arbitrary initial determination $u|_{x=x_0, y=y_0} = c_1, u_x|_{x=x_0, y=y_0} = c_2, u_y|_{x=x_0, y=y_0} = c_3$, $u_{xy}|_{x=x_0, y=y_0} = c_4, v|_{x=x_0} = \varphi_1(y), v_x|_{y=y_0} = \varphi_2(x)$ and $v_{xy}|_{y=y_0} = \varphi_3(x)$, where $x_0, y_0, c_1, c_2, c_3, c_4$ are complex numbers, $y_0 \neq 0$, and $\varphi_1, \varphi_2, \varphi_3$ are holomorphic functions.

Finally, we can distinguish two aspects of passive (complete) orthonomic systems of PDEs. Firstly, there is a dynamic aspect which consists in the algorithmic transformation of an arbitrary system of PDEs into passive (complete) orthonomic form. Secondly, there is a static aspect consisting in the determination of the solution space of a system of PDEs in passive (complete) orthonomic form.

Janet was the first who investigated the dynamic part from the point of view of efficiency and a crucial property of his algorithm is that it directly computes a complete system of PDEs (see (Janet , 1929)).

For an excellent overview on the basic ideas of the Riquier/Janet theory we refer to the article (Thomas , 1929) by Joseph M. Thomas.

3   Gröbner Bases of Polynomial Modules

Let $R = \mathbb{K}[X]$ be the polynomial ring over the field $\mathbb{K}$ in the variables $X = \{x_1, \ldots, x_n\}$, F a free R-module of rank k with basis $\{e_1, \ldots, e_k\}$ and $M \subseteq F$ an arbitrary submodule of F. An R-*term* is a monomial $x_1^{i_1} \cdots x_n^{i_n}$ without coefficient. By F-*term* we denote a module element te_i, where t is an R-term and e_i one of the basis elements of F. If no confusion is possible we will simply speak of terms instead of R- or F-terms. An admissible term order $\prec$ of F is a well-order of the set of all F-terms which is compatible with the multiplication by R-terms. By $\mathrm{lt}_\prec(f)$ we denote the leading term of a non-zero element $f \in F$ with respect to $\prec$, i.e. the largest (w.r.t. $\prec$) term appearing in f with non-zero coefficient. $\mathrm{in}_\prec(f)$ denotes the *initial term* of $f \in F$ with respect to $\prec$, that is $\mathrm{in}_\prec(f) = 0$ in case $f = 0$ and $\mathrm{in}_\prec(f) = c\, \mathrm{lt}_\prec(f)$, where c is the coefficient of $\mathrm{lt}_\prec(f)$ in f, otherwise. For arbitrary subsets $V \subseteq F$ we extend these notions by $\mathrm{lt}_\prec(V) = \{\mathrm{lt}_\prec(v) \mid v \in V \setminus \{0\}\}$ and $\mathrm{in}_\prec(V)$ to be the submodule of F which is generated by all initial terms of elements of V.

Then a (finite) subset $V \subset M$ is called a *Gröbner basis* of M with respect to $\prec$ if and only if $\mathrm{in}_\prec(V) = \mathrm{in}_\prec(M)$. One way to compute a Gröbner basis of M from an arbitrary given finite generating set of M is the application of Buchberger's famous algorithm (see (Buchberger , 1965, 1985); for the module case see also Möller and Mora (1986)). We will illustrate the notions and methods by an example.

Example 2 (Part 1). Consider the polynomial ring $R = \mathbb{C}[x, y]$ and the free R-module F with basis $\{e_1, e_2\}$. Moreover, let $\prec$ be the degreewise lexicographical order of the multiplicative monoid of R-terms which is defined by $x^\alpha y^\beta \prec x^{\alpha'} y^{\beta'}$ iff the first non-zero component of the integer vector $((\alpha + \beta) - (\alpha' + \beta'), \beta - \beta', \alpha - \alpha')$ is negative. $\prec$ is extended to F-terms by $te_i \prec se_j :\Leftrightarrow t \prec s$ or $t = s$ and $i < j$.

Consider the module elements $p = x^2 y e_1 - 2xy^2 e_2 + x^3 e_2$ and $q = 2x^2 y e_1 + xy e_1 + x^2 y e_2$. Then $\mathrm{in}_\prec(p) = -2xy^2 e_2$ since $x^3 \prec x^2 y \prec xy^2$ and, hence, $x^3 e_2 \prec x^2 y e_1 \prec xy^2 e_2$. Similarly, $\mathrm{in}_\prec(q) = x^2 y e_2$ because $xy \prec x^2 y$ and $x^2 y e_1 \prec x^2 y e_2$. Consequently, $\mathrm{in}_\prec(V) = R\,(-2xy^2 e_2, x^2 y e_2) \subset F$ is the initial module of the set $V = \{p, q\}$. By M we denote the submodule of F generated by V. Then we have $r := xp + 2yq = 4x^2 y^2 e_1 + x^3 y e_1 + x^4 e_2 + 2xy^2 e_1 \in M$ and since $\mathrm{in}_\prec(r) = 4x^2 y^2 e_1 \notin \mathrm{in}_\prec(V)$ we deduce that V is not a Gröbner basis of M with respect to $\prec$. (p, q) was the only critical pair of V (see Buchberger (1985)) and since the initial term of the new element r belongs to a different module generator also the extended set $V \cup \{r\}$ has no additional critical pairs. Hence,

$\text{in}_{\prec}(M) = R\ (-2xy^2e_2, x^2ye_2, 4x^2y^2e_1)$, i.e. $\{p, q, r\}$ is a Gröbner basis of M with respect to $\prec$.

For a comprehensive introduction to the theory of Gröbner bases we refer to the survey article Buchberger (1985) or one of the textbooks Adams and Loustaunau (19~~ Becker and Weispfenning (1993); Eisenbud (1995).

4   Relationships between Gröbner Bases and Passive Systems

The idea behind passive complete systems of PDEs can be translated in terms of distinguished Gröbner bases of submodules of a free module F of rank k over the polynomial ring $R = \mathbb{C}[X]$, the so-called involutive bases of polynomial modules. For this purpose let us consider the special case of systems of linear PDEs with constant coefficients. Then in view of the natural isomorphisms between the monoid $\mathcal{D}$ with differentiation, the multiplicative monoid $T = \langle X \rangle$ of R-terms, and the additive monoid $\mathbb{N}^n$, there is a natural isomorphism between the ring $\mathbb{C}[\mathcal{D}]$ of polynomial differential operators and the polynomial ring $R = \mathbb{C}[X]$. The left hand side of a linear partial differential equation $G = 0$ with constant coefficients can be considered as an element of the $\mathbb{C}[\mathcal{D}]$-(left)module $\mathcal{F}$ which is freely generated by the unknown functions $u_1, \ldots, u_k$ and scalar multiplication of $m \in \mathcal{F}$ by $\Delta \in \mathbb{C}[\mathcal{D}]$ means application of the differential operator Δ to m. The modules $\mathcal{F}$ and $F = R^k$ are isomorphic. There is a natural bijection between the equivalence classes of systems of PDEs of the form $G = 0$ with left hand sides in $\mathcal{F}$ and the submodules of F. Moreover, a passive system of PDEs is mapped to a Gröbner basis and vice versa, where the corresponding orders $\prec$ are the same under the natural isomorphism between $\mathcal{D}$ and T.

Let us outline the overlapping area of both theories a little more detailed. First of all, note, none of the theories is completely covered by the other. On the one hand the translation from systems of PDEs to R-modules over some ring R is restricted to linear PDEs. Nevertheless, the condition of constant coefficients can be omitted if one allows so-called D-modules as algebraic counterpart. On the other hand also Gröbner bases can be considered in a much more general context than polynomial modules, see e.g. Apel (2000) and the references therein for huge classes of rings allowing a generalized Gröbner theory. But PDE-analogues to the corresponding Gröbner bases can be considered only for graded structures originating from an $\mathbb{N}^n$-filtration of a ring of functions.

Example 1 (Part 4). The system considered in Example 1 does not fit in the above concept due to the non-constant coefficients. However, it has a natural algebraic translation in a free left module F of rank 2 over the Weyl algebra W_2 in the indeterminates $x, y, \partial_x, \partial_y$, where $\partial_x\, x = x\partial_x + 1, \partial_y\, y = y\partial_y + 1$ and all other pairs of indeterminates commute.

The input translates in the module elements $\partial_x^2 e_1 - y\partial_x\, e_2$ and $\partial_y^2 e_1 + x^2 e_1$, where e_1 and e_2 are the free generators of F. By M we denote the left submodule of F generated by these two elements. Applying the Gröbner basis techniques for finitely generated left modules over Weyl algebras (see Apel and Laßner (1988)) we obtain the left Gröbner basis $\{\partial_x^2 e_1 - y\partial_x\ e_2, \partial_y^2 e_1 + x^2 e_1, y\partial_x\partial_y^2\ e_2 + 2\partial_x\partial_y\ e_2 + x^2 y\partial_x\ e_2 + 4x\partial_x\ e_1 + 2e_1\}$. Using the methods introduced in (Apel , 1998) it is also possible to compute involutive bases in F (for the notion of an involutive basis see the forthcoming Definition 1). One particular such basis can be obtained by adding the element $\partial_x^2\partial_y\ e_1 - y\partial_x\partial_y\ e_2 - \partial_x\ e_2$ to the above Gröbner basis. Retranslation in the language of PDEs confirms the analogy to our previous results. There remains one difference, namely the division by y appearing in Equation 5 cannot take place in the context of the Weyl algebra W_2. But we could extend W_2 by allowing division by x and y and apply the Gröbner theory developed in (Apel , 2000) which does not require the variables and coefficients to commute.

Example 2 (Part 2). The module elements p and q from Example 2 correspond to the following system of linear PDEs with constant coefficients

$$u_{xxy} - 2v_{xyy} + v_{xxx} = 0$$
$$2u_{xxy} + u_{xy} + v_{xxy} = 0$$

in unknown functions u and v and independent variables x and y. Transformation into orthonomic form yields

$$v_{xyy} = \frac{1}{2}\left(u_{xxy} + v_{xxx}\right)$$
$$v_{xxy} = -2u_{xxy} - u_{xy}$$

Differentiating the first equation by x and the second by y leads to a non-contradicting integrability condition and adding the resulting equation

$$u_{xxyy} = -\frac{1}{4}\left(u_{xxxy} + v_{xxxx} + 2u_{xyy}\right)$$

makes the system passive. Moreover, the system is already complete as the decomposition

$$\mathcal{PR} = [v_{xyy}, \{x, y\}] \cup [v_{xxy}, \{x\}] \cup [u_{xxyy}, \{x, y\}] \tag{6}$$

of the set of principal derivative shows. The decomposition

$$\mathcal{PA} = [v, \{y\}] \cup [v_x, \{x\}] \cup [v_{xy}, \emptyset] \cup [u, \{y\}] \cup [u_x, \{y\}] \cup [u_{xx}, \{x\}] \cup [u_{xxy}, \{x\}]$$

of the set of parametric derivatives implies that the system has a unique holomorphic solution for an arbitrary initial determination of the type $v_{xy}|_{x=x_0, y=y_0} = c, v|_{x=x_0} = \varphi_1(y), v_x|_{y=y_0} = \varphi_2(x), u|_{x=x_0} = \varphi_3(y), u_x|_{x=x_0} = \varphi_4(y), u_{xx}|_{y=y_0} = \varphi_5(x), u_{xxy}|_{y=y_0} = \varphi_6(x)$.

Recall, that the basic property of a Gröbner basis V of the submodule $M \subset F$ with respect to $\prec$ consists in the equality $in_\prec(M) = in_\prec(V)$ or, equivalently, in the sum representation

$$in_\prec(M) = \sum_{v \in V} R\, in_\prec(v) . \tag{7}$$

In case V is the image of a minimal passive complete system of PDEs in F we have the stronger condition

$$\text{in}_{\prec}(M) = \bigoplus_{v \in V} \mathbb{C}[Y_v]\text{in}_{\prec}(v) . \tag{8}$$

Example 2 (Part 3). From Equation (6) we deduce that $\{p, q, r\}$ is not only a Gröbner basis but even an involutive basis providing the direct sum decomposition $\text{in}_{\prec}(M) = \mathbb{C}[x, y]\ xy^2 e_2 \oplus \mathbb{C}[x]\ x^2 ye_2 \oplus \mathbb{C}[x, y]\ x^2 y^2 e_1$ of the initial module $\text{in}_{\prec}(M) = \text{in}_{\prec}(\{p, q, r\})$.

Equation (8) is a good starting point for the definition of an involutive basis of the submodule $M \subset F$ with respect to $\prec$ which we wish to be the analogous notion to a passive complete system of PDEs. In fact, if we restrict our interests to only the static aspect then this equation would be already the correct analogue. However, it is slightly too general in order to reflect also Janet's dynamical aspects in a satisfactory way.

One method for the computation of a passive complete system is to first compute a passive system and to complete it in a subsequent step. Translated to the language of polynomial modules, that is, first to compute a Gröbner basis and afterwards to decompose the initial module in a direct sum of the required type. Actually, this is the idea which is implicitly suggested by Thomas' paper (Thomas , 1929). In contrast, the most important feature of Janet's algorithm consists in the fact, that it combines the completion of the set of principal derivatives and the transformation into passive form (see Janet (1929)). Also alternative definitions of multipliers and non-multipliers introduced in the past, e.g. by Thomas (Thomas , 1929, 1937) or by Pommaret (1978), show the nice behavior that an orthonomic system of PDEs is passive and complete if and only if replacing all principal derivatives in $\frac{\partial}{\partial x}(D_i u_{j_i} - G_i)$ yields zero for all equations $D_i u_{j_i} = G_i$ of the system and all non-multipliers x of $D_i u_{j_i}$.

In fact, this is the crucial condition which enables the application of a Janet like algorithm, see also Figure 1 and the explanations in Section 5.

5   Involutive Bases of Polynomial Modules

In what follows we will consider arbitrary coefficient fields $\mathbb{K}$. The only reason we had to restrict ourselves to the field $\mathbb{K} = \mathbb{C}$ of complex numbers in the previous section was the relation to systems of PDEs. Here we will also switch to the more frequently used right module notation. In our setting of a module over a commutative polynomial ring this makes of course no difference. Note, however, in the general case of PDEs or non-commutative algebraic situations such as the Weyl algebra case considered in Part 4 of Example 1 we would have to take care about the correct sides.

The demands on the notion of involutive basis as a "good" module theoretical analogue to passive complete systems resulting from the investigations made in the previous section can be summarized in two conditions:

1. a minimal involutive basis V of M has to satisfy Equation (8) and
2. there should exist a Janet type algorithm (see Figure 1) which transforms an arbitrary finite generating set of M in an involutive basis V of M.

Two different approaches satisfying both conditions have been introduced by Gerdt and Blinkov (1998) and by Apel (1998). While the Gerdt/Blinkov method saves almost all classical dynamic features our approach maintains only the principal idea behind Janet's algorithm explained at the end of the previous section.

Definition 1. *Let F be a free module of rank k over the polynomial ring $R = \mathbb{K}[X]$ in the indeterminates $X = \{x_1, \ldots, x_n\}$ and M a submodule of F. Moreover, let $\prec$ be an admissible term order of F. Then a subset $V \subset M \setminus \{0\}$ is called a* minimal involutive basis *of M with respect to $\prec$ if there exists a linear order $\sqsubset$ of $\mathrm{lt}_\prec(V)$ and a family $(Y_v)_{v \in V}$ of subsets of X such that*

1. $\mathrm{in}_\prec(M) = \bigoplus_{v \in V} \mathrm{in}_\prec(v)\mathbb{K}[Y_v]$ *and*
2. $\mathrm{in}_\prec(v')\mathbb{K}[X] \cap \mathrm{in}_\prec(v)\mathbb{K}[Y_v] = \{0\}$ *for all $v, v' \in V$ such that $\mathrm{lt}_\prec(v') \sqsubset \mathrm{lt}_\prec(v)$.*

A subset $V' \subseteq M$ containing a subset V which is a minimal involutive basis of M w.r.t. $\prec$ is called an involutive basis *of M w.r.t. $\prec$.*

It is an immediate consequence of Definition 1 that the initial terms of the elements of a minimal involutive basis are pairwise distinct. We will call a family $(Y_v)_{v \in V}$ of subsets of X *admissible* for the ordered set $(\mathrm{lt}_\prec(V), \sqsubset)$ if for all $v, v' \in V \setminus \{0\}$ such that $\mathrm{lt}_\prec(v') \sqsubseteq \mathrm{lt}_\prec(v)$ it holds one of the conditions $\mathrm{in}_\prec(v')\mathbb{K}[Y_{v'}] \subseteq \mathrm{in}_\prec(v)\mathbb{K}[Y_v]$ or $\mathrm{in}_\prec(v)\mathbb{K}[Y_v] \cap \mathrm{in}_\prec(v')\mathbb{K}[X] = \{0\}$. Moreover, by writing that $(Y_v)_{v \in V}$ is admissible for $\mathrm{lt}_\prec(V)$ we express the existence of a linear order $\sqsubset$ on $\mathrm{lt}_\prec(V)$ such that $(Y_v)_{v \in V}$ is admissible for $(\mathrm{lt}_\prec(V), \sqsubset)$. Note, in the language of Apel (1998) a family $(Y'_s)_{s \in S}$ defines an involutive division $\mathcal{M} = (s \cdot \langle Y'_s \rangle)_{s \in S}$. If $S = \mathrm{lt}_\prec(V)$ and $Y_v = Y'_{\mathrm{lt}_\prec(v)}$ for all $v \in V$ then the family $(Y_v)_{v \in V}$ of multiplier sets is admissible for $(\mathrm{lt}_\prec(V), \sqsubset)$ if and only if the involutive division $\mathcal{M} = (s \cdot \langle Y'_s \rangle)_{s \in S}$ is admissible for $(S, \sqsubset)$.

Using the admissibility notion we can say that $V \subset M$ is an involutive basis of M with respect to $\prec$ if and only if there exists a family $(Y_v)_{v \in V}$ of multiplier sets which is admissible for $\mathrm{lt}_\prec(V)$ and satisfies $\mathrm{in}_\prec(M) = \sum_{v \in V} \mathrm{in}_\prec(v)\mathbb{K}[Y_v]$.

Note, there is a close relationship between the notions of minimal involutive bases and *reduced* Gröbner bases. However, since we did not assume that the elements of a minimal involutive basis are tail reduced, the better analogue are the Gröbner bases which are minimal with respect to set inclusion. The following two examples will illustrate Definition 1.

Example 3 (Part 1). Consider the polynomial ring $R = \mathbb{K}[x, y, z, w]$ as a free R-module F of rank 1 and let M be the submodule of F, i.e. the ideal of R, which is generated by the set $V = \{xz, xw, yz, yw\}$ of monomials.

If we order V by $xz \sqsubset xw \sqsubset yz \sqsubset yw$ then there is a unique admissible family of maximal multiplier sets, namely, $Y_{xz} = \{x, y, z, w\}, Y_{xw} = \{x, y, w\}, Y_{yz} = \{y, z, w\}$ and $Y_{yw} = \{y, w\}$. It is easy to verify that the so-defined family $(Y_v)_{v \in V}$

is admissible for $(V, \sqsubset)$. That each of the sets Y_v, $v \in V$, is the only maximal set of multipliers which v can take follows immediately by the observation that for any $v \in V$ and any variable $s \notin Y_v$ there exists $v' \in V$ such that $v' \sqsubset v$ and $v\mathbb{K}[s] \cap v'R \supsetneq \{0\}$ in contradiction to the admissibility conditions. Moreover, one easily observes $M = \mathrm{in}_\prec(M) = xz\mathbb{K}[x, y, z, w] \oplus xw\mathbb{K}[x, y, w] \oplus yz\mathbb{K}[y, z, w] \oplus yw\mathbb{K}[y, w]$ and, therefore, V is an involutive basis of M.

Example 4. In general, the situation is more complicated even for monomial ideals of polynomial rings. Let us consider another example of this type with $R = \mathbb{K}[x, y, z, w]$ and $F = R$ being as in the previous example and $V = \{xy, xz, w\}$.

If V is ordered according to $xy \sqsubset xz \sqsubset w$ then there are two maximal admissible families of multipliers. In both cases we have $Y_{xy} = \{x, y, z, w\}$ and $Y_{xz} = \{x, z, w\}$. But there are two ways to fix a maximal set of multipliers for w, namely either $Y_w = \{y, z, w\}$ or $Y_w = \{x, w\}$. In both cases we have only the inclusion $xy\mathbb{K}[Y_{xy}] \oplus xz\mathbb{K}[Y_{xz}] \oplus w\mathbb{K}[Y_w] \subsetneq M$, hence, none of the two multiplier assignments makes V an involutive basis.

Nevertheless, V is an involutive basis of M. We just need to reorder V according to $w \sqsubset xy \sqsubset xz$. Then $Y_w = \{x, y, z, w\}$, $Y_{xy} = \{x, y, z\}$ and $Y_{xz} = \{x, z\}$ is the one and only maximal admissible choice of the sets of multipliers and one easily verifies the direct sum decomposition $M = w\mathbb{K}[x, y, z, w] \oplus xy\mathbb{K}[x, y, z] \oplus xz\mathbb{K}[x, z]$.

We remind that there are also sets V of monomials, e.g. $V = \{x^2, y^2\}$, which by no linear order $\sqsubset$ and by no admissible family of multiplier sets can be made an involutive basis. Also the question whether there is exactly one way to define a maximal admissible family of multiplier sets for a monomial set V is not related to V being an involutive basis or not. At the first glance Definition 1 seems to have no dynamical part. However, the following theorem shows that Condition 2, which is slightly more restrictive than the property $\mathrm{in}_\prec(v)\mathbb{K}[Y_v] \cap \mathrm{in}_\prec(v')\mathbb{K}[Y_{v'}] = \{0\}$ for all $v, v' \in V$ following already from Condition 1 of Definition 1 alone, enables the application of a Janet like algorithm for the computation of the so-defined involutive bases.

Theorem 1. *Let F be a free module of rank k over the polynomial ring $R = \mathbb{K}[X]$ in the variables $X = \{x_1, \ldots, x_n\}$ and $M \subset F$ a submodule with generating set V, where all elements of V are monic and have pairwise distinct leading terms. Moreover, let $\prec$ be an admissible term order of F, $\sqsubset$ a linear order of $\mathrm{lt}_\prec(V)$ and $(Y_v)_{v\in V}$ a family of multiplier sets which is admissible for the ordered set $(\mathrm{lt}_\prec(V), \sqsubset)$.*

Then with $V' = \{v' \in V \mid \mathrm{in}_\prec(v') \notin \mathrm{in}_\prec(v)\mathbb{K}[Y_v]$ for all $v \in V \setminus \{v'\}\}$ the following conditions are equivalent:

1. $\mathrm{in}_\prec(M) = \sum_{v\in V} \mathrm{in}_\prec(v)\mathbb{K}[Y_v]$,
2. $\mathrm{in}_\prec(M) = \bigoplus_{v'\in V'} \mathrm{in}_\prec(v')\mathbb{K}[Y_{v'}]$,
3. V *is a Gröbner basis of M with respect to $\prec$ and for all $w \in V$ and all variables $x \notin Y_w$ it holds $\mathrm{in}_\prec(xw) = x\,\mathrm{in}_\prec(w) \in \sum_{v\in V} \mathrm{in}_\prec(v)\mathbb{K}[Y_v]$,*

4. $(V \setminus V') \subset \sum_{v' \in V'} v' \mathbb{K}[Y_{v'}]$ *and for all* $w \in V'$ *and all variables* $x \notin Y_w$ *it holds* $xw \in \sum_{v' \in V'} v' \mathbb{K}[Y_{v'}]$.

Moreover, any of the above conditions implies that V *is an involutive basis of* M *with respect to* $\prec$.

Proof. Before we start to prove the equivalences let us make some remarks concerning the last statement. Obviously, V is an involutive basis of M with respect to $\prec$ if Conditions 1. or 2. are satisfied. But one might wonder why the opposite direction should fail to hold. The reason is simply that V could be an involutive basis according to the definition introduced here but the considered family $(Y_v)_{v \in V}$ is not suitable for providing the representations stated in 1. and 2. This problem can be avoided by defining the notion of an involutive basis V with respect to a given order $\sqsubset$ and a given family $(Y_v)_{v \in V}$ in the way it was done in (Apel , 1998).

In what follows, $\langle Y \rangle$ denotes the commutative monoid which is freely generated by some subset $Y \subseteq X$.

The equivalence of 1. and 2. follows immediately from the admissibility of $(Y_v)_{v \in V}$. The implication 1. $\rightarrow$ 3. is trivial. Let us sketch the main ideas behind the proof direction 3. $\rightarrow$ 1. The essential part consists in proving that for all $w \in V$ and all terms $t \in \langle X \rangle$ there is a representation $tw = a + b$, where $a = \sum_{v \in V} f_v v$ and $b = \sum_{v \in V} g_v v$ such that $f_v, g_v \in R$, only finitely many of the f_v and g_v are non-zero, $f_v \in \mathbb{K}[Y_v]$ for all $v \in V$, and $\mathrm{lt}_\prec(g_v v) \prec \mathrm{lt}_\prec(tw)$ for all $v \in V$ such that $g_v \neq 0$. In case $t \in \langle Y_w \rangle$ this is obvious, so let $x \in X \setminus Y_w$ be a non-multiplier occurring in t. Then by condition 3. we have $\mathrm{in}_\prec(xw) = x \, \mathrm{in}_\prec(w) \in \sum_{v \in V} \mathrm{in}_\prec(v) \mathbb{K}[Y_v]$ and, hence, there exist $w' \in V$ and $s \in \mathbb{K}[Y_{w'}]$ such that $\mathrm{lt}_\prec(xw) = \mathrm{lt}_\prec(sw')$. Since V is a Gröbner basis it follows the existence of a representation $xw - sw' = \sum_{v \in V} g'_v v$, where only finitely many g'_v are non-zero and $\mathrm{lt}_\prec(g'_v v) \prec \mathrm{lt}_\prec(xw)$ for all v such that $g'_v \neq 0$. Consequently, if one can show that $\frac{t}{x} sw'$ has a required representation $a' + b'$ then also $tw = \frac{t}{x} sw' + \sum_{v \in V} \frac{t}{x} g'_v v$ can be represented in the desired way. An obvious idea seems to be to repeat the process recursively and conclude termination by induction on the number of non-multipliers contained in t. But this approach will not work since a non-multiplier for w does not need to be also a non-multiplier for w'. Moreover, relaxing the admissibility condition of $(Y_v)_{v \in V}$ in such a way that only Condition 1 but not necessarily Condition 2 of Definition 1 holds can actually result in an infinite process. Note, there are simple counter examples showing that the implication 3. $\rightarrow$ 1. need not to hold in the relaxed case, cf. the forthcoming Example 3 (Part 2). However, the stronger notion of admissibility used here implies $\mathrm{lt}_\prec(w') \sqsubset \mathrm{lt}_\prec(w)$ which indeed ensures termination. Note, that by the finiteness of the number of terms dividing $t \, \mathrm{in}_\prec(w)$ this is even true if V is infinite and $\sqsubset$ is not a well-order. Application of simple inductive arguments on finite representations $m = \sum_{v \in V} h_v v$ of arbitrary elements $m \in M$ finishes the proof of direction 3. $\rightarrow$ 1.

The implication 3. $\rightarrow$ 4. is obvious and it remains to consider the branch 4. $\rightarrow$ 3. An arbitrary element $m \in M$ can be represented in the form $m = \sum_{v \in V} h_v v$, where only finitely many of the $h_v \in R$ are non-zero. According to the first part of 4. we can assume w.l.o.g. $h_v = 0$ for all $v \notin V'$. A rewriting process which successively replaces products of the form xv, where x is a non-multiplier for $v \in V'$, finally ends up with a finite representation $m = \sum_{v' \in V'} f_{v'} v'$, where $f_{v'} \in \mathbb{K}[Y_{v'}]$. Note, that here the strong definition of admissibility of $(Y_v)_{v \in V}$ is again essential. Now, we can deduce $\mathrm{lt}_{\prec}(f_{v'} v') \preceq \mathrm{lt}_{\prec}(m)$ for all $v' \in V'$ such that $f_{v'} \neq 0$ since no initial parts of the summands on the right hand side can cancel out[1]. This proves that V is a Gröbner basis and the second part of 3. follows easily by considering the special case $m = xw$ in our above investigations. □

Input:
   V $\cdots$ finite set of non-zero elements of a $\mathbb{K}[X]$-module M
   $\prec$ $\cdots$ admissible term ordering
Output:
   W $\cdots$ minimal involutive basis of M with respect to $\prec$

$U :=$ minimal subset of V such that $\mathrm{lt}_{\prec}(U) = \mathrm{lt}_{\prec}(V)$
$(Y_v)_{v \in U} :=$ admissible family for $\mathrm{lt}_{\prec}(U)$
$W := \left\{ w \in U \mid \mathrm{in}_{\prec}(w) \notin \bigoplus_{v \in U \setminus \{w\}} \mathrm{in}_{\prec}(v)\mathbb{K}[Y_v] \right\}$
$L := \{(1, v) \mid v \in V \setminus W\} \cup \{(y, w) \mid w \in W, y \in X \setminus Y_w\}$
while $L \neq \emptyset$ **do**
   choose $(t, u) \in L$
   $L := L \setminus \{(t, u)\}$
   $w :=$ involutive normal form of tu modulo W w.r.t. $(Y_v)_{v \in W}$ and $\prec$
   if $w \neq 0$ **then**
      $V := V \cup \{w\}$
      $U :=$ minimal subset of V such that $\mathrm{lt}_{\prec}(U) = \mathrm{lt}_{\prec}(V)$
      $(Y_v)_{v \in U} :=$ admissible family for $\mathrm{lt}_{\prec}(U)$
      $W := \left\{ w \in U \mid \mathrm{in}_{\prec}(w) \notin \bigoplus_{v \in U \setminus \{w\}} \mathrm{in}_{\prec}(v)\mathbb{K}[Y_v] \right\}$
      $L := \{(1, v) \mid v \in V \setminus W\} \cup \{(y, w) \mid w \in W, y \in X \setminus Y_w\}$
   end if
end while
return(W)

Fig. 1. Involutive basis construction method of Janet type

Figure 1 sketches the global structure of a Janet like method for computing an involutive basis of a given R-module. Some further details are necessary in order to make the presented frame algorithmic. A couple of ways to achieve this will be mentioned below.

First of all, we will explain the meaning of some objects used in our method. According to the specification it is allowed that various elements of the input basis V have the same leading term. U is a maximal subset of elements of V

---

[1] Note, by the same argument it will follow immediately that the sum $\sum_{v' \in V'} v'\mathbb{K}[Y_{v'}]$ is direct, i.e. $\sum_{v' \in V'} v'\mathbb{K}[Y_{v'}] = \bigoplus_{v' \in V'} v'\mathbb{K}[Y_{v'}]$.

whose leading terms are pairwise distinct. The family $(Y_v)_{v \in U}$ of multiplier sets which is admissible for $\mathrm{lt}_{\prec}(V) = \mathrm{lt}_{\prec}(U)$ is indexed by the subset $U \subseteq V$ since $Y_v = Y_{v'}$ must hold for all $v, v' \in V$ such that $\mathrm{lt}_{\prec}(v) = \mathrm{lt}_{\prec}(v')$ according to the admissibility property. After fixing $(Y_v)_{v \in U}$ we construct the set W as a maximal subset of U such that the sum $\sum_{v \in W} \mathrm{in}_{\prec}(v) \mathbb{K}[Y_v]$ is direct, i.e. $\mathrm{lt}_{\prec}(tv) \neq \mathrm{lt}_{\prec}(t'v')$ for all $v, v' \in W$, $t \in \langle Y_v \rangle$, $t' \in \langle Y_{v'} \rangle$. L is a set consisting of two types of critical pairs (t, u), where all of them have the characteristic property $\mathrm{in}_{\prec}(tu) \in \bigoplus_{v \in W} \mathrm{in}_{\prec}(v) \mathbb{K}[Y_v]$, i.e. tu is involutively reducible modulo W with respect to the admissible family $(Y_v)_{v \in W}$.

The global structure of the method displayed in Figure 1 is this of a critical element/completion algorithm. The main type of critical elements are the so-called *non-multiplicative prolongations* $(y, w) \in (X \setminus Y_w) \times W$. A second type are the pairs $(1, v)$, where the leading term of $v \in V$ is involutively reducible by some other element of V. Note, the second type of critical pairs does not occur when the leading terms of the input basis elements are pairwise distinct and Janet's definition of multipliers is used. In case the reduction of the polynomial tu resulting from the critical pair (t, u) yields a non-zero remainder w the basis is extended by w and the check is repeated. We remark the fact that completion of W by non-zero reduction remainders of non-multiplicative prolongations (and if applicable elements of $V \setminus W$) does not only complete the set of leading terms of W but at the same time completes W to a Gröbner basis of M without explicitly computing Buchberger's S-polynomials as the distinguishing feature of a Janet like algorithm.

If tu has involutive remainder 0 modulo W for all pairs $(t, u) \in L$ then W is an involutive basis of M according to Condition 4 of Theorem 1. Hence, the method displayed in Figure 1 works correctly provided that it terminates. A series of further restrictions is necessary in order to ensure termination for arbitrary inputs satisfying the specification.

The extension of V, and hence U, during the completion process requires the adaption of the admissible family $(Y_v)_{v \in U}$ of multiplier sets. Janet solved this problem by giving a general rule how to adapt the sets of multipliers and proved termination for his particular multiplier choice. But it turns out that the process will also terminate if one chooses the sets of multipliers in each step completely new, taking care only that the multiplier sets are "sufficiently large". More precisely, "sufficiently large" means $x \in Y_v$ for all $v \in U$ and $x \in X$ such that $\deg_x \mathrm{lt}_{\prec}(v) \geq \deg_x \mathrm{lt}_{\prec}(v')$ for all $v' \in U$. In the language of Apel (1998) these admissible families correspond exactly to the involutive divisions refining the *Thomas division* on $\mathrm{lt}_{\prec}(U)$, which is defined by the multiplier sets $Y_v^T = \{x \in X \mid \forall v' \in U : \deg_x v \geq \deg_x v'\}$, $v \in U$. The advantage of such an approach is the possibility of a better adjustment of the admissible family $(Y_v)_{v \in U}$ to the concrete intermediate basis. However, this freedom in choice causes a large overhead and some heuristical restrictions are advisable in order to find a compromise. A detailed analysis how to compute the families $(Y_v)_{v \in U}$ and a detailed discussion of the algorithm which completes an arbitrary finite generating set of M to a finite (minimal) involutive basis of M

can be found in (Apel , 1998). Nowadays, in view of the possibility to use Hemmecke's *sliced divisions* for initializing and adapting the family $(Y_v)_{v \in U}$ (see Hemmecke (2001)) in connection with the application of the criteria introduced by Apel and Hemmecke (2002) parts of the heuristics from (Apel , 1998) appear in a new light. In particular, the overhead caused by multiple reductions of non-multiplicative prolongations seems to be rather small for the sliced divisions strategy though the order $\sqsubset$ is changed during the adaption of the involutive division, or equivalently the admissible family of multiplier sets, to the intermediate basis.

Another generalization of Janet's method is the Gerdt/Blinkov approach which follows the classical method of fixing a priori all sets of multipliers, i.e. of fixing the families $(Y_s)_{s \in S}$ for all finite sets S of monomials. In this context a series of technical restrictions is necessary in order to ensure the termination of the algorithm. In (Apel , 1998) it is proved that there does not exist any family of multiplier assignments satisfying all conditions of Gerdt and Blinkov (1998) and having the additional property that for all finite sets of monomials the corresponding multiplier family is maximal. However, the Gerdt/Blinkov approach is also not subsumed by the method discussed here since it might include assignments of multiplier sets which violate Condition 2 of our Definition 1. It is an open question whether there are situations where one can take advantage of this fact.

Note, all methods discussed so far have in common that, in addition, the critical pairs $(u, h) \in L$ have to be processed according to the *normal selection strategy*, i.e. $\mathrm{lt}_{\prec}(uh)$ must be minimal with respect to $\prec$, in order to ensure termination for any specified input. Experiments show that the normal selection strategy is not preferable in Buchberger's algorithm for the computation of Gröbner bases. State of the art implementations frequently use *sugar strategy* instead (see Giovini et al. (1991)). In (Gerdt and Blinkov , 1998) Gerdt and Blinkov where the first who transfered Buchberger's criteria for detecting unnecessary critical pairs to the involutive method. Recently it could be shown by Apel and Hemmecke (2002) that the involutive basis algorithm terminates independent of the pair selection strategy if a series of criteria including those of Gerdt and Blinkov are applied in order to avoid the treatment of certain pairs $(u, h) \in L$.

6   "Good" Involutive Bases

There are theoretical as well as computational issues which pose further conditions on a "good" choice of an involutive basis V of a module $M \subset F$. Some of them are related to questions of the following type:

1. What is the minimal value of the sum $\sum_{v \in V} |X \setminus Y_v|$ which an involutive basis V of M can take?
2. What is the minimal number of elements of an involutive basis V of M?
3. What is the minimal highest degree of the elements of an involutive basis of M?

4. What is the maximal number d for which exist $V \subset M$ and a multiplier family $(Y_v)_{v \in V}$ making V an involutive basis of M such that $|Y_v| \geq d$ for all $v \in V$?

In case of generic coordinates and a degreewise reverse lexicographical term order $\prec$ Questions 1. - 3. have simple answers in terms of well-known algebraic invariants. In this case Bayer and Stillman (1987) could show that the crucial invariants of the module M coincide with those of its generic initial module, some generalizations of their results which are required here are due to Eisenbud (1995), and Mall (1998) proved that, under the same conditions, any set V such that $\mathrm{lt}_\prec(V)$ consists of the minimal monomial generators of $\mathrm{in}_\prec(M)$ becomes an involutive basis of M when the multipliers are assigned in a Pommaret like way. In case $rank\,F = 1$ the results of Bayer/Stillman, Eisenbud, and Mall answer also Question 4.

Let us consider the much more delicate but also much more interesting case of non-generic coordinates. We start with the investigation of involutive bases which are "good" in the sense that they can be computed in a minimal number of steps. The importance of the first three questions for a complexity analysis of the completion algorithm is obvious. In particular, the first question is highly interesting since the number $\sum_{v \in V} |X \setminus Y_v|$ provides an estimate for the costs of the involutive basis verification algorithm. In case of homogeneous modules M and the input generating set V' being a minimal homogeneous generating set of M the completion algorithm to involutive form has the same costs under the additional assumption that we have an oracle which tells us how to define the multipliers for the intermediate generating sets in the same way as they will appear in the final output. Note, the complexity analysis is harder than in the case of Buchberger's algorithm where analogous statements hold without an oracle. We can offer a heuristics for the multiplier assignment which in some sense is a static approximation of the oracle. While, in general, Question 1 is difficult to answer it is always easy to find a family $(Y_v)_{v \in V}$ of admissible multiplier assignments such that $\sum_{v \in V} |X \setminus Y_v|$ is minimal for a given finite set V, a given term order $\prec$, and a given linear order $\sqsubset$ of $\mathrm{lt}_\prec(V)$. Note, we neither require $M = \sum_{v \in V} v\mathbb{K}[Y_v]$ (which clearly would be unfulfillable in general) nor that $\sqsubset$ and $\prec$ are optimal in whatsoever sense. Let us discuss the objective of minimizing $\sum_{v \in V} |X \setminus Y_v|$ a little more detailed. In order to have a trivial intersection $v\mathbb{K}[Y_v] \cap v'\mathbb{K}[Y_{v'}] = \{0\}$ it is necessary and sufficient to ensure $\mathrm{lcm}(v, v') \notin v\mathbb{K}[Y_v] \cap v'\mathbb{K}[Y_{v'}]$, which is exactly the case if at least one of the conditions $\frac{\mathrm{lcm}(v,v')}{v} \notin \mathbb{K}[Y_v]$ or $\frac{\mathrm{lcm}(v,v')}{v'} \notin \mathbb{K}[Y_{v'}]$ holds. How this property is achieved by the different approaches applied in the past? Thomas uses the most rigid way which consists in requiring that for any distinct elements $v, v' \in V$ no variable of $\frac{\mathrm{lcm}(v,v')}{v}$ belongs to Y_v and no variable of $\frac{\mathrm{lcm}(v,v')}{v'}$ belongs to $Y_{v'}$. The method due to Janet is much more sophisticated. It enumerates the variables and requires that for any two distinct elements $v, v' \in V$ the first variable x in which v and v' have distinct degrees, without loss of generality let us assume $\deg_x v < \deg_x v'$, is not contained in Y_v and, therefore, Janet's method ensures $\frac{\mathrm{lcm}(v,v')}{v} \notin \mathbb{K}[Y_v]$ by introducing a single condition $x \notin Y_v$ for the pair (v, v').

Janet's approach sounds already very promising but still can be improved for two reasons. Firstly, it does not make any use of inclusions $v'\mathbb{K}[Y_{v'}] \subset v\mathbb{K}[Y_v]$ which might allow to waive the condition $x \notin Y_v$ for the pair (v, v') in case $v \mid v'$ and $Y_{v'} \subseteq Y_v \cup \{x\}$. Secondly, it always poses a condition on the first variable x in which v and v' have different degrees. But it can happen that the consideration of v and a third basis element $v'' \in V$ yields a condition $y \notin Y_v$ in such a way that by chance also $\deg_y v < \deg_y v'$. In this case the single condition $y \notin Y_v$ ensures already $v\mathbb{K}[Y_v] \cap v'\mathbb{K}[Y_{v'}] = v\mathbb{K}[Y_v] \cap v''\mathbb{K}[Y_{v''}] = \{0\}$ and the condition $x \notin Y_v$ can be refused again. By removing redundant admissibility conditions one obtains an admissible refinement of the Janet division. See (Apel , 1998, Figure 2) for an algorithm which computes a (sub-)maximal refinement for an arbitrary given admissible family of multiplier sets.

Example 5. Consider the ideal M generated by $V = \{x_2 x_3^2, x_2^2 x_3, x_1 x_2^2\}$ in $F = R = \mathbb{K}[x_1, x_2, x_3]$. Janet's sets of multipliers are $Y_{x_2 x_3^2} = \{x_3\}$, $Y_{x_2^2 x_3} = \{x_2, x_3\}$ and $Y_{x_1 x_2^2} = \{x_1, x_2, x_3\}$. A better assignment is $Y'_{x_2 x_3^2} = \{x_1, x_3\}$, $Y'_{x_2^2 x_3} = \{x_2, x_3\}$ and $Y'_{x_1 x_2^2} = \{x_1, x_2, x_3\}$. Not only that $\sum_{v \in V} |X \setminus Y'_v| = 2 < 3 = \sum_{v \in V} |X \setminus Y_v|$, we have even the proper inclusion $\{xv \mid v \in V \text{ and } x \in X \setminus Y'_v\} \subset \{xv \mid v \in V \text{ and } x \in X \setminus Y_v\}$ of the sets of elements which have to be reduced to zero during an involutive basis check. Hence, the running time in case of success is not only heuristically but provably faster for the second family of multipliers. Finally, we observe that the family $\left(Y'_{x_2 x_3^2}, Y'_{x_2^2 x_3}, Y'_{x_1 x_2^2}\right)$ yields the decomposition $M = x_2 x_3^2 \mathbb{K}[x_1, x_3] \oplus x_2^2 x_3 \mathbb{K}[x_2, x_3] \oplus x_1 x_2^2 \mathbb{K}[x_1, x_2, x_3]$ showing that V is an involutive basis of M. In contrast the original Janet assignment fails the check since $x_1 x_2 x_3^2 \in M \setminus (x_2 x_3^2 \mathbb{K}[x_3] \oplus x_2^2 x_3 \mathbb{K}[x_2, x_3] \oplus x_1 x_2^2 \mathbb{K}[x_1, x_2, x_3])$, i.e. V is not a Janet basis of M.

A mild variation of the above static approach which includes also dynamic features without essentially increasing the costs is to look for an admissible multiplier assignment $(Y_v)_{v \in V}$ such that the sum $\sum_{v \in V} |X \setminus (Y_v \cup Z_v)|$, where $Z_v \subseteq X$ denotes the set of all variables x for which xv had been reduced already during the previous completion process, becomes minimal. If a fixed linear order $\sqsubset$ of the set of all terms is used during the completion algorithm then this number is the better approximation for the remaining costs of the completion process since no repeated reductions are necessary under these conditions (see (Apel , 1998)). The basic idea behind the strategy suggested in (Apel , 1998, Remark 6.2) is to choose the multipliers according to this heuristics.

Now let us consider another notion of "good" involutive bases, namely, such which provide the maximal value d introduced in Question 4. The search for involutive bases which are "good" in such a sense is strongly related to the solution of combinatorial problems and the computation of invariants of the module. There is the following open conjecture due to Stanley (1982): an arbitrary monomial module M can be decomposed in the form (8) such that $|Y_v| \geq depth\, M$ for all $v \in V$. In (Apel , 2001 a) this conjecture could be proved for some huge classes of modules in the case $rank\, F = 1$ by proving the existence of "good" involutive bases, i.e. such which for a suitable multiplier definition satisfy $d \geq depth\, M$.

This property is even more restrictive than conjectured by Stanley since the additional Condition 2 of Definition 1 holds. So it is not surprising that the class of involutive bases defined in Definition 1 is not rich enough to prove Stanley's conjecture in general. The following simple counter examples illustrates that no definition of involutive bases allowing a Janet type algorithm need to provide the maximal possible d.

Example 3 (Part 2). In Part 1 we observed that $V = \{xz, xw, yz, yw\}$ is an involutive basis and one easily verifies that the largest element of V with respect to any linear order $\sqsubset$ will have at most two multipliers. However, M can be decomposed also in the form $M = xz\mathbb{K}[x, z, w] \oplus xw\mathbb{K}[x, y, w] \oplus yz\mathbb{K}[x, y, z] \oplus yw\mathbb{K}[y, z, w] \oplus xyzw\mathbb{K}[x, y, z, w]$ with vertices from the larger generating set $V' = \{xz, xw, yz, yw, xyzw\}$ of M. While the corresponding multiplier sets satisfy Equation (8) they do not allow a Janet like algorithm. Note, that V' and the above family of multipliers satisfy the last two but violate the first two conditions of Theorem 1. One easily observes that Condition 2 of Definition 1 is not satisfied for any linear order $\sqsubset$ of V'.

The above example shows that even in the case that the minimal generators of a monomial module M can be used as the vertices of a direct sum decomposition (8) of M such a decomposition need not be the "best" possible with respect to Question 4. Consequently, "good" in the sense of the first three questions can be quite distinct from "good" in the sense of the last question. The difficulties one meets in the study of Stanley's conjecture make clear that a relaxation of the assignment rules for the multipliers will not only disable Janet's algorithm but will have also impact on the costs of the alternative method which consists in first computing a Gröbner basis and doing the direct sum decomposition of the initial module subsequently. In fact, this shows that the sometimes underestimated task of computing an involutive basis of a given monomial module is not easy if additional conditions are posed on the decomposition. Note, there is an obvious but infeasible algorithm for solving this problem which is based on the simple observation that the number of decompositions of type (8) in maximal direct summands is finite. During the search for suitable initial determinations of a system of PDEs the demands on the decomposition of the set $\mathcal{PA}$ of parametric derivatives in disjoint cones are often similar to those of Stanley. Therefore, the investigation of Stanley's conjecture for modules R/I (see (Apel , 2001 b)) can contribute to this part of the Riquier/Janet theory which seems to be much less investigated in the past.

7 Concluding Remarks

The observation of Zharkov and Blinkov (1993) that there is something interesting at the junction of passive complete orthonomic systems of PDEs and Gröbner bases of polynomial ideals was the starting point of the investigation of involutive bases. Zharkov and Blinkov's motivation was to find a fast alternative algorithm for the computation of Gröbner bases in polynomial rings and a series of remarkable results and computer experiments in this direction have been reported in the past (c.f. Gerdt et al. (2001), Hemmecke (2001)). Each comparison depends strongly on the used implementations since Buchberger's algorithm for the computation of Gröbner bases and Janet's algorithm for the computation of involutive bases are of the same complexity order. Though, Gerdt/Blinkov/Yanovich compared to todays state of the art implementations of Buchberger's algorithm there still remains a little danger that at some day the original justification of (polynomial) involutive bases might disappear, for instance, when an essentially improvement in the strategies of Buchberger's algorithm will be found. This is the reason why one should also have in mind theoretical justifications resulting from the additional structure of involutive bases. This additional structure consists in a direct sum decomposition of the initial module and allows, for instance, to give a closed formula for the Hilbert function (see (Apel , 1998)) or to compute Stanley decompositions (see (Apel , 2001 a,b)).

The quite different approaches due to Gerdt and Blinkov (1998) and Apel (1998), respectively, show that there is no canonical generalization of Riquier's ideas to rings and modules. An objective of this article is to convince the reader that there is certainly no "best" algebraic generalization and that there is still a wide open field in finding "good" ones depending on the application one has in mind. The application might be one of the previously investigated as well as some completely new which probably will be discovered in the future.

References

W. W. Adams, P. Loustaunau. An Introduction to Gröbner Bases, *Graduate Studies in Mathematics*, Vol. **3**, AMS Press, Providence, (1994).

J. Apel. The Theory of Involutive Divisions and an Application to Hilbert Function Computations. *J. Symb. Comp.* **25**, 683-704 (1998).

J. Apel. Computational Ideal Theory in Finitely Generated Extension Rings. *Theoretical Computer Science*, **244/1-2**, 1–33 (2000).

J. Apel. On a Conjecture of R. P. Stanley; Part I – Monomial Ideals. *MSRI Preprint* 2001-004 (2001).

J. Apel. On a Conjecture of R. P. Stanley; Part II – Quotients modulo Monomial Ideals. *MSRI Preprint* 2001-009 (2001).

J. Apel. Stanley Decompositions and Riquier Bases. Talk presented at the Conference on Applications of Computer Algebra (ACA'01), Albuquerque (2001).

J. Apel, R. Hemmecke. Detecting Unnecessary Reductions in an Involutive Basis Computation. Work in Progress.

J. Apel, W. Laßner. An Extension of Buchberger's Algorithm and Calculations in Enveloping Fields of Lie Algebras. *J. Symb. Comp.* **6**, 361–370 (1988).

D. Bayer, M. Stillman. A Criterion for Detecting m-Regularity. *Invent. math.* **87**, 1–11 (1987).

T. Becker, V. Weispfenning, in cooperation with H. Kredel. Gröbner Bases, A Computational Approach to Commutative Algebra. *Springer*, New York, Berlin, Heidelberg, (1993).

B. Buchberger. Ein Algorithmus zum Auffinden der Basiselemente des Restklassenringes nach einem nulldimensionalen Polynomideal, *Ph.D. Thesis*, University of Innsbruck, (1965).

B. Buchberger. An Algorithmic Method in Polynomial Ideal Theory, Chapter 6 in: N.K. Bose, ed., Recent Trends in Multidimensional System Theory, *D.Reidel Publ.Comp.*, (1985).

D. Eisenbud. Commutative Algebra. With a view toward algebraic geometry. Graduate Texts in Mathematics, **150**. *Springer-Verlag*, New York, (1995).

V. P. Gerdt, Yu. A. Blinkov. Involutive Bases of Polynomial Ideals. *Mathematics and Computers in Simulation* **45**, 519–542 (1998).

V. P. Gerdt, Yu. A. Blinkov, D. A. Yanovich. Fast Computation of Polynomial Janet Bases. Talk presented at the Conference on Applications of Computer Algebra (ACA'01), Albuquerque (2001).

Giovini, A., Mora, T., Niesi, G., Robbiano, L., Traverso, C. (1991). "One sugar cube, please," or Selection Strategies in the Buchberger Algorithm. In: Watt, S.M. (ed.), Proc. ISSAC'91, ACM Press, New York, pp. 49–54.

R. Hemmecke. Dynamical Aspects of Involutive Bases Computations. SNSC'01, Linz (2001), This Volume.

M. Janet. Lecons sur les systèmes d'equations aux dérivées partielles. *Gauthier-Villars*, Paris (1929).

D. Mall, On the Relation Between Gröbner and Pommaret Bases. *AAECC* **9/2**, 117–123 (1998).

H. M. Möller, T. Mora. New Constructive Methods in Classical Ideal Theory, *J.Algebra* **100**, 138–178, (1986).

J. F. Pommaret. Systems of Partial Differential Equations and Lie Pseudogroups. *Gordan and Breach*, New York (1978).

C. H. Riquier. Les systémes d'equations aux dérivées partielles. *Gauthier-Villars*, Paris (1910).

R. P. Stanley. Linear Diophantine Equations and Local Cohomology. *Invent. math.* **68**, 175–193 (1982).

J. M. Thomas. Riquier's Existence Theorems. *Annals of Mathematics* **30/2**, 285–310 (1929).

J. M. Thomas. Differential Systems. *American Mathematical Society*, New York, (1937).

A. R. Tresse. Sur les invariants différentials des groupes continus de transformations. *Acta Mathematica* **18**, 1–8 (1894).

A. Yu. Zharkov, Yu. .A. Blinkov. Involution Approach to Solving Systems of Algebraic Equations. Proc. IMACS'93, 11–16 (1993).

Symmetries of Second- and Third-Order Ordinary Differential Equations

Fritz Schwarz

FhG, Institut SCAI, 53754 Sankt Augustin, Germany[*]
fritz.schwarz@gmd.de

Abstract. In order to apply Lie's symmetry theory for solving a differential equation it must be possible to identify the group of symmetries leaving the equation invariant. The answer is obtained in two steps. At first a classification of the possible symmetries of equations of the respective order is determined. Secondly a decision procedure is provided which allows to identify the symmetry type within this classification. For second-order equations the answer has been obtained by Lie himself. In this article the complete answer for quasilinear equations of order three is given. An important tool is the Janet base representation for the determining system of the symmetries.

1 Introduction

In general there are no solution algorithms available for solving nonlinear ordinary differential equations (ode's) in closed form. Usually several heuristics are applied in order to find a solution without any guarantee however to find one, or even to assure its existence. Stimulated by Galois' ideas for solving algebraic equations, Sophus Lie got interested in these problems around the middle of the 19th century. His most significant recognition was that the transformation properties of an ode under certain groups of continuous transformations play a fundamental role for answering this question, very much like the permutations of the solutions of an algebraic equation furnish the key to understanding its solution behavior.

With the insight gained by analysing simple examples of solvable equations, Lie developed a solution scheme for ode's that may be traced back to the following question: Are there any transformations leaving the form of the given equation invariant? If the answer is affirmative these transformations are called its *symmetries*. In general, equations that may be transformed into each other are called *equivalent*, they form an *equivalence class* and share the same *symmetry type*. The union of all equivalence classes sharing the same symmetry type is called a *symmetry class*. If the solution of a canonical representative within an equivalence class may be obtained, and moreover it is possible to determine the transformation of the given equation to this canonical form, the original problem is solved.

[*] Supported in part by INTAS Grant 99-0167

F. Winkler and U. Langer (Eds.): SNSC 2001, LNCS 2630, pp. 108–139, 2003.

Lie himself has worked out his theory in full detail for ode's of order two [Lie 1891]. However in many fields of application there occur ode's of order three or higher. Therefore it appears to be highly desirable to provide the full answer obtained by his theory at least for equations of order three. To this end a complete classification of the possible symmetries for these equations required. It is provided in this article. Furthermore it is shown how the various symmetry types may be obtained from the Janet base for its determining system.

Two fundamental discoveries that were made within about twenty years after Lie's death turned out to be of fundamental importance for applying his theories. The first is Loewy's theory of linear ode's [Loewy 1906], and its generalization to certain systems of linear pde's. The second is Janet's theory of linear pde's [Janet 1920] and in particular the canonical form he introduced for any such system which is known as a *Janet base* today. This concept is closely related to Buchbergers *Gröbner base* in polynomial ideal theory, and turned out to be fundamental in the realm of differential equations. The importance of these results for Lie's theory originates from the fact that the symmetries of any differential equation are obtained from its so-called *determining system*, a system of linear homogeneous pde's. Representing it as a Janet base allows to identify the type of its symmetry from its coefficients.

In this article ode's are polynomial in the derivatives, linear in the highest derivative, with coefficients that are rational functions of the dependent and the independent variables, it is called the *base field*. The overall scheme for solving ode's based on its possible symmetries decomposes in three major steps.

→ Determine the symmetry class of the given equation.
→ Transform the equation to a canonical form corresponding to its symmetry type.
→ Solve the canonical form and obtain the solution of the given equation from it.

It is the purpose of this article to provide a complete answer for the first step for equations of order two or three. The remaining two steps and many more details may be found in a forthcoming publication [Schwarz 2002].

The subsequent Section 2 deals with linear partial differential equations (pde's), and the concept of a Janet base is introduced for them. In Section 3, those results from the theory of continuous groups of the plane are presented that are relevant for the rest of this article. In Section 4 various geometric and algebraic properties of groups are expressed in terms of a Janet base for its defining equations. Section 5 contains the main result. Several theorems are proved allowing to identify the symmetry type to which a given quasilinear equation of order two or three belongs.

2   Linear Partial Differential Equations

The theory of systems of linear homogeneous pde's is of interest for its own right, independent of its applications for finding symmetries and invariants of differential equations. Any such system may be written in infinitely many ways by adding linear combinations of its members or derivatives thereof without changing its solution set. In general it is a difficult question whether there exist nontrivial solutions at all, or what the degree of arbitrariness of the general solution is. It may be a finite set of constants, in this case there is a finite dimensional solution space as it is true for the systems considered in this article. More general, one or more functions depending on a differing number of arguments may be involved.

These questions were the starting point for Maurice Janet. He introduced a unique representation for such systems of pde's similar to a Gröbner base representation of a system of algebraic equations, it is called a *Janet base*. There are two fundamental achievements of a Janet base representation for a system of linear homogeneous pde's. The most important quantity characterizing such systems, the degree of arbitrariness of its general solution, is obvious from its leading terms. Therefore the pattern of leading terms is called the *type of a Janet base*. Secondly due to the uniqueness of the coefficients of a Janet base, additional invariants may be obtained from them providing further information on the solutions of the corresponding system of pde's.

The systems of pde's considered in this article are linear and homogeneous in a finite set of differential indeterminates $u^1, \ldots, u^m$ over the ring of differential operators $\mathbf{Q}(x_1, \ldots, x_n)[\partial_{x_1}, \ldots, \partial_{x_n}]$. The u^α are also called *dependent variables* or *functions*, the x_i are called *independent variables*. A *ranking* of derivatives is a total ordering such that for any two derivatives u and v and any derivation δ there holds $u \leq v \Rightarrow \delta u \leq \delta v$. A term $t_{i,j}$ in a system of linear homogeneous pde's is a function u^k or a derivative of it, multiplied by an element of the base field $\mathbf{Q}(x_1, \ldots, x_n)$. The ranking of derivatives defined above generates a linear order for the terms. A *lex* order arranges derivatives according to some lexicographical order of the variables involved. In a *grlex* order, at first the order of the derivatives are compared. A linear homogeneous system of pde's with terms $t_{i,j}$ is always arranged in the form

$$\boxed{t_{1,1}} > t_{1,2} > \ldots > t_{1,k_1}$$
$$\wedge$$
$$\boxed{t_{2,1}} > t_{2,2} > \ldots > t_{2,k_2}$$
$$\wedge$$
$$\vdots$$
$$\wedge$$
$$\boxed{t_{N,1}} > t_{N,2} > \ldots > t_{N,k_N}$$

such that the ordering relations are valid as indicated. Each line of this scheme corresponds to an equation of the system of pde's, its terms are arranged in decreasing order from left to right. In order to save space, sometimes several equations are arranged into a single line. In these cases, in any line the leading terms *increase* from left to right. The terms in the rectangular boxes are the *leading terms* containing the *leading derivative*, i. e. the leading term is that term preceding any other term in the respective equation, its coefficient is usually assumed to be one. For any given system of pde's there is a finite number of term orderings. In any term ordering there is a *lowest term*, it is always one of the indeterminates itself.

The main steps for generating a Janet base for any system of pde's will be outlined now. The first fundamental operation is the *reduction*. Given any pair of equations e_1 and e_2 of the system, it may occur that the leading derivative of the latter equation e_2, or of a suitable derivative of it, equals some derivative in the former equation e_1. This coincidence may be applied to remove the respective term from e_1 by an operation which is called a *reduction step*. To this end, multiply the proper derivative of e_2 by the coefficient of the term in e_1 that is to be removed, and subtract it from e_1 multiplied by the leading coefficient of e_2. In general reductions may occur several times. However, due to the properties of the term orderings described above and the genuine lowering of terms in any reduction step it is assured that reductions always terminate after a finite number of iterations. After its completion there are no further reductions possible. In this case e_1 is called *completely reduced* with respect to e_2.

For any given system it will in general be possible to perform various reductions w.r.t. one or more equations of the system. It is required that all possible reductions be performed. A system with the property that there is no reduction possible of any equation of the system w.r.t. to any other equation is called an *autoreduced system*.

Autoreduced systems bear two major shortcomings which disqualify them from being considered as canonical form of a given system of pde's. In the first place, the result of autoreduction is non-unique in general. Furthermore, by proper differentiation of two equations of a system and subsequent reduction it may be possible to obtain two different equations with the same leading derivatives. Either of them may be considered as a representation of this leading derivative with a different answer. This apparent inconsistency holds already the clue how to avoid them. To this end, choose two equations with like leading function and solve both equations with respect to its leading derivative. By suitable cross differentiation it is always possible to obtain two new equations such that the derivatives at the left hand sides are identical to each other. Intuitively it is expected that the same should hold true for the right hand sides if all reductions w.r.t. to the remaining equations of the system are performed. If this is not true their difference is called an *integrability condition*. It has to be added to the system as a new equation. It can be shown that this proceeding terminates after a finite number of steps. An autoreduced system satisfying all integrability conditions is called a *Janet base* or a *coherent system*.

The following algorithm due to Janet [Janet 1920] accepts any linear homogeneous system of pde's and transforms it into a new system such that all integrability conditions are satisfied.

Algorithm. *JanetBase(S).* Given a linear homogeneous system of pde's $S = \{e_1, e_2, \ldots\}$, the Janet base corresponding to S is returned. As usual let v_i be the leading term of equation e_i.

$S1$: *Autoreduction.* Set $S := Autoreduce(S)$.

$S2$: *Completion.* Set $S := CompleteSystem(S)$.

$S3$: *Find Integrability Conditions.* Find all pairs of leading terms v_i and v_j such that differentiation w.r.t. a non-multiplier x_{i_k} and multipliers $x_{j_1}, \ldots x_{j_l}$ respectively leads to

$$\frac{\partial v_i}{\partial x_{i_k}} = \frac{\partial^{p_1 + \ldots + p_l} v_j}{\partial x_{j_1}^{p_1} \ldots \partial x_{j_l}^{p_l}}$$

and determine the integrability conditions

$$c_{i,j} = LeadingCoefficient(e_j) \cdot \frac{\partial e_i}{\partial x_{i_k}}$$
$$-LeadingCoefficient(e_i) \cdot \frac{\partial^{p_1 + \ldots + p_l} e_j}{\partial x_{j_1}^{p_1} \ldots \partial x_{j_l}^{p_l}}$$

$S4$: *Reduce Integrability Conditions.* For all $c_{i,j}$ set
$$c_{i,j} := Reduce(c_{i,j}, S)$$

$S5$: *Termination?* If all $c_{i,j}$ are zero return S otherwise set $S := S \cup \{c_{i,j} | c_{i,j} \neq 0\}$, reorder S properly and go to $S1$.

From now on any system of linear homogeneous pde's will be assumed to be a Janet base that has been generated by the algorithm described above. Furthermore it is assumed that the dimension of its solution space is finite, it is called the *order* of the system. Both the number dependent and independent variables are assumed to be not greater than two. The reason is that these constraints are *a priori* known to be true for the applications later in this article.

It turns out that systems with these properties allow a complete classification. For the corresponding Janet base types a unique notation $\mathcal{J}_{r,k}^{(m,n)}$ is introduced. Here m and n denote the number of dependent and independent variables respectively, r denotes the order, and k is a consecutive enumeration within this set. The symbol $\mathcal{J}^{(m,n)}$ without the lower indices denotes collectively all Janet base types corresponding to the given values of m and n.

In Section 5 Janet bases of type $\mathcal{J}^{(2,2)}$ represent the determining systems of the symmetries. Those types occuring in this analysis are introduced in the remaining part of this section. The two dependent variables are denoted by ξ and η, the independent ones by x and y. In *grlex* term ordering with $\eta > \xi$ and $y > x$, the derivatives of order not higher than two are arranged as follows.

$$\ldots \eta_{yy} > \eta_{yx} > \eta_{xx} > \xi_{yy} > \xi_{yx} > \xi_{xx} > \eta_y > \eta_x > \xi_y > \xi_x > \eta > \xi.$$

For a r-dimensional solution space, r derivatives must be selected from this arrangement such that none of it may be obtained by differentiation from any of the remaining ones. For $r = 1, 2$ or 3 the possible selections are as follows.

$r = 1 :$ $\{\eta\}, \{\xi\},$

$r = 2 :$ $\{\eta_x, \eta\}, \{\eta_y, \eta\}, \{\eta, \xi\}, \{\xi_y, \xi\}, \{\xi_x, \xi\},$

$r = 3 :$ $\{\eta_{xx}, \eta_x, \eta\}, \{\eta_y, \eta_x, \eta\}, \{\eta_{yy}, \eta_y, \eta\}, \{\eta_x, \eta, \xi\},$

$\{\eta_y, \eta, \xi\}, \{\xi_x, \eta, \xi\}, \{\xi_y, \eta, \xi\}, \{\xi_{yy}, \xi_y, \xi\}, \{\xi_y, \xi_x, \xi\}, \{\xi_{xx}, \xi_x, \xi\}.$

The corresponding classification of Janet bases is given next. There are two Janet base types of order 1.

$$\mathcal{J}_{1,1}^{(2,2)} : \quad \xi = 0, \ \eta_x + a\eta = 0, \ \eta_y + b\eta = 0. \tag{1}$$

$$\mathcal{J}_{1,2}^{(2,2)} : \quad \eta + a\xi = 0, \ \xi_x + b\xi = 0, \ \xi_y + c\xi = 0. \tag{2}$$

There are five Janet base types of order 2.

$$\mathcal{J}_{2,1}^{(2,2)} : \quad \xi = 0, \ \eta_y + a_1\eta_x + a_2\eta = 0, \ \eta_{xx} + b_1\eta_x + b_2\eta = 0 \tag{3}$$

$$\mathcal{J}_{2,2}^{(2,2)} : \quad \xi = 0, \ \eta_x + a_1\eta = 0, \ \eta_{yy} + b_1\eta_y + b_2\eta = 0. \tag{4}$$

$$\mathcal{J}_{2,3}^{(2,2)} : \quad \begin{aligned} &\xi_x + a_1\eta + a_2\xi = 0, \ \ \xi_y + b_1\eta + b_2\xi = 0, \\ &\eta_x + c_1\eta + c_2\xi = 0, \ \ \eta_y + d_1\eta + d_2\xi = 0. \end{aligned} \tag{5}$$

$$\mathcal{J}_{2,4}^{(2,2)} : \quad \eta + a_2\xi = 0, \ \xi_y + b_1\xi_x + b_2\xi = 0, \ \xi_{xx} + c_1\xi_x + c_2\xi = 0. \tag{6}$$

$$\mathcal{J}_{2,5}^{(2,2)} : \quad \eta + a_1\xi = 0, \ \xi_x + b_1\xi = 0, \ \xi_{yy} + c_1\xi_y + c_2\xi = 0. \tag{7}$$

There are ten Janet base types of order 3.

$$\mathcal{J}_{3,1}^{(2,2)} : \xi = 0, \ \eta_y + a_1\eta_x + a_2\eta = 0, \ \eta_{xxx} + b_1\eta_{xx} + b_2\eta_x + b_3\eta = 0. \tag{8}$$

$$\mathcal{J}_{3,2}^{(2,2)} : \begin{aligned} &\xi = 0, \ \eta_{xx} + a_1\eta_y + a_2\eta_x + a_3\eta = 0, \ \eta_{xy} + b_1\eta_y + b_2\eta_x + b_3\eta = 0, \\ &\eta_{yy} + c_1\eta_y + c_2\eta_x + c_3\eta = 0. \end{aligned} \tag{9}$$

$$\mathcal{J}_{3,3}^{(2,2)} : \xi = 0, \ \eta_x + a_1\eta = 0, \ \eta_{yyy} + b_1\eta_{yy} + b_2\eta_y + b_3\eta = 0. \tag{10}$$

$$\mathcal{J}_{3,4}^{(2,2)} : \begin{aligned} &\xi_x + a_1\eta + a_2\xi = 0, \ \ \xi_y + b_1\eta + b_2\xi = 0, \\ &\eta_y + c_1\eta_x + c_2\eta + c_3\xi = 0, \ \ \eta_{xx} + d_1\xi_x + d_2\eta + d_3\xi = 0. \end{aligned} \tag{11}$$

$$\mathcal{J}_{3,5}^{(2,2)} : \begin{aligned} &\xi_x + a_1\eta + a_2\xi = 0, \ \ \xi_y + b_1\eta + b_2\xi = 0, \\ &\eta_x + c_1\eta + c_2\xi = 0, \ \ \eta_{yy} + d_1\eta_y + d_2\eta + d_3\xi = 0. \end{aligned} \tag{12}$$

$$\mathcal{J}_{3,6}^{(2,2)} : \begin{aligned} &\xi_y + a_1\xi_x + a_2\eta + a_3\xi = 0, \ \ \eta_x + b_1\xi_x + b_2\eta + b_3\xi = 0, \\ &\eta_y + c_1\xi_x + c_2\eta + c_3\xi, \ \xi_{xx} + d_1\xi_x + d_2\eta + d_3\xi = 0. \end{aligned} \tag{13}$$

$$\mathcal{J}_{3,7}^{(2,2)} : \begin{aligned} &\xi_x + a_1\eta + a_2\xi = 0, \ \ \eta_x + b_1\xi_y + b_2\eta + b_3\xi = 0, \\ &\eta_y + c_1\xi_y + c_2\eta + c_3\xi = 0, \ \xi_{yy} + d_1\xi_y + d_2\eta + d_3\xi = 0. \end{aligned} \tag{14}$$

$$\mathcal{J}_{3,8}^{(2,2)} : \eta + a_1\xi = 0, \ \xi_x + b_1\xi = 0, \ \xi_{yyy} + c_1\xi_{yy} + c_2\xi_y + c_3\xi = 0. \tag{15}$$

$$\mathcal{J}_{3,9}^{(2,2)}: \quad \begin{array}{l} \eta + a_1\xi = 0, \ \ \xi_{xx} + b_1\xi_y + b_2\xi_x + b_3\xi = 0, \\ \xi_{xy} + c_1\xi_y + c_2\xi_x + c_3\xi = 0, \ \ \xi_{yy} + d_1\xi_y + d_2\xi_x + d_3\xi = 0. \end{array} \tag{16}$$

$$\mathcal{J}_{3,10}^{(2,2)}: \eta + a_1\xi = 0, \ \ \xi_y + b_1\xi_x + b_2\xi = 0, \xi_{xxx} + c_1\xi_{xx} + c_2\xi_x + c_3\xi = 0. \tag{17}$$

There are twenty Janet base types of order 4, only the following five types are required for the symmetry analysis in Section 5.

$$\mathcal{J}_{4,2}^{(2,2)}: \quad \begin{array}{l} \xi = 0, \ \ \eta_{xy} + a_1\eta_{xx} + a_2\eta_y + a_3\eta_x + a_4\eta = 0, \\ \eta_{yy} + b_1\eta_{xx} + b_2\eta_y + b_3\eta_x + b_4\eta = 0, \\ \eta_{xxx} + c_1\eta_{xx} + c_2\eta_y + c_3\eta_x + c_4\eta = 0. \end{array} \tag{18}$$

$$\mathcal{J}_{4,9}^{(2,2)}: \quad \begin{array}{l} \xi_y + a_2\xi_x + a_2\xi_x + a_3\eta + a_4\xi = 0, \ \ \eta_y + b_1\eta_x + b_2\xi_x + b_3\eta + b_4\xi = 0, \\ \xi_{xx} + c_1\eta_x + c_2\xi_x + c_3\eta + c_4\xi = 0, \ \ \eta_{xx} + d_1\eta_x + d_2\xi_x + d_3\eta + d_4\xi = 0. \end{array} \tag{19}$$

$$\mathcal{J}_{4,10}^{(2,2)}: \quad \begin{array}{l} \xi_y + a_1\eta_y + a_2\xi_x + a_3\eta + a_4\xi = 0, \ \ \eta_x + b_1\eta_y + b_2\xi_x + b_3\eta + b_4\xi = 0, \\ \xi_{xx} + c_1\eta_y + c_2\xi_x + c_3\eta + c_4\xi = 0, \ \ \eta_{yy} + d_1\eta_y + d_2\xi_x + d_3\eta + d_4\xi = 0. \end{array} \tag{20}$$

$$\mathcal{J}_{4,12}^{(2,2)}: \quad \begin{array}{l} \xi_x + a_1\eta_y + a_2\xi_y + a_3\eta + a_4\xi = 0, \ \ \eta_x + b_1\eta_y + b_2\xi_y + b_3\eta + b_4\xi = 0, \\ \xi_{yy} + c_1\eta_y + c_2\xi_y + c_3\eta + c_4\xi = 0, \ \ \eta_{yy} + d_1\eta_y + d_2\xi_y + d_3\eta + d_4\xi = 0. \end{array} \tag{21}$$

$$\mathcal{J}_{4,14}^{(2,2)}: \quad \begin{array}{l} \eta_x + a_1\xi_y + a_2\xi_x + a_3\eta + a_4\xi = 0, \ \ \eta_y + b_1\xi_y + b_2\xi_x + b_3\eta + b_4\xi = 0, \\ \xi_{xx} + c_1\xi_y + c_2\xi_x + c_3\eta + c_4\xi = 0, \ \ \xi_{xy} + d_1\xi_y + d_2\xi_x + d_3\eta + d_4\xi = 0, \\ \xi_{yy} + e_1\xi_y + e_2\xi_x + e_3\eta + e_4\xi = 0. \end{array} \tag{22}$$

$$\mathcal{J}_{4,17}^{(2,2)}: \quad \begin{array}{l} \eta + a_4\xi = 0, \ \ \xi_{xy} + b_1\xi_{xx} + b_2\xi_y + b_3\xi_x + b_4\xi = 0, \\ \xi_{yy} + c_1\xi_{xx} + c_2\xi_y + c_3\xi_x + c_4\xi = 0, \\ \xi_{xxx} + d_1\xi_{xx} + d_2\xi_y + d_3\xi_x + d_4\xi = 0. \end{array} \tag{23}$$

$$\mathcal{J}_{4,19}^{(2,2)}: \quad \begin{array}{l} \eta = 0, \ \ \xi_{xx} + a_2\xi_y + a_3\xi_x + a_4\xi = 0, \\ \xi_{xy} + b_2\xi_y + b_3\xi_x + b_4\xi = 0, \\ \xi_{yyy} + c_1\xi_{yy} + c_2\xi_y + c_3\xi_x + c_4\xi = 0. \end{array} \tag{24}$$

Finally there are three Janet base types of order five required.

$$\mathcal{J}_{5,1}^{(2,2)}: \begin{array}{l} \xi_x + a_1\xi = 0, \ \ \xi_y = 0, \\ \eta_{xy} + c_1\eta_x + c_2\eta + c_3\xi = 0, \ \ \eta_{yy} + d_1\eta_y + d_2\eta + d_3\xi = 0, \\ \eta_{xxx} + e_1\eta_{xx} + e_2\eta_y + e_3\eta_x + e_4\eta + e_5\xi = 0. \end{array} \tag{25}$$

$$\mathcal{J}_{5,2}^{(2,2)}: \begin{array}{l} \eta_x = 0, \ \ \eta_y + b_1\eta = 0, \\ \xi_{xx} + c_1\xi_x + c_2\eta + c_3\xi = 0, \ \ \xi_{xy} + d_1\xi_y + d_2\eta + d_3\xi = 0, \\ \xi_{yyy} + e_1\xi_{yy} + e_2\xi_y + e_3\xi_x + e_4\eta + e_5\xi = 0. \end{array} \tag{26}$$

$$\mathcal{J}_{5,3}^{(2,2)} : \quad \begin{aligned} &\eta_x + a_1\xi_x + a_2\eta + a_3\xi = 0, \quad \eta_y + b_1\xi_y + b_2\eta + b_3\xi = 0, \\ &\xi_{xy} + c_1\xi_{xx} + c_2\xi_y + c_3\xi_x + c_4\eta + c_5\xi = 0, \\ &\xi_{yy} + d_1\xi_{xx} + d_2\xi_y + d_3\xi_x + d_4\eta + d_5\xi = 0, \\ &\xi_{xxx} + e_1\xi_{xx} + e_2\xi_y + e_3\xi_x + e_4\eta + e_5\xi = 0. \end{aligned} \tag{27}$$

The coefficients $a_1, a_2, \dots$ occuring in these Janet bases have to obey certain differential constraints in order to assure coherence. They are rather complicated and are given explicitly in [Schwarz 2002].

3 Lie Groups and Lie Algebras

The discussion of groups in this article is limited to the finite groups of the plane. By definition their elements may be specified by a finite number of parameters. Let a set of transformations in the $x - y$-plane depending on a continuous parameter a be given by

$$\bar{x} = f(x, y, a), \quad \bar{y} = g(x, y, a). \tag{28}$$

Assume that the dependence on this parameter is such that a subsequent transformation

$$\bar{\bar{x}} = f(\bar{x}, \bar{y}, b), \quad \bar{\bar{y}} = g(\bar{x}, \bar{y}, b) \tag{29}$$

may be replaced by a single one of the same set (28)

$$\bar{\bar{x}} = f(x, y, c), \quad \bar{\bar{y}} = g(x, y, c) \tag{30}$$

such that the parameter c may be expressed through a and b alone, i.e. there is a relation $c = \phi(a, b)$ with ϕ *independent* of x and y. The functions f and g are supposed to be invertible such that the action (28) may be reversed, i. e. it is required that the functional determinant $\dfrac{D(f, g)}{D(x, y)} \neq 0$. Furthermore a parameter value corresponding to no action at all is supposed to exist. If not stated otherwise, it is always assumed that this latter value is $a = 0$, i. e. $f(x, y, 0) = x$ and $g(x, y, 0) = y$.

The definition of a one-parameter group may be generalized in a fairly straightforward manner to groups depending on any number $r > 1$ of parameters. The transformation (28) is replaced by the more general one

$$\bar{x} = f(x, y, a_1, a_2, \dots, a_r), \quad \bar{y} = g(x, y, a_1, a_2, \dots, a_r). \tag{31}$$

If

$$\bar{\bar{x}} = f(\bar{x}, \bar{y}, b_1, b_2, \dots, b_r), \quad \bar{\bar{y}} = g(\bar{x}, \bar{y}, b_1, b_2, \dots, b_r)$$

is a second transformation it is required that it may be represented in the form

$$\bar{\bar{x}} = f(x, y, c_1, c_2, \dots, c_r), \quad \bar{\bar{y}} = g(x, y, c_1, c_2, \dots, c_r)$$

such that the parameters c_k may be expressed as $c_k = \phi_k(a_1, \dots, a_r, b_1, \dots, b_r)$. The identity is supposed to correspond to $a_1 = a_2 = \dots = a_r = 0$, and the existence of the inverse requires the functional determinant of f and g with respect to x and y to be non-vanishing.

A first major result of Lie was to characterize those functions in (31) that define a group of transformations. The answer is essentially obtained in terms of a system of differential equations of a special form that must be satisfied by f and g, it is described next.

Theorem 1. *(Lie's First Fundamental Theorem). If the equations*

$$\bar{x} = f(x, y, a_1, a_2, \ldots, a_r), \quad \bar{y} = g(x, y, a_1, a_2, \ldots, a_r)$$

define a r-parameter group, $\bar{x}$ and $\bar{y}$, considered as functions of x, y and the parameters, satisfy a system of pde's of the form

$$\frac{\partial \bar{x}}{\partial a_k} = \sum_{j=1}^{r} \psi_{jk}(a_1, \ldots, a_r) \xi_j(\bar{x}, \bar{y}), \quad \frac{\partial \bar{y}}{\partial a_k} = \sum_{j=1}^{r} \psi_{jk}(a_1, \ldots, a_r) \eta_j(\bar{x}, \bar{y}) \quad (32)$$

for $k = 1, \ldots, r$. The determinant of the ψ_{jk} does not vanish identically. On the other hand, if two function f and g satisfy a system of differential equations of this form and there are parameter values $a_1^0, \ldots, a_r^0$ such that

$$f(x, y, a_1^0, \ldots, a_r^0) = x, \quad g(x, y, a_1^0, \ldots, a_r^0) = y$$

and the determinant of the ψ_{jk} is finite and different from zero for these parameter values, then they define a transformation group. The non-vanishing of the determinant of the ψ_{jk} allows solving the system (32) for the ξ's and the η's, i. e. there holds also

$$\xi_j = \sum_{k=1}^{r} \alpha_{jk} \frac{\partial \bar{x}}{\partial a_k}, \quad \eta_j = \sum_{k=1}^{r} \alpha_{jk} \frac{\partial \bar{y}}{\partial a_k}. \quad (33)$$

The proof of this theorem may be found in [Lie 1888], vol. I, Kapitel 2, page 27-34 and vol. III, Kapitel 25, page 545-564.

The functions ξ and η occuring at the right hand side of (32) have another important meaning, they occur as coefficients of the infinitesimal generators of the group (28) which are defined next. Let $\bar{x} = f(x, y, a)$, $\bar{y} = g(x, y, a)$ define a one-parameter group in the $x-y-$plane, $f(x, y, 0) = x$, $g(x, y, 0) = y$. Expanding f and g at the value $a = 0$ and retaining only the terms linear in a yields

$$\bar{x} = f(x, y, 0) + a \frac{\partial f(x, y, a)}{\partial a}\Big|_{a=0} + o(a^2) \equiv x + a\xi(x, y) + o(a^2),$$

$$\bar{y} = g(x, y, 0) + a \frac{\partial g(x, y, a)}{\partial a}\Big|_{a=0} + o(a^2) \equiv y + a\eta(x, y) + o(a^2).$$

$$(34)$$

The infinitesimal generator U is defined by

$$U = \xi(x, y)\partial_x + \eta(x, y)\partial_y. \quad (35)$$

For an r-parameter group, to each parameter a_i, with $a_j = 0$ for $j \neq i$, there corresponds an infinitesimal generator U_i with coefficients ξ_i and η_i

$$U_i = \xi_i(x, y)\partial_x + \eta_i(x, y)\partial_y, \qquad i = 1, \ldots, r. \tag{36}$$

An important operation for infinitesimal generators is to form its *commutator*. Let U_1 and U_2 be given by (36). Then the definition of its commutator

$$[U_1, U_2]f = U_1 U_2 f - U_2 U_1 f \equiv U_3 f$$

implies for the coefficients of U_3

$$\begin{aligned}
\xi_3 &= \xi_1 \xi_{2,x} - \xi_{1,x}\xi_2 + \eta_1 \xi_{2,y} - \xi_{1,y}\eta_2, \\
\eta_3 &= \xi_1 \eta_{2,x} - \eta_{1,x}\xi_2 + \eta_1 \eta_{2,y} - \eta_{1,y}\eta_2.
\end{aligned} \tag{37}$$

The importance of the infinitesimal generators of a group arises from the fact that a group is uniquely determined by them. This relation is established in the following Theorem due to Lie [Lie 1888].

Theorem 2. *(Lie's Second Fundamental Theorem). Any r-parameter group*

$$\bar{x} = f(x, y, a_1, a_2, \ldots, a_r), \qquad \bar{y} = g(x, y, a_1, a_2, \ldots, a_r)$$

of transformations contains r infinitesimal generators

$$U_i = \xi_i \partial_x + \eta_i \partial_y \quad for \ \ i = i, \ldots, r$$

which satisfy $r(r-1)$ relations of the form

$$[U_i, U_j] = \sum_{k=1}^{r} c_{ij}^k U_k \tag{38}$$

with constant coefficients c_{ij}^k. Conversely, r independent generators $U_1, \ldots, U_r$ satisfying relations (38) generate a r-parameter group of finite transformations.

Starting with the vector fields forming the infinitesimal generators of a group, a third representation in terms of the *defining equations* as Lie [Lie 1888], vol. I, page 185 called them may be obtained. By definition, these equations are such that its general solution generates the same vector space as the coefficients of the given generators. This representation is particularly important if the groups to be considered occur as symmetry groups of differential equations because they are usually defined in terms of systems of pde's. Let the infinitesimal generators of a group be

$$U_i = \xi_i(x, y)\partial_x + \eta_i(x, y)\partial_y, \quad i = 1, \ldots, r$$

and $\xi = c_1 \xi_1 + \ldots c_r \xi_r$, $\eta = c_1 \eta_1 + \ldots c_r \eta_r$ its general element. A system of linear homogeneous pde's which has ξ_i and η_i as a base for its solution space is called the system of *defining equtions*.

The existence of the defining equations has been shown by Lie and a constructive method for finding them has been given. A Janet base for these defining equations is called a Janet base for the group. In general it is a much harder problem to obtain the vector field coefficients from a given set of defining equations, because this comes down to determining the general solution of these latter equations. The relation between these representations of a group and how to turn from one to another is illustrated by the following diagram. The operations corresponding to the downward arrows at the left can always be performed algorithmically. Contrary to that, there is no generally valid algorithm for the operations at the right corresponding to the upward arrows.

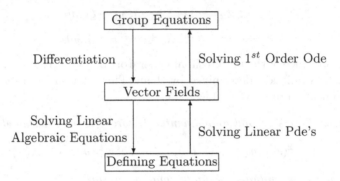

By abstraction from its geometric origin, the algebraic objects obeying commutation relations like (38) are studied as a subject of interest of its own right, they have been called *Lie algebras* by Hermann Weyl. First of all they have to be defined precisley.

Definition 1. *(Lie algebra). A Lie algebra $\mathcal{L}$ is a vector space over a field F equipped with a binary operation called the commutator satisfying the following constraints for any $U, V, W \in \mathcal{L}$ and $\alpha, \beta \in F$.*

i) $[U, V] = -[V, U]$ *(anti-commutativity)*

ii) $[\alpha U + \beta V, W] = \alpha[U, W] + \alpha[V, W]$ *(bilinearity)*

iii) $[U, [V, W]] + [V, [W, U]] + [W, [U, V]] = 0$ *(Jacobi identity)*

In this article always finite dimensional Lie algebras are considered, and it is always assumed that the field F has characteristic zero, usually it will be the rational number field $\mathbf{Q}$ or some algebraic extension $\bar{\mathbf{Q}}$ of it. The structure constants contain the complete information on the group from which they originate as the following result shows.

Theorem 3. *(Lie's Third Fundamental Theorem). The r^3 constants c_{ij}^k, i, j, $k = 1, \ldots, r$, determine an r-parameter continuous group if and only if they obey the relations $c_{ik}^j + c_{ki}^j = 0$ and*

$$\sum_{h=1}^{r} (c_{ik}^h c_{jh}^l + c_{kj}^h c_{ih}^l + c_{ji}^h c_{kh}^l) = 0.$$

The proof of this theorem may be found in [Lie 1888], vol. III, page 597 ff.

Fundamental problems are to classify all types of composition leading to non-isomorphic Lie algebras, and to determine simpler substructures within a given Lie algebra, i. e. to determine its invariant subalgebras. The *characteristic polynomial*

$$\Delta(\omega) \equiv \left| \sum E_i c_{ij}^k - \delta_{jk}\omega \right| =$$
$$\omega^r - \psi_1(E)\omega^{r-1} + \psi_2(E)\omega^{r-2} + \ldots + (-1)^{r-1}\psi_{r-1}(E)\omega \tag{39}$$

or correspondingly the *characteristic equation* $\Delta(\omega) = 0$ where $E_1, \ldots, E_r$ are r indeterminates turns out to be an important tool for dealing with these problems. It has been introduced by Killing [Killing 1887] in order to answer the question in which two-parameter groups a given one-parameter group is contained as an invariant subgroup. The number of independent coefficients of the characteristic polynomial is called the *rank of the Lie algebra*. If new basis elements are introduced in a Lie algebra by $\bar{U}_i = \sum \alpha_{i,j} U_j$, the coefficients ψ_k in (39) are changed according to

$$\bar{\psi}_k(\bar{E}_1, \ldots, \bar{E}_r) = \psi_k \left(\sum \alpha_{i,1} E_i, \ldots, \sum \alpha_{i,r} E_i \right),$$

i. e. the structure of the characteristic polynomial remains essentially unchanged. This feature is an important tool for identifying the isormorphy class of a Lie algebra.

Changing coordinates is one of the most important operations for groups and its infinitesimal generators. To this end it must be known in the first place how the new representation is obtained if a coordinate transformation is applied. Closely related to that is the problem how to decide whether two given groups or vector fields are related by a suitable variable transformation. Two groups are called *similar* if there is a point-transformation that transforms them into each other. This definition establishes an equivalence relation between groups. An equivalence class comprises all groups or its vector fields of infinitesimal generators that may be transformed into each other by a suitable variable change, it is called a *group type*. As soon as equivalence between any pair of groups may be decided, the objects of primary interest are the equivalence classes.

Usually groups of transformations are given in terms of their infinitesimal generators. Therefore the behavior of vector fields under a change of variables must be known in the first place. Let $U = \xi(x,y)\partial_x + \eta(x,y)\partial_y$ be a vector field in the $x - y$-plane. Furthermore let new coordinates u and v be defined by $u = \sigma(x,y)$ and $v = \rho(x,y)$ with the inverse $x = \phi(u,v)$ and $y = \psi(u,v)$. Then the coefficients of the transformed vector field $V = \alpha(u,v)\partial_u + \beta(u,v)\partial_v$ are

$$\alpha(u,v) = U\sigma|_{x=\phi,y=\psi}, \quad \beta(u,v) = U\rho|_{x=\phi,y=\psi}$$

Of particular importance is the question how the commutator of two vectorfields is transformed under a change of variables. The answer has been given by Lie [Lie 1888], part I, page 84. If three vector fields U_1, U_2 and U_3 in the

$x - y$−plane are transformed into V_1, V_2 and V_3 in the $u - v$−plane repectively, its commutator $[U_1, U_2] = U_3$ is transformed into $[V_1, V_2] = V_3$.

Let $\bar{x} = f(x, y, a)$ and $\bar{y} = g(x, y, a)$ be the finite equations of a one-parameter group in the plane with infinitesimal generator $U = \xi(x, y)\partial_x + \eta(x, y)\partial_y$. Under the action of this group an arbitrarily chosen point describes a curve the points of which are parametrized by a. Any two points on an individual curve may be transformed into each other by a suitable group element, they are *equivalent* w.r.t. this group. If they are located on different curves such a transformation does not exist. Consequently these curves form equivalence classes of points equivalent w.r.t. to transformations of the group. The curve that is generated by the action of a one-parameter group on an arbitrary point of the plane is called its *path curve* or its *orbit*. The orbits of a one-parameter group generate an *invariant decomposition* of the plane.

Given an r-parameter group of the plane, the following question may be raised: Is it possible to transform a given point of the $x - y$−plane into any other point that is arbitrarily chosen by a suitable group element? The answer leads to the following definition. An r-parameter group is called *simply transitive* or just *transitive* if amongst its operations at least one may be found that transforms a pair of arbitrarily chosen points of the plane into each other. If such a pair of points does not exist it is called intransitive.

Even if a group is transitive there may exist a more general kind of decomposition. This remarkable property is the basis for the new concept defined next. A transitive group of the plane is called *imprimitive* if there is a one-parameter family of curves that are transformed among themselves by the elements of the group. The corresponding curves are called a *system of imprimitivity* or *s.o.i.* for short. If such a system does not exist, the group is called *primitive*. The curves which are interchanged by the operations of an imprimitive group correspond to the blocks of elements which are permuted under the action of finite permutation groups.

The number of systems of imprimitivity is an important distinguishing feature of any transformation group. For a continuous group of the plane the complete answer has been obtained by Lie [Lie 1893], page 342, as follows. Either it is primitive, or it has exactly one or two systems of imprimitivity, or there is a system of imprimitivity depending on a parameter or a function of one variable.

For later applications a particularly well suited criterion in terms of the Janet base of a group may be described as follows [Schwarz 1995]. Let the infinitesimal generators of a symmetry group of the $x - y$−plane be of the form $U = \xi\partial_x + \eta\partial_y$ where the coefficients solve the corresponding determining system. In order for a system of imprimitivity described by $y' = \omega(x, y)$, different from $x = const.$ or $y = const$ to exist, an equation

$$\eta_y\omega + \eta_x - \eta\omega_y - \xi_y\omega^2 - \xi_x\omega - \xi\omega_x = 0 \tag{40}$$

with a suitable $\omega \equiv \omega(x, y)$ must be valid as a consequence of the determining system of this symmetry group. If $x = const$ or $y = const$ is a system of imprimitivity, (40) reduces to $\xi_y = 0$ or $\eta_x = 0$ respectively.

The subsequent classification of Lie-algebras of dimension not higher than three is complete without any duplication, i. e. any such Lie algebra is isomorphic to one and only one entry. In some cases there occur parameters. They have to be constraint suitably in order to guarantee this feature. In addition the following general rule is applied: Whenever an essential algebraic property like e. g. the dimension of the derived algebra changes for a special value of a parameter, the algebra corresponding to these parameter values is listed as a separate entry. For dimension higher than three only those algebras are included that occur in the symmetry analysis of second- and third-order ode's.

One-Dimensional Algebras

l_1: *commutative*, Characteristic equation: $\omega = 0$.

Two-Dimensional Algebras

$l_{2,1}$: $[U_1, U_2] = U_1$, $dim\, l'_{2,1} = 1$, $dim\, l''_{2,1} = 0$.
Characteristic equation: $\omega^2 - E_2\omega = \omega(\omega - E_2) = 0$.

$l_{2,2}$: *commutative*. Characteristic equation: $\omega^2 = 0$.

Three-Dimensional Algebras

$l_{3,1}$: $[U_1, U_2] = U_1$, $[U_1, U_3] = 2U_2$, $[U_2, U_3] = U_3$, $dim\, l'_{3,1} = dim\, l''_{3,1} = 3$.
Characteristic equation: $\omega^3 + (4E_3E_1 - E_2^2)\omega = \omega(\omega^2 + 4E_3E_1 - E_2^2) = 0$.

$l_{3,2}(c)$: $[U_1, U_3] = U_1$, $[U_2, U_3] = cU_2, c \neq 0$, $dim\, l'_{3,2} = 2$, $dim\, l''_{3,2} = 0$.
Characteristic equation: $\omega^3 - (c+1)E_3\omega^2 + cE_3^2\omega = \omega(\omega - E_3)(\omega - cE_3) = 0$.

$l_{3,3}$: $[U_1, U_3] = U_1$, $[U_2, U_3] = U_1 + U_2$, $dim\, l'_{3,3} = 2$, $dim\, l''_{3,3} = 0$.
Characteristic equation: $\omega^3 - 2E_3\omega^2 + E_3^2\omega = \omega(\omega - E_3)^2 = 0$.

$l_{3,4}$: $[U_1, U_3] = U_1$, $dim\, l'_{3,4} = 1$, $dim\, l''_{3,4} = 0$.
Characteristic equation: $\omega^3 - E_3\omega^2 = \omega^2(\omega - E_3) = 0$.

$l_{3,5}$: $[U_2, U_3] = U_1$, $dim\, l'_{3,5} = 1$, $dim\, l''_{3,5} = 0$.
Characteristic equation: $\omega^3 = 0$.

$l_{3,6}$: *commutative*. Characteristic equation: $\omega^3 = 0$.

Some Algebras with More than Three Dimensions

$l_{4,1}$: $[U_1, U_2] = U_1$, $[U_1, U_3] = 2U_2$, $[U_2, U_3] = U_3$, $dim\, l'_{4,1} = dim\, l''_{4,1} = 3$.
Characteristic equation: $\omega^4 + (4E_3E_1 - E_2^2)\omega^2 = (\omega^2 + 4E_3E_1 - E_2^2)\omega^2 = 0$.

$l_{4,4}$: $[U_1, U_4] = 2U_1$, $[U_2, U_3] = U_1$, $[U_2, U_4] = U_2$, $[U_3, U_4] = U_2 + U_3$,
$dim\, l'_{4,4} = 3$, $dim\, l''_{4,4} = 1$, $dim\, l'''_{4,4} = 0$ Characteristic equation:
$\omega^4 - 4E_4\omega^3 + 5E_4^2\omega^2 - 2E_4^3\omega = -\omega(\omega - 2E_4)(\omega - E_4)^2 = 0$.

$l_{4,5}(c)$: $[U_1, U_4] = cU_1$, $[U_2, U_3] = U_1$, $[U_2, U_4] = U_2$, $[U_3, U_4] = (c - 1)U_3$,
$c \neq 1$, $dim\, l'_{4,5} = 3$, $dim\, l''_{4,5} = 1$, $dim\, l'''_{4,4} = 0$. Characteristic equation:
$$\omega^4 - 2cE_4\omega^3 + (c^2 + c - 1)E_4^2\omega^2 - c(c - 1)E_4^3\omega.$$
$$= \omega(\omega - E_4)(\omega - cE_4)(\omega - (c - 1)E_4)) = 0.$$

$l_{4,10}$: $[U_1, U_2] = U_2$, $[U_3, U_4] = U_4$, $dim\, l'_{4,10} = 2$, $dim\, l''_{4,10} = 0$.
Characteristic equation:
$$\omega^4 + (E_3 + e_1)\omega^3 + E_3E_1\omega^2 = \omega^2(\omega + E_1)(\omega + E_3) = 0.$$

l_8: $[U_1, U_3] = U_1$, $[U_1, U_5] = U_2$, $[U_1, U_7] = U_4$, $[U_1, U_8] = U_6 + 2U_3$,
$[U_2, U_4] = U_1$, $[U_2, U_6] = U_2$, $[U_2, U_7] = 2U_6 + U_3$, $[U_2, U_8] = U_5$,
$[U_3, U_4] = -U_4$, $[U_3, U_5] = U_5$, $[U_3, U_8] = U_8$, $[U_4, U_5] = U_6 - U_3$,
$[U_4, U_8] = U_7$, $[U_5, U_6] = U_5$, $[U_6, U_7] = U_7$
Characteristic equation: $\psi_1 = -2e_6$, $\psi_2 = 3e_2e_7 + 6e_1e_8 + e_6^2 + e_6e_3 - 2e_3^2$.

The infinitesimal generators of those groups that occur later on as symmetry groups of differential equations are listed next. In addition the Janet bases for the determining systems are given in *grlex* term ordering with $\eta > \xi$ and $y > x$. The coefficients ϕ_k for $k = 1, \ldots, r$ occuring in the group types $\mathbf{g}_{15}$ and $\mathbf{g}_{16}$ are undetermined functions of x from some function field. The coefficients q_k occuring in the Janet base for these groups are defined by

$$q_k = -\frac{W_k^{(2)}(\phi_1, \phi_2)}{W^{(2)}(\phi_1, \phi_2)}$$

for $k = 1, \ldots, r$. The algebraic properties of groups with $r \leq 4$ may be obtained from their Lie algebra type which is always given.

One-Parameter Group

$\mathbf{g}_{27}$: $\{\partial_y\}$. Type: $\mathbf{l}_1$. Janet base: $\{\xi, \eta_x, \eta_y\}$

Two-Parameter Groups

$\mathbf{g}_4$: $\{\partial_y, y\partial_y\}$. Type: $\mathbf{l}_{2,1}$. Janet base: $\{\xi, \eta_x, \eta_{yy}\}$.

$\mathbf{g}_{15}(r = 2)$: $\{\phi_1\partial_y, \phi_2\partial_y\}$. Type: $\mathbf{l}_{2,2}$.
Janet base: $\{\xi, \eta_y, \eta_{xx} + q_1\eta_x + q_2\eta\}$.

$\mathbf{g}_{25}$: $\{\partial_x, x\partial_x + y\partial_y\}$. Type: $\mathbf{l}_{2,1}$. Janet base: $\{\xi_x - \frac{1}{y}\eta, \xi_y, \eta_x, \eta_y - \frac{1}{y}\eta\}$.

$\mathbf{g}_{26}$: $\{\partial_x, \partial_y\}$. Type: $\mathbf{l}_{2,2}$. Janet base: $\{\xi_x, \xi_y, \eta_x, \eta_y\}$.

Three-Parameter Groups

$\mathbf{g}_5$: $\{\partial_x, \partial_y, y\partial_y\}$. Type: $\mathbf{l}_{3,4}$. Janet base: $\{\xi_x, \xi_y, \eta_x, \eta_{yy}\}$.

$\mathbf{g}_7$: $\{\partial_x, \partial_y, x\partial_x + cy\partial_y\}$ with $c \neq 0, 1$. Type: $\mathbf{l}_{3,2}$.
Janet base: $\{\xi_y, \eta_x, \eta_y - c\xi_x, \xi_{xx}\}$.

$\mathbf{g}_8$: $\{\partial_y, y\partial_y, y^2\partial_y\}$. Type: $\mathbf{l}_{3,1}$. Janet base: $\{\xi, \eta_x, \eta_{yyy}\}$.

$\mathbf{g}_{10}$: $\{\partial_x + \partial_y, x\partial_x + y\partial_y, x^2\partial_x + y^2\partial_y\}$. Type: $\mathbf{l}_{3,1}$. Janet base:
$\{\xi_y, \eta_x, \eta_y + \xi_x + \frac{2}{x-y}(\eta - \xi), \xi_{xx} - \frac{2}{x-y}\xi_x - \frac{2}{(x-y)^2}(\eta - \xi)\}$.

$\mathbf{g}_{13}$: $\{\partial_x, 2x\partial_x + y\partial_y, x^2\partial_x + xy\partial_y\}$. Type: $\mathbf{l}_{3,1}$.
Janet base: $\{\xi_x - \frac{2}{y}\eta, \xi_y, \eta_y - \frac{1}{y}\eta, \eta_{xx}\}$.

$\mathbf{g}_{15}(r = 3)$: $\{\phi_1\partial_y, \phi_2\partial_y, \phi_3\partial_y\}$. Type: $\mathbf{l}_{3,6}$.
Janet base: $\{\xi, \eta_y, \eta_{xxx} + q_1\eta_{xx} + q_2\eta_x + q_3\eta\}$.

$\mathbf{g}_{16}(r = 2)$: $\{\phi_1\partial_y, \phi_2\partial_y, y\partial_y\}$. Type: $\mathbf{l}_{3,2}$ with $c = 1$.
Janet base: $\{\xi, \eta_{xx} + q_1\eta_x + q_2\eta, \eta_{xy}, \eta_{yy}\}$.

$\mathbf{g}_{17}(l = 1, \rho_1 = 1, \alpha_1 = 0)$: $\{\partial_x, \partial_y, x\partial_y\}$. Type: $\mathbf{l}_{3,5}$.
Janet base: $\{\xi_x, \xi_y, \eta_y, \eta_{xx}\}$.

$\mathbf{g}_{17}(l = 1, \rho_1 = 1, \alpha_1 = 1)$: $\{\partial_x, e^x\partial_y, xe^x\partial_y\}$. Type: $\mathbf{l}_{3,3}$.
Janet base: $\{\xi_x, \xi_y, \eta_y, \eta_{xx} - 2\eta_x + \eta\}$.

$\mathbf{g}_{17}(l = 2, \rho_1 = \rho_2 = 0, \alpha_1 = 0, \alpha_2 = 1)$: $\{\partial_x, \partial_y, e^x\partial_y\}$. Type: $\mathbf{l}_{3,4}$.
Janet base: $\{\xi_x, \xi_y, \eta_y, \eta_{xx} - \eta_x\}$.

$\mathbf{g}_{17}(l = 2, \rho_1 = \rho_2 = 0, \alpha_1 = 1, \alpha_2 = c)$: $\{\partial_x, e^x\partial_y, e^{cx}\partial_y\}$, $c \neq 0, 1$.
Type: $\mathbf{l}_{3,2}$. Janet base: $\{\xi_x, \xi_y, \eta_y, \eta_{xx} - (c + 1)\eta_x + c\eta\}$.

$\mathbf{g}_{20}(r = 1)$: $\{\partial_x, \partial_y, x\partial_x + (x + y)\partial_y\}$. Type: $\mathbf{l}_{3,3}$.
Janet base: $\{\xi_y, \eta_x - \xi_x, \eta_y - \xi_x, \xi_{xx}\}$.

$\mathbf{g}_{24}$: $\{\partial_x, \partial_y, x\partial_x + y\partial_y\}$. Type: $\mathbf{l}_{3,2}$ with $c = 1$.
Janet base: $\{\xi_y, \eta_x, \eta_y - \xi_x, \xi_{xx}\}$.

Selected Groups with More than Three Parameters

$\mathbf{g}_6$: $\{\partial_x, \partial_y, x\partial_x, y\partial_y\}$. Type $l_{4,10}$. Janet base: $\{\xi_y, \eta_x, \xi_{xx}, \eta_{yy}\}$.

$\mathbf{g}_{14}$: $\{\partial_x, x\partial_x, x^2\partial_x + xy\partial_y, y\partial_y\}$. Type $l_{4,1}$.

Janet base: $\{\xi_y, \eta_y - \frac{1}{y}\eta, \xi_{xx} - \frac{2}{y}\eta_x, \eta_{xx}\}$.

$\mathbf{g}_{19}$ with $r = 2$: $\{\partial_y, x\partial_y, \partial_x, x\partial_x + cy\partial_y\}$. Type: $l_{4,5}(c)$.

Janet base: $\{\xi_y, \eta_y - c\xi_x, \xi_{xx}, \eta_{xx}\}$.

$\mathbf{g}_{20}$ with $r = 2$: $\{\partial_x, \partial_y, x\partial_y, x\partial_x + (x^2 + 2y)\partial_y\}$. Type: $l_{4,4}$.

Janet base: $\{\xi_y, \eta_y - 2\xi_x, \xi_{xx}, \eta_{xx} - 2\xi_x\}$.

$\mathbf{g}_{18}(l = 2, \rho_1 = 1, \rho_2 = 0, \alpha_1 = 0, \alpha_2 = 1)$: $\{\partial_y, x\partial_y, e^x\partial_y, y\partial_y, \partial_x\}$.

Janet base: $\{\xi_x, \xi_y, \eta_{xy}, \eta_{yy}, \eta_{xxx} - \eta_{xx}\}$.

$\mathbf{g}_{18}(l = 3, \rho_1 = \rho_2 = \rho_3 = 0, \alpha_1 = 0, \alpha_2 = 1, \alpha_3 = c)$, $c \neq 0, 1$:

$\{\partial_y, x\partial_y, e^{cx}\partial_y, y\partial_y, \partial_x\}$.

Janet base: $\{\xi_x, \xi_y, \eta_{xy}, \eta_{yy}, \eta_{xxx} - (c + 1)\eta_{xx} + c\eta_x\}$.

$\mathbf{g}_{23}(r = 3) = \{\partial_y, x\partial_y, x^2\partial_y, y\partial_y, \partial_x, x\partial_x, x^2\partial_x + 2xy\partial_y\}$.

Janet base: $\{\xi_y, \eta_{xy} - \xi_{xx}, \eta_{yy}, \xi_{xxx}, \eta_{xxx}\}$.

$\mathbf{g}_3$: $\{\partial_x, \partial_y, x\partial_y, y\partial_y, x\partial_x, y\partial_x, x^2\partial_x + xy\partial_y, xy\partial_x + y^2\partial_y\}$.

Janet base: $\{\xi_{yy}, \eta_{xx}, \eta_{xy}, \eta_{yy}, \xi_{xxx}, \eta_{xxy}\}$.

4   Lie Systems

In Section 2 general systems of linear homogeneous pde's have been discussed without any assumptions on their origin or special properties of their solutions. The uniqueness of the Janet base representation makes it possible to express additional knowledge in terms of its coefficients. For applications in this article, the determining systems describing the symmetries of ordinary differential equations, this additional knowledge is of threefold origin.

→ *Lie's relations* are always true because they express the general properties of commutators of vector fields as given in (37).

→ *Geometric relations* express geometric properties of particular group types, e. g. the number of systems of imprimitivity.

→ *Algebraic relations* express algebraic properties of particular Lie algebra types, e. g. in terms of the structure of its characteristic polynomial.

The coefficient constraints following from Lie's relations (37) are satisfied for any determining system. Due to its importance a special term is introduced for them. A system of linear homogeneous pde's with the property that (ξ_3, η_3) defined by (37) is a solution if this is true for (ξ_1, η_1) and (ξ_2, η_2) is called a *Lie system*.

In order to distinguish groups by their geometric or algebraic properties, the properties must be expressed in terms of the Janet base coefficients of its defining system. For those Janet bases occuring in the symmetry analysis of second- and third order ode's this is provided by the theorems of this section. At first geometric constraints are considered.

Lemma 1. *The Lie systems of types $\mathcal{J}_{1,1}^{(2,2)}$ and $\mathcal{J}_{1,2}^{(2,2)}$ define one-parameter groups allowing a system of imprimitivity depending on an undetermined function.*

Proof. The systems of imprimitivity for the two Janet bases (1) and (2) are determined by

$$\omega_y + b\omega + a = 0 \quad \text{or} \quad \omega_x + a\omega_y + c\omega^2 + (a_y - b - c)\omega + a_x - ab = 0.$$

The general solution of either of these pde's for $\omega(x, y)$ contains an undetermined function of a single argument. Consequently they determine systems of imprimitivity depending on an undetermined function.   •

Lemma 2. *The Lie systems of types $\mathcal{J}_{2,1}^{(2,2)}, \ldots, \mathcal{J}_{2,5}^{(2,2)}$ define vector fields for two-parameter groups with the following systems of imprimitivity.*

Type $\mathcal{J}_{2,1}^{(2,2)}$. Lie system (3) does not allow a one-parameter systems of imprimitivity. It allows two system of imprimitivity iff $a_1 \neq 0$, and a single one iff $a_1 = 0$.

Type $\mathcal{J}_{2,2}^{(2,2)}$. Lie system (4) allows always two systems of imprimitivity.

Type $\mathcal{J}_{2,3}^{(2,2)}$. Lie system (5) defines always a one-parameter system of imprimitivity.

Type $\mathcal{J}_{2,4}^{(2,2)}$. Lie system (6) allows always two systems of imprimitivity.

Type $\mathcal{J}_{2,5}^{(2,2)}$. Lie System (7) does not allow a one-parameter system of imprimitivity. It allows two system of imprimitivity iff $a_1 \neq 0$ and a single one iff $a_1 = 0$.

Proof. Type $\mathcal{J}_{2,1}^{(2,2)}$. There is always the system of imprimitivity corresponding to $\xi_y = 0$. Any other system of imprimitivity is determined by the equations $a_1\omega - 1 = 0$, $\omega_y + a_2\omega = 0$. If $a_1 \neq 0$ there is one solution, if $a_1 = 0$ the first equation becomes inconsistent, consequently a solution does not exist.

Type $\mathcal{J}_{2,2}^{(2,2)}$. There is always the systems of imprimitivity corresponding to $\xi_y = 0$, and a second one that follows from the equations $\omega = 0$, $\omega_y + a_1 = 0$ with the constraint $a_1 = 0$ which is obtained as a consequence of (37).

Type $\mathcal{J}_{2,3}^{(2,2)}$. Any system of imprimitivity is determined by the coherent equations

$$\omega_x - b_2\omega^2 - (a_2 - d_2)\omega + c_2 = 0, \quad \omega_y - b_1\omega^2 - (a_1 - d_1)\omega + c_1 = 0.$$

They determine always a system of imprimitivity depending on a parameter.

Type $\mathcal{J}_{2,4}^{(2,2)}$. Any system of imprimitivity is determined by the equations

$$b_1\omega^2 + (a_2b_1 - 1)\omega = 0, \quad a_2\omega_y - \omega_x + b_2\omega^2 - (a_{2,y} - a_2b_2)\omega - a_{2,x} = 0.$$

If $b_1 \neq 0$, the first equation determines two systems of imprimitivity because $a_1b_1 + 1 \neq 0$. If $b_1 = 0$, the condition (37) requires $b_2 = 0$, i. e. there is one system of imprimitivity corresponding to $\xi_y = 0$, and a second one follows from $\omega = 0$.

Type $\mathcal{J}_{2,5}^{(2,2)}$. Any system of imprimitivity is determined by the equations

$$\omega(\omega + a_1) = 0, \quad a_1 \omega_y - \omega_x - (a_{1,y} + b_1)\omega - a_{1,x} + a_1 b_1 = 0.$$

There is always the solution $\omega = 0$. If $a_1 \neq 0$ there is a second system of imprimitivity corresponding to $\omega = -a_1$. •

There are three Janet base types that occur in the symmetry analysis of second- and third order ode's in Section 5, their systems of imprimitivity are considered next.

Lemma 3. *The Lie systems of types $\mathcal{J}_{3,4}^{(2,2)}$, $\mathcal{J}_{3,6}^{(2,2)}$ and $\mathcal{J}_{3,7}^{(2,2)}$ define vector fields for three-parameter groups with the following systems of imprimitivity.*

Type $\mathcal{J}_{3,4}^{(2,2)}$. *Lie system (11) defines a three-parameter group allowing two systems of imprimitivity iff $c_1 \neq 0$ and a single one iff $c_1 = 0$. There cannot be a system of imprimitivity depending on a parameter.*

Type $\mathcal{J}_{3,6}^{(2,2)}$. *Lie system (13) defines a three-parameter group allowing a system of imprimitivity depending on a parameter iff $a_1 = b_1 = 0$ and $c_1 = -1$, two systems of imprimitivity iff either $a_i = b_i = 0$ for $i = 1, 2, 3$ or $D \equiv 4a_1 b_1 + (c_1 + 1)^2 \neq 0$, and a single one iff $D = 0$ and $a_1 \neq 0$ or $b_1 \neq 0$ or $c_1 \neq -1$.*

Type $\mathcal{J}_{3,7}^{(2,2)}$. *Lie system (14) defines a two-parameter group which does not allow a system of imprimitivity depending on a parameter. It allows two systems of imprimitivity iff $D \equiv b_1 - \frac{1}{4}c_1^2 \neq 0$ and a single one iff $D = 0$.*

The proof of this lemma and the next one is similar to the proof of Lemma 2, it is therefore omitted. There are two Lie systems of order four for which the systems of imprimitivity have to be determined.

Lemma 4. *The Lie systems of types $\mathcal{J}_{4,9}^{(2,2)}$ and $\mathcal{J}_{4,14}^{(2,2)}$ define vector fields for four-parameter groups with the following systems of imprimitivity.*

Type $\mathcal{J}_{4,9}^{(2,2)}$. *Lie system (19) defines a four-parameter group which does not allow a system of imprimitivity depending on a parameter. It allows two systems of imprimitivity iff $a_1 = 0$, $a_2 = a_3 = a_4 = 0$, $b_2 + 1 = 0$ and $b_1 \neq 0$, and a single one iff $a_2 = a_3 = a_4 = b_1 = 0$.*

Type $\mathcal{J}_{4,14}^{(2,2)}$. *Lie system (22) defines a four-parameter group which does not allow a system of imprimitivity depending on a parameter. It allows two systems of imprimitivity iff $a_1 = 0$, $a_2 = a_3 = a_4 = 0$, $b_2 + 1 = 0$ and $b_1 \neq 0$ or $a_2 = 0$, $b_2 + 1 = 0$, $b_1^2 - 4a_1 \neq 0$ and $a_1 \neq 0$ or $a_3 \neq 0$ or $a_4 \neq 0$.*

The third set of constraints for the coefficients of Janet bases occuring in the symmetry analysis of ode's originates from the algebraic properties of their Lie algebras. They are based on the above mentioned property of the characterisitc polynomial w.r.t. to basis transformations of the respective Lie algebra.

Lemma 5. *The Lie system (5) of type $\mathcal{J}_{2,3}^{(2,2)}$ defines a Lie algebra of vector fields of a certain type as given in the classification on page 121 if its coefficients satisfy the following constraints.*

Lie algebra $\mathfrak{l}_{2,1}$ iff the coefficient obey $a_1 = b_2$ and $c_1 = d_2$.
Lie algebra $\mathfrak{l}_{2,2}$ iff the coefficient obey $a_1 \neq b_2$ or $c_1 \neq d_2$.

Proof. For the general Lie system of type $\mathcal{J}_{2,3}^{(2,2)}$ the characteristic polynomial has the form $\omega^2 - [(c_1 - d_2)E_2 - (a_1 - b_2)E_1]\omega$. The above coefficient constraints generate the structure of the two two-dimensional Lie algebras $\mathfrak{l}_{2,1}$ or $\mathfrak{l}_{2,2}$ respectively listed on page 121. ●

Lemma 6. *The Lie system (11) of type $\mathcal{J}_{3,4}^{(2,2)}$ defines a Lie algebra of vector fields of a certain type as given in the classification on page 121 if its coefficients satisfy the following constraints.*

Lie algebra $\mathfrak{l}_{3,1}$ iff the coefficient obey

$$c_1 = 0, \ a_3 - 2c_3 + d_1 = 0, \ a_2 \neq 0 \text{ or } c_{3,x} - c_3^2 + c_3 d_1 - d_2 \neq 0.$$

Lie algebra $\mathfrak{l}_{3,2}(c)$ iff

$$c_{3,x} + \frac{1}{4}a_3(a_3 - 4c_3 + 2d_1) + \frac{1}{4}d_1^2 - d_2 \neq 0$$

and $c \neq 1$ is a solution of

$$c^2 + \frac{2c_{3,x} + 2c_3^2 - 2c_3 d_1 - 4a_3 c_3 + a_1^2 + d_1^2 - 2d_2}{c_{3,x} - c_3^2 + c_3 d_1 - d_2}c + 1 = 0.$$

Lie algebra $\mathfrak{l}_{3,3}$ iff the coefficients obey

$$a_1 = 0, \ a_2 = 0, \ c_{3,x} + \frac{1}{4}a_3(a_3 - 4c_3 + 2d_1) + \frac{1}{4}d_1^2 - d_2 = 0, \ a_3 - 2c_3 + d_1 \neq 0.$$

Lie algebra $\mathfrak{l}_{3,4}$ iff the coefficients obey

$$a_2 = 0, \ c_{3,x} - c_3^2 + c_3 d_1 - c_1 d_3 - d_2 = 0, \ c_1 \neq 0 \text{ or } a_3 - 2c_3 + d_1 \neq 0.$$

Lie algebra $\mathfrak{l}_{3,5}$ iff the coefficients obey

$$c_1 = 0, \ a_2 = 0, \ a_3 - 2c_3 + d_1 = 0, \ c_{3,x} - c_3^2 + c_3 d_1 - c_1 d_3 - d_2 = 0.$$

Lie algebra $\mathfrak{l}_{3,6}$ not possible.

Proof. According to the definition (39) the characteristic polynomial for any three-dimensional Lie algebra has the form $\omega^2 + C_2\omega^2 + C_1\omega$. In terms of a Lie system of type $\mathcal{J}_{3,4}^{(2,2)}$ the coefficients may be expressed as follows.

$$C_2 = c_1 E_1 + (a_3 - c_3 + d_1)c_1 E_2 - (a_3 - 2c_3 + d_1)E_3,$$

$$C_1 = a_2 c_1 E_1 E_2 - a_2(a_2 - a_3 c_1 + c_1 c_3 - c_1 d_1)E_2^2 - 2a_2 E_1 E_3$$
$$+ [c_{3,x} - c_1 d_3 - c_3^2 + c_3 d_1 - d_2)c_1 - a_2(a_3 + d_1)]E_2 E_3$$
$$- (c_{3,x} - c_1 d_3 - c_3^2 + c_3 d_1 - d_2)E_3^2.$$

Imposing the constraints on C_1 and C_2 that follow from the structure given in the listing on page 121, the above conditions are obtained after some simlifications. The condition for $\mathfrak{l}_{3,6}$ follows from the fact that some structure constants are numbers.   ●

Lemma 7. *The Lie system (13) of type $\mathcal{J}_{3,6}^{(2,2)}$ defines a Lie algebra of vector fields of a certain type as given in the classification on page 121 if its coefficients satisfy the following constraints.*

Lie algebra $\mathfrak{l}_{3,1}$ iff the coefficients obey

$$c_1 = 1, \quad a_3 b_1 - a_1 b_3 + b_2 - c_3 + d_1 = 0$$

and

$$a_1 b_1 + 1 \neq 0 \ \text{ or } \ a_1(b_1 - c_3) + a_3 \neq 0 \ \text{ or } \ a_3 b_1 - b_2 + c_3 \neq 0$$
$$\text{or } \ a_{3,x} - a_3(a_1 b_3 + c_3 - d_1) - a_1 d_3 - d_2 \neq 0$$
$$\text{or } \ a_{3,x} b_1 + (a_1 b_3 + c_3 - d_1)(c_3 - b_2) - b_1 d_2 + d_3 \neq 0$$
$$\text{or } \ a_{3,x} + \tfrac{1}{2}(a_1 b_2 - a_1 c_3 - a_3)(a_1 b_3 + c_3 - d_1) \neq 0.$$

Lie algebra $\mathfrak{l}_{3,2}(c)$ iff $c \neq 1$ is a solution of

$$c^2 + \frac{2a_1 b_1 + c_1^2 + 1}{a_2 b_1 + c_1} c + 1 = 0.$$

Lie algebra $\mathfrak{l}_{3,3}$ iff the coefficients obey either

$$c_1 + 1 = 0, \ b_1 = 0, \ a_1 b_3 + b_2 - c_3 - d_1 = 0,$$
$$a_{3,x} + a_1(b_2 - c_3)^2 - a_3 b_2 - a_1 d_3 - d_2 = 0$$

or

$$c_1 + 1 = 0, \ b_1 = 0,$$
$$a_{3,x} b_1 - \tfrac{1}{4}a_3^2 b_1^2 - \tfrac{1}{4}(b_2 - c_1 - d_1)^2 - \tfrac{1}{2}a_3 b_1(b_2 + c_3 - d_1) - b_1 d_1 = 0$$

or

$$4a_1b_1 + (c_1 + 1)^2 = 0,$$

$$(c_1 + 3)(b_2 - c_3) + (c_1 - 1)(d_1 - a_1b_3) + (c_1 - 5)a_1b_1 = 0,$$

$$a_3(c_1 + 1)^2(c_1 - 5) + 4a_1(a_1b_3 - d_1)(c_1 - 1) - 4a_1(c_1 + 3)(b_2 - c_3) = 0,$$

$$a_{3,x}b_1 - \tfrac{1}{8}(a_3b_3 - 2d_3)(c_1 + 1)^2 - \tfrac{1}{4}(b_2 - d_1)^2 - \tfrac{1}{4}(a_1^2b_3^2 + a_3^2b_1^2) - b_1d_2$$

$$- \tfrac{1}{2}a_3b_1(b_2 + c_3 + d_1) - \tfrac{1}{2}a_1b_3(b_2 - c_3 - d_1) + \tfrac{1}{4}c_3(2b_2 - c_3 - 2d_1) = 0$$

and $b_1 \neq 0$.

Lie algebra $l_{3,4}$ iff the coefficients obey either

$$a_1 = 0, \ c_1 = 0, \ a_3b_1 = b_2 - c_3, \ a_{3,x} = a_3(c_3 - d_1) + d_2, \ c_3 - b_2 \neq \tfrac{1}{2}d_1$$

or

$$a_1b_1 + 1 = 0, \ a_3c_1 + a_1(b_2 - c_3) = 0, \ a_3b_1 = b_2 - c_3,$$

$$a_{3,x} - a_3(a_1b_3 + c_3 - d_1) - a_1d_3 - d_2 = 0,$$

$$a_1b_3 - 2(b_2 - c_3) \neq d_1, \ a_3 + a_1(a_1b_3 + c_3 - b_2 - d_1) \neq 0.$$

Lie algebra $l_{3,5}$ iff the coefficient obey

$$c_1 = 1, \ a_1b_1 + 1 = 0, \ a_1b_3 - 2(b_2 - c_3) - d_1 = 0, \ a_3 + a_1(b_2 - c_3) = 0,$$

$$(b_{2,x} - c_{3,x})a_1 + (b_2 - c_3)(a_1d_1 - a_3c_1) + a_1d_3 + d_2 = 0.$$

Lie algebra $l_{3,6}$ not possible.

The proof of this lemma and the next one is similar to the proof of Lemma 6 and is therefore skipped.

Lemma 8. *The Lie system (14) of type $\mathcal{J}_{3,7}^{(2,2)}$ defines a Lie algebra of vector fields of a certain type as given in the classification on page 121 if its coefficients satisfy the following constraints.*

Lie algebra $l_{3,1}$ iff the coefficient obey

$$c_1 = 0, \ a_2 - \tfrac{1}{2}(c_2 + d_1) = 0,$$

$$b_1 \neq 0 \text{ or } c_3 \neq 0 \text{ or } c_{2,y} + d_{1,y} - \tfrac{1}{2}c_2^2 + \tfrac{1}{2}d_1^2 - 2d_3 \neq 0.$$

Lie algebra $l_{3,2}(c)$ iff $c \neq 1$ is a solution of

$$c^2 + \frac{2b_1 - c_1^2}{b_1}c + 1 = 0.$$

Lie algebra $l_{3,3}$ *iff the coefficients obey*

$$b_1 - \tfrac{1}{4}c_1^2 = 0, \quad c_3 - \tfrac{1}{2}c_1(a_2 + 2c_2 + 2d_1) = 0,$$
$$c_{2,y} + d_{1,y} + c_1 d_2 - \tfrac{1}{2}c_2^2 + \tfrac{1}{2}d_1^2 - 2d_3 = 0.$$

Lie algebra $l_{3,4}$ *iff the coefficients obey*

$$b_1 = 0, \quad c_3 - a_2 c_1 = 0,$$
$$c_{2,y} + d_{1,y} - 2a_2(a_2 + c_2 + d_1) + c_1 d_2 - c_2 d_1 - c_2^2 - 2d_3 = 0,$$
$$c_1 \neq 0 \text{ or } a_2 - \tfrac{1}{2}(c_2 + d_1) \neq 0.$$

Lie algebra $l_{3,5}$ *iff the coefficients obey*

$$b_1 = 0, \ c_1 = 0, \ c_3 = 0, \ a_2 - \frac{1}{2}(c_2 + d_1) = 0, \ c_{2,y} + d_{1,y} - \frac{1}{2}c_2^2 + \frac{1}{2}d_1^2 - 2d_3 = 0.$$

Lie algebra $l_{3,6}$ *not possible.*

Finally two Lie systems of order four have to be considered.

Lemma 9. *The Lie system (19) of type* $\mathcal{J}_{4,9}^{(2,2)}$ *defines a Lie algebra of vector fields of a certain type as given in the classification on page 121 if its coefficients satisfy the following constraints.*

Lie algebra $l_{4,1}$ *iff the coefficient obey* $a_2 = b_1$, $b_2 = 0$ *and* $b_1 \neq 0$.
Lie algebra $l_{4,4}$ *iff the coefficients obey* $b_2 = -2$, $b_1 = a_2 = 0$ *and* $a_4 = \tfrac{1}{2}c_1$.
Lie algebra $l_{4,5}(c)$ *iff the coefficients obey* $b_2 \neq 0$, $b_2 + 1 \neq 0$. *The parameter is determined by* $c = -b_2$.

Lemma 10. *For a Lie system (22) of type* $\mathcal{J}_{4,14}^{(2,2)}$ *to define a Lie algebra of vector fields of type* $l_{4,4}$ *or* $l_{4,5}(c)$, *the following necessary constraints have to be satisfied.*

Lie algebra $l_{4,4}$: $b_2 = -\tfrac{1}{2}$.
Lie algebra $l_{4,5}(c)$: $b_2 = -\tfrac{1}{c}$, $c \neq 0$ *and* $c \neq -2$.

Lemmata 1 to 10 provide the interface between the geometric and algebraic properties of the Lie groups of the plane on the one hand, and the symmetries of ode's expressed in terms of the Janet base of its determining system.

5   Symmetries of Differential Equations

The behavior of ode's under various kinds of transformations will be investigated first. Let an ode of order n in the independent variable x and the dependent variable $y \equiv y(x)$ be given as

$$\omega(x, y, y', \ldots, y^{(n)}) = 0. \tag{41}$$

If not stated explicitly otherwise it will be assumed that ω is a polynomial in the derivatives with coefficients in some *base field* which is usually the field of rational functions in x and y, i. e. $\omega \in \mathbf{Q}(x, y)[y', \ldots, y^{(n)}]$. Any other field that occurs later on during the solution procedure is an extension of this base field.

A *point-transformation* between two planes with coordinates (x, y) and (u, v) and dependencies $y \equiv y(x)$ and $v \equiv v(u)$ respectively is considered in the form

$$u = \sigma(x, y), \qquad v = \rho(x, y). \tag{42}$$

A point-transformation is a diffeomorphism in the first place. Depending on the particular situation, the function field in which σ and ρ are contained has to be specified.

Let a curve in the $x - y$-plane described by $y = f(x)$ be transformed under (42) into $v = g(u)$. There arises the question how the derivative $y' = df/dx$ corresponds to $v' = dg/du$ under this transformation. A simple calculation leads to the *first prolongation*

$$v' = \frac{dv}{du} = \frac{\rho_x + \rho_y y'}{\sigma_x + \sigma_y y'} \equiv \chi_1(x, y, y'). \tag{43}$$

Similarly the transformation law for derivatives of second order is obtained as

$$v'' = \frac{dv'}{du} = \frac{\chi_{1,x} + \chi_{1,y} y' + \chi_{1,y'} y''}{\sigma_x + \sigma_y y'} \equiv \chi_2(x, y, y', y'').$$

Explicitly in terms of σ and ρ it is

$$\begin{aligned}
v'' = \frac{1}{(\sigma_x + \sigma_y y')^3} &\{ (\sigma_x \rho_y - \sigma_y \rho_x) y'' + (\sigma_y \rho_{yy} - \sigma_{yy} \rho_y) y'^3 \\
&+ [\sigma_x \rho_{yy} - \sigma_{yy} \rho_x + 2(\sigma_y \rho_{xy} - \sigma_{xy} \rho_y)] y'^2 \\
&+ [\sigma_y \rho_{xx} - \sigma_{xx} \rho_y + 2(\sigma_x \rho_{xy} - \sigma_{xy} \rho_x)] y' + \sigma_x \rho_{xx} - \sigma_{xx} \rho_x \}.
\end{aligned} \tag{44}$$

In general there holds

$$v^{(n)} = \frac{dv^{(n-1)}}{du} \equiv \chi_n(x, y, y', \ldots, y^{(n)})$$

and

$$v^{(n+1)} = \frac{dv^{(n)}}{du} = \frac{\chi_{n,x} + \chi_{n,y} y' + \cdots + \chi_{n,y^{(n-1)}} y^{(n)}}{\sigma_x + \sigma_y y'}.$$

The form of a differential equation is extremely sensitive to a variable change. The following definition introduces a particular term for the exceptional cases where this is not true. Equation (41) is said to be *invariant* under the transformation $x = \phi(u,v)$, $y = \psi(u,v)$ with $v \equiv v(u)$ if it retains its form under this transformation, i. e. if the functional dependence of the transformed equation $\omega(u,v,v',\dots,v^{(n)}) = 0$ on its arguments is the same as in the original equation (41). Such a transformation is called a *symmetry of the differential equation*. The totality of symmetry transformations of an ode form a continuous group, it is called the *symmetry group* of that equation. The type of this group is called its *symmetry type*. Lie [Lie 1891] has shown that equivalent equations have the same symmetry type. The reverse is not true, in general to any symmetry type there corresponds the union of one or more equivalence classes.

In order to utilize the symmetries of an ode for the solution procedure, the geometrical considerations above must be expressed in analytical terms. Of particular importance is the fact that this may be achieved in terms of the infinitesimal generators of a symmetry, the expressions for its finite transformations are not required. To this end the prolongation of a vector field in x and $y(x)$ is introduced next. Let an infinitesimal generator $U = \xi(x,y)\partial_x + \eta(x,y)\partial_y$ be given and $y \equiv y(x)$ depend on x. Its $n-th$ prolongation is

$$U^{(n)} = U + \zeta^{(1)}\frac{\partial}{\partial y'} + \zeta^{(2)}\frac{\partial}{\partial y''} + \dots + \zeta^{(n)}\frac{\partial}{\partial y^{(n)}}. \tag{45}$$

The functions $\zeta^{(k)}$ are recursively defined by

$$\zeta^{(1)} = D(\eta) - y'D(\xi), \quad \zeta^{(k)} = D(\zeta^{(k-1)}) - y^{(k)}D(\xi) \tag{46}$$

for $k \geq 2$, $D = \partial_x + y'\partial_y + y''\partial_{y'} \dots$ is the operator of total differentiation with respect to x. The three lowest $\zeta's$ are explicitly

$$\zeta^{(1)} = \eta_x + (\eta_y - \xi_x)y' - \xi_y y'^2,$$

$$\zeta^{(2)} = \eta_{xx} + (2\eta_{xy} - \xi_{xx})y' + (\eta_{yy} - 2\xi_{xy})y'^2 - \xi_{yy}y'^3$$

$$+(\eta_y - 2\xi_x)y'' - 3\xi_y y'y''$$

$$\zeta^{(3)} = \eta_{xxx} + (3\eta_{xxy} - \xi_{xxx})y' + 3(\eta_{xyy} - \xi_{xxy})y'^2 + (\eta_{yyy} - 3\xi_{xyy})y'^3$$

$$-\xi_{yyy}y'^4 + 3(\eta_{xy} - \xi_{xx})y'' + 3(\eta_{yy} - 3\xi_{xy})y'y'' - 6\xi_{yy}y'^2y''$$

$$+(\eta_y - 3\xi_x)y''' - 4\xi_y y'y''' - 3\xi_y y''^2. \tag{47}$$

The classification problem of second-order equations has been solved by Lie himself [Lie 1883]. The next theorem and the corresponding results for third order equations are completely contained in part I of [Lie 1883]. They are just a different view of Lie's listing of differential invariants of the groups of the plane.

Theorem 4. *(Lie 1883). Any symmetry generator of a second order quasilinear ode is similar to one in canonical variables u and v as given in the following listing. In addition the corresponding Janet base is given where $\alpha(u,v)$ and $\beta(u,v)$ are the coefficients of ∂_u and ∂_v respectively.*

One-Parameter Group
  $S_1^2 : \mathbf{g}_{27} = \{\partial_v\}$. *Janet base* $\{\alpha, \beta_u, \beta_v\}$.
Two-Parameter Groups
  $S_{2,1}^2 : \mathbf{g}_{26} = \{\partial_u, \partial_v\}$. *Janet base* $\{\alpha_u, \alpha_v, \beta_u, \beta_v\}$.
  $S_{2,2}^2 : \mathbf{g}_{25} = \{\partial_v, u\partial_u + v\partial_v\}$. *Janet base* $\{\alpha_v, \beta_u, \beta_v - \frac{1}{v}\beta, \alpha_{uu}\}$.
Three-Parameter Groups
  $S_{3,1}^2 : \mathbf{g}_{10} = \{\partial_u + \partial_v, u\partial_u + v\partial_v, u^2\partial_u + v^2\partial_v\}$. *Janet base*
  $\{\alpha_v, \beta_u, \beta_v + \alpha_u + \frac{2}{u-v}(\beta - \alpha), \alpha_{uu} - \frac{2}{u-v}\alpha_u - \frac{2}{(u-v)^2}(\beta - \alpha)\}$.
  $S_{3,2}^2 : \mathbf{g}_{13} = \{\partial_u, 2u\partial_u + v\partial_v, u^2\partial_u + uv\partial_v\}$. *Janet base*
  $\{\alpha_u - \frac{2}{v}\beta, \alpha_v, \beta_v - \frac{1}{v}\beta, \beta_{uu}\}$.
  $S_{3,3}^2(c) : \mathbf{g}_7(c) = \{\partial_u, \partial_v, u\partial_u + cv\partial_v\}, c \neq 1$.
  Janet base $\alpha_v, \beta_u, \beta_v - c\alpha_u, \alpha_{uu}\}$.
  $S_{3,4}^2 : \mathbf{g}_{20}(r = 1) = \{\partial_u, \partial_v, u\partial_u + (u + v)\partial_v\}$.
  Janet base $\{\alpha_v, \beta_u - \alpha_u, \beta_v - \alpha_u, \alpha_{uu}\}$.
Eight-Parameter Group
  $S_8^2 : \mathbf{g}_3 = \{\partial_u, \partial_v, u\partial_v, v\partial_v, u\partial_u, v\partial_u, u^2\partial_u + uv\partial_v, uv\partial_u + v^2\partial_v\}$.
  Janet base $\{\alpha_{vv}, \beta_{uu}, \beta_{uv}, \beta_{vv}, \alpha_{uuu}, \beta_{uuv}\}$.

This listing shows in particular that there does not exist any second order ode allowing a group of point symmetries with 4, 5, 6 or 7 parameters.

In Section 3 and the above Theorem 4 the groups have been characterized by its Janet bases in *canonical variables*. In applications, the symmetry generators of differential equation and accordingly its Janet base occur in *actual*, i. e. non-canonical variables. Therefore the relation between these two descriptions must be found. This is based on the following observation. Take the Janet base for any group in canonical variables u and v, and transform it into actual variables by means of (42). In general this change of variables will destroy the Janet base property. In order to reestablish it, the algorithm *JanetBase* has to be applied. During this process it may occur that a leading coefficient of an equation that is applied for reduction vanishes due to a special choice of the transformation (42). This has the consequence that alternatives occur which may lead to different types of Janet bases. In order to obtain the complete answer, each of these alternatives has to be investigated separately. Finally all Janet bases have to be combined such that a minimal number of *generic cases* is retained, and all those that are discarded may be obtained from them by specialization. It turns out that up to a few exceptional cases all alternatives are due to vanishing first order partial derivatives of the transformation functions σ and ρ. Based on these observations the following result has been obtained [Schwarz 1996].

Theorem 5. *(Schwarz 1996) The following criteria provide a decision procedure for the symmetry type of a second order ode if its Janet base in a grlex term ordering with $\eta > \xi$, $y > x$ is given.*
One-Parameter Group

$S_1^2 :$ *Group* $\mathbf{g}_{27}$, *Janet base type* $\mathcal{J}_{1,1}^{(2,2)}$ *or* $\mathcal{J}_{1,2}^{(2,2)}$.

Two-Parameter Groups

$S_{2,1}^2$: *Group* $\mathbf{g}_{26}$, *Lie algebra* $\mathbf{l}_{2,2}$, *Janet base type* $\mathcal{J}_{2,3}^{(2,2)}$.

$S_{2,2}^2$: *Group* $\mathbf{g}_{27}$, *Lie algebra* $\mathbf{l}_{2,1}$, *Janet base type* $\mathcal{J}_{2,3}^{(2,2)}$.

Three-Parameter Groups

$S_{3,1}^2$: *Group* $\mathbf{g}_{10}$, *two s.o.i., Lie algebra* $\mathbf{l}_{3,1}$, *Janet base type* $\mathcal{J}_{3,6}^{(2,2)}$ *or* $\mathcal{J}_{3,7}^{(2,2)}$.

$S_{3,2}^2$: *Group* $\mathbf{g}_{13}$, *one s.o.i., Lie algebra* $\mathbf{l}_{3,1}$, *Janet base type* $\mathcal{J}_{3,4}^{(2,2)}$, $\mathcal{J}_{3,6}^{(2,2)}$ *or* $\mathcal{J}_{3,7}^{(2,2)}$.

$S_{3,3}^2(c)$: *Group* $\mathbf{g}_7(c)$, *two s.o.i., Lie algebra* $\mathbf{l}_{3,2}(c)$, *Janet base type* $\mathcal{J}_{3,6}^{(2,2)}$ *or* $\mathcal{J}_{3,7}^{(2,2)}$.

$S_{3,4}^2$: *Group* $\mathbf{g}_{20}(r=1)$, *one s.o.i., Lie algebra* $\mathbf{l}_{3,3}$, *Janet base type* $\mathcal{J}_{3,6}^{(2,2)}$ *or* $\mathcal{J}_{3,7}^{(2,2)}$.

Eight-Parameter Group

S_8^2 : *Group* $\mathbf{g}_3$, *Janet base type* $\{\xi_{yy}, \eta_{xx}, \eta_{xy}, \eta_{yy}, \xi_{xxx}, \xi_{xxy}\}$.

Proof. One-parameter group. A one-parameter symmetry is uniquely determined by the Janet base type.

Two-parameter groups. The two groups are identified by their Lie algebra via Lemma 5.

Three-parameter groups. At first the groups are distinguished by the number of imprimitivity systems they allow by Lemma 3. The two pairs corresponding to one or two systems of imprimitivity are separated w.r.t. to their Lie algebras by Lemma 6, 7 and 8.

The projective group. The eight-parameter symmetry is uniquly identified by its Janet base type. ●

These results may be extended to higher order equations, although the larger number of symmetry types leads to a much larger number of alternatives that have to be considered. There is another feature that distinguishes equations of order three or higher from second-order equations. Whereas for the latter linear equations form a single equivalence class with canonical form $v'' = 0$, this is not true any more if the order is higher than two. Therefore it appears to be appropriate to distinguish equations that are equivalent to a linear one from those that are not. These latter equations are simply called nonlinear third-order equations, they are the subject of the next two theorems, whereas the linearizable equations are considered thereafter. The equivalent of Theorem 4 for nonlinear third order equations is given first.

Theorem 6. *Any symmetry generator of a third order quasilinear ode is similar to one in canonical variables u and v as given in the following listing. In* *addition the corresponding Janet base is given where $\alpha(u,v)$ and $\beta(u,v)$ are the coefficients of ∂_u and ∂_v respectively.*

One-Parameter Group

$\mathcal{S}_1^3$: $\mathbf{g}_{27} = \{\partial_v\}$. *Janet base* $\{\alpha, \beta_u, \beta_v\}$.

Two-Parameter Groups

$\mathcal{S}_{2,1}^3$: $\mathbf{g}_{26} = \{\partial_u, \partial_v\}$. *Janet base* $\{\alpha_u, \alpha_v, \beta_u, \beta_v\}$.

$\mathcal{S}_{2,2}^3$: $\mathbf{g}_{25} = \{u\partial_u + v\partial_v, \partial_v\}$. *Janet base* $\{\alpha_v, \beta_u, \beta_v - \frac{1}{v}\beta, \alpha_{uu}\}$.

$\mathcal{S}_{2,3}^3$: $\mathbf{g}_4 = \{v\partial_v, \partial_v\}$. *Janet base* $\{\alpha, \beta_u, \beta_{vv}\}$.

$\mathcal{S}_{2,4}^3$: $\mathbf{g}_{15}(r=2) = \{\partial_v, u\partial_v\}$. *Janet base* $\{\alpha, \beta_v, \beta_{uu}\}$.

Three-Parameter Groups

$\mathcal{S}_{3,1}^3$: $\mathbf{g}_{10} = \{\partial_u + \partial_v, u\partial_u + v\partial_v, u^2\partial_u + v^2\partial_v\}$. *Janet base*
$\{\alpha_v, \beta_u, \beta_v + \alpha_u + \frac{2}{u-v}(\beta - \alpha), \alpha_{uu} - \frac{2}{u-v}\alpha_u - \frac{2}{(u-v)^2}(\beta - \alpha)\}$.

$\mathcal{S}_{3,2}^3$: $\mathbf{g}_{13} = \{\partial_u, 2u\partial_u + v\partial_v, u^2\partial_u + uv\partial_v\}$. *Janet base*
$\{\alpha_u - \frac{2}{v}\beta, \alpha_v, \beta_v - \frac{1}{v}\beta, \beta_{uu}\}$.

$\mathcal{S}_{3,3}^3(c)$: $\mathbf{g}_7 = \{\partial_u, \partial_v, u\partial_u + cv\partial_v\}$, $c \neq 0, 1$.
Janet base $\alpha_v, \beta_u, \beta_v - c\alpha_u, \alpha_{uu}\}$.

$\mathcal{S}_{3,4}^3$: $\mathbf{g}_{20}(r=1) = \{\partial_u, \partial_v, u\partial_u + (u+v)\partial_v\}$. *Janet base*
$\{\alpha_v, \beta_u - \alpha_u, \beta_v - \alpha_u, \alpha_{uu}\}$.

$\mathcal{S}_{3,5}^3$: $\mathbf{g}_5 = \{\partial_u, \partial_v, v\partial_v\}$. *Janet base* $\{\alpha_u, \alpha_v, \beta_u, \beta_{vv}\}$.

$\mathcal{S}_{3,6}^3$: $\mathbf{g}_{17}(l=1, \rho_1=1, \alpha_1=0) = \{\partial_u, \partial_v, u\partial_v\}$. *Janet base* $\{\alpha_u, \alpha_v, \beta_v, \beta_{uu}\}$.

$\mathcal{S}_{3,7}^3$: $\mathbf{g}_{17}(l=1, \rho_1=1, \alpha_1=1) = \{e^u\partial_v, ue^u\partial_v, \partial_u\}$.
Janet base $\{\alpha_u, \alpha_v, \beta_v, \beta_{uu} - 2\beta_u + \beta\}$.

$\mathcal{S}_{3,8}^3$: $\mathbf{g}_{17}(l=2, \rho_1=\rho_2=0, \alpha_1=0, \alpha_2=1) = \{e^u\partial_v, \partial_v, \partial_u\}$.
Janet base $\{\alpha_u, \alpha_v, \beta_v, \beta_{uu} - \beta_u\}$.

$\mathcal{S}_{3,9}^3(c)$:
$\mathbf{g}_{17}(l=2, \rho_1=\rho_2=0, \alpha_1=1, \alpha_2=c) = \{e^u\partial_v, e^{cu}\partial_v, \partial_u\}$, $c \neq 0, 1$.
Janet base $\{\alpha_u, \alpha_v, \beta_v, \beta_{uu} - (a+1)\beta_u + a\beta\}$.

$\mathcal{S}_{3,10}^3$: $\mathbf{g}_{24} = \{\partial_u, \partial_v, u\partial_u + v\partial_v\}$. *Janet base* $\{\alpha_v, \beta_u, \beta_v - \alpha_u, \alpha_{uu}\}$.

Four-Parameter Groups

$\mathcal{S}_{4,1}^3$: $\mathbf{g}_6 = \{\partial_u, \partial_v, v\partial_v, u\partial_u\}$. *Janet base* $\{\alpha_v, \beta_u, \alpha_{uu}, \beta_{vv}\}$.

$\mathcal{S}_{4,2}^3$: $\mathbf{g}_{14} = \{\partial_u, u\partial_u, v\partial_v, u^2\partial_u + uv\partial_v\}$.
Janet base $\{\alpha_v, \beta_v - \frac{1}{v}\beta, \alpha_{uu} - \frac{2}{v}\beta_u, \beta_{uu}\}$.

$\mathcal{S}_{4,3}^3(c)$: $\mathbf{g}_{19}(r=2) = \{\partial_u, \partial_v, u\partial_v, u\partial_u + cv\partial_v\}$, $c \neq 2$.
Janet base $\{\alpha_v, \beta_v - c\alpha_u, \alpha_{uu}, \beta_{uu}\}$.

$\mathcal{S}_{4,4}^3$: $\mathbf{g}_{20}(r=2) = \{\partial_u, \partial_v, u\partial_v, u\partial_u + (u^2 + 2v)\partial_v\}$.
Janet base $\{\alpha_v, \beta_v - 2\alpha_u, \alpha_{uu}, \beta_{uu} - 2\alpha_u\}$.

Groups with Six Parameters

$\mathcal{S}_6^3$: $\mathbf{g}_{12} = \{\partial_u, \partial_v, u\partial_u, v\partial_v, u^2\partial_u, v^2\partial_v\}$. *Janet base* $\{\alpha_v, \beta_u, \alpha_{uuu}, \beta_{vvv}\}$.

As it has been mentioned before, this theorem is an immediate consequence of Lie's classification of differential invariants [Lie 1883]. The next theorem allows to determine the symmetry type of an actually given equation from its Janet base, it is the equivalent of Theorem 5.

Theorem 7. *The following criteria provide a decision procedure for the symmetry type of a third order ode if its Janet base in a grlex term ordering with* $\eta > \xi$, $y > x$ *is given.*

One-Parameter Group

S_1^3 : *Janet base of type* $\mathcal{J}_{1,1}^{(2,2)}$ *or* $\mathcal{J}_{1,2}^{(2,2)}$.

Two-Parameter Groups

$S_{2,1}^3$: *Group* g_{26}, *one-parameter s.o.i., Lie algebra* $l_{2,2}$, *Janet base type* $\mathcal{J}_{2,3}^{(2,2)}$.

$S_{2,2}^3$: *Group* g_{27}, *one-parameter s.o.i., Lie algebra* $l_{2,1}$, *Janet base type* $\mathcal{J}_{2,3}^{(2,2)}$.

$S_{2,3}^3$: *Group* g_4, *two s.o.i., Lie algebra* $l_{2,1}$, *Janet base type* $\mathcal{J}_{2,1}^{(2,2)}$, $\mathcal{J}_{2,2}^{(2,2)}$ *or* $\mathcal{J}_{2,4}^{(2,2)}$.

$S_{2,4}^3$: *Group* $g_{15}(r = 2)$, *one s.o.i., Lie algebra* $l_{2,1}$, *Janet base type* $\mathcal{J}_{2,1}^{(2,2)}$ *or* $\mathcal{J}_{2,5}^{(2,2)}$.

Three-Parameter Groups

$S_{3,1}^3$: *Group* g_{10}, *two s.o.i., Lie algebra* $l_{3,1}$, *Janet base type* $\mathcal{J}_{3,6}^{(2,2)}$ *or* $\mathcal{J}_{3,7}^{(2,2)}$.

$S_{3,2}^3$: *Group* g_{13}, *one s.o.i., Lie algebra* $l_{3,1}$, *Janet base type* $\mathcal{J}_{3,4}^{(2,2)}$, $\mathcal{J}_{3,6}^{(2,2)}$ *or* $\mathcal{J}_{3,7}^{(2,2)}$.

$S_{3,3}^3(c)$: *Group* $g_7(c)$, *two s.o.i., Lie algebra* $l_{3,2}(c)$, *Janet base type* $\mathcal{J}_{3,6}^{(2,2)}$ *or* $\mathcal{J}_{3,7}^{(2,2)}$.

$S_{3,4}^3$: *Group* $g_{20}(r = 1)$, *one s.o.i., Lie algebra* $l_{3,3}$, *Janet base type* $\mathcal{J}_{3,6}^{(2,2)}$ *with* $c_1 \neq$ *or* $\mathcal{J}_{3,7}^{(2,2)}$ *with* $c_1 \neq 0$.

$S_{3,5}^3$: *Group* g_5, *two s.o.i., Lie algebra* $l_{3,4}$, *Janet base type* $\mathcal{J}_{3,4}^{(2,2)}$, $\mathcal{J}_{3,5}^{(2,2)}$, $\mathcal{J}_{3,6}^{(2,2)}$ *or* $\mathcal{J}_{3,7}^{(2,2)}$.

$S_{3,6}^3$: *Group* $g_{17}(l = 1, \rho_1 = 1, \alpha_1 = 0)$, *one s.o.i., Lie algebra* $l_{3,5}$, *Janet base type* $\mathcal{J}_{3,4}^{(2,2)}$, $\mathcal{J}_{3,6}^{(2,2)}$ *or* $\mathcal{J}_{3,7}^{(2,2)}$.

$S_{3,7}^3$: *Group* $g_{17}(l = 1, \rho_1 = 1, \alpha_1 = 1)$, *one s.o.i., Lie algebra* $l_{3,3}$, *Janet base type* $\mathcal{J}_{3,4}^{(2,2)}$, $\mathcal{J}_{3,6}^{(2,2)}$ *with* $c_1 = 1$ *or* $\mathcal{J}_{3,7}^{(2,2)}$ *with* $c_1 = 0$.

$S_{3,8}^3$: *Group* $g_{17}(l = 2, \rho_1 = \rho_2 = 0, \alpha_1 = 0, \alpha_2 = 1)$, *one s.o.i., Lie algebra* $l_{3,4}$, *Janet base type* $\mathcal{J}_{3,4}^{(2,2)}$, $\mathcal{J}_{3,6}^{(2,2)}$ *or* $\mathcal{J}_{3,7}^{(2,2)}$.

$S_{3,9}^3(c)$: *Group* $g_{17}(l = 2, \rho_1 = \rho_2 = 0, \alpha_1 = 1, \alpha_2 = c)$, *one s.o.i., Lie algebra* $l_{3,2}(c)$, *Janet base type* $\mathcal{J}_{3,4}^{(2,2)}$, $\mathcal{J}_{3,6}^{(2,2)}$ *or* $\mathcal{J}_{3,7}^{(2,2)}$.

$S_{3,10}^3$: *Group* g_{24}, *one-parameter s.o.i., Lie algebra* $l_{3,2}(c = 1)$, *Janet base type* $\mathcal{J}_{3,6}^{(2,2)}$.

Four-Parameter Groups

$\mathcal{S}_{4,1}^3$: Group $\mathbf{g}_6$, two s.o.i., Lie algebra $\mathbf{l}_{4,10}$, Janet base type $\mathcal{J}_{4,9}^{(2,2)}$, $\mathcal{J}_{4,10}^{(2,2)}$ or $\mathcal{J}_{4,14}^{(2,2)}$.

$\mathcal{S}_{4,2}^3$: Group $\mathbf{g}_{14}$, one s.o.i., Lie algebra $\mathbf{l}_{4,1}$, Janet base type $\mathcal{J}_{4,9}^{(2,2)}$ or $\mathcal{J}_{4,12}^{(2,2)}$.

$\mathcal{S}_{4,3}^3(c)$: Group $\mathbf{g}_{19}(r = 2)$, one s.o.i., Lie algebra $\mathbf{l}_{4,5}(c)$, Janet base type $\mathcal{J}_{4,9}^{(2,2)}$ or $\mathcal{J}_{4,14}^{(2,2)}$.

$\mathcal{S}_{4,4}^3$: Group $\mathbf{g}_{20}(r = 2)$, one s.o.i., Lie algebra $\mathbf{l}_{4,4}$, Janet base type $\mathcal{J}_{4,9}^{(2,2)}$ or $\mathcal{J}_{4,14}^{(2,2)}$.

Six-Parameter Group

$\mathcal{S}_6^3$: Janet base of type $\{\xi_y, \eta_x, \xi_{xxx}, \eta_{yyy}\}$, type $\{\xi_y, \eta_y, \xi_{xxx}, \eta_{xxx}\}$, type $\{\eta_x, \eta_y, \xi_{xy}, \xi_{xxx}, \xi_{yyy}\}$ or type $\{\eta_x, \eta_y, \xi_{yy}, \xi_{xxx}, \xi_{xxy}\}$.

Proof. One-parameter group. A one-parameter symmetry is uniquely determined by the Janet base type.

Two-parameter groups. The first two symmetry classes are singled out by their one-parameter system of imprimitivity by Lemma 2. The further distiction between the two groups is obtained through their Lie algebra by Theorem 5. For Janet base type $\mathcal{J}_{2,1}^{(2,2)}$ the symmetry class is uniquely identified by Lemma 2 through their systems of imprimitivity. If the Janet base type is $\mathcal{J}_{2,2}^{(2,2)}$, $\mathcal{J}_{2,4}^{(2,2)}$ or $\mathcal{J}_{2,5}^{(2,2)}$ the symmetry class is unique.

Three-parameter groups. If there is a type $\mathcal{J}_{3,5}^{(2,2)}$ Janet base, the symmetry class $\mathcal{S}_{3,5}^3$ is uniquely identified. For Janetbase type $\mathcal{J}_{3,4}^{(2,2)}$, by Lemma 3 again symmetry class $\mathcal{S}_{3,5}^3$ is uniquely identified by its two systems of imprimitivity. The remaining cases allow a single system of imprimitivity, the symmetry type is identified from its Lie algebra by means of Lemma 6. For Janetbase type $\mathcal{J}_{3,6}^{(2,2)}$, Lemma 3 gives the unique answer $\mathcal{S}_{3,10}^3$ if a one-parameter system of imprimitivity is found, otherwise it combines the symmetry classes into those allowing two systems of imprimitivity, i. e. $\mathcal{S}_{3,1}^3$, $\mathcal{S}_{3,3}^3(c)$ and $\mathcal{S}_{3,5}^3$, and the remaining ones. In the former case Lemma 7 leads to a unique identification by its Lie algebra. In the latter the same is true except for the two symmetry classes $\mathcal{S}_{3,4}^3$ and $\mathcal{S}_{3,7}^3$, both with a $\mathbf{l}_{3,3}$ Lie algebra. The are distinguished by their different connectivity properties which is identified by $c_1 \neq 1$ of $c_1 = 1$ respectively. For Janet base type $\mathcal{J}_{3,7}^{(2,2)}$ the discussion is identical, Lemmata 3 and 8 are applied now. The distinction between symmetry classes $\mathcal{S}_{3,4}^3$ and $\mathcal{S}_{3,7}^3$ is obtained now by the condition $c_1 \neq 0$ or $c_1 = 0$ respectively.

Four-parameter groups. The symmetry classes $\mathcal{S}_{4,1}^3$ and $\mathcal{S}_{4,2}^3$ are uniquely identified from the Janet base types $\mathcal{J}_{4,10}^{(2,2)}$ and $\mathcal{J}_{4,12}^{(2,2)}$. For Janet base type $\mathcal{J}_{4,9}^{(2,2)}$, by Lemma 4 symmetry class $\mathcal{S}_{4,1}^3$ is uniquely identified by its two systems of imprimitivity. The remaining three symmetry types are distinguished by their Lie algebra applying Lemma 9. For Janetbase type $\mathcal{J}_{4,14}^{(2,2)}$, by Lemma 4 again symmetry class $\mathcal{S}_{4,1}^3$ is uniquely identified through its two systems of imprimitivity. The distinction between the two remaining cases is obtained from Lemma 10.

Six-parameter group. The only symmetry class $\mathcal{S}_6^3$ comprising six parameters is identified from any of the Janet bases of order six without constraints on its coefficients. ●

As already mentioned before, for order higher than two, linearizable equations comprise more than a single equivalence class. Moreover they combine into several symmetry classes. As usual by now at first a complete survey of all possible symmetry types is provided.

Theorem 8. *Any symmetry generator of a third order quasilinear ode is similar to one in canonical variables u and v as given in the following listing. In addition the corresponding Janet base is given where $\alpha(u, v)$ and $\beta(u, v)$ are the coefficients of ∂_u and ∂_v respectively.*

Four-Parameter Symmetry

$\mathcal{S}_{4,5}^3$: $\mathbf{g}_{16}(r = 3) = \{\partial_v, u\partial_v, \phi(u)\partial_v, v\partial_v\}$.

Janet base $\{\alpha, \beta_{uv}, \beta_{vv}, \beta_{uuu} - \dfrac{\phi'''}{\phi''}\beta_{uu}\}$.

Five-Parameter Symmetry

$\mathcal{S}_{5,1}^3$: $\mathbf{g}_{18}(l = 2, \rho_1 = 1, \rho_2 = 0, \alpha_1 = 0, \alpha_2 = 1) = \{\partial_v, u\partial_v, e^u\partial_v, v\partial_v, \partial_u\}$.
Janet base: $\{\alpha_u, \alpha_v, \beta_{uv}, \beta_{vv}, \beta_{uuu} - \beta_{uu}\}$.
$\mathcal{S}_{5,2}^3(a) \equiv \mathbf{g}_{18}(l = 3, \rho_1 = \rho_2 = \rho_3 = 0, \alpha_1 = 0, \alpha_2 = 1, \alpha_3 = a) = \{\partial_v, e^u\partial_v, e^{au}\partial_v, v\partial_v, \partial_u\}$, $a \neq 0, 1$.
Janet base: $\{\alpha_u, \alpha_v, \beta_{uv}, \beta_{vv}, \beta_{uuu} - (a + 1)\beta_{uu} + a\beta_u\}$.

Seven-Parameter Symmetry

$\mathcal{S}_7^3$: $\mathbf{g}_{23}(r = 3) = \{\partial_v, u\partial_v, u^2\partial_v, v\partial_v, \partial_u, u\partial_u, u^2\partial_u + 2uv\partial_v\}$.
Janet base: $\{\alpha_v, \beta_{uv} - \alpha_{uu}, \beta_{vv}, \alpha_{uuu}, \beta_{uuu}\}$.

Based on the above classification of symmetries, the subsequent theorem provides algorithmic means for obtaining the symmetry class of any given quasilinear third-order ode.

Theorem 9. *The following criteria provide a decision procedure for the symmetry class of a third order ode that is equivalent to a linear one if its Janet base in a grlex term ordering with $\eta > \xi$, $y > x$ is given.*
Four-Parameter Symmetry

$\mathcal{S}_{4,5}^3$: *Janet base of type* $\mathcal{J}_{4,2}^{(2,2)}$, *type* $\mathcal{J}_{4,17}^{(2,2)}$ *or type* $\mathcal{J}_{4,19}^{(2,2)}$.

Five-Parameter Symmetries

$\mathcal{S}_{5,1}^3$ *and* $\mathcal{S}_{5,2}^3(a)$: *Three Janet base types may occur. In all three cases the parameter a determining the symmetry type is a solution of the equation*

$$P^2(a^2 - a + 1)^3 + Q^2 R(a - \frac{1}{2})^2(a + 1)^2(a - 2)^2) = 0. \tag{48}$$

If $a \neq 0$ the symmetry class is $\mathcal{S}_{5,2}^3(a)$ and $\mathcal{S}_{5,1}^3$ otherwise. Its coefficients P, Q and R for the various cases are given below.

a) *Janet base type* $\mathcal{J}_{5,1}^{(2,2)}$.

$$P = -\tfrac{9}{2}(a_{1,x}a_1 + \tfrac{1}{3}a_{1,x}e_1 + c_{3,x} - a_1^3 - \tfrac{2}{3}a_1^2 e_1 - 2a_1 c_3$$
$$-\tfrac{1}{3}a_1 e_1^2 + a_1 e_3 + c_1 e_2 - \tfrac{2}{27}e_1^3 + \tfrac{1}{3}e_1 e_3 - e_4),$$

$$Q = a_{1,x} - 2a_1^2 - a_1 e_1 - 3c_3 - \frac{1}{3}e_1^2 + e_3, \quad R = 3Q.$$

b) *Janet base type* $\mathcal{J}_{5,2}^{(2,2)}$.

$$P = -\tfrac{9}{2}(b_{1,y}b_1 + \tfrac{1}{3}b_{1,y}e_1 + d_{2,y} - b_1^3 - \tfrac{2}{3}b_1^2 e_1 - 2b_1 d_2$$
$$-\tfrac{1}{3}b_1 e_1^2 + b_1 e_2 + c_1 e_3 - \tfrac{2}{27}e_1^3 + \tfrac{1}{3}e_1 e_2 - e_5),$$

$$Q = b_{1,y} - 2b_1^2 - b_1 e_1 - 3d_2 - \frac{1}{3}e_1^2 + e_2, \quad R = 3Q.$$

c) *Janet base type* $\mathcal{J}_{5,3}^{(2,2)}$.

$$P = \tfrac{9}{2}(a_{3,x}a_3 c_1^3 - \tfrac{1}{3}a_{3,x}c_1^2 e_1 + a_{3,x}c_1 c_3 + c_{5,x} + a_2 a_3^2 c_1^3 - \tfrac{1}{3}a_2 a_3 c_1^2 e_1$$
$$+a_2 a_3 c_1 c_3 - a_2 c_5 + 2a_3^3 c_1^4 - a_3^2 c_1^3 e_1 + 3a_3^2 c_1^2 c_3 + a_3 c_1^3 e_2$$
$$+\tfrac{1}{3}a_3 c_1^2 e_1^2 - a_3 c_1^2 e_3 + 2a_3 c_1 c_5 + a_3 c_4 - \tfrac{1}{3}c_1^2 e_1 e_2$$
$$+c_1 c_3 e_2 - \tfrac{2}{27}c_1 e_1^3 + \tfrac{1}{3}c_2 e_1 e_3 - c_1 e_5 - e_4),$$

$$Q = a_{3,x}c_1^2 + a_2 a_3 c_1^2 + 3a_3^2 c_1^3 - a_3 c_1^2 e_1 + 3a_3 c_1 c_3 + c_1^2 e_2$$
$$+\tfrac{1}{3}c_1 e_1^2 - c_1 e_3 + 3c_5, \quad R = \tfrac{3}{c_1}Q.$$

Seven-Parameter Symmetry

$\mathcal{S}_7^3$: Janet base of type $\{\xi_y, \eta_{xy}, \eta_{yy}, \xi_{xxx}, \eta_{xxx}\}$, type $\{\eta_x, \xi_{xx}, \eta_{yy}, \xi_{xyy}, \xi_{yyy}\}$ or type $\{\eta_y, \xi_{yy}, \eta_{xx}, \xi_{xxx}, \xi_{xxy}\}$.

Proof. A Janet base of type $\mathcal{J}_{4,2}^{(2,2)}$, $\mathcal{J}_{4,17}^{(2,2)}$ or $\mathcal{J}_{4,19}^{(2,2)}$ uniquely identifies symmetry class $\mathcal{S}_{4,5}^3$ because these Janet base types do not occur for any other four-parameter symmetry as it may be seen from Theorem 7. If the equations relating the generic coefficients of a Janet base of type $\mathcal{J}_{5,1}^{(2,2)}$, $\mathcal{J}_{5,2}^{(2,2)}$ or $\mathcal{J}_{5,3}^{(2,2)}$ and the transformation functions $\sigma(x, y)$ and $\rho(x, y)$ are transformed into a Janet base, one of the equations is (48) from which the value of a and thereby the symmetry class $\mathcal{S}_{5,2}^3(a)$ may be obtained if $a \neq 0$ or $\mathcal{S}_{5,1}^3$ if $a = 0$. A Janet base allowing a seven-parameter symmetry group identifies uniquely symmetry class $\mathcal{S}_7^3$.   •

6   Concluding Remarks

The results described in this article are the basis for extending Lie's symmetry analysis of differential equations and the solution procedures based on it to equations of third order. It should be emphasized that the application of the theorems of Section 5 for determining the symmetry type are completely algorithmic. They do not require to extend the base field determined by the given differential equation. Only the next step, when the transformation to canonical form is considered, in general a base field extension will be required. This is discussed in detail in [Schwarz 2002].

References

Bluman 1990. Bluman G. W., S. Kumei S.: Symmetries of Differential Equations, Springer, Berlin (1990).

Janet 1920. Janet M.: Les systèmes d'équations aux dérivées partielles, Journal de mathématiques **83**, 65–123 (1920).

Kamke 1961. Kamke E.: Differentialgleichungen: Lösungsmethoden und Lösungen, I. Gewöhnliche Differentialgleichungen. Akademische Verlagsgesellschaft, Leipzig (1961).

Killing 1887. Killing W.: Die Zusammensetzung der stetigen endlichen Transformationsgruppen. Mathematische Annalen **31**, 252-290, **33**, 1-48, **34**, 57-122, **36**, 161-189 (1887).

Lie 1883. Lie S.: Klassifikation und Integration von gewöhnlichen Differentialgleichungen zwischen x, y, die eine Gruppe von Transformationen gestatten I, II, III and IV. Archiv for Mathematik **VIII**, page 187-224, 249-288, 371-458 and **IX**, page 431-448 respectively (1883) [Gesammelte Abhandlungen, vol. V, page 240-281, 282-310, 362-427 and 432-446].

Lie 1888. Lie S.: Theorie der Transformationsgruppen I, II and III. Teubner, Leipzig (1888). [Reprinted by Chelsea Publishing Company, New York (1970)].

Lie 1891. Lie S.: Vorlesungen über Differentialgleichungen mit bekannten infinitesimalen Transformationen. Teubner, Leipzig (1891). [Reprinted by Chelsea Publishing Company, New York (1967)].

Lie 1893. Lie S.: Vorlesungen über continuierliche Gruppen. Teubner, Leipzig (1893). [Reprinted by Chelsea Publishing Company, New York (1971)].

Loewy 1906. Loewy A.: Über vollständig reduzible lineare homogene Differentialgleichungen. Mathematische Annalen **56**, 89-117 (1906).

Olver 1986. Olver P.: Application of Lie Groups to Differential Equations. Springer, Berlin (1986).

Schwarz 1995. Schwarz F.: Symmetries of 2nd and 3rd Order ODE's. In: Proceedings of the ISSAC'95, ACM Press, A. Levelt, Ed., page 16-25 (1995).

Schwarz 1996. Schwarz F.: Janet Bases of 2nd Order Ordinary Differential Equations. In: Proceedings of the ISSAC'96, ACM Press, Lakshman, Ed., page 179-187 (1996)

Schwarz 2002. Schwarz F.: Algorithmic Lie Theory for Solving Ordinary Differential Equations. To appear.

Symbolic Methods
for the Equivalence Problem for Systems
of Implicit Ordinary Differential Equations

Kurt Schlacher[1], Andreas Kugi[2], and Kurt Zehetleitner[3]

[1] Department of Automatic Control and Control Systems Technology, Johannes
Kepler University of Linz, Austria; Christian Doppler Laboratory for Automatic
Control of Mechatronic Systems in Steel Industries, Linz, Austria.
kurt.schlacher@jku.at
[2] Chair of System Theory and Automatic Control, University of Saarland, Germany.
andreas.kugi@lsr.uni-saarland.de
[3] Department of Automatic Control and Control Systems Technology, Johannes
Kepler University of Linz, Austria.
kurt.zehetleitner@jku.at

Abstract. This contribution deals with the equivalence problem for systems of implicit ordinary differential equations. Equivalence means that
every solution of the original set of equations is a solution of a given
normal form and vice versa. Since we describe this system as a submanifold in a suitable jet-space, we present some basics from differential and
algebraic geometry and give a short introduction to jet-theory and its
application to systems of differential equations. The main results of this
contribution are two solutions for the equivalence problem, where time
derivatives of the input are admitted or not. Apart from the theoretical
results we give a sketch for computer algebra based algorithms necessary
to solve these problems efficiently.

1 Introduction

This contribution deals with the equivalence problem for dynamic systems described by a set of n_e nonlinear implicit ordinary differential equations of the
type

$$f^{i_e}\left(t, z, \tfrac{\mathrm{d}}{\mathrm{d}t}z\right) = 0 , \quad i_e = 1, \ldots, n_e , \tag{1}$$

with the independent variable t and the dependent variables z^{α_z}, $\alpha_z = 1, \ldots, q$.
We assume that the coordinates (t, z) of (1) are local coordinates of a smooth
manifold $\mathcal{E}$ with $\dim \mathcal{E} = q + 1$. The equivalence problem under consideration
is the following one: Given the system (1) and a prescribed normal form, does
there exist transformations such that these transformations map each solution
of (1) to a solution of the normal form and vice versa? The class of systems (1)
covers implicit systems, DAE-systems (differential algebraic equation)

$$\begin{aligned}
\tfrac{\mathrm{d}}{\mathrm{d}t}x^{\alpha_x} &= f^{\alpha_x}(t, x, u) , \alpha_x = 1, \ldots, n_x \\
0 &= f^{\alpha_s}(t, x, u) , \alpha_s = n_x + 1, \ldots, n_e
\end{aligned} \tag{2}$$

F. Winkler and U. Langer (Eds.): SNSC 2001, LNCS 2630, pp. 140–151, 2003.

and descriptor systems like

$$n_{\alpha_z}^{\alpha_e}(t, z) \tfrac{\mathrm{d}}{\mathrm{d}t} z^{\alpha_z} = m^{\alpha_e}(t, z) , \quad \alpha_e = 1, \ldots, n_e \tag{3}$$

in the unknowns z^{α_z}, $\alpha_z = 1, \ldots, n_z$. Of course, (3) contains (2) as a subclass, if one puts the input u, the set of functions that can be chosen freely, and the state x on an equal footing by setting $z = (x, u)$.

The investigation of nonlinear systems like (1), (2), (3) requires new mathematical tools. A modern approach considers differential equations as submanifolds of new spaces, called jet manifolds. One may state that the theory of jet manifolds has been invented for dealing with calculations involving any arbitrary order of differentiation, a prerequisite to solve problems concerning linear or nonlinear partial differential equations. This paper is organized as follows: In Section 2 we present some mathematical preliminaries which include manifolds and submanifolds, as well as certain rings over smooth functions defined on the manifolds under consideration and ideals of these rings. The goal is to combine methods from differential and algebraic geometry. The Section 3 deals with jet manifolds and their application to systems of ordinary or partial differential equations. The main results of this contribution are presented in Section 4. Finally, this contribution ends with some conclusions. Let us finish, the introduction with some technical remarks. Throughout this contribution, we consider only generic problems, because we are interested in methods based on computer algebra to derive algorithms for their investigation. Therefore, we assume that all functions are smooth and that several rank conditions are fulfilled. Roughly speaking, these conditions guarantee that the dynamic systems under consideration describe regular submanifolds of certain jet-spaces. Furthermore, we use the tensor-notation to keep the formulas short and readable.

2   Some Preliminaries

Before we start to repeat some mathematical preliminaries, we introduce some useful notations. Let $\mathcal{M}$ be a smooth finite-dimensional manifold, then $\mathcal{T}\mathcal{M}$ denotes the tangent bundle of $\mathcal{M}$, $\wedge_k^*\mathcal{M}$ the exterior k bundle and $\wedge^*\mathcal{M}$ the exterior algebra over $\mathcal{M}$. d : $\wedge_k^*\mathcal{M} \to \wedge_{k+1}^*\mathcal{M}$ is the exterior derivative and i : $\mathcal{T}\mathcal{M}\times\wedge_{k+1}^*\mathcal{M} \to \wedge_k^*\mathcal{M}$ is the interior product written as $i_X(\omega)$ with $X \in \mathcal{T}\mathcal{M}$ and $\omega \in \wedge_{k+1}^*\mathcal{M}$. $\wedge$ denotes the exterior product of the exterior algebra $\wedge^*\mathcal{M}$. Furthermore, the Lie derivative of $\omega \in \wedge^*\mathcal{M}$ along the tangent vector field $f \in \mathcal{T}\mathcal{M}$ is written as $f(\omega)$.

Although the methods being presented are independent of the special choice of the coordinates for a manifold $\mathcal{M}$, any computer algebra implementation relies on coordinates. The following theorem shows that the choice of the coordinates influences the representation of certain maps significantly.

Theorem 1. Let $\mathcal{M}$, $\mathcal{N}$ be two m- and n-dimensional smooth manifolds and $\varphi : \mathcal{M} \to \mathcal{N}$ a smooth map of maximal rank at $p \in \mathcal{M}$. Then there exist *neighborhoods $\mathcal{U}_p \subset \mathcal{M}$, $\mathcal{V}_{\varphi(p)} \subset \mathcal{N}$ and local coordinates $(x) = (x^1, \ldots, x^m)$ for $\mathcal{U}_p$ and $(y) = (y^1, \ldots, y^n)$ for $\mathcal{V}_{\varphi(p)}$ such that φ takes the simple form*

$$
\begin{aligned}
(y) &= (x^1, \ldots, x^m, 0, \ldots 0) && \text{if } n > m \\
(y) &= (x^1, \ldots, x^n) && \text{if } n \leq m
\end{aligned}
\tag{4}
$$

in these coordinates.

This theorem is an easy consequence of the implicit function theorem [1]. Later on we identify dynamic systems with submanifolds of certain manifolds. To avoid subtleties, we will assume that these submanifolds are at least locally regular. The following definition tells us that these manifolds have a very simple representation in special coordinates [1], [3].

Definition 1. *Let $\mathcal{M}$ be an m-dimensional smooth manifold. An n-dimensional submanifold $\mathcal{S} \subset \mathcal{M}$ is regular iff for each $p \in \mathcal{S}$ there exist local coordinates $(x) = (x^1, \ldots, x^m)$ defined on a neighborhood $\mathcal{U}_p \subset \mathcal{M}$ of p such that*

$$
\mathcal{S} \cap \mathcal{U}_p = \{x : x^{n+1} = \cdots = x^m = 0\}
\tag{5}
$$

is met. This coordinate chart is called a flat coordinate chart.

Often, we can derive only local results, which are valid in a sufficiently small neighborhood $\mathcal{U}_p \subset \mathcal{M}$ of $p \in \mathcal{S}$ only. To shorten the notation, we will write $\mathcal{S}_p$ for $\mathcal{S} \cap \mathcal{U}_p$. Regular submanifolds follow often from systems of implicit equations. The general concept is presented by the following theorem, which follows from Theorem 1.

Theorem 2. *Let $\mathcal{M}$ be an m-dimensional smooth manifold with local coordinates (x) and let the $f^i : \mathcal{M} \to \mathbb{R}$, $i = 1, \ldots, m - s$ denote smooth functions, which meet*

$$
\bigwedge_{i=1}^{m-s} \mathrm{d} f^i \neq 0
\tag{6}
$$

on the subset $\mathcal{S} = f^{-1}(\{0\})$, then $\mathcal{S}$ is a closed regular s-dimensional submanifold of $\mathcal{M}$.

A crucial observation for the problems discussed later on is that the functions f^i of Theorem 2 have no intrinsic meaning in contrast to $\mathcal{S}$. Let $\mathcal{F}_p(\mathcal{M})$ denote the set of smooth functions $f : \mathcal{M} \cap \mathcal{U}_p \to \mathbb{R}$ then it is easy to see that $\mathcal{F}_p$ has the structure of a unitary commutative local ring denoted by $R(\mathcal{F}_p(\mathcal{M}))$. Let $\mathcal{S}_p \subset \mathcal{M}$ be a regular submanifold, then the subset of all functions of $\mathcal{F}_p(\mathcal{M})$, $p \in \mathcal{S}$ vanishing on $\mathcal{S}_p$ forms an ideal denoted by $I(\mathcal{S}_p) \subset R(\mathcal{F}_p(\mathcal{M}))$. On the other hand, we can start with a set of functions $B = \{f\} = \{f^1, \ldots, f^{m-s}\}$, $f^i(\mathcal{S}_p) = \{0\}$ and study the ideal $\langle f \rangle = \langle f^1, \ldots, f^{m-s} \rangle$ generated by the functions f. The following theorem tells us, whether the ideals $\langle f \rangle$ and $I(\mathcal{S}_p)$ coincide.

Theorem 3. *Let $\mathcal{M}$ be a smooth m-dimensional manifold and $\mathcal{S}_p$ be a regular s-dimensional submanifold. Let the functions $f^i \in \mathcal{F}_p(\mathcal{M})$, $i = 1, \ldots, m - s$, $p \in \mathcal{S}$ meet $f^i(\mathcal{S}_p) = \{0\}$ and (6) on $\mathcal{S}_p$, then*

$$\langle f \rangle = I(\mathcal{S}_p) \tag{7}$$

is fulfilled. Furthermore, the set $B = \{f\} = \{f^1, \ldots, f^{m-s}\}$ forms a minimal basis of $I(\mathcal{S}_p)$, or given any function $g \in I(\mathcal{S}_p)$, then one can find functions $\lambda_i \in R(\mathcal{F}_p(\mathcal{M}))$ such that the relation

$$g = \lambda_i f^i \tag{8}$$

is met.

To prove this theorem, we choose flat coordinates $(x^1, \ldots, x^m)$ according to Definiton 1 such that p is mapped to 0. Obviously, the set $B = \{x^1, \ldots, x^{m-s}\}$ is a basis for $I(\mathcal{S}_p)$ in these coordinates. Let g be any function of $\mathcal{F}_p(\mathcal{M})$, which vanishes on $\mathcal{S}_p$, then from

$$g(x_1, x_2) - g(0, x_2) = \int_0^1 \partial_j g(x_1 \tau, x_2) x^j d\tau = x^j \int_0^1 \partial_j g(x_1 \tau, x_2) d\tau$$

with $x_1 = (x^j)$, $j = 1, \ldots, m - s$, $x_2 = (x^k)$, $k = m - s + 1, \ldots, m$ it follows $g \in \langle x^1, \ldots, x^{m-s} \rangle$ because of $g(0, x_2) = 0$. This shows that B is a minimal basis.

Let the two manifolds $\mathcal{M}$, $\mathcal{N}$ and the map φ be like in Theorem 1 with $n < m$. The pullback $g \circ \varphi \in \mathcal{F}_p(\mathcal{M})$ of $g \in \mathcal{F}_{\varphi(p)}(\mathcal{N})$ by φ is denoted by $\varphi^*(g)$. Since $\varphi^* : \mathcal{F}_{\varphi(p)}(\mathcal{N}) \to \mathcal{F}_p(\mathcal{M})$ is a ring homomorphism, we obtain $\varphi^*\left(R\left(\mathcal{F}_{\varphi(p)}(\mathcal{N})\right)\right) \subset R(\mathcal{F}_p(\mathcal{M}))$. The following theorem characterizes the intersection of this subring with an ideal of $R(\mathcal{F}_p(\mathcal{M}))$.

Theorem 4. *Let the two manifolds $\mathcal{M}$, $\mathcal{N}$ and the map φ be like in Theorem 1 with $n < m$. Let $\mathcal{S}_p \subset \mathcal{M}$ and $\varphi(\mathcal{S}_p) \subset \mathcal{N}$ be regular submanifolds and let $I(\mathcal{S}_p)$, $I(\varphi(\mathcal{S}_p))$ denote the corresponding ideals, then $\pi_\varphi(I(\mathcal{S}_p))$*

$$\pi_\varphi(I(\mathcal{S}_p)) = I(\mathcal{S}_p) \cap \varphi^*\left(R\left(\mathcal{F}_{\varphi(p)}(\mathcal{N})\right)\right) = \varphi^*(I(\varphi(\mathcal{S}_p))) \tag{9}$$

is an ideal such that its elements are functions of φ only. The corresponding manifold will be denoted by $\pi_\varphi(\mathcal{S}_p)$.

To prove this theorem, we choose any function $g \in \mathcal{F}_{\varphi(p)}(\mathcal{N})$, and get $\varphi^*(g) \in \varphi^*\left(R\left(\mathcal{F}_{\varphi(p)}(\mathcal{N})\right)\right)$. Now, $g \in I(\varphi(\mathcal{S}_p))$ implies $\varphi^*(g)(\mathcal{S}_p) = \{0\}$, or $\varphi^*(g) \in \pi_\varphi(I(\mathcal{S}_p))$ is met. Now, we choose any function $f \in \pi_\varphi(I(\mathcal{S}_p))$. It follows from $f \in \varphi^*\left(R\left(\mathcal{F}_{\varphi(p)}(\mathcal{N})\right)\right)$ that there exists a function $g \in \mathcal{F}_{\varphi(p)}(\mathcal{N})$ such that $f = \varphi^*(g)$ is met, and $f = g \circ \varphi \in I(\mathcal{S}_p)$ implies $g \in I(\varphi(\mathcal{S}_p))$, since φ is onto.

To give a geometric interpretation of Theorem 4, we choose a map $\psi : \mathcal{N} \to \mathcal{M}$ of maximal rank such that $\psi \circ \varphi(p) = p$ is met, and that $\mathcal{P}_p = \psi(\mathcal{N}) \cap \mathcal{U}_p \subset \mathcal{M}$ is a regular submanifold. Since the function $\pi_p^{\mathcal{M}} = \psi \circ \varphi : \mathcal{M}_p \to$

$\psi\left(\mathcal{N}_p\right)$ maps all points of the set $\left(\pi_{\mathcal{P}}^{\mathcal{M}}\right)^{-1}\left(\{q\}\right)$ to $q \in \mathcal{P}_p$, we may consider $\pi_{\mathcal{P}}^{\mathcal{M}}$ as a projection of $\mathcal{U}_p$ on $\mathcal{P}_p$. In particular, the map $\pi_{\mathcal{P}}^{\mathcal{M}}$ projects $\mathcal{S}_p$ on $\mathcal{P}_p$. Therefore, the ideal $\pi_\varphi\left(I\left(\mathcal{S}_p\right)\right)$ may be identified with the ideal $\pi_{\mathcal{P}}^{\mathcal{M}}\left(I\left(\mathcal{S}_p\right)\right)$. Obviously, this interpretation depends on the construction of $\mathcal{P}_p$. The situation becomes simpler, if we choose the special coordinates of Theorem 1. Since the ring $\varphi^*\left(R\left(\mathcal{F}_{\varphi(p)}\left(\mathcal{N}\right)\right)\right)$ contains the functions, which depend only on $x^1, \ldots, x^n$, the ideal $\pi_\varphi\left(I\left(\mathcal{S}_p\right)\right)$ contains all functions of $I\left(\mathcal{S}_p\right)$, which are independent of the coordinates $x^{n+1}, \ldots, x^m$. Additionally, let the point p have the coordinates $x = 0$, then $\mathcal{P}_p$ is the plane $x^{n+1} = \cdots = x^m = 0$ and $\pi_{\mathcal{P}}^{\mathcal{M}}$ takes the form $\left(x^1, \ldots, x^m\right) \to \left(x^1, \ldots, x^n, 0, \ldots, 0\right)$.

Given a minimal basis f^i, $i = 1, \ldots, m-s$ for $I\left(\mathcal{S}_p\right)$ one can construct a minimal basis for $\pi_\varphi\left(I\left(\mathcal{S}_p\right)\right) \neq \{0\}$. We form a new basis

$$\hat{f}^l = \lambda_i^l f^i , \quad l = 1, \ldots, r \tag{10a}$$

$$\check{f}^{\bar{l}} = \lambda_i^{\bar{l}} f^i , \quad \bar{l} = r+1, \ldots, m-s \tag{10b}$$

with a regular matrix $[\lambda_i^{\bar{i}}]$, $\bar{i} = 1, \ldots, m-s$ such that the equations

$$\lambda_i^l v_k\left(f^i\right) = \lambda_i^l \mathrm{i}_{v_k}\left(\mathrm{d}f^i\right) = 0 , \quad k = 1, \ldots, m-n \tag{11a}$$

$$v_k\left(\varphi^j\right) = \mathrm{i}_{v_k}\left(\mathrm{d}\varphi^j\right) = 0 , \quad j = 1, \ldots, n \tag{11b}$$

are met, for all linearly independent solutions of (11b). The set of functions $\hat{f}^l \bmod \left\{\check{f}^{\bar{l}} = 0\right\}$ form a basis of $\pi_\varphi\left(I\left(\mathcal{S}_p\right)\right)$. To show our claim, we choose a function $\hat{f}^l$. From $\hat{f}^l \in \varphi^*\left(R\left(\mathcal{F}_p\left(\mathcal{N}\right)\right)\right)$ and $\hat{f}^l \in I\left(\mathcal{S}_p\right)$ it follows

$$\mathrm{d}\hat{f}^l = \mu_j^l \mathrm{d}\varphi^j = \lambda_i^l \mathrm{d}f^i.$$

This equation is solvable, iff (11a) is met for all linearly independent solutions of (11b). Let us denote the inverse of $[\lambda_i^{\bar{i}}]$ by $[\bar{\lambda}_i^{\bar{i}}]$. Then the relations

$$v_k\left(\lambda_i^l f^i\right) = v_k\left(\lambda_i^l\right) f^i + \lambda_i^l v_k\left(f^i\right)$$
$$= v_k\left(\lambda_i^l\right) \bar{\lambda}_l^i \hat{f}^l + \underline{v_k\left(\lambda_i^l\right) \bar{\lambda}_{\bar{l}}^i \check{f}^{\bar{l}} + \lambda_i^l v_k\left(f^i\right)}$$

show that $v_k\left(\lambda_i^l f^i\right) \in \pi_\varphi\left(I\left(\mathcal{S}_p\right)\right)$ since the underbraced term vanishes on the solution set $\check{f}^{\bar{l}} = 0$ because of (11a). Now $v_k\left(\lambda_i^l f^i\right) \in \pi_\varphi\left(I\left(\mathcal{S}_p\right)\right)$ implies that $v_k\left(\hat{f}^l\right) \overset{\pi_\varphi(\mathcal{S}_p)}{=} 0$ or $\left\langle \hat{f} \right\rangle = \pi_\varphi\left(I\left(\mathcal{S}_p\right)\right) \bmod \left\{\check{f}^{\bar{l}} = 0\right\}$ is met. It is worth mentioning that in general the computation of the functions needs the implicit function theorem, but for polynomial systems we can use Groebner bases for the determination of the corresponding elimination ideal [2].

3   Jet Manifolds and Differential Equations

Throughout this contribution we use the concept of fibered manifolds and jet-bundles. A fibered manifold is a triple $\left(\mathcal{E}, \pi, \mathcal{B}\right)$ with the total manifold $\mathcal{E}$, the base manifold $\mathcal{B}$ and the surjective submersion $\pi : \mathcal{E} \to \mathcal{B}$ ([4], [5]). For each point $x \in \mathcal{B}$, the subset $\pi^{-1}(x) = \mathcal{E}_x$ is called the fiber over x. If all fibers $\mathcal{E}_x$ are diffeomorphic to a manifold $\mathcal{F}$, then we call $(\mathcal{E}, \pi, \mathcal{B})$ a bundle. In the finite dimensional case with $\dim \mathcal{E} = p + q$, $\dim \mathcal{B} = p$ we can introduce adapted coordinates (x^i, u^α) at least locally, with the independent coordinates x^i, $i = 1, \ldots, p$ and the dependent ones u^α, $\alpha = 1, \ldots, q$. We will write $\mathcal{E}$ instead of $(\mathcal{E}, \pi, \mathcal{B})$, whenever the projection π and the base manifold $\mathcal{B}$ follow from the context. A section σ of $\mathcal{E}$ is a map $\sigma : \mathcal{B} \to \mathcal{E}$ such that $\pi \circ \sigma = \mathrm{id}_\mathcal{B}$ is met, where $\mathrm{id}_\mathcal{B}$ denotes the identity map on $\mathcal{B}$. From now on we use Latin indices for the independent and Greek indices for the dependent variables.

Although jet manifolds and systems of differential equations admit a coordinate free description, we will us adapted coordinates from now on to simplify the following consideration. Let f be a smooth section of $(\mathcal{E}, \pi, \mathcal{B})$. The kth order partial derivatives of f^α will be denoted by

$$\frac{\partial^{l_0}}{\partial_1^{j_1} \cdots \partial_p^{j_p}} f^\alpha = \partial_J f^\alpha = f_J^\alpha , \quad \partial_i = \frac{\partial}{\partial x^i}$$

with $J = j_1, \ldots, j_p$, and $k = \#J = \sum_{i=1}^p j_i$. J is nothing else than an ordered multi-index ([4], [5]). The special index $J = j_1, \ldots, j_p$, $j_i = \delta_{ik}$ will be denoted by 1_k and $J+1_k$ is a shortcut for $j_i + \delta_{ik}$ with the Kronecker symbol δ_{ik}. Let $(\mathcal{E}, \pi, \mathcal{B})$ be a bundle with adapted coordinates (x^i, u^α) and let f be a smooth section of $\mathcal{E}$. We can prolong f to a map $j^1 f : x \to (x, f(x), \partial_i f(x))$, then the first jet $j_x^1 f$ of f at x is the equivalence class of sections f, g with $j^1 f(x) - j^1 g(x)$. Roughly speaking, the first jet manifold $J^1(\mathcal{E})$ may be considered as a container for all first order jets of smooth sections of $\mathcal{E}$. An adapted coordinate system of $\mathcal{E}$ induces an adapted system on $J^1(\mathcal{E})$, which is denoted by $(x^i, u^\alpha, u_{1_i}^\alpha)$ with the pq new coordinates $u_{1_i}^\alpha$. The manifold $J^1(\mathcal{E})$ has two natural projections, $\pi : J^1(\mathcal{E}) \to \mathcal{B}$ and $\pi_0^1 : J^1(\mathcal{E}) \to \mathcal{E}$ with $\pi(j^1 f(x)) = x$ and $\pi_0^1(j^1 f(x)) = f(x)$. Furthermore, $(J^1(\mathcal{E}), \pi_0^1, \mathcal{E})$ is an affine bundle and $(J^1(\mathcal{E}), \pi, \mathcal{B})$ is bundle, if $\mathcal{E}$ is one. Analogously to the first jet of a section f, we define the nth-jet $j^n f$ of f by $j^n f = (x, f(x), \partial_J f(x))$, $\#J = 1, \ldots, n$. The nth-jet manifold $J^n(\mathcal{E})$ of $\mathcal{E}$ may be considered as a container for n^{th}-jets of sections of $\mathcal{E}$ ([3], [4], [5]). Furthermore, an adapted coordinate system of $\mathcal{E}$ induces an adapted system on $J^n(\mathcal{E})$ with $(x^i, u^{(n)})$ and $u^{(n)} = u_J^\alpha$, $\alpha = 1, \ldots, q$, $\#J = 0, \ldots, n$. The natural projections are given by $\pi : J^n(\mathcal{E}) \to \mathcal{B}$ and $\pi_m^n : J^n(\mathcal{E}) \to J^m(\mathcal{E})$, $m = 1, \ldots, n-1$ with $\pi(j^n f(x)) = x$ and $\pi_m^n(j^n f(x)) = j^m f(x)$. To simplify certain formulas later on, we set $J^0(\mathcal{E}) = \mathcal{E}$.

Let us consider a bundle $(\mathcal{E}, \pi, \mathcal{B})$ with adapted coordinates (x^i, u^α), $i = 1, \ldots, p$, $\alpha = 1, \ldots, q$ and its nth-order jet manifold $J^n(\mathcal{E})$. Because of Definition 1, a regular submanifold $\mathcal{S}_p^n \subset J^n(\mathcal{E})$, $p \in \mathcal{S}^n$ corresponds locally to a system of equations

$$f^i\left(x, u^{(n)}\right) = 0 , \quad i = 1, \ldots, n_e \tag{12}$$

that involves jet coordinates $u^{(n)}$ of u up to the order n. If we substitute the nth-jet $j^n \sigma$ of a smooth section σ into (12), then we get a system of ordinary

$(p = 1)$ or partial differential equations $(p > 1)$ of the type

$$f^i \left(j^n \sigma \left(x \right) \right) = 0 \ , \quad i = 1, \ldots, n_e \ . \tag{13}$$

From now on, we will deal exclusively with the submanifold S_p^n, and the ideal $I \left(S_p^n \right)$ instead of the differential equations (12). There exist two natural operations for S_p^n, $I \left(S_p^n \right)$, their projection and their prolongation [4]. The projection of $I \left(S_p^n \right)$ is simply given by $I \left(S_p^n \right) \cap R \left(J^m \left(\mathcal{E} \right) \right)$, see Theorem 4 with $\varphi = \pi_m^n$ and will be denoted by $\pi_m^n \left(I \left(S_p^n \right) \right)$. We assume that the ideal $\pi_m^n \left(I \left(S_p^n \right) \right)$ describes a regular submanifold of $J^m \left(\mathcal{E} \right)$ denoted by $\pi_m^n \left(S_p^n \right)$. To define the prolongation of S_p^n, we introduce the total derivative d_i,

$$d_i = \partial_i + u_{J+1_i}^\alpha \partial_\alpha^J \ , \quad \partial_\alpha^J = \frac{\partial}{\partial u_J^\alpha} \ , \quad \#J \geq 0 \tag{14}$$

with respect to the independent variables x^i [3], [4], [5]. It is straightforward to see that for any section σ of $\mathcal{E}$ and $f \in \mathcal{F} \left(J^n \left(\mathcal{E} \right) \right)$ the following relations

$$d_i f \circ j^{k+1} \sigma = \partial_i f \left(j^k \sigma \right) \tag{15}$$

are met. Therefore, d_i is a map $d_i : \mathcal{F} \left(J^n \left(\mathcal{E} \right) \right) \to \mathcal{F} \left(J^{n+1} \left(\mathcal{E} \right) \right)$. The ideal $J \left(I \left(S_p^n \right) \right) = I \left(S_p^n \right) + \langle d_i f, \ i = 1, \ldots, q, \ f \in I \left(S_p^n \right) \rangle$ is the first prolongation of $I \left(S_p^n \right)$. Obviously, the ideal $J^1 \left(I \left(S_p^n \right) \right)$ contains all functions of $I \left(S_p^n \right)$, as well as their total derivatives with respect to the independent variables. Again, we assume that $J \left(I \left(S_p^n \right) \right)$ defines a regular submanifold, which we denote by $J \left(S_p^n \right)$ by some abuse of notation. The nth prolongation of $I \left(S_p^n \right)$ is, as usual defined by $J^n \left(I \left(S_p^n \right) \right) = J \left(J^{n-1} \left(I \left(S_p^n \right) \right) \right)$ and $J^0 \left(I \left(S_p^n \right) \right) = I \left(S_p^n \right)$. Given a minimal basis $B \left(S_p^n \right) = \left\{ f^1, \ldots, f^k \right\}$ of $I \left(S_p^n \right)$ one can construct a minimal basis of $J \left(I \left(S_p^n \right) \right)$ in a straightforward manner. We determine the set $B \cup \left\{ d_i f^j : j = 1, \ldots, k, \ i = 1, \ldots, q \right\}$ and take out a minimal number of functions such that the members of the remaining set $B \left(J \left(S_p^n \right) \right)$ are functionally independent.

Let σ be a section $\mathcal{E}$ such that $j_x^n \sigma \in S_p^n$ is met. By construction any section of (12) is a solution of the prolonged and projected systems. Unfortunately, this is not true for points in general, since a point $j_x^{n+r} \sigma$ with $\pi_n^{n+r} \left(j_x^{n+r} \sigma \right) \in J^n \left(\mathcal{E} \right)$ may fail to meet $j_x^{n+r} \sigma \in J^r \left(S_p^n \right)$. This is a result of the fact that in general $I \left(S_p^n \right) \subset \pi_n^{n+r} \left(J^r \left(I \left(S_p^n \right) \right) \right)$ or $\pi_n^{n+r} \left(S_p^n \right) \subset S_p^n$ is met only. Formally integrable systems do not show this unpleasant behavior.

Definition 2. *Let S_p^n be a locally regular submanifold of the nth jet manifold $J^n \left(\mathcal{E} \right)$ of the bundle $\mathcal{E}$. We call the system S_p^n formally integrable, iff its rth prolongations $J^r \left(S_p^n \right)$ are regular submanifolds of $J^{n+r} \left(\mathcal{E} \right)$ and*

$$\pi_{n+r}^{n+r+s} \left(J^{r+s} \left(S_p^n \right) \right) = J^r \left(S_p^n \right)$$

is met for all $r, s \geq 0$.

Roughly speaking, formally integrable systems do not contain hidden constraints. These constraints are often called integrability conditions. It is an extremely hard job to check a general system for formal integrability. Fortunately, this is not true for systems of ordinary differential equations and systems of linear partial differential equations of the Frobenius-type.

4 Main Results

Let $(\mathcal{E}, \pi, \mathcal{B})$ be a smooth manifold with adapted coordinates (t, z^{α_z}) in one independent variable t and q dependent variables z^{α_z}, $\alpha_z = 1, \ldots, q$. In the case $\dim \mathcal{B} = 1$, we denote the single independent coordinate by t and choose x or z for the dependent coordinates, as customary. We study the following set of n_e nonlinear ordinary differential equations

$$f^{i_e}(t, z, z_1) = 0 , \quad i_e = 1, \ldots, n_e , \quad n_e \leq q \tag{16}$$

with smooth functions $f^{i_e} : J^1(\mathcal{E}) \to \mathbb{R}$. To simplify the following considerations we assume that

$$\mathrm{d}t \wedge \bigwedge_{i_e=1}^{n_e} \mathrm{d}f^{i_e}(p) \neq 0 \tag{17}$$

is met for a point $p \in J^1(\mathcal{E})$ with $f^{i_e}(p) = 0$. This condition ensures that the system (16) defines a regular submanifold $\mathcal{S}_p^1 \subset J^1(\mathcal{E})$ of $J^1(\mathcal{E})$ near p, which contains an open interval of the independent variable t.

Let us choose a point $p \in \mathcal{S}^1$ and a neighborhood $\mathcal{U}_p \subset J^1(\mathcal{E})$. Then we are ready to pose the following problem. Is it possible to construct a transformation near p that converts the implicit system (16) to an explicit system of the form

$$x_1^{\alpha_x} = f^{\alpha_x}\left(t, x, u^{(n)}\right) , \quad \alpha_x = 1, \ldots, n_x . \tag{18}$$

Here, x denotes the state and $u^{(n)}$ the nth jet of the input u^{α_u}, $\alpha_u = 1, \ldots, n_u$, the functions, which we can choose freely.

A crucial observation for the problem above is that the system (16) has no intrinsic meaning in contrast to $\mathcal{S}_p^1$ or the ideal $I(\mathcal{S}_p^1)$. Taking into account the considerations of Section 2, this problem takes the form of the ideal membership problem, whether the relations

$$(x_1^{\alpha_x} - f^{\alpha_x}) \circ j^n \varphi \in I(\mathcal{S}_p^1)$$

can be fulfilled for suitable functions f^{α_x} and a suitable map φ or not. Obviously, we have to construct a suitable basis to perform the ideal membership test.

Let us assume that the system (16) is rewritten in the following form

$$f^{\alpha_x}(t, z, z_1) = 0 , \quad \alpha_x = 1, \ldots, n_x$$
$$f^{\alpha_s}(t, z, z_1) = 0 , \quad \alpha_s = n_x + 1, \ldots, n_e ,$$

where the functions f^{α_s} meet the relations

$$\partial^1_{z^{\alpha_z}} f^{\alpha_s} = 0$$

on the solution set $f^{\alpha_x}(t, z, z_1) = 0$. This implies that we can eliminate the variable z_1 from the equations f^{α_s}. Therefore, we will write $f^{\alpha_s} = f^{\alpha_s}(t, z)$ and consider the special system

$$f^{\alpha_x}(t, z, z_1) = 0 \ , \quad \alpha_x = 1, \ldots, n_x \tag{19a}$$

$$\begin{aligned} d_1 f^{\alpha_s}(t, z) &= 0 \\ f^{\alpha_s}(t, z) &= 0 \end{aligned} , \quad \alpha_s = n_x + 1, \ldots, n_e \ . \tag{19b}$$

Additionally, we assume that the system (19) is formally integrable or

$$\mathrm{d}t \wedge \bigwedge_{\alpha_z} \mathrm{d}z^{\alpha_z} \wedge \bigwedge_{\alpha_x} \mathrm{d}f^{\alpha_x} \wedge \bigwedge_{\alpha_s} \mathrm{dd}_1 f^{\alpha_s} \overset{\mathcal{S}^1_p}{\neq} 0 \ , \tag{20}$$

is fulfilled.

Now, we claim that the formally integrable system (19) with $q \geq n_e$ is equivalent to the explicit form (18)

$$x^{\alpha_x}_1 - \bar{f}^{\alpha_x}\left(t, x, s, u^{(1)}\right) = 0 \ , \quad \alpha_x = 1, \ldots, n_x \tag{21a}$$

$$\begin{aligned} s^{\alpha_s}_1 &= 0 \\ s^{\alpha_s} &= 0 \end{aligned} , \quad \alpha_s = n_x + 1, \ldots, n_e \tag{21b}$$

with $n = 1$. To prove this claim, we introduce the new bundle $\bar{\mathcal{E}}$ with adapted coordinates (t, x, s, u) and the bundle diffeomorphism $\varphi : \mathcal{E} \to \bar{\mathcal{E}}$,

$$\begin{aligned} x^{\alpha_x} &= \varphi^{\alpha_x}(t, z) \ , & \alpha_x &= 1, \ldots, n_x \\ s^{\alpha_s} &= \varphi^{\alpha_s}(t, z) = f^{\alpha_s} \ , & \alpha_s &= n_x + 1, \ldots, n_e \\ u^{\alpha_u} &= \varphi^{\alpha_u}(t, z) \ , & \alpha_u &= n_e + 1, \ldots, q \end{aligned} \tag{22a}$$

$$z^{\alpha_z} = \psi^{\alpha_z}(t, x, s, u) \tag{22b}$$

that does not change the independent variable t. Following the considerations above, we have to check, whether the functions of (21) are contained in the ideal $I\left(\mathcal{S}^1_p\right)$, or

$$\left(x^{\alpha_x}_1 - \bar{f}^{\alpha_x}\right) \circ j\varphi = \lambda^{\alpha_x}_{\beta_x} f^{\beta_x} + \lambda^{\alpha_x,1}_{\beta_s} d_1 f^{\beta_s} + \lambda^{\alpha_x}_{\beta_s} f^{\beta_s} \tag{23a}$$

$$s^{\alpha_s}_1 \circ j\varphi = \lambda^{\alpha_s}_{\beta_x} f^{\beta_x} + \lambda^{\alpha_s,1}_{\beta_s} d_1 f^{\beta_s} + \lambda^{\alpha_s}_{\beta_s} f^{\beta_s} \tag{23b}$$

is met. It is straightforward to solve (23b) and the solution is

$$\varphi^{\alpha_s} = f^{\alpha_s} \ , \quad \lambda^{\alpha_s}_{\beta_x} = 0 \ , \quad \lambda^{\alpha_s,1}_{\beta_s} = \delta^{\alpha_s,1}_{\beta_s} \ , \quad \lambda^{\alpha_s}_{\beta_s} = 0 \ . \tag{24}$$

To solve (23a) we consider its total derivative with respect to t,

$$\partial_{\alpha_z} \varphi^{\alpha_x} z^{\alpha_z}_2 - \partial^1_{\alpha_u} \bar{f}^{\alpha_x} \partial_{\alpha_z} \varphi^{\alpha_u} z^{\alpha_z}_2 + \ldots \overset{\mathcal{S}^1_p}{\doteq} \lambda^{\alpha_x}_{\beta_x} \partial^1_{\alpha_z} f^{\beta_x} z^{\alpha_z}_2 + \lambda^{\alpha_x,1}_{\beta_s} \partial_{\alpha_z} \varphi^{\beta_s} z^{\alpha_z}_2 + \ldots \ ,$$

where the terms not involving z_2 are omitted. Comparing the coefficients of $z_2^{\alpha_z}$, we derive

$$\partial_{\alpha_z}\varphi^{\alpha_x} - \partial^1_{\alpha_u}\bar{f}^{\alpha_x}\partial_{\alpha_z}\varphi^{\alpha_u} = \lambda^{\alpha_x}_{\beta_x}\partial^1_{\alpha_z}f^{\beta_x} + \lambda^{\alpha_x,1}_{\beta_s}\partial_{\alpha_z}\varphi^{\beta_s}$$

and rewrite these relations as

$$\left(d\varphi^{\alpha_x} - \partial^1_{\alpha_u}\bar{f}^{\alpha_x}\circ j\varphi d\varphi^{\alpha_u}\right) \wedge dt \overset{S^1_p}{=} \left(\lambda^{\alpha_x}_{\beta_x}\partial^1_{\alpha_z}f^{\beta_x}dz^{\alpha_z} + \lambda^{\alpha_x,1}_{\beta_s}d\varphi^{\beta_s}\right) \wedge dt \ . \quad (25)$$

Let us choose functions φ^{α_x}, φ^{α_u} such that φ is a diffeomorphism and

$$\bigwedge_{\alpha_x}\partial^1_{\alpha_z}f^{\alpha_x}dz^{\alpha_z} \wedge \bigwedge_{\alpha_u}d\varphi^{\alpha_u} \wedge \bigwedge_{\alpha_s}d\varphi^{\alpha_s} \overset{S^1_p}{\neq} 0$$

is met, this is always possible, then it is easy to see that one derives the remaining functions $\lambda^{\alpha_x}_{\beta_x}$, $\lambda^{\alpha_x,1}_{\beta_s}$ immediately from (25). Finally, the functions $\bar{f}^{\alpha_x}$ follow from (23a). This proves our claim and leads to the following theorem.

Theorem 5. *Let (19) describe a formally integrable system, then there exists locally a diffeomorphism (22) such that the equivalent system takes the form (21).*

An interesting problem is, whether we can choose $\bar{f}^{\alpha_x}(t,x,s,u)$ in (22a) instead of $\bar{f}^{\alpha_x}(t,x,s,u^{(1)})$. It is straightforward to see that this restriction has no impact on (25), but the terms involving $\bar{f}^{\alpha_x}$ in (25) vanish. This implies that we cannot use the simple strategy from above to determine the functions φ^{α_x}, φ^{α_u}. Now we have to find functions φ^{α_x} such that

$$d\varphi^{\alpha_x} \wedge dt \overset{S^1_p}{=} \left(\lambda^{\alpha_x}_{\beta_x}\partial^1_{\alpha_z}f^{\beta_x}dz^{\alpha_z} + \lambda^{\alpha_x,1}_{\beta_s}d\varphi^{\alpha_s}\right) \wedge dt \quad (26)$$

is met. The compatibility conditions for this problem are

$$d\left(\partial^1_{\alpha_z}f^{\beta_x}dz^{\alpha_z}\right) \wedge dt \overset{S^1_p}{=} \left(\theta^{\alpha_x}_{\beta_x} \wedge \partial^1_{\alpha_z}f^{\beta_x}dz^{\alpha_z} + \theta^{\alpha_x}_{\alpha_s} \wedge d\varphi^{\alpha_s}\right) \wedge dt \quad (27)$$

for suitable 1-forms $\theta^{\alpha_x}_{\beta_x}$, $\theta^{\alpha_x}_{\alpha_s}$. Now, the theorem of Frobenius [1], [3] tells us that the relation (27) implies locally the existence of the functions φ^{α_x}. Unfortunately, here we have to solve a system of partial differential equations. The next theorem summarizes this result.

Theorem 6. *Let (19) describe a formally integrable system that meets (27) additionally. Then there exists locally a diffeomorphism (22) such that the equivalent system takes the form (21), where the functions $\bar{f}^{\alpha_x}$ are independent of u_1.*

It is worth mentioning that the compatibility conditions are always met by DAE-systems like (2), see [6]. The Theorems 5, 6 deal with the two cases where derivatives of all inputs or no derivatives of the inputs are admitted. An open problem, at least according to the knowledge of the authors, is to minimize the number of inputs, whose derivatives appear in the normal form (21a). Unfortunately, this problem is beyond the scope of this contribution, since it leads to a system of nonlinear partial differential equations. Of course, the case of one input is solved, which follows from a trivial argument.

Finally, we have to show that we can transform the system (16) to the formally integrable one (19), if some regularity conditions are met. We choose $\varphi = \pi_0^1 : J(\mathcal{E}) \to \mathcal{E}$ or in coordinates $\varphi : (t, z, z_1) \to (t, z)$ and apply Theorem 4 and the relations (10), (11) to rewrite (16) as

$$I\left(\mathcal{S}_p^1\right) = \left\langle f^{\alpha_x^0}, f^{\alpha_s^0} \right\rangle \ , \quad f^{\alpha_s^0} \in \pi_0^1 \left(I\left(\mathcal{S}_p^1\right) \right)$$

with $\alpha_x^0 = 1, \ldots, n_x^0$, $\alpha_s^0 = n_x^0 + 1, \ldots, n_e$. We set $r = 0$ and run the following algorithm.

1. Using Theorem 4 and the relations (10), (11), we calculate N^r,

$$N^r = \left\langle f^{\alpha_x^r}, d_1 f^{\alpha_s^r} \right\rangle \cap R\left(\mathcal{F}_p\left(\mathcal{E}\right)\right) \ .$$

   If $N^r = \langle 0 \rangle$ or equivalently (20) is met then we are done.
2. Eliminate a minimal number of functions from the set $\{f^{\alpha_x^r}\}$, $\alpha_x^{r+1} = 1, \ldots, n_x^{r+1}$ such that the set $\{f^{\alpha_x^{r+1}}\}$, $\alpha_x^{r+1} = 1, \ldots, n_x^{r+1}$ meets (20).
3. Determine a minimal number of functions $f^{\alpha_s^{r+1}}$, $\alpha_s^{r+1} = n_x^{r+1} + 1, \ldots, n_e^{r+1}$ such that $\left\langle f^{\alpha_s^{r+1}} \right\rangle = N^r + \left\langle f^{\alpha_s^r} \right\rangle$ is fulfilled. Set $r = r + 1$. If the solution set of $f^{\alpha_x^r} = d_1 f^{\alpha_s^r} = f^{\alpha_s^r} = 0$, $\alpha_x^r = 1, \ldots, n_x^r$, $\alpha_s^r = n_x^r + 1, \ldots, n_e^r$ describes a regular manifold, then go to 1, otherwise stop.

Two facts are worth mentioning. If all manifolds are regular, then this algorithm stops reliably after a finite number of steps. Carrying out the elimination step 1, one has to deal with nonlinear algebraic equations. The general case is a hard problem for any computer algebra system, but two cases admit a significant simplification. Linear algebra is enough to solve this problem for descriptor systems (3), see [6], and one can use Groebner basis [2], if the equations are polynomial ones.

5   Conclusions

This contribution is devoted to the equivalence problem for systems of nonlinear implicit ordinary differential equations, where equivalence means that the solution set of the original system and a given normal form coincide. Since regular submanifolds in certain jet spaces were chosen for the mathematical models of these dynamic systems, one can identify geometric objects like submanifolds with algebraic objects like ideals over certain rings. This combination of methods from differential and algebraic geometry allows us to reduce the equivalence problem to an ideal membership problem. This problem has been solved for the two cases where the derivatives of all inputs or no derivatives of the inputs are admitted, but there is still the open problem to minimize the number of inputs, whose derivatives appear in the normal form. Unfortunately, the latter problem leads to the test, whether a system of nonlinear partial differential equations admits a solution or not. To the best of the authors´ knowledge a test for this problem exists only for the analytic scenario. Therefore, the future research will be devoted to this problem.

References

1. Boothby, W.J. (1986): An Introduction to Differentiable Manifolds and Riemannian Geometry, Academic Press, Orlando.
2. Cox, D., Little, J., O'Shea, D. (1998): Using Algebraic Geometry, Springer, New York.
3. Olver, P.J. (1995): Equivalence, Invariants and Symmetry, Cambridge University Press, Cambridge.
4. Pommaret, J.-F. (2001): Partial Differential Control Theory, Kluwer Academic Publishers, Dordrecht.
5. Saunders, D.J. (1989): The Geometry of Jet Bundles, London Mathematical Society Lecture Note Series 142, Cambridge University Press, Cambridge.
6. Schlacher, K., Kugi, A. (2001): Control of Nonlinear Descriptor Systems, A Computer Algebra Based Approach, Lecture Notes in Control and Information Sciences, **259**, 397–395, Springer, London.

On the Numerical Analysis of Overdetermined Linear Partial Differential Systems*

Marcus Hausdorf and Werner M. Seiler

Lehrstuhl für Mathematik I, Universität Mannheim,
68131 Mannheim, Germany,
{hausdorf,werner.seiler}@math.uni-mannheim.de
http://www.math.uni-mannheim.de/~wms

Abstract. We discuss the use of the formal theory of differential equations in the numerical analysis of general systems of partial differential equations. This theory provides us with a very powerful and natural framework for generalising many ideas from differential algebraic equations to partial differential equations. We study in particular the existence and uniqueness of (formal) solutions, the method of an underlying system, various index concepts and the effect of semi-discretisations.

1 Introduction

The majority of the literature on differential equations is concerned with so-called normal systems or systems in Cauchy–Kovalevskaya form. However, many important systems arising in applications are not of this form. As examples we may mention Maxwell's equations of electrodynamics, the incompressible Navier–Stokes equations of fluid dynamics, the Yang–Mills equations describing the fundamental interactions in particle physics or Einstein's equations of general relativity.

For ordinary differential equations, the importance of non-normal systems has been recognised for about twenty years; one usually speaks of *differential algebraic equations*. Introductions into their theory can be found e. g. in [2,8]. Recently, the extension to partial differential systems has found some interest, see e. g. [5,15]. However, this is non-trivial, as new phenomena appear.

About a century ago the first methods for the analysis of general partial differential systems were designed; by now, a number of different approaches exists. Some of them have already been applied in a numerical context [14,16,20,25]. We use the formal theory [17,23] with its central notion of an *involutive system*. In contrast to our earlier works [10,22,24], we take a more algebraic point of view closely related to the theory of involutive bases [3,6]. For simplicity, we concentrate on linear systems, although many results remain valid in the non-linear case.

* Supported by Deutsche Forschungsgemeinschaft, Landesgraduiertenförderung Baden-Württemberg and INTAS grant 99-1222.

2   Involutive Systems

We consider differential equations in n independent variables $\boldsymbol{x} = (x_1, \ldots, x_n)$ and m dependent variables $\boldsymbol{u}(\boldsymbol{x}) = (u_1(\boldsymbol{x}), \ldots, u_m(\boldsymbol{x}))$, using a multi index notation for the derivatives: $p_{\alpha,\mu} = \partial^{|\mu|} u_\alpha / \partial x_\mu = \partial^{|\mu|} u_\alpha / \partial x_{\mu_1} \cdots \partial x_{\mu_n}$ for each multi index $\mu = [\mu_1, \ldots, \mu_n] \in \mathbb{N}_0^n$. The dependent variable u_α is identified with the derivative $p_{\alpha,[0,\ldots,0]}$. We fix the following ranking $\prec$ on the set of all derivatives: $p_{\alpha,\mu} \prec p_{\beta,\nu}$ if $|\mu| < |\nu|$ or if $|\mu| = |\nu|$ and the rightmost non-vanishing entry in $\nu - \mu$ is negative; if $\mu = \nu$, we set $p_{\alpha,\mu} \prec p_{\beta,\nu}$, if $\alpha < \beta$. The *class* of a derivative is the leftmost non-vanishing entry of its multi index: $\mathrm{cls}(p_{\alpha,\mu}) := \min\{i \mid \mu_i > 0\}$ and $\mathrm{cls}(u_\alpha) := n$. The *order* of $p_{\alpha,\mu}$ is the length of its multi index $|\mu| = \sum \mu_i$. The ranking defined above respects classes: if the derivatives $p_{\alpha,\mu}, p_{\beta,\nu}$ are of the same order but $\mathrm{cls}(p_{\alpha,\mu}) < \mathrm{cls}(p_{\beta,\nu})$, then $p_{\alpha,\mu} \prec p_{\beta,\nu}$. The independent variables x_i with $i \leq \mathrm{cls}(p_{\alpha,\mu})$ are called *multiplicative* for $p_{\alpha,\mu}$, the remaining ones *non-multiplicative*.

We consider a *linear (homogeneous) differential system* $\boldsymbol{\Phi}(\boldsymbol{x}, \boldsymbol{p}) = 0$ where each of the p component functions Φ_τ has the form

$$\Phi_\tau(\boldsymbol{x}, \boldsymbol{p}) = \sum_{\alpha=1}^{m} \sum_{0 \leq |\mu| \leq q} a_{\tau\alpha\mu}(\boldsymbol{x}) p_{\alpha,\mu} = 0. \tag{1}$$

The *leader* of an equation is the highest occurring derivative with respect to the ranking $\prec$. Concepts like class and (non-)multiplicative variables are transfered to equations by defining them in terms of their leaders. We denote by $\beta_j^{(k)}$ the number of equations of order j and class k contained in the system. Equations obtained by differentiating with respect to a (non-)multiplicative variable are called *(non-)multiplicative prolongations*.

We introduce *involutive systems* via a normal form and its properties. For simplicity, we present it only for a first order system. This poses no real restriction, as every system can be transformed into an equivalent first order one. We may thus write down (after some algebraic manipulations and, possibly, coordinate transformations) each system in its *Cartan normal form*:

$$p_{\alpha,n} - \phi_{\alpha,n}(\boldsymbol{x}, \boldsymbol{u}, p_{\gamma,j}, p_{\delta,n}) = 0, \qquad \begin{cases} 1 \leq \alpha \leq \beta_1^{(n)}, \\ 1 \leq j < n, \ \beta_1^{(n)} < \delta \leq m, \end{cases} \tag{2a}$$

$$p_{\alpha,n-1} - \phi_{\alpha,n-1}(\boldsymbol{x}, \boldsymbol{u}, p_{\gamma,j}, p_{\delta,n-1}) = 0, \qquad \begin{cases} 1 \leq \alpha \leq \beta_1^{(n-1)}, \\ 1 \leq j < n-1, \ \beta_1^{(n-1)} < \delta \leq m, \end{cases} \tag{2b}$$

$$\vdots$$

$$p_{\alpha,1} - \phi_{\alpha,1}(\boldsymbol{x}, \boldsymbol{u}, p_{\delta,1}) = 0, \qquad \begin{cases} 1 \leq \alpha \leq \beta_1^{(1)}, \\ \beta_1^{(1)} < \delta \leq m, \end{cases} \tag{2c}$$

$$u_\alpha - \phi_\alpha(\boldsymbol{x}, u_\beta) = 0, \qquad \begin{cases} 1 \leq \alpha \leq \beta_0^{(n)} \leq m;, \\ \beta_0^{(n)} < \beta \leq m. \end{cases} \tag{2d}$$

Here, the functions $\phi_{\alpha,k}$ and ϕ_α are linear in the dependent variables and derivatives. The system is in a triangular form where the subsystem in the first line comprises all equations of class n, the one in the second line all of class $n-1$ and so on. The derivatives on the right hand side have always a class lower than or equal to the one on the left hand side. The subsystem in the last line collects all algebraic constraints; their number is denoted by $\beta_0^{(n)}$. For an involutive system, we must have $0 \leq \beta_0^{(n)} \leq \beta_1^{(1)} \leq \cdots \leq \beta_1^{(n)} \leq m$, so that the subsystems may be empty below a certain class.

We deal with a *normal system* or a system in *Cauchy–Kovalevskaya form*, if all equations are of class n, i.e. if $\beta_1^{(n)} = m$ and and all other $\beta_1^{(k)}$ vanish. An existence and uniqueness theory (in the real analytic category) for such systems is provided by the famous Cauchy–Kovalevskaya theorem [21]. More generally, the system is underdetermined, if and only if $\beta_1^{(n)} < m$. If the system is not underdetermined, then the subsystem (2a) is always normal. In the sequel we will exclusively study such systems.

These purely structural aspects of the normal form (2) do not yet capture that our differential system is involutive; they only express that we have chosen a local representation in triangular form.[1] Any differential system can be brought into such a form. The important point about the Cartan normal form is that involution implies certain relations between prolonged equations. First of all, we require that any *non-multiplicative* prolongation can be written as a linear combination of *multiplicative* ones. Thus, if D_k denotes the total differentiation with respect to x_k, then functions $A_{\beta ij}(\boldsymbol{x})$, $B_{\beta i}(\boldsymbol{x})$, and $C_\beta(\boldsymbol{x})$ must exist such that whenever $1 \leq \ell < k \leq n$

$$
D_k(p_{\alpha,\ell} - \phi_{\alpha,\ell}) = \sum_{i=1}^{k} \sum_{\beta=1}^{\beta_1^{(i)}} \left\{ \sum_{j=1}^{i} A_{\beta ij} D_j(p_{\beta,i} - \phi_{\beta,i}) + B_{\beta i}(p_{\beta,i} - \phi_{\beta,i}) \right\}
$$
$$
+ \sum_{\beta=1}^{\beta_0^{(n)}} C_\beta(u_\beta - \phi_\beta). \tag{3}
$$

Furthermore, no prolongation of the algebraic equations in (2d) may lead to a new equation. This implies the existence of functions $\bar{C}_\beta(\boldsymbol{x})$ such that

$$
\frac{\partial \phi_\alpha}{\partial x_k} - \phi_{\alpha,k} + \sum_{\beta=\beta_0^{(n)}+1}^{m} \frac{\partial \phi_\alpha}{\partial u_\beta} \phi_{\beta,k} = \sum_{\beta=1}^{\beta_0^{(n)}} \bar{C}_\beta(u_\beta - \phi_\beta). \tag{4}
$$

We cannot go into details here, but this second set of conditions has a geometric interpretation and is only partially present in purely algebraic approaches like involutive bases. We will see in Sect. 5 that this geometric ingredient is crucial for obtaining correct index values. Involution comprises *formal integrability*: an involutive system is always consistent and possesses at least formal power series solutions. They can be computed order by order in a straightforward manner via the Cartan normal form.

---

[1] In a more algebraic terminology one may say that the equations are head autoreduced.

The analysis of the *compatibility conditions* will later play an important role. They define (differential) relations between the equations in (2) and correspond to *syzygies* in commutative algebra. We introduce for each equation in our system (2) an inhomogeneity $\epsilon(x)$ on the right hand side. The relations (3) and (4) imply that the inhomogeneous system possesses solutions, if and only if the functions ϵ satisfy the homogeneous linear system

$$\frac{\partial \epsilon_{\alpha,\ell}}{\partial x_k} - \sum_{i=1}^{k} \sum_{\beta=1}^{\beta_1^{(i)}} \left\{ \sum_{j=1}^{i} A_{\beta ij} \frac{\partial \epsilon_{\beta,i}}{\partial x_j} + B_{\beta i} \epsilon_{\beta,i} \right\} - \sum_{\beta=1}^{\beta_0^{(n)}} C_\beta \epsilon_\beta = 0, \tag{5a}$$

$$\frac{\partial \epsilon_\alpha}{\partial x_k} - \epsilon_{\alpha,k} + \sum_{\beta=\beta_0^{(n)}+1}^{m} \frac{\partial \phi_\alpha}{\partial u_\beta} \epsilon_{\beta,k} - \sum_{\beta=1}^{\beta_0^{(n)}} \bar{C}_\beta \epsilon_\beta = 0. \tag{5b}$$

Example 1. In vacuum, *Maxwell's equations* of electrodynamics are

$$E_t = \operatorname{curl} B, \qquad B_t = -\operatorname{curl} E, \tag{6a}$$
$$0 = \operatorname{div} E, \qquad 0 = \operatorname{div} B. \tag{6b}$$

We have six dependent variables $(E_1, E_2, E_3, B_1, B_2, B_3)$, the components of the electric and magnetic field, and four independent variables (x, y, z, t). The system is almost in Cartan normal form; only the Gauß laws (6b) have not been solved for their leaders. (6a) corresponds to (2a); (6b) to (2b). Thus we find $\beta_0^{(4)} = \beta_1^{(1)} = \beta_1^{(2)} = 0$, $\beta_1^{(3)} = 2$ and $\beta_1^{(4)} = 6$.

The relations (3) follow from adding the (non-multiplicative) t-prolongation of (6b) to the (multiplicative) divergence of (6a). Due to the identity div $\circ$ curl $= 0$, they are satisfied and Maxwell's equations are involutive. Introducing inhomogeneities j_e, j_m, ρ_e, ρ_m, we get as compatibility equations the familiar continuity equations relating charge density and current

$$(\rho_e)_t + \operatorname{div} j_e = 0, \qquad (\rho_m)_t + \operatorname{div} j_m = 0. \tag{7}$$

Example 2. Involution is more than the absence of integrability conditions. Consider the following system for two unknown functions $v(x,t)$, $w(x,t)$:

$$v_t = w_x, \qquad w_t = 0, \qquad v_x = 0. \tag{8}$$

It arises, if we transform the second order system $u_{tt} = u_{xx} = 0$ to a first order one. Obviously, no integrability conditions are hidden in this simple system. Thus it is formally integrable, but it is *not* involutive. Differentiating the last equation with respect to the non-multiplicative variable t yields (after a simplification) a new second order equation: $w_{xx} = 0$. Such an equation is called *obstruction to involution*; we will see later why it is rather important.

3   Existence and Uniqueness of Solutions

An important notion in the theory of differential algebraic equation is that of an *underlying equation*. It refers to an unconstrained ordinary differential system such that any solution of the given system is also a solution of it. We may straightforwardly extend this notion to partial differential equations.

Definition 1. *An* underlying system *of a given differential system is any normal system that is solved by any solution of the original system.*

An underlying system exists, if and only if the given system is not underdetermined. It is of course not unique. For an involutive system in Cartan normal form (2), an underlying system is given by the subsystem (2a). Thus an underlying system for Maxwell's equations is (6a). Although (8) is not involutive, the first two equations form an underlying system.

The generalisation of the Cauchy–Kovalevskaya theorem from normal to arbitrary involutive systems is provided by the *Cartan–Kähler theorem*. It guarantees the existence of a unique analytic solution for the system (2) provided the functions $\phi_{\alpha,k}$ and the initial data are analytic. For a proof and for more information on the choice of the initial conditions we refer to [23].

In order to sketch some of the basic ideas behind the proof of the Cartan–Kähler theorem and to demonstrate how they may be used to prove the existence and uniqueness of more general solutions than only analytic ones, we consider a special class of linear systems.

Definition 2. *An involutive differential system with Cartan normal form (2) is* weakly overdetermined, *if* $\beta_1^{(n)} = m$, $\beta_1^{(n-1)} > 0$ *and* $\beta_1^{(k)} = 0$ *else.*

In the sequel, we are only interested in equations that can be interpreted in some sense as evolution equations, as we concentrate on initial value problems. We slightly change our notation and denote the independent variables by $(x_1, \ldots, x_n, t)$, i.e. we have $n + 1$ variables and write x_{n+1} as t. We study linear systems with smooth coefficients of the following form:

$$\boldsymbol{u}_t = \sum_{i=1}^{n} A_i(\boldsymbol{x},t)\boldsymbol{u}_{x_i} + B(\boldsymbol{x},t)\boldsymbol{u}, \tag{9a}$$

$$0 = \sum_{i=1}^{n} C_i(\boldsymbol{x},t)\boldsymbol{u}_{x_i} + D(\boldsymbol{x},t)\boldsymbol{u}. \tag{9b}$$

Here $\boldsymbol{u}$ is again the m-dimensional vector of dependent variables. The square matrices $A_i(\boldsymbol{x},t)$ and $B(\boldsymbol{x},t)$ have m rows and columns; the rectangular matrices $C_i(\boldsymbol{x},t)$ and $D(\boldsymbol{x},t)$ have r rows and m columns. The system is weakly overdetermined, if for at least one i we have rank $C_i(\boldsymbol{x},t) = r$; without loss of generality, we may assume that this is the case for C_n.

A straightforward computation shows that (9) is involutive, if and only if $r \times r$ matrices $H_i(\boldsymbol{x}, t)$, $K(\boldsymbol{x}, t)$ exist such that for all values $1 \leq i, j \leq n$

$$H_i C_j + H_j C_i = C_i A_j + C_j A_i, \tag{10a}$$

$$H_i D + K C_i + \sum_{k=1}^{n} H_k C_{i,x_k} = C_i B + D A_i + \sum_{k=1}^{n} C_k A_{i,x_k} + C_{i,t}, \tag{10b}$$

$$K D + \sum_{k=1}^{n} H_k D_{x_k} = D B + \sum_{k=1}^{n} C_k B_{x_k} + D_t. \tag{10c}$$

Because of our assumption on rank C_n, it is not difficult to see that if such matrices H_i, K exist, they are uniquely determined by (10). We derive the compatibility conditions of the linear system (9) under the assumption that it is involutive. We add on the right hand side of (9a) a "perturbation" $\boldsymbol{\delta}$ and on the right hand side of (9b) a "perturbation" $-\boldsymbol{\epsilon}$. The inhomogeneous system admits (at least formal) solutions, if and only if these functions satisfy the compatibility conditions

$$\boldsymbol{\epsilon}_t - \sum_{i=1}^{n} H_i \boldsymbol{\epsilon}_{x_i} - K \boldsymbol{\epsilon} = \sum_{i=1}^{n} C_i \boldsymbol{\delta}_{x_i} + D \boldsymbol{\delta}. \tag{11}$$

Recall that the system (9a) is *hyperbolic in t-direction*, if for any vector $\boldsymbol{\xi} \in \mathbb{R}^n$ the matrix $A_{\boldsymbol{\xi}}(\boldsymbol{x}, t) := \sum \xi_i A_i(\boldsymbol{x}, t)$ has at every point $(\boldsymbol{x}, t)$ only real eigenvalues and an eigenbasis. It is *strongly hyperbolic*, if there exists for any $A_{\boldsymbol{\xi}}(\boldsymbol{x}, t)$ a symmetric, positive definite matrix $P_{\boldsymbol{\xi}}(\boldsymbol{x}, t)$, a *symmetriser*, depending smoothly on $\boldsymbol{\xi}$, $\boldsymbol{x}$ and t such that $P_{\boldsymbol{\xi}} A_{\boldsymbol{\xi}} - A_{\boldsymbol{\xi}}^t P_{\boldsymbol{\xi}} = 0$. The system (9b) is *elliptic*, if the matrix $C_{\boldsymbol{\xi}}(\boldsymbol{x}, t) := \sum \xi_i C_i(\boldsymbol{x}, t)$ defines for any vector $0 \neq \boldsymbol{\xi} \in \mathbb{R}^n$ a surjective mapping.

Example 3. Maxwell's equations are weakly overdetermined. Their evolution part (6a), corresponding to (9a), forms a symmetric hyperbolic system, whereas the constraints (6b), corresponding to (9b), are elliptic. The compatibility conditions (11) are given by (7) with $\boldsymbol{\epsilon} = (\rho_e, \rho_m)$ and $\boldsymbol{\delta} = (\boldsymbol{j}_e, \boldsymbol{j}_m)$.

A classical result in the theory of hyperbolic systems [12] states that the smooth, strongly hyperbolic system

$$\boldsymbol{u}_t = \sum_{i=1}^{n} A_i(\boldsymbol{x}, t) \boldsymbol{u}_{x_i} + B(\boldsymbol{x}, t) \boldsymbol{u} + \boldsymbol{F}(\boldsymbol{x}, t) \tag{12}$$

possesses for periodic boundary conditions and smooth initial conditions $\boldsymbol{u}(\boldsymbol{x}, 0) = \boldsymbol{f}(\boldsymbol{x})$ a unique smooth solution. Furthermore, at any time $t \in [0, T]$ this solution can be estimated in the weighted Sobolev norm $\|\cdot\|_{P,H^p}$ (the weight depends on the symmetriser $P_{\boldsymbol{\xi}}$, see [12] for details) with $p \geq 0$ by

$$\|\boldsymbol{u}(\cdot, t)\|_{P,H^p} \leq K_p \left[\|\boldsymbol{f}\|_{P,H^p} + \int_0^t \|\boldsymbol{F}(\cdot, \tau)\|_{P,H^p} \, d\tau \right]. \tag{13}$$

We exploit this to obtain an existence and uniqueness theorem for smooth solutions of (9). If the subsystem (9a) is strongly hyperbolic, then the cited theorem ensures that it has a unique smooth solution for arbitrary smooth initial conditions. Let us assume that the initial data $f(x)$ has been chosen such that they satisfy the constraints (9b) for $t = 0$. The question is whether our solution of (9a) satisfies the constraints also for $t > 0$.

The answer to this question lies in the compatibility condition (11). Entering our solution into (9b) yields residuals $\epsilon(x, t)$. As we are dealing with an exact solution of (9a), the residuals ϵ satisfy (11) with $\delta \equiv 0$ so that the system becomes homogeneous. Furthermore, the choice of our initial data implies $\epsilon(x, 0) = 0$, hence $\epsilon \equiv 0$ is obviously a smooth solution. We are done, if we can show that it is the only one.

Thus we need a uniqueness result for (11). If the matrices H_i, K were analytic, we could apply Holmgren's theorem [21]. However, our coefficients are only smooth. Some linear algebra shows that, provided the constraints (9b) are elliptic, the compatibility system (11), viewed as system for ϵ only, inherits the strong hyperbolicity of the underlying system (9a) with a symmetriser Q_ξ determined by $Q_\xi^{-1} = C_\xi P_\xi^{-1} C_\xi^t$ [24]. Thus we may again apply the above cited theorem to prove the needed uniqueness.

These considerations lead to a simple approach to the numerical integration of the overdetermined system (9): we consider the equations (9b) only as constraints on the initial data and otherwise ignore them, i.e. we simply solve numerically the initial value problem for (9a) with initial data satisfying (9b). This integration is a standard problem in numerical analysis.

We must expect a *drift* off the constraints, i.e. the numerical solution ceases to satisfy the constraints (9b). Some discussions of this problem for Maxwell's equations are contained in [11]. For a numerical solution the residuals δ do not vanish but lead to a "forcing term" in the compatibility condition (11). Thus the residuals ϵ do not vanish either. Their growth depends on the properties of (11). In the particular case of a strongly hyperbolic system with elliptic constraints we may estimate the size of the drift via (13):

$$\|\epsilon(\cdot, t)\|_{Q, H^p} \leq K_p \int_0^t \left\| \sum_{i=1}^n C_i \delta_{x_i}(\cdot, \tau) + D\delta(\cdot, \tau) \right\|_{Q, H^p} d\tau \tag{14}$$

This estimate depends not only on the residuals δ but also on their spatial derivatives. While any reasonable numerical method controls the size of δ, it is difficult to control the size of the derivatives.

Example 4. In the case of Maxwell's equations, the estimate (13) depends only on the divergence of δ and not on δ itself. Thus a good numerical method for them should be constructed such that this divergence vanishes.

4   Completion to Involution

If a given system is not involutive, one should *complete* it to an equivalent involutive one. "Equivalent" means here that both systems possess the same (formal) solution space. In the case of a linear system with constant coefficients, the completion is in fact equivalent to the determination of a Gröbner basis for the corresponding polynomial module. We present now a completion algorithm for linear systems that combines algebraic and geometric ideas. More details on the algorithm can be found in [9]; an implementation in the computer algebra system *MuPAD* is briefly described in [1].

In order to formulate our algorithm, we need some further notations. We suppress the zero in (1) and call the remaining left hand side a *row*. The basic idea is to restrict computations as much as possible to non-multiplicative prolongations; multiplicative prolongations are only used for determining a triangular form of the system. In order to indicate which multiplicative prolongations are present, we introduce a *global level* λ, initialised to zero and denoting the number of the current iteration, and assign to each row one or two non-negative integers, its *initial level* and its *phantom level*. The set of such indexed rows is the *skeleton* $\mathcal{S}_\lambda$ of the system and we reproduce the full system $\bar{\mathcal{S}}_\lambda$ by replacing each indexed row by a set of multiplicative prolongations determined by its indices. For a single indexed row $f_{(k)}$ these are

$$\{D_\mu f \mid 0 \leq |\mu| \leq (\lambda - k); \ \forall i > \mathrm{cls}(f) : \mu_i = 0\} \tag{15}$$

and for a double indexed row $f_{(k,l)}$

$$\{D_\mu f \mid (\lambda - l) < |\mu| \leq (\lambda - k); \ \forall i > \mathrm{cls}(f) : \mu_i = 0\}. \tag{16}$$

Without loss of generality, we assume that the given system of order q is already in triangular form. We turn it into the skeleton $\mathcal{S}_0$ by setting the initial level of each row to 0. Furthermore, the numbers $e_{\mathcal{S}_\lambda,i}$ count how many rows of order i are present in the system $\bar{\mathcal{S}}_\lambda$. Since $\mathcal{S}_0 = \bar{\mathcal{S}}_0$, the starting values $e_{\mathcal{S}_0,i}$ are obtained at once. Finally, we initialise the counter $r := 0$. Each iteration step with $\mathcal{S}_\lambda$ being the current skeleton proceeds as follows.

Prolongation. The global level λ is increased by one. This automatically adds all new multiplicative prolongations to the system $\bar{\mathcal{S}}_\lambda$ defined by the skeleton $\mathcal{S}_\lambda$. Concerning the non-multiplicative prolongations, only those single indexed rows $f_{(k)}$ with $k = \lambda - 1$ (these have been created in the last iteration) are computed and become part of the new skeleton with initial level λ. These changes necessitate to recompute the numbers $e_{\mathcal{S}_\lambda,i}$: for each row in $\mathcal{S}_{\lambda-1}$ they are modified as follows:

 - If $f_{(k)}$ is a single indexed row of order t with $k = \lambda - 1$, prolongations in the direction of all independent variables are computed, so $e_{\mathcal{S}_\lambda,t+1} = e_{\mathcal{S}_\lambda,t+1} + n$.
 - If the initial level of $f_{(k)}$ with order t and class j is less than λ, $\binom{(\lambda-k)+j}{j-1}$ new rows of order $t + (\lambda - k) + 1$ enter the system.
 - If $f_{(k,l)}$ is a double indexed row of order t and class j, there are $\binom{(\lambda-k)+j}{j-1}$ new rows of order $t+(\lambda-k)+1$ and $\binom{(\lambda-l)+j}{j-1}$ rows of order $t+(\lambda-l)+1$ are removed.

Triangulation. Next, the skeleton S_λ is algebraically transformed such that $\bar{S}_\lambda$ is in triangular form, i. e. all rows possess different leaders. Starting with the row with the highest ranked leader, one searches through the rows of the skeleton and the allowed multiplicative prolongations for a row with the same leader. If this is the case, reductions are carried out until none are possible. Then the process is repeated with the next leader.

One slight subtlety has to be watched: if a row is reduced which has already produced multiplicative prolongations, removing it would mean to lose all these rows. This is the whole reason behind the introduction of double indexed rows: by adding a phantom level $l = \lambda$ to $f_{(k)}$, we ensure that in future iterations new prolongations are still taken into account and prolongations becoming reducible are removed.

The changes of the $e_{S_\lambda, i}$ are trivial: If a row is reduced, the corresponding value is decreased by 1 and, if the reduction has not yielded zero, the value at the appropriate order is increased by 1.

Involution Analysis. We now check whether we have reached an involutive system. Two conditions must be fulfilled for this:
  - No single indexed rows of order $q + r + 1$ may exist in S_λ.
  - The values of $e_{S_{\lambda-1}, i}$ and $e_{S_\lambda, i}$ coincide for $1 \le i \le q + r$.

If the first condition is violated, we set $r := r + 1$ and proceed with the next iteration. Otherwise, the new value for r is $q' - q$, where q' is the highest order at which a row occurs in S_λ. If the second condition was not satisfied, we continue with the next iteration. Otherwise, we have finished.

Note that the last iteration only serves to check the involution of the system obtained in the last but one step. Obviously, nothing "new" can happen here, since otherwise our termination conditions would not hold.

Given the triangulised skeleton S_λ at the end of each iteration step, one easily determines the numbers $\beta_{q+r}^{(i)}$ of the highest order part of the Cartan normal form of the corresponding system. A single indexed row $f_{(k)}$ of order t and class c contributes to them, if and only if $\lambda - k + t \ge q + r$ and $1 \le i \le c$; for a double indexed row with phantom level l, it is additionally required that $\lambda - l + t < q + r$. The contribution is then given by $B(c - i + 1, q + r - t, 1)$, where $B(n, c, q) = \binom{n-c+q-1}{q-1}$ denotes the number of multi indices of length n, order q and class c.

Example 5. For the system (8) our algorithm needs only two iterations. In the first one the equation $w_{xx} = 0$ is added to the skeleton, as it cannot be reduced by a multiplicative prolongation. After the second iteration the algorithm stops with $r = 1$ as the final system contains a second order equation.

Our algorithm also works for quasi-linear systems provided they remain quasi-linear during the completion. The application to fully non-linear systems leads to a number of serious issues which we cannot discuss here.

Example 6. In order to demonstrate an important difference between our combined algebraic-geometric completion algorithm and purely algebraic methods like involutive bases we analyse the planar pendulum. The dependent variables are the positions (x, y), the velocities (u, v) and a Lagrange multiplier λ. We set all parameters like mass or length to 1. This yields the equations of motion:

$$\dot{x} = u\,, \quad \dot{y} = v\,, \quad \dot{u} = -x\lambda\,, \quad \dot{v} = -y\lambda - 1\,, \quad 0 = x^2 + y^2 + 1\,. \tag{17}$$

Our algorithm needs four iterations. The first two produce the new equations $0 = xu + yv$ and $0 = u^2 + v^2 - y - \lambda$. In the third step, we find no algebraic constraint, but an equation for $\dot{\lambda}$ enters the system. Finally, in the fourth step nothing further happens, so we have arrived at an involutive system.

A purely algebraic approach would need one iteration less. It would not determine an equation for the derivative $\dot{\lambda}$, as the system contains already an algebraic equation for λ. The next section shows that this additional iteration of geometric origin is important for obtaining the correct index values.

5   Indices for Differential Equations

Indices play an important role for differential algebraic equations. They serve as indicators for the difficulties one has to expect in the numerical integration of the given system: the higher the index, the more problems arise. Typical problems are that classical methods exhibit suddenly a lower order of convergence or the already mentioned drift off the constraint manifold.

The literature abounds with definitions of indices; a partial survey may be found in [4]. A recent trend is their extension to partial differential equations. One often speaks of *partial differential algebraic equations*, but this terminology is misleading, as in many cases (like Maxwell's equations) such systems do not contain any algebraic equations: the non-normality is due to the presence of equations of lower class and not of lower order.

We distinguish two classes of indices. *Differentiation indices* count the number of prolongations needed until the system possesses certain properties. In the case of "the" differentiation index, the property is that an underlying system has been found. *Perturbation indices* are based on estimates on the difference of solutions of the original system and of a perturbation of it.

Two differentiation indices follow naturally from our completion algorithm. Recall that it produces a sequence[2] of linear systems in triangular form $\bar{\mathcal{S}}_0 \longrightarrow \bar{\mathcal{S}}_1 \longrightarrow \cdots \longrightarrow \bar{\mathcal{S}}_{\lambda_f}$.

Definition 3. *The* determinacy index ν_D *of a differential system in n independent and m dependent variables is the first λ such that we have $\beta_{q+r}^{(n)} = m$ for the system $\bar{\mathcal{S}}_\lambda$ and the corresponding value of the counter r. The* involution index ν_I *is the first λ for which $\bar{\mathcal{S}}_\lambda$ is involutive and thus $\lambda_f - 1$.*

The determinacy index corresponds to "the" differentiation index. Its name reflects that $\bar{\mathcal{S}}_\lambda$ is not an underdetermined system. ν_D has no finite value for

---

[2] The systems are taken at the *end* of each iteration step.

an underdetermined system, as then $\beta_{q+r}^{(n)} < m$ for any value of λ. In contrast to the involution index, the determinacy index cannot be used as a basis of an existence theory for solutions, as it is does not require the full completion. Thus it is well possible that the system is in fact inconsistent. The involution index is equivalent to the strangeness index [13]; one also obtains the same result with the geometric approaches presented in [18,19].

Example 7. It follows from Ex. 6 that both the determinacy and the involution index of the pendulum are $\nu_I = \nu_D = 3$, as an underlying system is obtained only after an equation for $\dot{\lambda}$ is present. A purely algebraic completion would yield too low values for the indices.

In system (8), we find different values for ν_I and ν_D. The determinacy index is obviously 0, as the first two equations form already an underlying system. But $\lambda_f = 2$ and hence $\nu_I = 1$.

The perturbation index was introduced by Hairer et al. [7] for differential algebraic equations and extended to partial differential equations by Campbell and Marszalek [5]. We assume that the system lives on a compact domain $\Omega \subset \mathbb{R}^n$ and choose a norm $\| \cdot \|$ on some function space $\mathcal{F}$ over Ω which is usually either the maximum norm $\| \cdot \|_{L^\infty}$ or the uniform norm $\| \cdot \|_{L^1}$. In addition, we define on $\mathcal{F}$ for each integer $k > 0$ a kind of Sobolev norm

$$\| f \|_k = \sum_{0 \leq |\mu| \leq k} \left\| \frac{\partial^{|\mu|} f}{\partial x_\mu} \right\|, \tag{18}$$

i. e. we sum the norms of f and its partial derivatives up to order k. Of course, the function space $\mathcal{F}$ must be such that these norms make sense on it.

Partial differential equations are usually accompanied by initial or boundary conditions. In order to accommodate for this, we take the following simple approach. Let $\Omega' \subseteq \partial\Omega$ be a subdomain of the boundary of Ω and introduce on the restriction $\mathcal{F}'$ of $\mathcal{F}$ to Ω' similar norms denoted by $\| \cdot \|_k'$. The conditions are assumed to be of the form $\boldsymbol{\Psi}(\boldsymbol{x}, \boldsymbol{p})\big|_{\Omega'} = 0$. This comprises most kinds of initial or boundary conditions in applications. The highest derivative in $\boldsymbol{\Psi}$ determines the order ℓ of the conditions.

We consider again a linear partial differential system with p equations (1). We do not require that $p = m$ or that the Jacobian $\partial\boldsymbol{\Phi}/\partial\boldsymbol{p}$ has any special properties. Assume that we are given a smooth solution $\boldsymbol{u}(\boldsymbol{x})$ defined on the whole domain Ω and satisfying our initial or boundary conditions on Ω'. We compare it with solutions of the perturbed equation $\boldsymbol{\Phi}(\boldsymbol{x}, \boldsymbol{p}) = \boldsymbol{\delta}(\boldsymbol{x})$ with a smooth right hand side $\boldsymbol{\delta}$.

Definition 4. *Let $\boldsymbol{u}(\boldsymbol{x})$ be a smooth solution satisfying some initial or boundary conditions of order ℓ on Ω'. The system has* perturbation index ν_P *along this solution, if ν_P is the smallest integer such that for any solution $\tilde{\boldsymbol{u}}(\boldsymbol{x})$ of the perturbed equation at every point $\boldsymbol{x} \in \Omega$ the estimate*

$$|\boldsymbol{u}(\boldsymbol{x}) - \tilde{\boldsymbol{u}}(\boldsymbol{x})| \leq C \big(\| \boldsymbol{f} - \tilde{\boldsymbol{f}} \|_\ell' + \| \boldsymbol{\delta} \|_{\nu_P - 1} \big) \tag{19}$$

holds, *whenever the right hand side is sufficiently small. Here f and $\tilde{f}$ represent the restrictions of the solutions u and $\tilde{u}$, respectively, to Ω'. The constant C may depend only on the domains Ω, Ω' and on the function Φ.*

In the case of initial value problems for differential algebraic equations this definition coincides with the one given by Hairer et al. The term $\|f - \tilde{f}\|'_k$ takes there the simple form $|u(x_0) - \tilde{u}(x_0)|$ where $\Omega' = \{x_0\}$. For partial differential systems our definition is almost identical with the one given by Campbell and Marszalek; the only difference is that we do not include the order ℓ of the initial or boundary conditions in the index.

Whereas it is not so clear why differentiation indices should indicate the difficulty of the numerical integration of the system, this is rather obvious for the perturbation index. If we take for $\tilde{u}(x)$ an approximate solution, we may interpret $\delta(x)$ as the residual obtained by entering it into the system. The estimate (19) tells us that for an equation with $\nu_P > 1$ it does not suffice to keep this residual as small as possible, since also some of its derivatives enter it. Strictly speaking, this implies that the considered initial or boundary value problem is ill-posed in the sense of Hadamard!

Obviously, it is much harder to obtain estimates of the form (19) than to compute a differentiation index, but the perturbation index contains more useful information. Hence there is much interest in relating the two concepts.

Conjecture 1. For any linear differential system $\nu_D \leq \nu_P \leq \nu_D + 1$.

For differential algebraic equations, a rigorous proof of this conjecture can be found in [10].[3] One first shows that a normal ordinary differential system has the perturbation index $\nu_P \leq 1$ (a consequence of Gronwall's Lemma). For a general system, one follows the completion algorithm until an underlying system is reached. One can prove that its right hand side contains derivatives of the perturbations of order ν_D. This yields the estimate above.

For partial differential systems, the situation is much more complicated, as the perturbation index will depend in general on the chosen norm. A simple case arises, if the underlying equation can be treated with semi-group theory [21]. Then we may consider our overdetermined system as a differential algebraic equation on an infinite-dimensional Banach space and the same argument as above may be applied. Thus we obtain the same estimate.

6   Semi-discretisation

For simplicity, we restrict to first order linear homogeneous systems with constant coefficients in $n + 1$ independent variables of the form

$$\sum_{i=1}^{n+1} M_i u_{x_i} + N u = 0. \qquad (20)$$

---

[3] In that article non-linear systems are treated where one must introduce *perturbed* differentiation indices. For linear systems they are identical with the above defined indices.

The matrices M_i and N are here completely arbitrary. Now we discretise the derivatives with respect to n of the independent variables by some finite difference method. This yields a differential algebraic equation in the one remaining independent variable. We are interested in the relation between the involution indices of the original partial differential system and the obtained differential algebraic equation, respectively.

Instead of (20) we complete to involution a *perturbed* system with a generic right hand side $\gamma(x)$. For a linear constant coefficients system, the perturbation does not affect the completion. It only serves as a convenient mean of "bookkeeping" which prolongations are needed during the completion.

Definition 5. *The involution index ν_ℓ in direction x_ℓ of the system (20) is the maximal number of differentiations with respect to x_ℓ in a γ-derivative contained in the involutive completion of the perturbed system.*

For the semi-discretisation we proceed as follows. Assume that x_ℓ is the "surviving" independent variable; in other words, afterwards we are dealing with an ordinary differential system containing only x_ℓ-derivatives. In order to simplify the notation, we rewrite (20). We denote x_ℓ by t and renumber the remaining independent variables as x_i with $1 \leq i \leq n$. Then we solve as many equations as possible for a t-derivative, as we consider these as the derivatives of highest class. This yields a system of the form

$$Eu_t = \sum_{i=1}^n A_i u_{x_i} + Bu + \delta, \qquad 0 = \sum_{i=1}^n C_i u_{x_i} + Du + \epsilon. \tag{21}$$

Here we assume that the (not necessarily square) matrix E is of maximal rank, i. e. every equation in the first subsystem really depends on a t-derivative. But we do not pose any restrictions on the ranks of the matrices C_i.

In (21) we have introduced perturbations δ and ϵ which are related to the original perturbations γ by a linear transformation with constant coefficients. As we are only interested in the number of differentiations applied to them during the completion, such a transformation has no effect.

We discretise the "spatial" derivatives, i. e. those with respect to the x_i, on a grid where the points x_k are labelled by integer vectors $k = [k_1, \ldots, k_n]$. $u_k(t)$ denotes the value of the function u at the point x_k at time t. We approximate in (21) the spatial derivative $u_{x_i}(x_k, t)$ by the finite difference

$$\delta_i u_k(t) = \sum_{\ell_i = -a_i}^{b_i} \alpha_{\ell_i}^{(i)} u_{[k_1, \ldots, k_i + \ell_i, \ldots, k_n]}(t) \tag{22}$$

with some real coefficients $\alpha_{\ell_i}^{(i)}$. Thus different discretisations are allowed for different values of i, but u_{x_i} is everywhere discretised in the same way. Entering the approximations (22) into (21) yields

$$E\dot{u}_k = \sum_{i=1}^n A_i \delta_i u_k + Bu_k, \qquad 0 = \sum_{i=1}^n C_i \delta_i u_k + Du_k. \tag{23}$$

Theorem 1. *The involution index of the differential algebraic equation (23) obtained in the described semi-discretisation of (20) with respect to x_ℓ is ν_ℓ.*

The proof is given in [24]; we outline here only the basic idea. We compare what happens during the completion of the perturbed system (21) and the differential algebraic equation (23). One can show that each new equation in the discretised system corresponds to either an integrability condition or an obstruction to involution in the original system. Further examination shows that this only happens for prolongations in t-direction; mere spatial differentiations do not lead to new equations in the differential algebraic system. Since ν_l counts only the number of differentiations with respect to t, the involution index of (23) must coincide with ν_l.

This result is somewhat surprising. While one surely expects that integrability conditions of the original partial differential system induce integrability conditions in the differential algebraic equation obtained by semi-discretisation, Theorem 1 says that obstructions to involution also turn into integrability conditions upon semi-discretisation. Thus even if the original partial differential system is formally integrable, the differential algebraic equation might contain integrability conditions.

Example 8. A semi-discretisation with backward differences for the spatial derivatives of the linear system (8) leads to the differential algebraic equation

$$\dot{v}_n = (w_n - w_{n-1})/\Delta x\,, \qquad \dot{w}_n = 0\,, \qquad v_n - v_{n-1} = 0\,. \tag{24}$$

It hides the integrability condition $w_n - 2w_{n-1} + w_{n-2} = 0$ obviously representing a discretisation of the obstruction to involution $w_{xx} = 0$ by centred differences. The involution index of (8) in direction t is one which is also the involution index of (24) in agreement with Theorem 1.

For weakly overdetermined systems, Theorem 1 can be strengthened. For them the differential algebraic equation obtained by a semi-discretisation with respect to t is formally integrable, *if and only if* the original partial differential system is involutive [24]. This result does not only hold for semi-discretisations by finite differences but also for spectral methods: assuming periodic boundary conditions, we make the Fourier ansatz

$$u(x,t) \approx \sum_{k \in \mathcal{G}} [a_k(t) + i b_k(t)]\, e^{ikx}\,. \tag{25}$$

Here $\mathcal{G}$ is a finite grid of wave vectors, and we split the complex Fourier coefficients into their real and imaginary part. Entering this ansatz into (9) yields the following differential algebraic equation where the vectors A and C consist of the matrices A_i and C_i, respectively:

$$\begin{pmatrix} \dot{a}_k \\ \dot{b}_k \end{pmatrix} = \begin{pmatrix} B & -k \cdot A \\ k \cdot A & B \end{pmatrix} \begin{pmatrix} a_k \\ b_k \end{pmatrix}, \qquad 0 = \begin{pmatrix} D & -k \cdot C \\ k \cdot C & D \end{pmatrix} \begin{pmatrix} a_k \\ b_k \end{pmatrix}\,. \tag{26}$$

One can show that this differential algebraic equation is formally integrable, *if and only if* the weakly overdetermined system (9) is involutive [24].

References

1. J. Belanger, M. Hausdorf, and W.M. Seiler. A *MuPAD* library for differential equations. In V.G. Ghanza, E.W. Mayr, and E.V. Vorozhtsov, editors, *Computer Algebra in Scientific Computing — CASC 2001*, pages 25–42. Springer-Verlag, Berlin, 2001.
2. K.E. Brenan, S.L. Campbell, and L.R. Petzold. *Numerical Solution of Initial-Value Problems in Differential-Algebraic Equations*. Classics in Applied Mathematics 14. SIAM, Philadelphia, 1996.
3. J. Calmet, M. Hausdorf, and W.M. Seiler. A constructive introduction to involution. In R. Akerkar, editor, *Proc. Int. Symp. Applications of Computer Algebra — ISACA 2000*, pages 33–50. Allied Publishers, New Delhi, 2001.
4. S.L. Campbell and C.W. Gear. The index of general nonlinear DAEs. *Numer. Math.*, 72:173–196, 1995.
5. S.L. Campbell and W. Marszalek. The index of an infinite dimensional implicit system. *Math. Model. Syst.*, 1:1–25, 1996.
6. V.P. Gerdt and Yu.A. Blinkov. Involutive bases of polynomial ideals. *Math. Comp. Simul.*, 45:519–542, 1998.
7. E. Hairer, C. Lubich, and M. Roche. *The Numerical Solution of Differential-Algebraic Equations by Runge-Kutta Methods*. Lecture Notes in Mathematics 1409. Springer-Verlag, Berlin, 1989.
8. E. Hairer and G. Wanner. *Solving Ordinary Differential Equations II*. Springer Series in Computational Mathematics 14. Springer-Verlag, Berlin, 1996.
9. M. Hausdorf and W.M. Seiler. An efficient algebraic algorithm for the geometric completion to involution. *Appl. Alg. Eng. Comm. Comp.*, 13:163–207, 2002.
10. M. Hausdorf and W.M. Seiler. Perturbation versus differentiation indices. In V.G. Ghanza, E.W. Mayr, and E.V. Vorozhtsov, editors, *Computer Algebra in Scientific Computing — CASC 2001*, pages 323–337. Springer-Verlag, Berlin, 2001.
11. B.N. Jiang, J. Wu, and L.A. Povelli. The origin of spurious solutions in computational electrodynamics. *J. Comp. Phys.*, 125:104–123, 1996.
12. H.-O. Kreiss and J. Lorenz. *Initial-Boundary Value Problems and the Navier-Stokes Equations*. Pure and Applied Mathematics 136. Academic Press, Boston, 1989.
13. P. Kunkel and V. Mehrmann. Canonical forms for linear differential-algebraic equations with variable coefficients. *J. Comp. Appl. Math.*, 56:225–251, 1994.
14. G. Le Vey. Some remarks on solvability and various indices for implicit differential equations. *Num. Algo.*, 19:127–145, 1998.
15. W. Lucht, K. Strehmel, and C. Eichler-Liebenow. Indexes and special discretization methods for linear partial differential algebraic equations. *BIT*, 39:484–512, 1999.
16. Y.O. Macutan and G. Thomas. Theory of formal integrability and DAEs: Effective computations. *Num. Algo.*, 19:147–157, 1998.
17. J.F. Pommaret. *Systems of Partial Differential Equations and Lie Pseudogroups*. Gordon & Breach, London, 1978.
18. P.J. Rabier and W.C. Rheinboldt. A geometric treatment of implicit differential algebraic equations. *J. Diff. Eq.*, 109:110–146, 1994.
19. S. Reich. On an existence and uniqueness theory for nonlinear differential-algebraic equations. *Circ. Sys. Sig. Proc.*, 10:343–359, 1991.
20. G.J. Reid, P. Lin, and A.D. Wittkopf. Differential elimination-completion algorithms for DAE and PDAE. *Stud. Appl. Math.*, 106:1–45, 2001.

21. M. Renardy and R.C. Rogers. *An Introduction to Partial Differential Equations*. Texts in Applied Mathematics 13. Springer-Verlag, New York, 1993.
22. W.M. Seiler. Indices and solvability for general systems of differential equations. In V.G. Ghanza, E.W. Mayr, and E.V. Vorozhtsov, editors, *Computer Algebra in Scientific Computing — CASC 1999*, pages 365–385. Springer-Verlag, Berlin, 1999.
23. W.M. Seiler. Involution – the formal theory of differential equations and its applications in computer algebra and numerical analysis. Habilitation thesis, Dept. of Mathematics, Universität Mannheim, 2001.
24. W.M. Seiler. Completion to involution and semi-discretisations. *Appl. Num. Math.*, 42:437–451, 2002.
25. J. Tuomela and T. Arponen. On the numerical solution of involutive ordinary differential systems. *IMA J. Num. Anal.*, 20:561–599, 2000.

Dynamical Aspects
of Involutive Bases Computations

Ralf Hemmecke

Research Institute for Symbolic Computation
Johannes Kepler University
A-4040 Linz, Austria
ralf@hemmecke.de

Abstract. The article is a contribution to a more efficient computation
of involutive bases. We present an algorithm which computes a 'sliced
division'. A sliced division is an admissible partial division in the sense
of Apel. Admissibility requires a certain order on the terms. Instead of
ordering the terms in advance, our algorithm additionally returns such
an order for which the computed sliced division is admissible. Our algorithm gives rise to a whole class of sliced divisions since there is some
freedom to choose certain elements in the course of its run. We show that
each sliced division refines the Thomas division and thus leads to ter-
minating completion algorithms for the computation of involutive bases.
A sliced division is such that its cones 'cover' a relatively 'big' part of
the term monoid generated by the given terms. The number of prolongations that must be considered during the involutive basis algorithm is
tightly connected to the dimensions and number of the cones. By some
computer experiments, we show how this new division can be fruitful for
the involutive basis algorithm.
We generalise the sliced division algorithm so that it can be seen as
an algorithm which is parameterised by two choice functions and give
particular choice functions for the computation of the classical divisions
of Janet, Pommaret, and Thomas.

1 Introduction

Involutive bases have their origin in the Riquier-Janet Theory of partial differential equations at the beginning of the 20th century. In Riquier's Existence
and Uniqueness Theorem, a separation of variables into multiplicative and non-
multiplicative ones is essential. Janet was the first to give an algorithm to transform a system of partial differential equations (PDEs) into a canonical form
(passive system) required for Riquier's Theorem. He also described a particular
separation of variables. Zharkov and Blinkov [8] translated Janet's algorithm
into a polynomial context where its form is strikingly similar to Buchberger's
well known Gröbner basis algorithm [3]. In fact, the polynomial Janet algorithm
computes a Gröbner basis, though not a reduced one, but it does this in a
slightly different fashion. Basically, it is also a critical element/completion algorithm, however, instead of S-polynomials, it considers *prolongations* as critical
elements, and the usual polynomial reduction is applied in a restricted sense.

F. Winkler and U. Langer (Eds.): SNSC 2001, LNCS 2630, pp. 168–182, 2003.

Later Apel [1] and Gerdt/Blinkov [4] introduced the notion of involutive division in order to generalise the separation of variables that could be used in Janet's algorithm. Unfortunately, two slightly differing notions for involutive division are now available and it remains to be seen which one is more appropriate. Our paper follows the ideas of Apel.

The article is structured as follows. We first recall the basic definitions. In Section 5 we present an algorithm to compute a sliced division and prove termination and correctness of this algorithm. In Section 6 we show that each sliced division refines the Thomas division on the same set of terms. Section 7 deals with a generalisation of the sliced division algorithm. We show that every generalised sliced division is an admissible partial division. We present the classical partial divisions as generalised sliced divisions. In Section 8 we show by some computer experiments how the involutive basis algorithm can benefit from using the sliced division sub-algorithm.

Parts of the material presented in this article have already been given in a talk at the ACA 2001 conference, Albuquerque, New Mexico.

2   Preliminaries

For the whole paper let $X = \{x_1, \ldots, x_n\}$ be a set of indeterminates and $\mathbb{K}[X]$ be the set of polynomials in X over the field $\mathbb{K}$ of coefficients. By T we denote the set of power products of $\mathbb{K}[X]$. Since we are not interested in this paper in computations with respect to different term orders, we fix a certain admissible term order. We denote for $0 \neq f \in \mathbb{K}[X]$ by $\operatorname{supp} f \subset T$ the set of terms of f that appear with a non-zero coefficient and by $\operatorname{lt} f$ the biggest term of $\operatorname{supp} f$ with respect to the term order. Furthermore, if $F \subseteq \mathbb{K}[X]$, let $\operatorname{lt}(F) := \{\operatorname{lt} f \mid 0 \neq f \in F\}$.

By $\deg_x t$ we denote the degree of $t \in T$ in $x \in X$, whereas $\deg t$ is the total degree of t.

Definition 1. *For any set $U \subseteq T$ we define $\langle U \rangle$ as the monoid generated by U, i. e., $\langle U \rangle := \{t \in T \mid \exists u_1, \ldots, u_m \in U : t = u_1 u_2 \cdots u_m\}$. By definition, the product of $m = 0$ elements is 1, in particular, $\langle \emptyset \rangle = \{1\}$.*

Definition 2. *Let $Y \subseteq X$, $t \in T$, $U \subseteq T$. By tU we denote the set $\{tu \mid u \in U\}$. The set $t\langle Y \rangle$ is called a **cone**, t its **base point**, and the elements of Y its **multipliers**. The number of elements of Y is called the **dimension** of the cone.*

Definition 3. *The lexicographical order on T is defined by $u <_{\text{lex}} v$ iff there exists an $i \in \{1, \ldots, n\}$ such that $\deg_{x_j} u = \deg_{x_j} v$ for all $1 \leq j < i$ and $\deg_{x_i} u < \deg_{x_i} v$. The term u is inverse lexicographically smaller than v (denoted $u <_{\text{invlex}} v$) iff $v <_{\text{lex}} u$.*

3 Partial Divisions

After Zharkov and Blinkov [8] brought the Janet algorithm to a polynomial context (see Chapter 8), Apel and Gerdt/Blinkov realised that the particular choice of multiplicative and non-multiplicative variables that was used by Janet in his algorithm could be generalised. They came up with the concept of involutive division [1], [4], and is called admissible partial division in [2].

In order to get an intuitive picture of a partial division, one should read the ordinary divisibility relation $u \mid v$ of two terms $u, v \in T$ as $v \in u\langle X \rangle$. That is to say: to each term $t \in T$ we assign the cone of all its multiples. If we restrict these cones to sub-cones and define divisibility in terms of these restricted cones, we arrive at the concept of partial division. The word 'partial' should suggest that two terms which are ordinarily divisible might not be divisible with respect to some partial division. The idea behind 'admissible partial division' is that cones should be disjoint whereas 'complete' suggests that a certain region is completely covered by disjoint cones.

Apel and Gerdt/Blinkov introduced slightly different notions of involutive division. Since this paper is based on Apel's version, the definitions below mainly follow the definitions in [2].

Definition 4. *Let $(Y_t)_{t \in T}$ be a family of subsets of X. We call the family $\mathcal{M} = (t\langle Y_t \rangle)_{t \in T}$ a **partial division**. If $v \in u\langle Y_u \rangle$, then v is called an $\mathcal{M}$-**multiple** of u, and u is an $\mathcal{M}$-**divisor** of v.*

*Let $U \subseteq T$. Each family $\mathcal{N} = (t\langle Y_t \rangle)_{t \in U}$ **induces** a partial division $\mathcal{M} = (t\langle Y_t \rangle)_{t \in T}$ by setting $Y_t := X$ for $t \notin U$. We will also call $\mathcal{N}$ a partial division and mean its induced partial division.*

*Let $\sqsubset$ be a total order on $U \subseteq T$. A partial division $\mathcal{M} = (t\langle Y_t \rangle)_{t \in T}$ is called **admissible on** $(U, \sqsubset)$ if for all $u, v \in U$ with $u \sqsubset v$, one of the conditions*

$$u\langle X \rangle \cap v\langle Y_v \rangle = \emptyset \ or \tag{1}$$

$$u\langle Y_u \rangle \subset v\langle Y_v \rangle \tag{2}$$

*holds. $\mathcal{M}$ is **admissible on** U if there exists a total order $\sqsubset$ on U such that $\mathcal{M}$ is admissible on $(U, \sqsubset)$.*

*The set U will be called **complete** if there exists an admissible partial division $(t\langle Y_t \rangle)_{t \in T}$ on U such that*

$$\bigcup_{t \in U} t\langle Y_t \rangle = \bigcup_{t \in U} t\langle X \rangle.$$

Remark 1. The first condition of admissibility involves the set X. So it would have been more precise to speak of X-admissibility. Because of the following observation, this additional qualification is not needed. Suppose a partial division $(u\langle Y_u \rangle)_{u \in U}$ is X-admissible on $(U, \sqsubset)$. Then it is also X'-admissible for all $X' \supseteq \bigcup_{u \in U} Y_u$.

Let $u, v \in U$ with $u \sqsubset v$ and suppose the relation

$$u\langle X \rangle \cap v\langle Y_v \rangle = \emptyset \tag{3}$$

holds. Then also

$$u\langle X'\rangle \cap v\langle Y_v\rangle = \emptyset \qquad (4)$$

trivially holds for all $X' \subset X$. From (3) it follows that $\deg_x u > \deg_x v$ for some $x \in X \setminus Y_v$. Therefore, (4) holds for $X' \supset X$, as well.

The following definition introduces a partial order on the set of all partial divisions.

Definition 5. *Let* $\mathcal{M} = (M_t)_{t\in T}$ *and* $\mathcal{N} = (N_t)_{t\in T}$ *be two partial divisions. If* $M_t \subseteq N_t$ *for all* $t \in T$ *we say that* $\mathcal{N}$ **refines** $\mathcal{M}$. *Let* $U \subseteq T$. $\mathcal{M}$ *and* $\mathcal{N}$ *are* U**-equivalent** *if* $M_u = N_u$ *for all* $u \in U$ *(denoted* $\mathcal{M} \equiv_U \mathcal{N}$*).*

4   Classical Partial Divisions

The following three divisions are called 'classical' in the literature. We include them here as examples of admissible partial divisions. Whereas the Thomas division is important for the termination of the involutive basis algorithm, we will refer to Janet and Pommaret division in Section 7.

Janet [5] used a particular partial division which is defined as follows.

Example 1 (Janet Division). Let $U \subseteq T$ be a finite set of power products. Define $Y_t = X$ for all $t \notin U$ and

$$Y_t := \left\{ x_i \in X \,\middle|\, \nexists u \subset U : \deg_{x_i} u > \deg_{x_i} t \wedge \forall 1 \le j < i : \deg_{x_j} u = \deg_{x_j} t \right\}$$

for all $t \in U$. The division $(t\langle Y_t\rangle)_{t\in T}$ is called **Janet division on** U.

The Thomas division [7] is mainly of theoretical interest. Apel showed in [1] that the involutive basis algorithm will terminate if one chooses a partial division which refines the Thomas division.

Example 2 (Thomas Division). Let $U \subseteq T$ be a set of power products. Define $Y_t = X$ for all $t \notin U$ and $Y_t := \{x \in X \mid \forall u \in U : \deg_x u \le \deg_x t\}$ for all $t \in U$. The division $(t\langle Y_t\rangle)_{t\in T}$ is called **Thomas division on** U.

Another division is due to Pommaret [6].

Example 3 (Pommaret Division). Let $Y_1 := X$. If $t \in \langle x_1, \ldots, x_k \rangle$ and $\deg_{x_k} t > 0$ for some $k \in \{1, \ldots, n\}$ then let $Y_t := \{x_k, \ldots, x_n\}$. The division $(t\langle Y_t\rangle)_{t\in T}$ is called **Pommaret division.**

All of the above divisions are admissible with respect to the inverse lexicographical order on T (cf. [1]).

As one easily recognises, Janet and Pommaret divisions can be defined for any permutation of the variables.

Gerdt/Blinkov [4] gave further divisions whereas Apel [1] considers the whole lattice of partial divisions and argues to prefer a maximal (with respect to the refinement relation, cf. Definition 5) admissible division in each step of the involutive basis algorithm.

5   Sliced Division Algorithm

In this section we present an algorithm which not only computes an admissible partial division for a given set U of terms but also a particular order on U so that the resulting division is admissible on U with respect to that order. This approach is different from Apel's sub-maximal refinement of a division with respect to some given order.

The basic objective for the sliced division algorithm was to find an admissible partial division on a set U of terms such that the cones attached to elements of U cover a relatively 'big' part of the monoid ideal generated by U. This is achieved by assigning all variables as multiplicative to a term of minimal total degree and then applying such an assignment recursively (with fewer variables) to the remaining terms.

Algorithm 1 SLICEDDIVISION

Call: $M := $ SLICEDDIVISION(Y, U)

Input: $Y \subseteq X$ finite set of indeterminates
$U = \{u_1, \ldots, u_r\} \subseteq w\langle Y \rangle$ for some $w \in \langle X \setminus Y \rangle$ where $r = \text{card}\, U$

Output: A list $[v_1\langle Y_{v_1}\rangle, \ldots, v_r\langle Y_{v_r}\rangle]$ of Y-cones such that $U = \{v_1, \ldots, v_r\}$ and $\forall i, j = 1, \ldots, r : i < j \implies (v_i\langle Y\rangle \cap v_j\langle Y_{v_j}\rangle = \emptyset \vee v_i\langle Y_{v_i}\rangle \subset v_j\langle Y_{v_j}\rangle)$

1: $L := []$ *# empty list*
2: <u>if</u> $U = \emptyset$ <u>then</u> <u>return</u> L
3: <u>if</u> $Y = \emptyset$ <u>then</u> <u>return</u> $[w\langle Y\rangle]$ *# Note that here $U = w\langle Y\rangle = \{w\}$ if $Y = \emptyset$.*
4: Choose $m \in U$ with $\deg m = \min\{\deg u \mid u \in U\}$
5: $C := (U \cap m\langle Y\rangle) \setminus \{m\}$; $L_C := $ SLICEDDIVISION(Y, C)
6: $V := U \setminus m\langle Y\rangle$ *# same as $V := U \setminus (C \cup \{m\})$*
7: <u>while</u> $V \neq \emptyset$ <u>do</u>
8:     Choose $x \in Y$ such that $\deg_x(\gcd V) < \deg_x m$
9:     $d := \deg_x(\gcd V)$
10:    $S := \{v \in V \mid \deg_x v = d\}$
11:    $V := V \setminus S$
12:    $L_S := $ SLICEDDIVISION$(Y \setminus \{x\}, S)$
13:    $L := $ CONCAT(L_S, L)
14: <u>return</u> CONCAT$(L_C, [m\langle Y\rangle], L)$

5.1   Some Comments to the Algorithm

Informally speaking, one should associate the set U of terms with its monoid ideal represented in some $\mathbb{N}^n$. As the first step, some element m of minimal total degree is taken from U and gets the full-dimensional cone assigned to it (a shifted copy of $\mathbb{N}^n$). All elements which are divisible by m are treated recursively. Any of the other elements will get a cone which is at most of dimension $n - 1$. By taking hyperplanes perpendicular to the direction of some variable x (chosen in each iteration step), slices are cut from the 'outermost' part of the ideal until the cone of m is reached. The elements inside such a slice correspond to elements of V (cf. Algorithm 1) having minimal degree in x. Such slices are of dimension $n - 1$ and thus can be treated recursively.

Definition 6. *Let $U \subseteq T$ be a finite set of terms and $[u_1\langle Y_{u_1}\rangle, \ldots, u_r\langle Y_{u_r}\rangle]$ be an output of* SLICEDDIVISION(X, U). $(u\langle Y_u\rangle)_{u \in U}$ *is called a **sliced division on** U.*

The condition $U \subseteq w\langle Y\rangle$, $w \in \langle X \setminus Y\rangle$ in the specification of Algorithm 1 just states that elements of U agree in the 'non-multiplicative part'.

5.2   Termination and Correctness

Theorem 1. *The algorithm* SLICEDDIVISION *(page 172) terminates and meets its specification.*

Proof. One easily verifies that the parameters of recursive calls of the algorithm fulfil the input specification.

Termination. Because of lines 3, 6, and 7, the choice of x in line 8 is possible. There can be only finitely many while iterations (lines 7–13), since S contains at least one element and thus the cardinality of V decreases. In each recursive call the sum of the cardinality of the input parameters is strictly smaller than $\text{card}\,Y + \text{card}\,U$. Thus the recursion depth is bounded, and hence SLICEDDIVISION terminates.

Correctness. Let Y and U be as specified in the input of Algorithm 1. The proof will be an induction on the number of elements in the input sets. Obviously, the algorithm is correct if it is started with empty input parameters. Let us assume that the algorithm is correct if the sum of the cardinalities of the input parameters is smaller than $\text{card}\,Y + \text{card}\,U$. Let $[v_1\langle Y_{v_1}\rangle, \ldots, v_{r'}\langle Y_{v_{r'}}\rangle]$ be the output of the algorithm for Y and U. Consider the particular run where this output was produced. Let m, C, and V be as specified before the while loop, and let s be the number of actual iterations.

In the following let σ range over $1, \ldots, s$. The values of x, and S after the σ-th iteration will be denoted by y_σ and S_σ.

First, we must show that $r' = r$. By induction hypothesis, we have $\text{card}\,C = \text{card}\,L_C$ and $\text{card}\,S_\sigma = \text{card}\,L_{S_\sigma}$ for all σ. From lines 1, 13, and 14 and the fact that U is equal to the disjoint union $C \cup \{m\} \cup S_s \cup \cdots \cup S_1$, we can conclude $r = r'$.

Lines 5, 12, and 14 together with the induction hypothesis then imply $U = \{v_1, \ldots, v_r\}$.

Take $i, j \in \{1, \ldots, r\}$ with $i < j$. It remains to show that

$$v_i\langle Y\rangle \cap v_j\langle Y_{v_j}\rangle = \emptyset \text{ or} \tag{5}$$

$$v_i\langle Y_{v_i}\rangle \subset v_j\langle Y_{v_j}\rangle \tag{6}$$

holds. There are several cases to consider.

1. $v_i, v_j \in C$: (5) or (6) follows by induction hypothesis.
2. $v_i \in C$, $v_j = m$: We have $Y_{v_j} = Y$ and (6) by construction of C.
3. $v_i \in C \cup \{m\}$, $v_j \in S_\sigma$ for some $\sigma \in \{1, \ldots, s\}$: By construction of m, C, and S_σ, we have $m\langle Y \rangle \cap v_j \langle Y_{v_j} \rangle = \emptyset$, because $\deg_{y_\sigma} v_j < \deg_{y_\sigma} m$ and $y_\sigma \in X \setminus Y_{v_j}$. Since $c\langle Y_c \rangle \subset m\langle Y \rangle$ for all $c \in C$, relation (5) holds.
4. $v_i \in S_\sigma$, $v_j \in S_{\sigma'}$ for some $\sigma, \sigma' \in \{1, \ldots, s\}$: If $\sigma = \sigma'$ then (5) or (6) follows from the induction hypothesis and Remark 1. Otherwise, $i < j$ implies $\sigma > \sigma'$ by line 14 and construction of L in line 13. The inequality $\deg_{y_{\sigma'}}(v_i t) > \deg_{y_{\sigma'}} v_j$ holds for $t = 1$ by construction of $S_{\sigma'}$ and remains true for any $t \in \langle Y \rangle$. Since $y_{\sigma'} \notin Y_{v_j}$, we can conclude (5).

By line 14 there cannot be other cases. Thus, the correctness of SLICEDDIVISION is proved. $\square$

Corollary 1. *Let $W \subseteq T$ be a finite set of terms, and $[w_1\langle Y_{w_1}\rangle, \ldots, w_r\langle Y_{w_r}\rangle]$ an output of* SLICEDDIVISION(X, W). *Define $\sqsubset$ on W by $w_i \sqsubset w_j \iff i < j$, $i, j = 1, \ldots, r$. Then $\mathcal{M} = (t\langle Y_t\rangle)_{t \in W}$ is admissible on $(W, \sqsubset)$.*

We want to emphasise the fact that instead of giving an order $\sqsubset$ in advance and then compute an admissible partial division with respect to that order, our algorithm only starts with a set of terms and returns a division and an order for which this division is admissible.

One easily verifies that SLICEDDIVISION never calls itself with an empty set of variables if it is started with a non-empty set Y. Line 3 is only important if the algorithm is initially called with an empty set of variables or for a generalisation of the algorithm. In fact, algorithm SLICEDDIVISION also meets its specification if an arbitrary element $m \in U$ in line 4 is chosen. For such a generalisation, however, it can happen that the algorithm calls itself with an empty set Y. One easily verifies that the proof of Theorem 1 does not depend on the particular choice of m. For an extensive treatment of the generalised algorithm see Section 7.

6   Sliced Division Refines Thomas Division

As Apel showed in [1], termination of the involutive basis algorithm (Algorithm 2) is guaranteed if the division used in its while loop refines the Thomas division.

The following property of the Thomas division is easy to prove.

Lemma 1. *Let $U, V \subseteq T$ be two sets with $U \supseteq V$ and $(t\langle Y_t\rangle)_{t \in T}$, $(t\langle Z_t\rangle)_{t \in T}$ be the Thomas divisions on U and V, respectively. Then $Y_t \subseteq Z_t$ for all $t \in T$.*

Proof. For all $t \in T \setminus U$ the equality $Y_t = Z_t = X$ holds. Trivially, $Y_t \subseteq Z_t = X$ is true for all $t \in U \setminus V$. For all $t \in V$ and all $x \in X \setminus Z_t$ there exists $t' \in V$ with $\deg_x t' > \deg_x t$ and, therefore $x \in X \setminus Y_t$. Hence, $Y_t \subseteq Z_t$. $\square$

Theorem 2. *Let $Y \subseteq X$ be a set of variables, $U \subset T$ be a finite set of terms, and $[u_1 \langle Y_{u_1} \rangle, \dots, u_r \langle Y_{u_r} \rangle] = \text{SLICEDDIVISION}(Y, U)$. Then for the Thomas division $(t \langle Z_t \rangle)_{t \in T}$ on U the relation $Z_u \cap Y \subseteq Y_u$ holds for all $u \in U$.*

Proof. Since for $U = \emptyset$ there is nothing to prove, assume that $\text{card } U = r > 0$ and that the theorem is correct if the number of terms is less than r. Consider the run of SLICEDDIVISION where $[u_1 \langle Y_{u_1} \rangle, \dots, u_r \langle Y_{u_r} \rangle]$ was computed. Let s be the number of iterations of the while loop, and let C and m be as defined in the algorithm. After the σ-th iteration ($\sigma = 1, \dots, s$) of the while loop the values of x and S are denoted by x_σ and S_σ, respectively. Then U is the disjoint union $C \cup \{m\} \cup S_s \cup \cdots \cup S_1$.

Trivially, $Z_m \cap Y \subseteq Y_m = Y$. From the induction hypothesis it follows that $Z_u \cap Y \subseteq Y_u$ for all $u \in C$. Now consider S_σ for some $\sigma \in \{1, \dots, s\}$. Take any $u \in S_\sigma$. From the induction hypothesis we get $Z_u \cap (Y \setminus \{x_\sigma\}) \subseteq Y_u$ (see line 12 of Algorithm 1). By definition of the Thomas division, we have $x_\sigma \notin Z_u$ since $\deg_{x_\sigma} u < \deg_{x_\sigma} m$. Hence, $Z_u \cap Y \subseteq Y_u$ as claimed. $\square$

The following corollary is Theorem 2 for the special case $X = Y$.

Corollary 2. *Let $U \subseteq T$ be a finite set of terms. Then a sliced division on U refines the Thomas division on U.*

7   Generalised Sliced Division

The algorithm SLICEDDIVISION was designed to construct an admissible partial division on a set U of terms such that the cones attached to elements of U cover a relatively 'big' part of the monoid ideal generated by U. For theoretical considerations it is interesting to generalise the algorithm such that it can compute other known admissible divisions.

As we have already seen, the particular choice of m in line 4 of Algorithm 1 does not affect the correctness of Theorem 1. By a further generalised form of SLICEDDIVISION, not only one can compute the Janet division, but also Pommaret and Thomas division as we shall see in Section 7.2.

Let us first generalise the sliced division algorithm.

7.1   Every Generalised Sliced Division
Is an Admissible Partial Division

Let $U \subseteq T$ be a finite set of terms. SLICEDDIVISION(X, U) always returns a partial division where at least one term of U has a full-dimensional cone. It cannot compute, for example, the Pommaret division for the terms $\{x_1 x_2^2, x_1^2 x_2\}$ since x_1 is not Pommaret-multiplicative for both terms. To produce admissible divisions on a set U where no element of U will get a cone corresponding to all possible variables, we introduce a distinct symbol $\infty \notin U$ and allow to choose in line 4 of SLICEDDIVSION either an element of U or this symbol depending on whether or not the final division should contain a full-dimensional cone.

In order to simplify notation, for $m = \infty$, $Y \subseteq X$, and $U \subseteq T$, we require that the set $U \cap m\langle Y \rangle$ is empty (no matter how $m\langle Y \rangle$ is actually defined), and we declare $\deg_x t < \deg_x m$ for all $x \in X$, $t \in T$.

By using the same specification as that of SLICEDDIVISION, the generalised version of Algorithm 1 is obtained by replacing lines 4, 8, and 14 by

    4: $m := \text{CHOOSEFIRST}(Y, U)$
    8: $x := \text{CHOOSENONMULTIPLIER}(Y, V, m)$
  14: <u>if</u> $m \in U$ <u>then</u> <u>return</u> $\text{CONCAT}(L_C, [m\langle Y \rangle], L)$ <u>else</u> <u>return</u> L

with two sub-algorithms CHOOSEFIRST and CHOOSENONMULTIPLIER. These algorithms should have the following properties.

S1 If $\emptyset \neq U \subseteq T$ is a finite set of terms, $Y \subseteq X$, and $m = \text{CHOOSEFIRST}(Y, U)$ then $m \in U \cup \{\infty\}$.

S2 Let $\emptyset \neq V \subseteq T$ be a finite set of terms, $Y \subseteq X$, $m \in T \cup \{\infty\}$,

$$\forall v \in V \, \exists y \in Y : \deg_y v < \deg_y m$$

and $x = \text{CHOOSENONMULTIPLIER}(Y, V, m)$. Then

$$x \in Y \wedge \exists v \in V : \deg_x v < \deg_x m.$$

One easily verifies that the input specifications are always satisfied when these functions are called in the generalised sliced division algorithm. By lines 6 and 7 of Algorithm 1 it follows $\forall v \in V \, \exists y \in Y : \deg_y v < \deg_y m$.

Theorem 3. *The generalised version of the algorithm* SLICEDDIVISION *terminates and meets its specification.*

Proof. The proof of termination and correctness of the generalised version is similar to the proof of Theorem 1. In fact, our proof of Theorem 1 has been formulated in such a way that it can even be taken literally. One should note that the cases 2 and 3 at the end of the proof cannot occur if $m = \infty$, since in the proof $v_i, v_j \in U$, but $\infty \notin U$. $\square$

Remark 2. Given U and Y, one may think of ∞ as the term $m = \text{lcm}(U) \prod_{y \in Y} y$ as it fulfils all conditions that we posed on ∞. Since the cone $m\langle Y \rangle$ is never returned and $\deg_y v < \deg_y m$ will be true for all $y \in Y$ and $v \in V$, such an assignment would not cause any problem.

We define a **generalised sliced division** to be a division which is induced by the output of a generalised sliced division algorithm.

7.2 Janet, Pommaret, and Thomas Divisions as Generalised Sliced Divisions

In this section, let $W \subseteq T$ be a finite set of terms and let $\sqsubset = <_{\text{invlex}}$ be the inverse lexicographical order on W. Let $\mathcal{J}$, $\mathcal{P}$, and $\mathcal{T}$ be the Janet, Pommaret, and Thomas division on W, respectively. They are admissible on $(W, \sqsubset)$.

We are going to present particular sub-algorithms such that the generalised form of Algorithm 1 computes one of the classical divisions if these sub-algorithms are used.

Definition 7. *Let $\emptyset \neq Y \subseteq X$ and $\emptyset \neq U, V \subseteq W$ finite sets of terms. Let $m \in T \cup \{\infty\}$ and assume $\forall v \in V \, \exists y \in Y : \deg_y v < \deg_y m$. We define the following functions.*

Janet case: *Let $\mathrm{CF}_{\mathcal{J}}(Y, U) := \max_{<_{\mathrm{lex}}} U$. Let $\mathrm{CNM}_{\mathcal{J}}(Y, V, m) := x_i$ where i is the smallest index such that $\deg_{x_i} v < \deg_{x_i} m$ for some $v \in V$.*
Pommaret case: *Let x_k be such that k is the smallest index of all variables of Y and let $U' := U \cap \langle (X \setminus Y) \cup \{x_k\} \rangle$ then $\mathrm{CF}_{\mathcal{P}}(Y, U)$ returns ∞ if $U' = \emptyset$, otherwise it returns the element of U' of smallest degree. Let $\mathrm{CNM}_{\mathcal{P}} := \mathrm{CNM}_{\mathcal{J}}$.*
Thomas case: *Let $\mathrm{CF}_{\mathcal{T}}(Y, U)$ return $l := \mathrm{lcm}(U)$ if $l \in U$ and $\deg_y t \leq \deg_y l$ for all $y \in Y$ and $t \in W$, otherwise let it return ∞. Let $\mathrm{CNM}_{\mathcal{T}}(Y, V, m) := x_i$ where i is the smallest index such that $\deg_{x_i} v < \deg_{x_i} m$ and $\deg_{x_i} v < \deg_{x_i}(\mathrm{lcm}\, W)$ for some $v \in V$.*

Definition 8. *Let $\mathcal{M}$ be a partial division and $\mathrm{CF}_{\mathcal{M}}$ and $\mathrm{CNM}_{\mathcal{M}}$ be functions satisfying **S1** and **S2** in place of CHOOSEFIRST and CHOOSENONMULTIPLIER, respectively. The algorithm which is obtained from Algorithm 1 by replacing lines 4, 8, and 14 by*

> *4:* $m := \mathrm{CF}_{\mathcal{M}}(Y, U)$
> *8:* $x := \mathrm{CNM}_{\mathcal{M}}(Y, U, m)$
> *14:* <u>*if*</u> $m \in U$ <u>*then*</u> <u>*return*</u> $\mathrm{CONCAT}(L_C, [m\langle Y \rangle], L)$ <u>*else*</u> <u>*return*</u> L

is called SLICEDDIVISION$_{\mathcal{M}}$.

It is trivial to check that the functions $\mathrm{CF}_{\mathcal{J}}$, $\mathrm{CF}_{\mathcal{P}}$, and $\mathrm{CF}_{\mathcal{T}}$ fulfil the property **S1**. $\mathrm{CNM}_{\mathcal{J}}$ and $\mathrm{CNM}_{\mathcal{P}}$ fulfil **S2**. For $\mathrm{CNM}_{\mathcal{T}}$ we must show that there is some $y \in Y$ and some $v \in V$ in line 8 such that $\deg_y v < \deg_y m$ and $\deg_y v < \deg_y(\mathrm{lcm}\, W)$. If $m \in W$, this follows by the input specification given in Definition 8. If $m = \infty$, it follows from the definition of $\mathrm{CF}_{\mathcal{T}}$.

Since SLICEDDIVISION$_{\mathcal{J}}$, SLICEDDIVISION$_{\mathcal{P}}$, and SLICEDDIVISION$_{\mathcal{T}}$ are generalised sliced division algorithms, by Theorem 3, they return admissible partial divisions.

Theorem 4. *Let $Y \subseteq X$, $U = \{u_1, \ldots u_r\} \subseteq W$. Let $\mathcal{M} = (t\langle X_t \rangle)_{t \in T}$ stand for the Janet division $\mathcal{J}$, the Pommaret division $\mathcal{P}$, or the Thomas division $\mathcal{T}$ on W. Let $\forall u \in U : X_u \subseteq Y$ and $[v_1\langle Y_{v_1} \rangle, \ldots, v_r \langle Y_{v_r} \rangle] = \mathrm{SLICEDDIVISION}_{\mathcal{M}}(Y, U)$. If $\mathcal{M} = \mathcal{P}$ is the Pommaret division then additionally assume that there exists $k \in \{1, \ldots, n\}$ such that $Y = \{x_k, \ldots, x_n\}$. Then $X_u = Y_u$ for all $u \in U$.*

Proof. The proof goes by induction on the cardinality of the input parameters Y and U. All line numbers will refer to SLICEDDIVISION$_{\mathcal{M}}$. For empty input parameters there is nothing to show. Now consider $Y \subseteq X$ and $U \subseteq W$ and assume that the theorem is proved for all calls of SLICEDDIVISION$_\mathcal{M}$ where the sum of the cardinalities of the input parameters is strictly less than card Y + card U. Let m, C, and V be as defined in lines 4–6. Let s be the number of iterations of the <u>while</u> loop, and let y_σ, d_σ, S_σ, and V_σ be the values of x, d, S, and V after the σ-th iteration ($\sigma = 1, \ldots, s$). By the input specification of SLICEDDIVISION$_\mathcal{M}$, there exists an element $w \in \langle X \setminus Y \rangle$ such that $U \subseteq W \cap w\langle Y \rangle$. For line 12 we find that $S_\sigma \subseteq y_\sigma^{d_\sigma} w\langle Y \setminus \{y_\sigma\}\rangle$, thus the input specification of SLICEDDIVISION$_\mathcal{M}$ is satisfied. There are two cases to consider.

$m \in U$: By line 14, $Y_m = Y$ will be returned, it remains to show $X_m = Y$. Since card C, card $V <$ card U, the correct assignment of multiplicative variables for the elements of $U \setminus \{m\}$ follows from the induction hypothesis if we can show that the assumptions of the theorem are satisfied in lines 5 and 12. For line 5 this is trivial, since $C \subsetneq U$ and Y does not change. Note that $C = \emptyset$ if $\mathcal{M} = \mathcal{J}$.

For the remaining considerations, we treat the various cases separately.

$\mathcal{M} = \mathcal{J}$: From the assumption of the theorem, $X_m \subseteq Y$. By definition of $\mathcal{J}$ and CF$_\mathcal{J}$, we must have $X_m = Y$. By $m = \max_{<_{\text{lex}}} U$, line 6, the choice of y_σ through CNM$_\mathcal{J}$, and from line 10 we conclude $\deg_{y_\sigma} v < \deg_{y_\sigma} m$ for all $v \in S_\sigma$. By definition of the Janet division, y_σ is non-multiplicative for all elements of S_σ. Hence, the theorem is applicable for $(Y \setminus \{y_\sigma\}, S_\sigma)$ in line 12.

$\mathcal{M} = \mathcal{T}$: From the assumption of the theorem, $X_m \subseteq Y$. By definition of $\mathcal{T}$ and CF$_\mathcal{T}$, we must have $X_m = Y$. By $m = \text{lcm}(U)$, line 6, the choice of y_σ through CNM$_\mathcal{T}$, and from line 10 we conclude $\deg_{y_\sigma} v < \deg_{y_\sigma} m$ for all $v \in S_\sigma$. By definition of the Thomas division, y_σ is non-multiplicative for all elements of S_σ. Hence, the theorem is applicable for $(Y \setminus \{y_\sigma\}, S_\sigma)$ in line 12.

$\mathcal{M} = \mathcal{P}$: Let $U' = U \cap \langle (X \setminus Y) \cup \{x_k\}\rangle$. By definition of the Pommaret division, $X_u = Y$ for all $u \in U'$. By definition of CF$_\mathcal{P}$, $m = wx_k^a \in U'$ with the smallest possible a. Note that $U' \subseteq m\langle Y \rangle$. By the choice of m and line 6, it follows that for all $v \in V$ $\deg_{x_k} v < a$ and $\exists j > k : \deg_{x_j} v > 0$. Therefore $y_\sigma = x_k$ and $x_k \notin X_v$. Hence, the theorem is applicable for $(Y \setminus \{x_k\}, S_\sigma) = (\{x_{k+1}, \ldots, x_n\}, S_\sigma)$ in line 12.

$m = \infty$: This case does not occur if $\mathcal{M}$ is the Janet division. From lines 5 and 6 follows $C = \emptyset$ and $V = U$. Let $\sigma \in \{1, \ldots, s\}$. Correct assignment of the multipliers follows from the induction hypothesis applied to $(Y \setminus \{y_\sigma\}, S_\sigma)$ in line 12. We must show that the assumptions of the theorem are applicable. If $\mathcal{M} = \mathcal{T}$ then $X_v \subseteq Y \setminus \{y_\sigma\}$ for all $v \in V$ follows from the definition of CNM$_\mathcal{T}$ and the definition of the Thomas division on W. If $\mathcal{M}$ is the Pommaret division, we observe that $y_\sigma = x_k$ by CNM$_\mathcal{P}$ and that for all $v \in V$ there exists $j > k$ such that $\deg_{x_j} v > 0$, i.e., $X_v \subseteq \{x_{k+1}, \ldots, x_n\}$.

Hence, all assumptions of the theorem are satisfied in line 12 and we get $X_u = Y_u$ for all $u \in S_\sigma$ and consequently for all $u \in U$.

The theorem is proved. $\square$

The following corollary states that Janet, Pommaret, and Thomas divisions are generalised sliced divisions.

Corollary 3. *Let $\mathcal{M}$ be the Janet, the Pommaret, or the Thomas division on W. Let $[v_1 \langle Y_{v_1} \rangle, \ldots, v_r \langle Y_{v_r} \rangle] = \text{SLICEDDIVISION}_{\mathcal{M}}(X, W)$. $\mathcal{M}$ and the induced division of $(v \langle Y_v \rangle)_{v \in W}$ are W-equivalent.*

8 Computer Experiments

The involutive basis algorithm appears in various flavours in the literature, for example, [8], [1], [4]. Let us give here only a basic treatment which is necessary to understand our computer experiments. For more details we refer to the papers above.

In order to make the paper as self-contained as possible, we repeat here several definitions.

Definition 9. *Let $G \subseteq \mathbb{K}[X]$ be a set of polynomials. G is called an **involutive basis** if G is a Gröbner basis and $\mathrm{lt}(G)$ is complete (cf. Def. 4).*

A list of equivalent conditions can be found in [1].

Definition 10. *Let $G \subseteq \mathbb{K}[X]$ be a set of polynomials and let $\mathcal{M} = (M_t)_{t \in T}$ be a partial division. We say that $f \in \mathbb{K}[X]$ is in $\mathcal{M}$-**normal form** with respect to G if $\mathrm{supp}\, f \cap M_{\mathrm{lt}\, g} = \emptyset$ for every $g \in G \setminus \{0\}$.*

Obviously, this leads to a concept of restricted reduction; elements in G can only be multiplied by their respective multiplicative variables in order to reduce some term of f. The sub-algorithm $\mathrm{NF}_{\mathcal{M}}$ that computes an $\mathcal{M}$-normal form is similar to a normal form algorithm in a Gröbner basis computation where divisor and multiple is replaced by $\mathcal{M}$-divisor and $\mathcal{M}$-multiple. For example, if $G = \{x^2, y^2\}$ and $\mathcal{M}$ denotes the Janet division on $\mathrm{lt}\, G$. Then $\mathrm{NF}_{\mathcal{M}}(xy^2, G) = xy^2$ since there is no $\mathcal{M}$-divisor of xy^2 in $\mathrm{lt}\, G$.

The basic form of the involutive basis algorithm is given in Algorithm 2. The different approaches for a definition of an admissible partial division [2] and an involutive division [4] are reflected in the sub-algorithm PARTIALDIVISION. Termination of INVOLUTIVEBASIS is ensured by Apel through the requirement that $\mathcal{M}$ be an admissible partial division on $\mathrm{lt}(G)$ that refines the Thomas division on $\mathrm{lt}(G)$. Gerdt and Blinkov require that PARTIALDIVISION returns a constructive involutive division (cf. definition in [4]). Our Algorithm 2 incorporates already some statements (reflected by the set P in lines 1, 6, and 11) to avoid multiple reductions of the same prolongation. We refer to [1] for conditions on PARTIALDIVISION such that Algorithm 2 will be correct. While the Janet division fulfils these conditions, the algorithm SLICEDDIVISION does not. In order to yield an involutive basis, Algorithm 2 is iterated until all elements of C from line 3 immediately reduce to zero. Termination follows from the fact that SLICEDDIVISION refines the Thomas division (cf. [1]). We denote this iterated algorithm by IBSD.

Algorithm 2 INVOLUTIVEBASIS

Call: $G = \text{INVOLUTIVEBASIS}(F)$

Input: $F \subset K[X] \setminus \{0\}$ is a finite set of polynomials.

Output: G is an involutive basis of the ideal of F in $\mathbb{K}[X]$.

1: $G := F$; $P := \emptyset$
2: $\mathcal{M} := (t\langle Y_t\rangle)_{t\in\text{lt}(G)} := \text{PARTIALDIVISION}(\text{lt } G)$
3: $C := \{xg \mid g \in G \wedge x \in X \setminus Y_{\text{lt } g}\}$
4: while $C \neq \emptyset$ do
5:     Choose smallest (w.r.t. the term order) p from C
6:     $C := C \setminus \{p\}$; $P := P \cup \{p\}$
7:     $h := \text{NF}_{\mathcal{M}}(p, G)$
8:     if $h \neq 0$ then
9:        $G := G \cup \{h\}$
10:       $\mathcal{M} := (t\langle Y_t\rangle)_{t\in\text{lt}(G)} := \text{PARTIALDIVISION}(\text{lt } G)$
11:       $C := \{xg \mid g \in G \wedge x \in X \setminus Y_{\text{lt } g}\} \setminus P$
12: return G

If in Algorithm 2 PARTIALDIVISION returns the Janet division, it is denoted by JB. If PARTIALDIVISION returns a sub-maximally refined (cf. [1]) Janet division, we denote Algorithm 2 by JBR.

The algorithm INVOLUTIVEBASIS presented in [4] where the Janet division is used, is denoted by JBC. This algorithm incorporates some improvements to Algorithm 2, in particular a variant of Buchberger's chain criterion is applied. Since the other, above mentioned, algorithms do not use criteria, we denote by JBW the algorithm JBC without application of criteria.

We have programmed a package CALIX[1] in the programming language AL-DOR[2] to investigate all the above algorithms.

We have selected the well-known Cyclic4 and Cyclic6 examples as they appear in many collections of polynomial data[3]. They can also be found in [4].

We computed here involutive bases with respect to the degree inverse lexicographical order where first degrees are compared and ties are broken by $<_{\text{invlex}}$. With respect to that term order, the (minimal) Janet bases of the above examples have much more elements than the reduced Gröbner bases which are 7 and 45, respectively.

Let us start with Cyclic4.

---

[1] http://www.hemmecke.de/calix
[2] http://www.aldor.org
[3] for example: http://www.symbolicdata.org

Cyclic4	Length	Prolong.	Reductions	Zeros
JBW	37	91	253	51
JB	37	82	254	49
JBR	29	61	198	35
JBC	37	47	**63**	**7**
IBSD	**7**	**34** (27)	194 (146)	28 (21)

The first column denotes the algorithm, the second contains the length of the involutive basis, the third contains the number of prolongations that have actually been reduced. The fourth column presents the number of elementary reductions, i.e., head reductions of the form $f \longrightarrow f - ctg$ for some coefficient c, some term t and a simplifier polynomial g. The last column counts the number of reductions to zero.

As can be seen from the JBR line, a submaximal refinement of the Janet division can result in a smaller involutive basis. The involutive basis is even smaller for IBSD. The numbers for IBSD include the fact that 2 iterations of Algorithm 2 have been performed. The values in parentheses incorporate the fact that one need not redo zero reductions in the next iteration after the last polynomial has been added to the basis.

For Cyclic6 the number of involutive basis elements differs even more between IBSD and the other algorithms.

Cyclic6	Length	Prolong.	Reductions	Zeros
JBW	385	20572	280696	11964
JB	840	3769	200522	2935
JBR	385	2912	106568	2160
JBC	385	9947	66910	1339
IBSD	**46**	**1104** (1044)	**48959** (45985)	**923** (863)

Surprisingly, the fact that by means of SLICEDDIVISION fewer polynomials are necessary to complete the basis than by using the Janet division, IBSD behaves better even against the Janet basis algorithm JBC which incorporates a variant of the chain criterion.

9   Conclusion

Apel suggested in [1] to use a maximally or sub-maximally refined division in an involutive basis algorithm. Our experiments seem to support this. However, instead of requiring a $\sqsubset$ ordering from the beginning and computing a sub-maximally refined partial division admissible with respect to that order, we compute with SLICEDDIVISION a partial division and an order $\sqsubset$ for which this division is admissible.

In our approach we use a sliced division in order to try to minimise the number of (non-multiplicative) prolongations that must be considered in the course of the involutive basis algorithm. Take, for example, x^7, xy, y^7. This is already an involutive basis with respect to a sliced division. No matter which order of the variables we use, in a Janet basis xy will not get the full cone and thus the basis will contain 8 elements.

One drawback of using SLICEDDIVISION is that Algorithm 2 must be iterated. Since sliced division is actually a whole class of admissible partial divisions, we will try to impose further conditions such that it is enough to consider each prolongation only once during an involutive basis computation or to find a connection between the division and the prolongations that must be reconsidered.

Acknowledgement

This work was supported by the Austrian Science Fund (FWF), SFB F013, project 1304. I am grateful to Joachim Apel for valuable discussions on the subject of involutive bases. I also thank Raymond Hemmecke for his helpful feedback on earlier versions of this article.

References

1. Joachim Apel. The theory of involutive divisions and an application to Hilbert function computations. *Journal of Symbolic Computation*, 25(6):683–704, June 1998.
2. Joachim Apel. *Zu Berechenbarkeitsfragen der Idealtheorie*. Habilitationsschrift, Universität Leipzig, Fakultät für Mathematik und Informatik, Augustusplatz 10–11, 04109 Leipzig, 1998.
3. Bruno Buchberger. Gröbner bases: An algorithmic method in polynomial ideal theory. In N. K. Bose, editor, *Recent Trends in Multidimensional Systems Theory*, chapter 6, pages 184–232. D. Reidel Publishing Company, Dordrecht, The Netherlands, 1985.
4. Vladimir P. Gerdt and Yuri A. Blinkov. Involutive bases of polynomial ideals. *Mathematics and Computers in Simulation*, 45:519–541, 1998.
5. Maurice Janet. Les systèmes d'équations aux dérivées partielles. *Journal de Mathematique*. 8ᵉ série, 3:65–151, 1920.
6. Jean-François Pommaret. *Systems of Partial Differential Equations and Lie Pseudogroups*, volume 14 of *Mathematics and Its Applications*. Gordon and Breach Science Publishers, Inc., One Park Avenue, New York, NY 10016, 1978.
7. Joseph Miller Thomas. *Differential Systems*. American Mathematical Society, New York, 1937.
8. A. Yu. Zharkov and Yuri A. Blinkov. Involution approach to solving systems of algebraic equations. In G. Jacob, N. E. Oussous, and S. Steinberg, editors, *Proceedings of the 1993 International IMACS Symposium on Symbolic Computation*, pages 11–16. IMACS, Laboratoire d'Informatique Fondamentale de Lille, France, 1993.

Congestion and Almost Invariant Sets in Dynamical Systems*

Michael Dellnitz[1] and Robert Preis[1,2]

[1] Department of Mathematics and Computer Science, Universität Paderborn,
Paderborn, Germany, {dellnitz,robsy}@uni-paderborn.de
[2] Computer Science Research Institute, Sandia National Laboratories***,
Albuquerque, United States of America, rpreis@sandia.gov

Abstract. An *almost invariant set* of a dynamical system is a subset of state space where typical trajectories stay for a long period of time before they enter other parts of state space. These sets are an important characteristic for analyzing the macroscopic behavior of a given dynamical system. For instance, recently the identification of almost invariant sets has successfully been used in the context of the approximation of so-called chemical *conformations* for molecules.

In this paper we propose new numerical and algorithmic tools for the identification of the number and the location of almost invariant sets in state space. These techniques are based on the use of *set oriented* numerical methods by which a graph is created which models the underlying dynamical behavior. In a second step graph theoretic methods are utilized in order to both identify the number of almost invariant sets and for an approximation of these sets. These algorithmic methods make use of the notion of *congestion* which is a quantity specifying bottlenecks in the graph. We apply these new techniques to the analysis of the dynamics of the molecules Pentane and Hexane. Our computational results are compared to analytical bounds which again are based on the congestion but also on spectral information on the transition matrix for the underlying graph.

1   Introduction

The dynamical behavior of a real world system can consist of motions on different time scales. For instance, a molecule can stay close to a certain geometric configuration in position space for quite a long time while exhibiting high frequency oscillations around this configuration. In this case the high frequency oscillations are irrelevant and it is desirable to identify the corresponding macroscopic structures for the given molecule. In general – that is, for an arbitrary dynamical

* Research is supported by the Priority Program 1095 of the Deutsche Forschungsgemeinschaft.
*** Sandia is a multiprogram laboratory operated by Sandia Corporation, a Lockheed Martin Company, for the United States Department of Energy under Contract DE-AC04-94AL85000.

F. Winkler and U. Langer (Eds.): SNSC 2001, LNCS 2630, pp. 183–209, 2003.
© Springer-Verlag Berlin Heidelberg 2003

system – such macroscopic structures in state space are called *almost invariant sets*. In this article we propose a new combination of numerical and graph theoretic tools for both the identification of the number and the location of almost invariant sets in state space.

The crucial observation for the developments in this article is the following. If the underlying dynamics can be represented by a graph then the almost invariant sets are subgraphs which show a high coupling inside them and only a low coupling to other parts of the graph. Thus, paths beginning in a subgraph representing an almost invariant set will, with high probability, remain in the set for a long period of time. Having this point of view the identification of almost invariant sets boils down to the identification of corresponding appropriate cuts in the representing graph. Using this idea the topic of this paper is to develop a combination of numerical and graph theoretical methods for the identification of decompositions of phase space into regions which are almost invariant. Moreover it is in general a non-trivial task to identify the number of almost invariant sets and we will also attack this problem in this paper.

As indicated above the main application we have in mind is the identification of chemical *conformations* for molecules, see e.g. [9,40]. However, in this paper we are not going to address the specific analytical and numerical problems which arise due to the fact that molecular dynamics is modeled via Hamiltonian systems. Rather this present work has to be viewed as a first step towards the efficient combination of graph theory and the numerical analysis for general dynamical systems.

Our topic is closely related to the graph partitioning problem in a natural way. In general this problem consists of the task to partition the vertices of a graph into a prescribed number of parts such that certain conditions on the number of edges between the parts are optimized. An enormous amount of work has been done to analyze the graph partitioning problem, to develop efficient methods for solving the problem and to design efficient code implementations in order to be used in several different applications. One important field of application is the efficient load balancing of graph-structured tasks on parallel computer systems.

It is our goal to use results from graph theory and graph algorithms and apply them in the context of dynamical systems. In this context a crucial question is to choose adequate cost functions which makes it necessary to modify some of the analytical results and to extend several existing algorithms. A first step in this direction has been made in [18] where greedy algorithms have been proposed for the identification of almost invariant sets. In our paper we extend this work significantly, in particular by proposing an algorithm which allows to approximate all the almost invariant sets by using the *congestion* of a graph.

A more detailed outline of the paper is as follows. In Section 2 we set the scene by stating the problem and by introducing the basic definitions. Moreover we give an overview of analytic bounds on the cost functions which are adequate for the underlying problem.

In Section 3 we briefly describe state-of-the-art graph partitioning methods. There are freely available graph partitioning tools such as our own tool PARTY. The problem we address here is different to the standard graph partitioning tools. Thus, we modify and extend some of the algorithms in PARTY. Furthermore, we implement some new algorithms in order to optimize the cost functions defined in Section 2.

In Section 4 we introduce the congestion of a graph. This quantity has been used in graph theory in order to specify bottlenecks within the graph. Since the computation of the congestion can be very costly we propose a couple of alternative approaches for approximating this quantity in a faster way.

Also the congestion of a graph can be used for bounding the magnitude of almost invariance of a subset of state space. In Section 5 we compare the bound derived by the congestion with the bound based on the spectrum of the graph and show that the congestion is highly suitable for identifying the quality of a specific decomposition. Moreover we quantify the increase in the congestion which is going to be observed if two invariant sets are loosely coupled and become almost invariant.

Finally, in Section 6 we use the congestion of a graph for the identification of an appropriate number of almost invariant sets for the given dynamical system. More precisely the congestion of a graph is used in combination with recursive bisection methods in order to decide whether or not a subgraph should be further partitioned. All the algorithms and in particular this partitioning strategy are illustrated by a number of examples.

2   Background

In this section we present the basic definitions and state the problem of finding almost invariant sets. The approach that we review here involves the optimization of certain cost functions and has already been presented in [18]. In addition to this review we state well known results on bounds of these cost functions by using spectral information on the corresponding transition matrix.

2.1   Basic Definitions and Problem Formulations

Dynamical System: Let $T : X \to X$ be a continuous map on a compact manifold $X \subset \mathbb{R}^n$. Then the transformation T defines a discrete time dynamical system of the form

$$x_{k+1} = T(x_k), \quad k = 0, 1, 2, \dots$$

on X. In typical applications T is an explicit numerical discretization scheme for an ordinary differential equation on a subset of $\mathbb{R}^n$.

Measure: In the following we assume that μ is an **invariant measure** for the transformation T describing the statistics of long term simulations of the dynamical system. That is, the probability measure μ satisfies

$$\mu(A) = \mu(T^{-1}(A)) \quad \text{for all measurable } A \subset X.$$

Moreover we assume that μ is a unique so-called **SRB-measure** in the sense that this is the only invariant measure which is robust under small random perturbations. In other words μ is the only physically relevant invariant measure for the dynamical system T. For a precise definition of SRB-measures see e.g. [7].

For two sets $A_1, A_2 \subset X$ we define the **transition probability** ρ from A_1 to A_2 as

$$\rho(A_1, A_2) := \frac{\mu(A_1 \cap T^{-1}A_2)}{\mu(A_1)}$$

whenever $\mu(A_1) \neq 0$.

Almost Invariance: The transition probability $\rho(A, A)$ from a set $A \subset X$ into itself is called the **invariant ratio** of A. Heuristically speaking a set A is **almost invariant** if $\rho(A, A)$ is close to one, that is, if almost all preimages of the points in A are in A itself. Using this observation we state the optimization problem for the identification of almost invariant sets as follows.

Problem 1 (Almost Invariant). Let $p \in \mathbb{N}$. Find p pairwise disjoint sets $A_1, \ldots, A_p \subset X$ with $\bigcup_{1 \leq i \leq p} A_i = X$ such that

$$\frac{1}{p} \sum_{j=1}^{p} \rho(A_j, A_j) = \max!$$

Observe that in this optimization problem we have assumed that the number p of almost invariant sets is known. In this article we are going to address the question of how to identify an appropriate p in Section 6.

Box Collection: In general the infinite dimensional optimization problem 1 cannot be solved directly. Therefore we have to discretize the problem. Following [18] we suppose that we have a (fine) box covering of X consisting of d boxes $B_1, \ldots, B_d$ such that

$$X = \bigcup_{i=1}^{d} B_i.$$

In practice this box covering can be created by using a subdivision scheme as described in [6]. The optimization problem 1 is now reduced to all subsets which are finite unions of boxes, that is, on subsets of

$$\mathcal{C}_d = \left\{ A \subset X : A = \bigcup_{i \in I} B_i, I \subset \{1, \ldots, d\} \right\}.$$

For two sets $A_1^d, A_2^d \in \mathcal{C}_d$ the **box transition probability** ρ_d from A_1^d to A_2^d is nothing else but

$$\rho_d(A_1^d, A_2^d) := \rho(A_1^d, A_2^d) = \frac{\mu(A_1^d \cap T^{-1}A_2^d)}{\mu(A_1^d)}$$

assuming that $\mu(A_1^d) \neq 0$.

Transition Matrix / Markov Chain / Undirected Graph: The box collection $\mathcal{C}_d$ can be used to define a weighted transition matrix for our dynamical system. This is easily accomplished by defining the transition matrix $P \in \mathbb{R}^{d \times d}$ with $P_{ij} = \rho_d(B_i, B_j)$, $1 \leq i, j \leq d$. By this procedure we have constructed an approximation of our dynamical system via a finite state Markov chain: the boxes are the vertices and two vertices i and j are connected by an edge if the image of box i under the transformation T has a nontrivial intersection with box j. The transition probability is given by P_{ij}.

By construction the matrix P is stochastic, that is $\sum_{1 \leq j \leq n} P_{ij} = 1$, and by our assumption on the SRB-measure μ it is reasonable to assume that P is also irreducible. Denote by $\mu^d \in \mathbb{R}^d$ the corresponding unique **stationary distribution** of P and by M the matrix which has the components of μ^d on its diagonal. Then it appears very unlikely that P is **reversible** in the sense that for all $1 \leq i, j \leq d$ we have

$$\mu_i^d P_{ij} = \mu_j^d P_{ji} \ .$$

However, as shown in [19] the value of the cost functions that we are going to consider remains unchanged if we are going to consider the reversibilization R of P given by the transformation

$$R := \frac{1}{2}(P + M^{-1} P^T M) \ .$$

From now on we denote by P a (not necessarily) reversible stochastic matrix and by R its corresponding reversible matrix.

The transition matrix R of a reversible Markov chain can be viewed as the adjacency matrix A of an undirected, edge-weighted graph $G = (V, E)$ with vertex set $V = \{v_1, \ldots, v_d\}$. Since R is reversible it follows that A is symmetric. The entry $A_{ij} = \mu_i^d R_{ij} = \mu_j^d R_{ji}$ denotes the edge weight between two vertices v_i and v_j. It is $\{v_i, v_j\} \in E$ if and only if $A_{ij} > 0$.

Partition of the Index Set / Graph: Let $I = \{1, \ldots, d\}$ be a set of indices. For two (not necessarily disjoint) subsets $S_1, S_2 \subset I$ let

$$\tilde{E}_{S_1, S_2}(P) = \sum_{i \in S_1, j \in S_2} \mu_i P_{ij}$$

and

$$E_{S_1, S_2}(P) = \frac{\tilde{E}_{S_1, S_2}(P) + \tilde{E}_{S_2, S_1}(P)}{2} \ .$$

Clearly, it is $\tilde{E}_{S_1, S_2}(R) = E_{S_1, S_2}(R)$ for a reversible matrix R. Furthermore, it is quite easy to see that

$$E_{S_1, S_2}(P) = E_{S_1, S_2}(R).$$

In this paper we focus on the two cases when S_1 is equal to S_2 and when S_1 is the complement of S_2. Clearly, for any $S \subset I$ it is $\tilde{E}_{S,S}(P) = E_{S,S}(P)$. For a subset $S \subset I$ let $\bar{S} = I \backslash S$ be the complement of S with respect to I.

We summarize the observations by defining

$$E_{S,S} := \tilde{E}_{S,S}(P) = E_{S,S}(P) = \tilde{E}_{S,S}(R) = E_{S,S}(R)$$
$$E_{S,\bar{S}} := E_{S,\bar{S}}(P) = \tilde{E}_{S,\bar{S}}(R) = E_{S,\bar{S}}(R) \ .$$

As discussed above, the transition matrix can be viewed as an undirected, edge-weighted graph $G = (V, E)$. Thus, a partition of the index set I can be viewed as a partition of $V = \{v_1, \ldots, v_d\}$, i.e. a set $S \subset I$ represents a subset $S_V \subset V$ with $S_V = \{v_i \in V : i \in S\}$. It is an easy task to rewrite the edge sets defined above and the cost functions defined below in graph notation.

We have now turned the optimization problem 1 into a problem of finding a minimal cut in an undirected graph G.

Cost Functions: By a slight abuse of notation we define $\mu(S) = \sum_{i \in S} \mu_i^d$ for a set $S \subset I$. For a set $S \subset I$ with $S \neq \emptyset$ and $S \neq I$ let

$$C_{int}(S) := \frac{E_{S,S}}{\mu(S)} \tag{1}$$

be the **internal cost of** S and let

$$C_{ext}(S) := \frac{E_{S,\bar{S}}}{\mu(S) \cdot \mu(\bar{S})} \tag{2}$$

be the **external cost of** S.

For $p \in \mathbb{N}$ let $\pi : I \to \{1, \ldots, p\}$ be a partition of I into $p \geq 2$ disjoint sets $I = S_1 \uplus \ldots \uplus S_p$ with $S_i \neq \emptyset$ for all $1 \leq i \leq p$. Define

$$C_{int}(\pi) := \frac{1}{p} \sum_{i=1}^{p} C_{int}(S_i) \tag{3}$$

to be the **internal cost of** π and let

$$C_{ext}(\pi) := \frac{\sum_{1 \leq i < j \leq p} E_{S_i, S_j}}{\prod_{i=1}^{p} \mu(S_i)} \tag{4}$$

be the **external cost of** π.

Remark 1. Both the internal and external costs are relevant cost functions. The following discussion shows that the minimization of the external cost is not equivalent to the maximization of the internal cost. In fact, the maximization of the internal cost favors in general parts that are on average very loosely coupled to the rest of the system. However, the size of these parts can in principle become very small. On the other hand the minimization of the external cost favors balanced weighting of the components.

One can easily construct examples for which the decomposition provided by the optimization of the internal cost is very different to the one given by the external cost and vice versa, but these examples have a rather artificial character. We have observed that in our examples, a decomposition with an optimal or almost optimal internal cost will also have an optimal or almost optimal external cost.

We can now state two optimization problems with respect to the internal and external costs. These have to be viewed as discrete versions of the continuous optimization problem 1.

Problem 2 (Internal Cost). Let $p \in \mathbb{N}$ and $I = \{1, \ldots, d\}$ be a set of indices. Find a partition $\pi : I \to \{1, \ldots, p\}$ of I into p disjoint sets $I = S_1 \uplus \ldots \uplus S_p$ such that

$$C_{int}(\pi) = \max!$$

Problem 3 (External Cost). Let $p \in \mathbb{N}$ and $I = \{1, \ldots, d\}$ be a set of indices. Find a partition $\pi : I \to \{1, \ldots, p\}$ of I into p disjoint sets $I = S_1 \uplus \ldots \uplus S_p$ such that

$$C_{ext}(\pi) = \min!$$

2.2   Spectral Bounds on the Cost Functions

In this section we state some bounds on the internal and external costs of a partition. These bounds are obtained via part of the spectrum of the underlying reversible Markov chain R.

Sinclair [43] introduces the **conductance** of an ergodic, reversible Markov chain. The bounds on the conductance can be transformed to bounds of the internal cost of a set. For a set $S \subset I$ let

$$\Phi(S) := \frac{E_{S,\bar{S}}}{\mu(S)} \; .$$

With $\mu(S) = E_{S,S} + E_{S,\bar{S}}$ we have

$$C_{int}(S) = 1 - \frac{E_{S,\bar{S}}}{\mu(S)} = 1 - \Phi(S).$$

The conductance Φ of a Markov chain is defined as

$$\Phi = \min_{0 < |S| < d, \mu(S) \leq 1/2} \Phi(S) = \min_{0 < |S| < d} \max\{\Phi(S), \Phi(\bar{S})\} \; .$$

This value is sometimes also called the **isoperimetric value** (see e.g. [1]).

Based on the eigenvalues $1 = \beta_1 \geq \beta_2 \geq \ldots \geq \beta_d \geq -1$ of R, Sinclair gave a proof for the bounds

$$1 - 2\Phi \leq \beta_2 \leq 1 - \frac{\Phi^2}{2} \; .$$

These bounds can be transformed into bounds on the internal cost of a set S, namely

$$1 - 2\sqrt{1 - \beta_2} \leq \max_{0 < |S| < d} \min\{C_{int}(S), C_{int}(\bar{S})\} \leq \frac{\beta_2 + 1}{2} \; .$$

Remark 2. Recently Huisinga [25] has obtained very similar estimates in the situation where the underlying Markov process is induced by different stochastic models for molecular dynamics. The crucial difference to the material presented here lies in the fact that the estimates we are using are valid just for finite approximations of the underlying process.

A discussion of these bounds can also be found in [19].

Now consider the graph $G = (V, E)$ with adjacency matrix A as described in Section 2.1, that is in that notation $A = M \cdot R$. For a vertex $v_i \in V$ denote by

$$\deg(v_i) := \sum_{j=1}^{d} A_{ij}$$

the **degree** of v_i and let D be a diagonal matrix with $D_{ii} = \deg(v_i)$. Then the matrix $L = D - A$ is called the **Laplacian** of G. The **generalized Laplacian** of G is $\mathcal{L} = D^{-1/2}LD^{-1/2}$ with the entries

$$\mathcal{L}_{i,j} = \begin{cases} \frac{\deg(v_i) - A_{ii}}{\mu_i^d}, & \text{if } i = j \\ -\frac{A_{ij}}{\sqrt{\mu_i^d \cdot \mu_j^d}}, & \text{if } i \neq j \text{ and } \mu_i^d \cdot \mu_j^d > 0 \\ 0, & \text{otherwise.} \end{cases}$$

It can be shown that the eigenvalues of $\mathcal{L}$ are nonnegative. We denote them by $0 = \lambda_1 \leq \lambda_2 \leq \ldots \leq \lambda_d$. In fact, it is $\lambda_i = 1 - \beta_i$ for $1 \leq i \leq d$. The multiplicity of 0 as an eigenvalue is equal to the number of connected components in the graph. By our irreducibility assumption on the underlying Markov chain we can conclude that $\lambda_2 > 0$.

We conclude, based on the work by Fiedler [16], that it can be shown that

$$\lambda_2 \leq C_{ext}(S) \leq \lambda_d . \tag{5}$$

Further discussion on spectral graph partitioning is to be found in e.g. [2,4] [13,31,37,43].

3   Algorithms for Graph Partitioning

In this section we focus on algorithms for partitioning the vertex set of a graph. A number of methods and tools for graph partitioning exist. However, they do not exactly optimize our cost functions mentioned in the previous section. Thus, in this section we give an overview of the most successful graph partitioning methods and implementations and point out the necessary modifications to such tools to optimize our cost functions.

For the remainder of this section we assume that we partition into a given number of p parts. In Section 6 we will show how to identify this number.

We want to calculate a partition of the vertex set V of a graph $G = (V, E)$ into p parts $V = S_1 \cup \ldots \cup S_p$ such that either one of our cost functions of

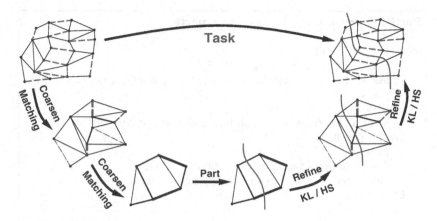

Fig. 1. The multilevel approach with several coarsening steps, a partitioning of the smallest graph and the local refinements.

equations (3) or (4) is optimized. However, the calculation of an optimal solution of either cost functions is **NP**-complete. Another widely discussed partitioning problem is to minimize the cut size $cut(\pi) = \sum_{1 \leq i < j \leq p} E_{S_i, S_j}$ of the partition π under the constraint that all parts have an equal (or almost equal) number of vertices. This problem is sometimes called *Balanced Partitioning Problem* and is **NP**-complete, even in the simplest case when a graph with constant vertex and edge weights is to be partitioned into two parts [20].

In recent years, efficient graph partitioning strategies have been developed for a number of different applications. Efficiency and generalizations of graph partitioning methods strongly depend on specific implementations. There are several software libraries, each of which provides a range of different methods. Examples are CHACO [22], JOSTLE [44], METIS [26], SCOTCH [35] or PARTY [38]. The goal of the libraries is to both provide efficient implementations and to offer a flexible and universal graph partitioning interface to applications. These libraries are designed to create solutions to the balanced partitioning problem. We first describe the most significant methods of these tools and then describe how we modified parts of the code in order to optimize the cost functions in the previous section.

The tool PARTY has been developed by the second author and we have used it for partitioning the graphs in this paper. PARTY, like other graph partitioning tools, follows the Multilevel Paradigm (Figure 1). The multilevel graph partitioning strategies have been proven to be very powerful approaches to efficient graph-partitioning [3,21,23,27,28,34,36,39]. The efficiency of this paradigm is dominated by two parts: graph coarsening and local improvement.

The graph is coarsened down in several levels until a graph with a sufficiently small number of vertices is constructed. A single coarsening step between two levels can be performed by the use of graph matchings. A matching of a graph is a subset of the edges such that each vertex is incident to at most one matching

---

Partition graph $G_0 = (V_0, E_0)$ **into** p **parts**

$i = 0$;
WHILE $(|V_i| > p)$
   calculate a graph matching $M_i \subset E_i$;
   use M_i to coarse graph $G_i = (V_i, E_i)$ to a graph $G_{i+1} = (V_{i+1}, E_{i+1})$;
   $i := i + 1$;
END WHILE
let π_i be a p-partition of G_i such that each vertex is one part;
WHILE $(i > 0)$
   $i := i - 1$;
   use M_i to project π_{i+1} to a p-partition π_i of G_i;
   modify the partition π_i on G_i locally to optimize $C_{int}(\pi_i)$;
END WHILE
output π_0.

Fig. 2. Calculating almost invariant sets with the multilevel approach.

edge. A matching of the graph is calculated and the vertices incident to a matching edge are contracted to a super-vertex. Experimental results reveal that it is important to contract those vertices which are connected via an edge of a high weight, because it is very likely that this edge does not cross between parts in a partition with a low weight of crossing edges. PARTY uses a fast approximation algorithm which is able to calculate a good matching in linear time [39].

PARTY stops the coarsening process when the number of vertices is equal to the desired number of parts. Thus, each vertex of the coarse graph is one part of the partition. However, it is also possible to stop the coarsening process as soon as the number of vertices is sufficiently small. Then, any standard graph partitioning method can be used to calculate a partition of the coarse graph.

Finally, the partition of the smallest graph is projected back level-by-level to the initial graph. The partition is locally refined on each level. Standard methods for local improvement are Kernighan/Lin [29] type of algorithms with improvement ideas from Fiduccia/Mattheyses [15]. An alternative local improvement heuristic is the Helpful-Set method [11] which is derived from a constructive proof of upper bounds on the bisection width of regular graphs [24,32,33].

As mentioned above, the tools are designed for solving the balanced graph partitioning problem. Thus, the optimization criteria is different from our cost functions of Section 2. The coarsening step of the multilevel approach does not consider the balancing of the weights of the super-vertices. It is the local refinement step which not only improves the partition locally but also balances the weights of the parts. Thus, we have to modify the local improvement part of the multilevel approach. We therefore modified the Kernighan/Lin implementation in PARTY such that it optimizes the cost-function C_{int}.

Overall, we use the algorithm of Figure 2 to calculate almost invariant sets.

As an example to illustrate the partitioning we take the graph of a Pentane molecule from [9]. Figure 3 (left) shows the box collection and all transitions. As

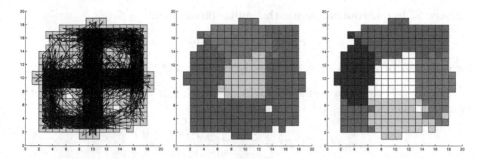

Fig. 3. Partition of the Pentane300. Left: Graph with all transitions. Center: Partition into 5 parts with $C_{int} = .980$. Right: Partition into 7 parts with $C_{int} = .963$.

we will see in section 6, it is adequate to partition the graph into 5 or 7 parts. These are shown in Figures 3 (center/right). The resulting internal costs are .980 for the 5-partition and 0.963 for the 7-partition.

4 The Congestion

The congestion of a graph can be used to spot bottlenecks in the graph. It is based on an all-to-all routing or a multi-commodity flow on the graph (see e.g. [30,43]). We are going to make use of the congestion both for bounding the external cost of subgraphs S (Section 5) and for the identification of the number p of almost invariant sets (Section 6).

4.1 The Definition

Let $G = (V, E)$ be an undirected graph with vertex weights $vw(v_i) = \mu_i^d$, for $v_i \in V$ and edge weights $ew(\{v_i, v_j\}) = A_{ij}$ for $\{v_i, v_j\} \in E$ $(i, j = 1, 2, \ldots, d)$.

Let $v_s \in V$ be a source and $v_t \in V$ be a target vertex in G. Let $c \in \mathbb{R}$ be a **single-commodity** which has to flow from v_s to v_t along the edges in G. A **single-commodity flow** f of the commodity c on G is a function $f : V \times V \to \mathbb{R}$ such that

- i. $f(v, w) = 0$ for all $\{v, w\} \notin E$ (flow on edges only),
- ii. $f(v, w) = -f(w, v)$ for all $v, w \in V$ (symmetry),
- iii. $\sum\limits_{w \in V} f(v, w) = 0$ for all $v \in V \setminus \{v_s, v_t\}$ (flow conservation) and
- iv. $\sum\limits_{w \in V} f(v_s, w) = \begin{cases} -\sum\limits_{w \in V} f(v_t, w) = c & \text{if } v_s \neq v_t \text{ (from source to target)} \\ 0 & \text{if } v_s = v_t \text{ (within source/target)} \end{cases}$

Now let $c_{s,t} \in \mathbb{R}$, $1 \leq s, t \leq d$, be **multi-commodities**. Each commodity $c_{s,t}$ has to flow from the source $v_s \in V$ to the target $v_t \in V$ along the edges in G. A **multi-commodity flow** F of the commodities $c_{s,t}$ on G is a function $F : V \times V \times V \times V \to \mathbb{R}$ such that $F(s, t, v, w)$ is a single-commodity flow of the commodity $c_{s,t}$ on G for each fixed pair $s, t \in V$.

Remark 3. In our context we use the multi-commodities

$$c_{s,t} = vw(v_s) \cdot vw(v_t)$$

for each pair of vertices $v_s, v_t \in V$ for the definition of the congestion.

The flow F causes a traffic on the edges in G. For an edge $\{v, w\} \in E$ the **edge congestion of $\{v, w\}$ in F on G** is

$$cong(\{v, w\}, F) = \frac{\sum_{1 \leq s,t \leq d} |F(s, t, v, w)|}{ew(\{v, w\})} .$$

We will see in the following that high edge congestion occurs at bottlenecks of the graphs.

The **flow congestion of F on G** is

$$cong(F) = \max_{\{v,w\} \in E} cong(\{v, w\}, F) ,$$

and finally the **congestion** of the graph G is

$$cong(G) = \min_{F \text{ is multi-commodity flow}} cong(F) .$$

Intuitively it is clear that a low congestion of the graph ensures that the graph is very compact. For illustration we list the congestion for a couple of specific graphs. The proofs can be found in [30].

Example 1.

a. For the specific case where the graph G is a line consisting of n vertices connected by $n - 1$ edges it is obvious that the congestion is given by

$$cong(G) = 2 \left\lceil \frac{n}{2} \right\rceil \cdot \left\lfloor \frac{n}{2} \right\rfloor .$$

b. Suppose that the graph G is given by a z-dimensional torus with n^z grid points (n even). Then the congestion is

$$cong(G) = \frac{n^{z+1}}{4} .$$

In particular for the circle we have $z = 1$ and obtain $cong(G) = \frac{n^2}{4}$.

c. For the hypercube G of dimension k the congestion is given by

$$cong(G) = 2^k .$$

d. For the complete graph with n vertices – that is a graph where there is an all-to-all coupling between the vertices – the congestion is given by

$$cong(G) = 2 .$$

Further discussion of the congestion and a listing of congestions of further graph classes can be found in e.g. [30,39].

4.2 Heuristics for Calculating a Flow

The computation of the congestion can be costly. We will see in Section 5.1 that the congestion can be used to bound the external cost of a partition. However, that bound holds for the congestion of any flow. Thus, a sub-optimal flow produces a sub-optimal, but still valid bound. In this section we discuss heuristics for calculating a flow with a small flow congestion.

There are some hints of how to construct a flow with a small flow congestion. Clearly, cycles in the flow should be avoided. Furthermore, the flow should - not always, but primarily - go along shortest paths between the pairs of vertices. Here, the length of a path is the sum of the reciprocal values of the edge weights along the path.

A straightforward method is to send the flow along shortest paths only. If more than one shortest paths exist, the flow value can be split among them. If the edge weights are constant, all shortest paths can be calculated in time $O(|V| \cdot |E|)$. This can be done by $|V|$ independent Breath-First searches in time $O(|E|)$ each. If the edge weights are non-negative, all shortest paths can be calculated in time $O(|V| \cdot (|V| \cdot \log |V| + |E|))$, e.g. with $|V|$ runs of the single-source shortest path Dijkstra algorithm using Fibonacci heaps. We refer to [5] for a deeper discussion of shortest paths algorithms.

A different method is to consider n commodities at a time. For each vertex v_s, $1 \le s \le n$, consider the commodities $c_{s,t}$, $1 \le t \le n$, i.e. all commodities with v_s as the source. For each source v_s we calculate a flow F_s which transports all commodities from v_s to all other vertices, i.e. it replaces n single-commodity flows such that $F_s(v, w) = \sum_{t=1}^{n} F(s, t, v, w)$. F_s is a single-source, multiple-destination commodity flow. Definitions (i.) and (ii.) for the single commodity flow (Section 4.1) remain unchanged whereas the definitions of (iii.) and (iv.) are replaced by

v. $\sum_{w \in V} F_s(v_t, w) = -c_{s,t}$ for all $t \ne s$ (from source s to target t) and

vi. $\sum_{w \in V} F_s(v_s, w) = \sum_{t=1}^{n} c_{s,t} - c_{s,s}$ (from source s to all targets t except to source s itself).

It is left to show how we calculate a single-source flow F_s. We use algorithms from diffusion load balancing on distributed processor networks for this task, see e.g. [10,12,14]. Here, the problem is to balance the work load in a distributed processor network such that the volume of data movement is as low as possible. We use these algorithms in the setting that the vertex v_s models a processor with load $\sum_{t=1}^{n} c_{s,t}$ and all other vertices model a processor with no load. Furthermore, the processors are heterogeneous with a capacity of $c_{s,t}$ for processor v_t [14]. The diffusion algorithms calculate a balancing flow such that each vertex/processor v_t gets a load of $c_{s,t}$. That is exactly what we need in our context. The resulting balancing flow has a nice property: it is minimal in the l_2-norm, i.e. the diffusion algorithms minimize the value $\sqrt{\sum_{1 \le v,w \le n} |F_s(v, w)|}$. This ensures that there are no cycles in any flow F_s. Furthermore, the flows are not

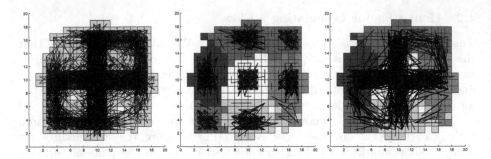

Fig. 4. Congestion of the Pentane. Left: all transitions. Center: only transitions with a low congestion. Right: only transitions with a high congestion.

restricted along shortest paths and can avoid high traffic along shortest paths. However, the flows are still favored to be along reasonably short paths. Thus, it is expected that the overall edge congestion of the resulting flow is reasonably small and that the flow congestion is close to the congestion of the graph.

The PARTY library includes efficient code of a variety of diffusion algorithms. We use them to calculate the single-source, multiple-destination flows for each source s and then add up the values to get the multi-commodity flow. We will see in Section 5.2 that the resulting flow congestion is very small.

4.3   An Example: Pentane

We now illustrate the meaning of the congestion in the context of dynamical systems. For this we use data from [9] in which a reversible stochastic matrix has been generated by simulations of an ODE-based model of the n-Pentane molecule. Those computations made use of two dihedral angles as the essential variables. The graph corresponding to this dynamical system is illustrated in Figure 4 (left) and the edges with low (middle) resp. high (right) congestion are also shown.

Furthermore, the boxes in Figure 4 (center) and Figure 4 (right) are colored according to a partition into seven almost invariant sets. It will be shown in Section 6, in which way the number of almost invariant sets has been determined. This decomposition has been calculated using the graph partitioning tool PARTY. Note that the partitioning was calculated independently from the congestion.

It can be observed that – as expected – edges with low congestion can mainly be found inside the almost invariant sets. On the other hand edges between different almost invariant sets have a large congestion. Thus, a high congestion indicates that there are at least two regions in the phase space which are only loosely coupled. As we will see in Section 6, this observation is the basis for using the congestion as an identifier for the number of almost invariant sets which have to be approximated.

5 The Congestion and Almost Invariant Sets

In this section we show how the congestion can be used in order to obtain quantitative estimates on the magnitude of almost invariance for a given set. Moreover we derive estimates on the decrease of the congestion if two almost invariant sets are decoupled. Obviously these estimates have to depend on the structure of the graph inside the almost invariant sets. These results have to be viewed as a first step in using graph theoretic notions for the identification of dynamical properties in multiscale dynamical systems.

5.1 The Congestion Bound

The congestion can be used to derive a lower bound on $C_{ext}(S)$ for any $S \subset I$ – recall the definitions (1) and (2) of the two different cost functions. As before, see Section 4.1, we use multi-commodities $c_{s,t} = vw(v_s) \cdot vw(v_t)$ for each source $v_s \in V$ and each destination $v_t \in V$.

Clearly, at least $\mu(S) \cdot \mu(\bar{S})$ units have to cross the cut from S to $\bar{S}$ and the same amount has to cross in the opposite direction. However, a congestion of $cong$ ensures that there are at most $cong \cdot E_{S,\bar{S}}$ units crossing the cut. Thus, $2 \cdot \mu(S) \cdot \mu(\bar{S}) \leq cong \cdot E_{S,\bar{S}}$, and we have shown the following inequality

$$C_{ext}(S) \geq \frac{2}{cong} . \tag{6}$$

Obviously, a high and tight lower bound can only be achieved with a small congestion. Although the computation of the congestion can be done in polynomial time, it remains to be very costly. Nevertheless, the congestion can be approximated by the congestion of any flow. Heuristics for calculating a small congestion were discussed in Section 4.2 above.

Remark 4. It is known that an optimal setting for the lower bound (6) occurs if all flows are along shortest paths and the amount of flow is equal on all edges (with respect to their weights). Unfortunately, this 'optimal' case does not exist for arbitrary graphs. Moreover, it can be shown that there are graphs for which such an 'optimal' setting exists, but for which the equation (6) is not tight [39]. Nevertheless, the comparisons in the following section show that the congestion is highly suitable for the type of graphs which are induced by the underlying dynamical systems.

Further discussion of lower bounds based on different variations of multi-commodity flows can be found in [41].

5.2 Comparison of Bounds

In this section we experimentally compare the lower bounds on the external costs of a partition. We compare the bound based on the congestion (see (6)) and the bound based on the second eigenvalue (see (5)).

As mentioned in the previous section, there are examples of graphs for which the congestion bound is not tight although there exists an optimal setting for the calculation of the multi-commodity flow (Remark 4). Furthermore, the examples for which the spectral bound is tight are also very limited, see e.g. [2]. A comparison of the two bounds on some structured graphs can be found in [39]. It is very interesting to know how the two bounds perform on the examples resulting from dynamical systems. We use the heuristic described in the previous section to calculate a flow f which should result in a very low congestion. A comparison of bounds for some examples is shown in Table 1.

Table 1. Example Graphs. The number $|V|$ of vertices is given as the total number of vertices and the number of vertices with nonvanishing discrete measure. The *selfloops* indicate transitions within a box and are listed with total number and total weight. The congestion $cong(f)$ is calculated from a heuristic flow f as described in Section 4.2.

Graph	$|V|$	$|E|$	selfloops	λ_2	$cong(f)$	$\frac{2}{cong(f)}$
Pentane200	400/181	1127	56 .244	.001591	1168.441628	.001712
Pentane300	400/255	3328	96 .170	.014069	139.939697	.014292
Pentane600	400/330	13016	171 .068	.103959	23.558954	.084893
Hexane	729/397	4258	295 .625	.010174	208.911692	.009573
Logistic	512/512	1501	1 .002	.067301	20.728386	.096486
Lorenz	5025/4842	37284	8 .000	.019535	58.842596	.033989
E32	82/80	676	41 .021	.003164	487.721222	.004101
E33	90/90	731	39 .022	.011117	113.961640	.017550
E34	92/92	760	37 .022	.015016	73.138238	.027345
E35	92/92	766	38 .022	.017796	62.112404	.032200
E36	96/96	795	34 .022	.020236	46.650151	.042872
E37	102/102	858	34 .022	.021510	38.152007	.052422
E38	104/104	892	34 .020	.023039	44.240662	.045207
E39	104/104	917	34 .020	.023933	42.678191	.046862
E40	106/106	937	37 .020	.024319	44.689798	.044753
E41	106/106	948	36 .020	.024771	44.692308	.044750
E42	116/116	1024	35 .020	.024545	45.899904	.043573
E43	124/124	1077	33 .019	.025073	46.176803	.043312
E44	128/128	1139	33 .019	.025813	44.759473	.044683
E45	128/128	1174	32 .019	.025643	44.865978	.044577
Butterfly	888/888	4752	16 .001	.003175	168.883748	.011842

The names in the first column in Table 1 refer to the following dynamical systems:

- The first three examples *Pentane200, Pentane300* and *Pentane600* are different models of the n-Pentane molecule with two dihedral angels [9]. An example of *Pentane300* can be found in Figure 3 and in Figure 4.
- *Hexane* is created from data for a model of the Hexane molecule. Again this data is taken from [9]. (An illustration of *Hexane* can be found in Figure 10 below.)

- The example *Logistic* refers to the *Logistic Map* which is given by the transformation

$$T(x) = bx(1-x)$$

on the interval $[0,1]$. We have chosen the parameter value $b = 4$.
- *Lorenz* represents the celebrated *Lorenz system* which is given by the ordinary differential equation

$$\dot{x} = -\sigma x + \sigma y$$
$$\dot{y} = \rho x - y - xz$$
$$\dot{z} = -\beta z + xy.$$

We have chosen the standard set of parameter values $\sigma = 10$, $\rho = 28$ and $\beta = \frac{8}{3}$.
- All the examples *Exx* refer to simulations for different energy levels of the Hamiltonian system with the potential

$$V(q) = \left(\frac{3}{2}q_1^4 + \frac{1}{4}q_1^3 - 3q_1^2 - \frac{3}{4}q_1 + 3\right) \cdot (2q_2^4 - 4q_2^2 + 3) .$$

For instance, E45 refers to the case where the sum of the kinetic and the potential energy is equal to $xx = 45$. This example has previously been analyzed in [8]. An illustration of the potential function is shown in Figure 5 (left) and the invariant measure for the corresponding dynamical behavior is shown in Figure 5 (right).

In Figure 6 we illustrate computations of the congestion for the example E45.

- Finally the example *Butterfly* is based on a simulation of the Hamiltonian system given in (7). An illustration of the corresponding graph can be found in Figure 7.

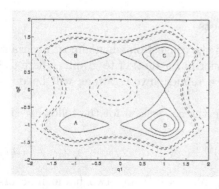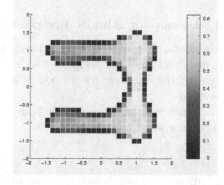

Fig. 5. Left: Contour plot for the potential function of *Exx*. Right: Typical invariant measure for *Exx*.

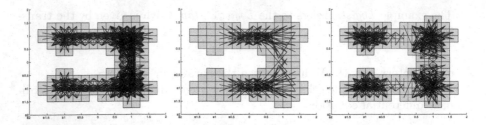

Fig. 6. Example E45 with all edges (left), edges with congestion at least 20 (center) and edges with congestion at most 10 (right).

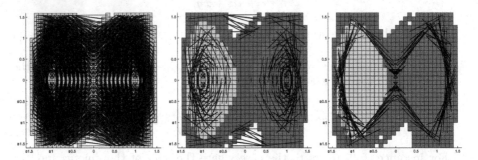

Fig. 7. Partition and congestion of Butterfly. Left: all transitions. Center: only transitions with a low congestion. Right: transitions with a high congestion. The phase space is partitioned into 3 parts.

The results of Table 1 show that the congestion bound $\frac{2}{cong(f)}$ is superior to the spectral bound λ_2 in most of the examples. Furthermore, we approximated the congestion $cong(G)$ with a flow congestion $cong(f)$ calculated by the heuristic described above. The computational results indicate that our heuristic approach produces a flow congestion which is at least close to the correct congestion of the graph.

5.3   Decoupling Almost Invariant Sets

In this section we investigate the change in the congestion when almost invariant sets are decoupled. For simplicity we restrict our attention to graphs where the weight of the vertices are all equal. More refined investigations will be done elsewhere.

For motivational purposes we compute the congestion for the following Hamiltonian system

$$\begin{aligned}
\dot{q} &= p \\
\dot{p} &= -4q(q^2 - 1) \ .
\end{aligned} \tag{7}$$

This system has a saddle point in $(q, p) = (0, 0)$ and two homoclinic orbits connecting this saddle point with itself. Figure 7 (left) shows the discretized system with all the transitions in the corresponding graph. Figure 7 (center/right) shows a partition of phase space into three parts. We calculated a multi-commodity flow f in the graph with the heuristics described in Section 4.2. The congestion of our flow is $cong(f) = 166.31$. Figure 7 (center) shows only the transitions with a low congestion and Figure 7 (right) shows only the transitions with a high congestion.

A more detailed inspection shows that the edges with the highest edge congestion are precisely those which are closest to the saddle point. In fact, this is precisely what has to be expected. Thus, roughly speaking we have the following rule of thumb:

High congestion identifies saddle points.

On the other hand one has to expect that the partition of phase space into almost invariant sets – in particular for Hamiltonian systems – takes place at precisely such saddle points. We now examine the change of the magnitude of the congestion if two invariant sets exhibiting complicated dynamics are coupled and become almost invariant. For this the following observation is crucial.

Proposition 1. *Let G be a graph and denote by H the graph consisting of two copies of G which are connected by m additional edges. Let n be the number of vertices in G. Then we have the following estimate on the congestion of H:*

$$cong(H) \geq \frac{2n^2}{m}$$

Proof. The result follows from the fact that at least $2n^2$ commodities have to cross the m connecting edges.   □

Let us apply Proposition 1 to the particular cases which have already been listed in the Examples 1

Corollary 1.

a. *Suppose that G is a z-dimensional torus with n vertices. Then $cong(G) = \frac{n^{(z+1)/z}}{4}$ and the coupling of two such graphs leads to an increase of the congestion by a factor of at least $\frac{8}{m} n^{(z-1)/z}$.*
b. *Suppose that G is a k-dimensional hypercube with $n = 2^k$ vertices. Then $cong(G) = n$ and the coupling leads to an increase of the congestion by a factor of at least $\frac{2n}{m}$.*
c. *If G is a complete graph with n vertices then $cong(G) = 2$. Thus, the congestion of the coupled graph will increase by a factor of at least $\frac{n^2}{4m}$.*

A couple of remarks are in order.

Remark 5.

a. Observe that in applications typically m will be very small in comparison to the number of vertices of G.

b. From the application point of view Corollary 1 can be interpreted in the following way. Suppose that the dynamics inside two almost invariant sets can adequately be modeled by a z-dimensional torus, a k-dimensional hypercube or a complete graph. Then the congestion will roughly drop by a factor of $\frac{8}{m}n^{(z-1)/z}$, by a factor of $\frac{2n}{m}$ or by a factor of $\frac{n^2}{4m}$ if two almost invariant sets are separated by a cut through m edges. Thus, this result has to be viewed as a tool for the identification of the internal dynamics inside almost invariant sets.

c. Observe that the torus and the hypercube are realistic models for the description of complicated dynamics inside the almost invariant sets. However, in the computations in the following section we sometimes even observe a more drastic decrease in the congestion than predicted by Corollary 1. This indicates that in those cases a model with an (almost) all-to-all coupling for the dynamical behavior, i.e. the complete graph, would be even more appropriate.

6   Identification of the Number of Almost Invariant Sets

We want to partition the system into almost invariant sets. In this section we discuss the problem of identifying the number of almost invariant sets.

Again, we cover the relevant dynamics by a box collection and deal with the graph $G = (V, E)$ represented by the transition matrix as described in Section 2. We want to determine a number $p \in \mathbb{N}$ such that there is a partition π of V into p parts $V = S_1 \cup \ldots \cup S_p$ with the internal cost of π being high. In order to explore the macroscopic dynamics of the system, π should not contain a part which could further be partitioned into sub-parts while maintaining or improving the internal cost. In other words, any further partition of any part of π should lower the internal cost of π. We will refer to this characteristic of a part as being *compact*, i.e. we want to find all compact almost invariant parts in the graph. This leads us to a strategy of how to determine the number of compact parts. It can be phrased as a general method:

Recursively bisect the graph into 2 parts until all parts are compact.

One needs to solve two tasks in order to follow this strategy. Firstly, one has to define the compactness of a graph and has to be able to decide whether a part is compact or not. Secondly, one needs a bisection method which bisects a non-compact part into two sub-parts such that the bisection preserves compact areas within a sub-part, i.e. the cut does not intersect compact areas of the graph.

Many graph partitioning methods were originally designed as graph bisection heuristics. Some of them are easy generalizable to partition the graph into any number p of parts. However, some methods are very difficult to generalize. It is a common approach to use bisection heuristics recursively to get a partition into more than two parts. Several experiments have been performed to compare the direct p-partitioning approach with the recursive bisection approach. An analyses on this comparison is to be found in [42]. We will use the methods described in Section 3 to recursively calculate bisections of a graph.

There are several graph theoretical areas which define some sort of compactness of a part. Thus, there are several criteria which can be used as indicators for a compact part. As some examples, a part is called compact if

1. it has a large internal cost (equation 1) and there is no bisection of the part into 2 sub-parts with a similar or higher internal cost.
2. the sub-graph representing the part has a large gap between the first and the second eigenvalue (e.g. [17]).
   In this case, equation (5) indicates that the external cost of any further cut is large and, thus, a further bisection would decrease the internal cost.
3. the Markov chain representing the part has a high conductance.
   The conductance is defined in Section 2.2. It is also referred to as the *isoperimetric value* of the graph (see e.g. [1]). Clearly, if the part has a high conductance, any further bisection of the graph would lead to a high connectivity of the two sub-parts.
4. **small congestion**.
   As stated in Section 5.1, a small congestion ensures a high external cost of any sub-part. Thus, any further bisection will decrease the internal cost of the partition.

Clearly, if a graph is considered compact with respect to one measure, it is also considered compact with respect to the other measures.

These characteristics are not only important in the analyses of dynamical systems, but they are also important in the explicit construction of compact graphs. Compact graphs have been studied for the construction of efficient topologies of communication networks for multi-processor systems. The main objective here is to construct a network such that the information flow in the network is as fast as possible and that there is no bottleneck within the network which could slow down the information flow.

We now illustrate the identification of the number of almost invariant sets on some examples.

Figure 8 illustrates the partitioning of the Pentane molecule from [9] which we already used in Section 3. As we can observe in the left most picture, the first bisection results in one part of 43 boxes and a very low congestion of 0.88. However, the other part consisting of 212 boxes has a high congestion of 168.82 . We continue to bisect parts with a congestion of more than 5. Thus, after a total of 4 bisection levels we get a partition into 7 parts and the highest congestion of any part is 3.67 .

A different method to derive the number of reasonable parts is to look at the sorted list of the eigenvalues of the Laplacian. A large gap in the value of two neighbored eigenvalues λ_i and λ_{i+1} is an indication to partition the graph into i parts. Figure 9 (left) shows the 20 smallest eigenvalues of the Pentane200, Pentane300 and Pentane600 graphs. As one can observe, there is a large gap between λ_7 and λ_8. This suggests to partition the Pentane300 into 7 parts and is conform with our calculation above using the congestion.

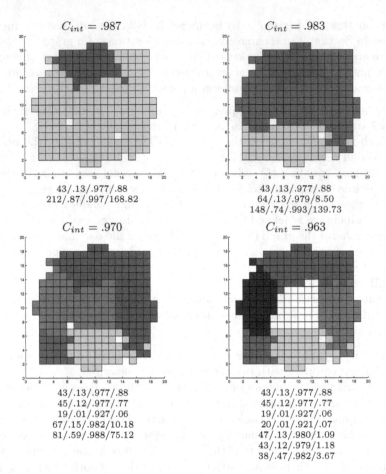

$C_{int} = .987$

43/.13/.977/.88
212/.87/.997/168.82

$C_{int} = .983$

43/.13/.977/.88
64/.13/.979/8.50
148/.74/.993/139.73

$C_{int} = .970$

43/.13/.977/.88
45/.12/.977/.77
19/.01/.927/.06
67/.15/.982/10.18
81/.59/.988/75.12

$C_{int} = .963$

43/.13/.977/.88
45/.12/.977/.77
19/.01/.927/.06
20/.01/.921/.07
47/.13/.980/1.09
43/.12/.979/1.18
38/.47/.982/3.67

Fig. 8. Recursive bisection of the Pentane300. The values indicate: number of boxes / invariant measure / internal cost C_{int} / congestion of subgraph. The graph has 255 vertices and a congestion of 139.67.

Figure 9 (right) shows the 20 smallest eigenvalues of the Laplacian of 'Hexane'. The large gap between λ_{17} and λ_{18} suggests to partition the graph into 17 parts. Table 2 shows the recursive bisection of Hexane by using the congestion as bisection criteria. Again, we recursively bisected the parts of the partition until all parts have a congestion of at most 5. This was achieved after 5 levels of recursion. The result is a partition into 17 parts with 1.65 being the highest congestion of any part. As we can see, the lowest internal cost of any of the 17 parts is .93.

Figure 10 shows a partition of Hexane into 17 parts. The example is 3-dimensional and the 3-dimensional boxes are drawn with a smaller size in order to be able to get a better visualization.

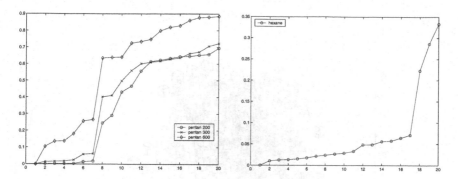

Fig. 9. Laplacian eigenvalues of Pentane200, Pentane300 and Pentane600 (left) and Hexane (right).

Table 2. Partitioning of Hexane into 17 parts with recursive bisection using the congestion as bisection criteria. The values in the first 6 columns indicate: number of boxes / congestion. The last two columns list the invariant measure and the internal cost of each of the resulting parts.

level 0	level 1	level 2	level 3	level 4	level 5	$\mu(S)$	$C_{int}(S)$
397/208.91	162/174.51	84/145.80	47/7.61	15/.05		.009	.95
				32/.63		.086	.98
			37/.71			.087	.98
		78/5670.31	51/1044.50	19/.12		.023	.98
				32/216.01	24/.21	.021	.98
					8/.01	.001	.93
			27/1.65			.320	.98
	235/506.16	90/139.35	38/214.00	10/.02		.002	.94
				28/.68		.092	.98
			52/8.23	17/.05		.010	.94
				35/.73		.090	.98
		145/141.11	46/5.88	29/.76		.085	.98
				17/.04		.009	.94
			99/30.13	59/16.68	26/.18	.026	.97
					33/1.32	.100	.98
				40/569.80	21/.12	.010	.94
					19/.15	.030	.98

7 Conclusion

The macroscopic behavior of a dynamical system can be analyzed by an identification of almost invariant sets in phase space. The calculation of almost invariant sets requires an appropriate partitioning of the phase space. We have addressed this topic by an application of graph partitioning theory in this context and use graph partitioning algorithms to calculate almost invariant sets of the system. We have shown that the congestion of the system is an appropriate measure of

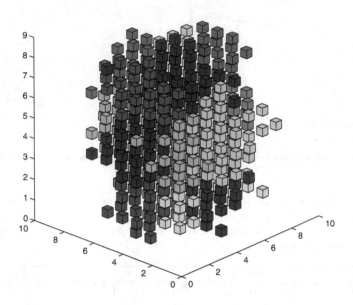

Fig. 10. Partition of Hexane into 17 parts

the quality of a partition. Furthermore, the congestion can be used to identify the number of almost invariant sets.

We view the results of this paper as an important step towards the application of graph theoretical results and algorithms in the field of dynamical systems exhibiting multiscale dynamics. So far, only a limited number of aspects are considered. For instance, it would be an interesting question to develop methods for the identification of appropriate graph models representing the dynamics inside the almost invariant sets. Then one could obtain even stronger statements about the change in congestion by an application of Proposition 1.

Finally the algorithmic methods presented eventually have to be combined with multilevel subdivision techniques which are used for the construction of the box coverings. This way a more efficient identification of the macroscopic structure of the dynamical behavior can be accomplished.

Acknowledgments

We thank Christof Schütte and his group at the Freie Universität Berlin for the very fruitful cooperation. In particular our treatment of the examples of the Pentane and Hexane molecules would not have been possible if Alexander Fischer and Wilhelm Huisinga would not have provided the corresponding data. We also thank Anthony Baker for patient discussions on the contents of this article.

References

1. N. Alon. On the edge-expansion of graphs. *Combinatorics, Probability and Computing*, 6:145–152, 1997.
2. S.L. Bezroukov, R. Elsässer, B. Monien, R. Preis, and J.-P. Tillich. New spectral lower bounds on the bisection width of graphs. In *Workshop on Graph-Theoretic Concepts in Computer Science (WG)*, LNCS 1928, pages 23–34, 2000.
3. N. Bouhmala. Impact of different graph coarsening schemes on the quality of the partitions. Technical Report RT98/05-01, University of Neuchatel, Department of Computer Science, 1998.
4. F.R.K. Chung. *Spectral Graph Theory*, volume 92 of *CBMS Regional conference series in mathematics*. American Mathematical Society, 1997.
5. T.H. Cormen, C.E. Leiserson, and R.L. Rivest. *Introduction to Algorithms*. MIT Press, 1990.
6. M. Dellnitz and A. Hohmann. A subdivision algorithm for the computation of unstable manifolds and global attractors. *Numerische Mathematik*, 75:293–317, 1997.
7. M. Dellnitz and O. Junge. On the approximation of complicated dynamical behavior. *SIAM J. Numer. Anal.*, 36(2):491–515, 1999.
8. P. Deuflhard, M. Dellnitz, O. Junge, and Ch. Schütte. Computation of essential molecular dynamics by subdivision techniques. In Deuflhard et al., editor, *Computational Molecular Dynamics: Challenges, Methods, Ideas*, LNCSE 4, 1998.
9. P. Deuflhard, W. Huisinga, A. Fischer, and Ch. Schütte. Identification of almost invariant aggregates in reversible nearly uncoupled markov chains. *Lin. Alg. Appl.*, 315:39–59, 2000.
10. R. Diekmann, A. Frommer, and B. Monien. Efficient schemes for nearest neighbor load balancing. *Parallel Computing*, 25(7):789–812, 1999.
11. R. Diekmann, B. Monien, and R. Preis. Using helpful sets to improve graph bisections. In D.F. Hsu, A.L. Rosenberg, and D. Sotteau, editors, *Interconnection Networks and Mapping and Scheduling Parallel Computations*, volume 21 of *DIMACS Series in Discrete Mathematics and Theoretical Computer Science*, pages 57–73. AMS, 1995.
12. R. Elsässer, A. Frommer, B. Monien, and R. Preis. Optimal and alternating-direction loadbalancing schemes. In P. Amestoy et al., editor, *Euro-Par'99 Parallel Processing*, LNCS 1685, pages 280–290, 1999.
13. R. Elsässer, T. Lücking, and B. Monien. New spectral bounds on k-partitioning of graphs. In *Proc. of the Symposium on Parallel Algorithms and Architectures (SPAA)*, pages 255–270, 2001.
14. R. Elsässer, B. Monien, and R. Preis. Diffusive load balancing schemes on heterogeneous networks. In *12th ACM Symp. on Parallel Algorithms and Architectures (SPAA)*, pages 30–38, 2000.
15. C.M. Fiduccia and R.M. Mattheyses. A linear-time heuristic for improving network partitions. In *Proc. IEEE Design Automation Conf.*, pages 175–181, 1982.
16. M. Fiedler. A property of eigenvectors of nonnegative symmetric matrices and its application to graph theory. *Czechoslovak Mathematical J., Praha*, 25(100):619–633, 1975.
17. A. Fischer, Ch. Schütte, P. Deuflhard, and F. Cordes. Hierarchical uncoupling-coupling of metastable conformations. In Schlick and Gan, editors, *Proc. of the 3rd International Workshop on Algorithms for Macromolecular Modeling*, LNCSE 24, 2002.

18. G. Froyland and M. Dellnitz. Detecting and locating near-optimal almost-invariant sets and cycles. Technical report, DFG-Priority Program 1095: "Analysis, Modeling and Simulation of Multiscale Problems", 2001.
19. G. Froyland and M. Dellnitz. μ almost-invariant sets and adaptive boundary refinement. manuscript, 2003.
20. M.R. Garey and D.S. Johnson. *Computers and Intractability - A Guide to the Theory of NP-Completeness.* Freemann, 1979.
21. A. Gupta. Fast and effective algorithms for graph partitioning and sparse matrix reordering. *IBM J. of Research and Development,* 41:171–183, 1997.
22. B. Hendrickson and R. Leland. The chaco user's guide: Version 2.0. Technical Report SAND94-2692, Sandia National Laboratories, Albuquerque, NM, 1994.
23. B. Hendrickson and R. Leland. A multilevel algorithm for partitioning graphs. In *Proc. Supercomputing '95.* ACM, 1995.
24. J. Hromkovič and B. Monien. The bisection problem for graphs of degree 4 (configuring transputer systems). In Buchmann, Ganzinger, and Paul, editors, *Festschrift zum 60. Geburtstag von Günter Hotz,* pages 215–234. Teubner, 1992.
25. W. Huisinga. *Metastability of Markovian systems.* Phd thesis, Freie Universität Berlin, 2001.
26. G. Karypis and V. Kumar. *METIS Manual, Version 4.0.* University of Minnesota, Department of Computer Science, 1998.
27. G. Karypis and V. Kumar. Multilevel k-way partitioning scheme for irregular graphs. *J. of Parallel and Distributed Computing,* 48:96–129, 1998.
28. G. Karypis and V. Kumar. A fast and high quality multilevel scheme for partitioning irregular graphs. *SIAM J. on Scientific Computing,* 20(1), 1999.
29. B.W. Kernighan and S. Lin. An effective heuristic procedure for partitioning graphs. *The Bell Systems Technical J.,* pages 291–307, 1970.
30. F.T. Leighton. *Introduction to Parallel Algorithms and Architectures: Arrays, Trees, Hypercubes.* Morgan Kaufmann Publishers, 1992.
31. B. Mohar. Some applications of laplace eigenvalues of graphs. In *Graph Symmetry: Algebraic Methods and Applications,* NATO ASI Ser. C 497, pages 225–275, 1997.
32. B. Monien and R. Diekmann. A local graph partitioning heuristic meeting bisection bounds. In *8th SIAM Conf. on Parallel Processing for Scientific Computing,* 1997.
33. B. Monien and R. Preis. Bisection width of 3- and 4-regular graphs. In *26th International Symposium on Mathematical Foundations of Computer Science (MFCS),* LNCS 2136, pages 524–536, 2001.
34. B. Monien, R. Preis, and R. Diekmann. Quality matching and local improvement for multilevel graph-partitioning. *Parallel Computing,* 26(12):1609–1634, 2000.
35. F. Pellegrini. SCOTCH 3.1 user's guide. Technical Report 1137-96, LaBRI, University of Bordeaux, 1996.
36. R. Ponnusamy, N. Mansour, A. Choudhary, and G.C. Fox. Graph contraction for mapping data on parallel computers: A quality-cost tradeoff. *Scientific Programming,* 3:73–82, 1994.
37. A. Pothen, H.D. Simon, and K.P. Liu. Partitioning sparse matrices with eigenvectors of graphs. *SIAM J. on Matrix Analysis and Applications,* 11(3):430–452, 1990.
38. R. Preis. *The PARTY Graphpartitioning-Library, User Manual - Version 1.99.* Universität Paderborn, Germany, 1998.
39. R. Preis. *Analyses and Design of Efficient Graph Partitioning Methods.* Heinz Nixdorf Institut Verlagsschriftenreihe, 2000. Dissertation, Universität Paderborn, Germany.

40. Ch. Schütte. *Conformational Dynamics: Modelling, Theory, Algorithm, and Application to Biomolecules.* Habilitation thesis, Freie Universität Berlin, 1999.
41. N. Sensen. Lower bounds and exact algorithms for the graph partitioning problem using multicommodity flows. In *Proc. European Symposium on Algorithms (ESA),* LNCS 2161, pages 391–403, 2001.
42. H.D. Simon and S.-H. Teng. How good is recursive bisection? *SIAM J. on Scientific Computing,* 18(5):1436–1445, 1997.
43. A. Sinclair. *Algorithms for Random Generation & Counting: A Markov Chain Approach.* Progress in Theoretical Computer Science. Birkhäuser, 1993.
44. C. Walshaw. *The Jostle user manual: Version 2.2.* University of Greenwich, 2000.

Datagraphs in Algebraic Geometry and K3 Surfaces

Gavin Brown

Mathematics Institute, University of Warwick, Coventry, CV4 7AL, UK
gavinb@maths.warwick.ac.uk

Abstract. *Datagraphs* are combinatorial graphs having database items at their vertices and geometric relationships along their edges. I describe their applicability to lists of examples in algebraic geometry generated by computer algebra, and illustrate this with a list of K3 surfaces as the database items. The main point is that when analysing a single surface during construction of the database, the datagraph makes available its close relations, and that very often these provide extra information required to complete the analysis.

1 Introduction: Lists in Algebraic Geometry

Beautiful lists of examples pervade algebraic geometry. Sometimes they comprise genuine classifications, such as the list of Du Val singularities in Table 1, see [D] for instance. At other times they are simply a list of known examples making no pretence of being comprehensive. In either case one expects there to be relationships between items on the list. A datagraph is a device to record geometric objects together with these relationships as separate pieces of data supporting one another. There are a lot of mathematical issues, but here I focus on the symbolic calculations that make the general method work.

Perhaps the best list of all is that which constitutes the statement of the classification of surfaces: if S is a nonsingular projective algebraic surface, defined over $\mathbb{C}$ say, then S is birational to a nonsingular surface $\tilde{S}$, a *minimal model* of S, in exactly one of the classes listed in Table 2. The Kodaira dimension κ is the order of growth of sections of nK_S with n, that is $h^0(nK_S) \sim n^\kappa$ with a conventional $\kappa = -\infty$ when no multiple has any sections. See Beauville [B] or Reid [R2] for an account of this classification.

The key example studied here is a list of K3 surfaces, described in detail below. But the general methods used can be applied to many similar situations

Table 1. Du Val surface singularities

(A_n)	$x^2 + y^2 + z^{n+1} = 0$	for $n \geq 1$
(D_n)	$x^2 + y^2z + z^{n-1} = 0$	for $n \geq 4$
(E_6)	$x^2 + y^3 + z^4 = 0$	
(E_7)	$x^2 + y^3 + yz^3 = 0$	
(E_8)	$x^2 + y^3 + z^5 = 0$	

F. Winkler and U. Langer (Eds.): SNSC 2001, LNCS 2630, pp. 210–224, 2003.

in algebraic geometry. Certainly many aspects of each class of surface in Table 2 have appeared in lists. For example, Reid [R1] worked out the case of a particular type of surface of general type. In fact, the approach described here has evolved from Reid's work over 20 years starting with this calculation. Lists of subcanonical curves and Enriques surfaces are in progress. Szendrői [Sz] has lists of Calabi–Yau 3-folds, three-dimensional analogues of K3 surfaces, and physicists have many more [KS]. And there are plenty of other lists of K3 surfaces from different points of view, the extremal elliptic fibrations of Shimada and Zhang [SZ] for instance, or [KS] again.

Table 2. The classification of surfaces

κ	class of minimal surfaces
2	surfaces of general type
1	certain elliptic fibrations
0	Abelian, K3, Enriques or bielliptic surfaces
$-\infty$	ruled or rational surfaces

As well as being enjoyable in their own right, these lists are very useful. Belcastro [BC] computes the Picard lattices of a list of K3 hypersurfaces known as the famous 95, described below, in a study related to mirror symmetry. Corti, Pukhlikov and Reid [CPR] use the same famous 95 as a hunting ground for examples of *birational links*, the basic birational maps between 3-folds. Although their results are for 3-folds, much of the calculation occurs on a hyperplane section, which in this case is a K3 surface.

Let me clarify the main point. The aim is to construct databases, exploiting relationships between objects of the final database during the construction. First one accumulates a lot of plausible but rather raw data that is implicit in the problem: collections of singularities that are expected to appear on the objects is a typical example. Then refinement steps apply geometrical constructions to the raw data to construct the desired objects, or perhaps to prove that they don't exist. If X and Y appear in the database and we have a map $X \rightarrow Y$ between them, then the construction of X may be able to inherit calculations used in the construction of Y. In fact, this is exactly what happens in the K3 case discussed here. Section 4 below illustrates this point with a naive example, while Section 5 works out the K3 case. In the final section we glimpse the Magma [Ma] implementation, at least enough to see how it follows the model.

As an application, the main theorem, discussed in Section 2, uses a trivial computer analysis of the datagraphs to identify K3s that are 'primitive' from the point of view of projection. It is worth remarking that although many computer-generated lists exist in the geometry literature, I cannot think of a single one that is available in a format to be analysed seriously using a computer, rendering them pointless if they are at all large.

2   K3 Surfaces and Polarisation

A *K3 surface* is a simply-connected projective surface having only A_n singularities and trivial canonical class. (This is a slight modification of the usual definition, but it suits the discussion here and does no harm.) If S is a K3 surface in this sense, then the minimal resolution of singularities $\widetilde{S}$ of S appears in the list of Table 2. But the definition of K3 is not so important here: we don't use it directly at all. It appears only implicitly in the Riemann–Roch formula (1) of Section 5. A favourite example of K3 is the Kummer surface of a curve of genus 2, a surface of degree 4 in $\mathbb{P}^3$ with sixteen A_1 singularities coming from the fixed points of the involution -1 on the Jacobian.

In 1979, Reid wrote down the complete list of 95 weighted hypersurfaces that are K3 surfaces, the 'famous 95' (see them in [Fl] or [K3]). An example is

$$X_{30} \subset \mathbb{P}(5,6,8,11)$$

that is, the locus of zeros of a general homogeneous polynomial of degree 30 in variables x, y, z, t, which themselves have weights $5, 6, 8, 11$ respectively. After a change of coordinates, any such surface can be written as

$$f_{30} = 0 \quad \text{where} \quad f_{30} = x^6 + y^5 + yz^3 + zt^2 + \lambda xyzt$$

for some $\lambda \in \mathbb{C}$. Notice then that although one refers to *the* K3 surface X_{30}, it actually denotes a whole family of K3 surfaces parametrised by λ. Implicitly one usually has in mind only the general elements of this family, although special members of the family will be interesting too.

Notice that such K3 surfaces are automatically *polarised*. That is, they come as a pair X, A of a K3 surface X together with an ample Weil divisor A. The class A contains the divisors of rational functions of weight 1. Consider the example X_{30}. The divisor of y/x lies in the class of A but it is clearly not an effective divisor. In fact, the smallest monomial weight is 5 so $H^0(nA) = 0$ for $n = 1, 2, 3, 4$ while $H^0(5A)$ is 1-dimensional with basis x. Nonetheless, A is ample since multiples of it really do embed X and we still refer to A as the *hyperplane section*, even though there is no effective divisor in its class.

One calculates that the general X_{30} has singularities A_1, A_7 and A_{10}. These singularities are also polarised. For example, $A_7 \in X_{30}$ is the singularity at the point $p_z = (0{:}0{:}1{:}0)$. Since $z = 1$ near p_z, f_{30} describes y as an implicit function and we can eliminate y there, realising the A_7 singularity as a $\mathbb{Z}/8$ quotient of $\mathbb{C}^2$. Such a quotient can happen in several ways. In this case, the global functions x, t restrict to the local axes u, v of the quotient and, modulo 8, the weights of x, t are $5, 3$. This polarised singularity is often denoted $\frac{1}{8}(3, 5)$. As a reasonable shorthand for computer input we denote such singuarities by a pair of coprime integers $[r, a]$ with $1 \le a \le r/2$ where r is the order of the cyclic group, called the *index* of the polarised singularity, and a is the smaller of its weights. So the A_7 singularity on X_{30} is denoted $[8, 3]$. Fletcher's paper [Fl] contains a tutorial on these calculations.

One can project away from either of the singularities A_7 and A_{10}. After projection from A_7 the resulting surface is again a K3 surface, this time

$$X_{66} \subset \mathbb{P}(5, 6, 22, 33).$$

Such a projection is called a *Gorenstein projection*. The definition is rather involved so suffice it to say that a Gorenstein projection from a singularity $\frac{1}{r}(a, b)$ where $a + b = r$ results in a surface with two singularities $\frac{1}{a}(r, -r)$ and $\frac{1}{b}(r, -r)$. (If $a = 1$ then the $\frac{1}{a}$ point is nonsingular so we omit it; similarly for b.) The singularities of X_{66} are $\frac{1}{2}(1, 1)$, $\frac{1}{3}(1, 2)$, $\frac{1}{5}(2, 3)$ and $\frac{1}{11}(5, 6)$ so this projection calculus works, breaking up the $\frac{1}{8}(3, 5)$ into two singularities.

It is often the case that Gorenstein projection is simply elimination of a single variable, as is the case with usual projection. This is called Type I Gorenstein projection. It occurs when the local polarising weights a, b are also the weights of global variables, that is, when a, b appear among the weights of the ambient weighted projective space. The A_{10} singularity above is actually $\frac{1}{11}(5, 6) \in X_{30}$ — the monomial zt^2 eliminates z at this point — so projection from there is Type I, whereas projection from $\frac{1}{8}(3, 5) \in X_{30}$ is not. Projections of other types are also defined.

Since the famous 95 were computed, a number of Reid's students have extended the list to higher codimension: Fletcher [Fl] found 84 codimension 2 cases and Altınok [A] found more in codimension 3 and 4. In total these lists contain 391 K3 surfaces. (A tentative list of 162 codimension 5 K3s is now at [K3].)

There are interesting questions of verification for these lists. Reid's list of hypersurfaces is known to be complete — for instance, Johnson and Kollár [JK] have a proof (and a list of several thousand Fano 4-folds). This means that if $X \subset \mathbb{P}(a, b, c, d)$ is a K3 hypersurface then it is one of the famous 95 (although not necessarily a general element of the family). In higher codimension there is no such proof, although the lists have survived for several years and are probably correct if one rules out certain degenerations which were deliberately overlooked. A more immediate problem is to prove that every element on these lists actually exists. This is easy for the famous 95 since one can write down equations and compute. Similar explicit calculations work in codimensions 2 and 3. But in codimension 4 one has no structure theorem explaining how to write down the equations. So Altınok [A] resorts to a clever trick, called *unprojection*, to justify much of her list. The idea is that if a variety exists one can make a Gorenstein projection from one of its singularities, as in the example $X_{30} \dashrightarrow X_{66}$ above. With luck the result will be a K3 surface in lower codimension. If one can identify that K3 in advance and make certain polynomial calculations it is sometimes possible to reverse the projection process and view it as a construction. Thus while some items on Altınok's lists are not described as explicitly as X_{30}, they certainly all exist and carry implicit methods for their construction.

I will describe a model for constructing such lists as part of a computer algebra package. This model has been realised to build lists of K3 surfaces in the Magma system [Ma]. The current version 2.9 of Magma contains the database of 391 known K3 surfaces in codimension at most 4. I will discuss a substantial revision of this which already exists in prototype. (See [K3] for the most recent information.) Of course one benefits at every step from extra mathematical knowledge, but even if that is hard to come by one can make progress. This is typified by the amazing examples of [BR] which were found even before we knew about the unprojections involved. Or consider the K3 datagraphs themselves. Anyone who knows about K3 surfaces will have noticed that there is no mention of Picard lattices, one of the basic pieces in the theory of K3 surfaces. Indeed they play no role in the initial construction of the database, although it would be possible to use them. Subject specific details like these can be applied to basic datagraphs at a later stage.

As an application of the K3 datagraphs I prove the following theorem.

Theorem 1. *Let X, A be a polarised K3 surface having Du Val singularities and no Gorenstein projection to another polarised K3 surface. Let $g = \dim H^0(A) - 1$ and assume in addition that $g \leq 2$. Then X, A has the same Hilbert series as one of the following:*

g	X
-1	codim $X \leq 4$ *and* X *is one of the 36 cases in Table 4*
0	codim $X = 1$ *and* X *is one of the 6 cases in Table 3*
1	$X = X_{12} \subset \mathbb{P}(1,1,4,6)$
2	$X = X_6 \subset \mathbb{P}(1,1,1,3)$

In Tables 3 and 4, each K3 surface X is identified by its number in the K3 datagraph of the same genus: $\frac{1}{r}$ is shorthand for $\frac{1}{r}(1, r - 1)$ and equation (2) in Section 5 determines A^2. More detailed Magma output is at [K3].

Table 3. K3 surfaces with $g = 0$ and no Gorenstein projections

No.	K3	Basket $\mathcal{B}$	Degree A^2
1	$X_{42} \subset \mathbb{P}(1,6,14,21)$	$\frac{1}{2}, \frac{1}{3}, \frac{1}{7}$	$1/42$
12	$X_{36} \subset \mathbb{P}(1,5,12,18)$	$\frac{1}{5}(2,3), \frac{1}{6}$	$1/30$
23	$X_{30} \subset \mathbb{P}(1,4,10,15)$	$\frac{1}{2}, \frac{1}{4}, \frac{1}{5}$	$1/20$
77	$X_{24} \subset \mathbb{P}(1,3,8,12)$	$2 \times \frac{1}{3}, \frac{1}{4}$	$1/12$
249	$X_{18} \subset \mathbb{P}(1,2,6,9)$	$3 \times \frac{1}{2}, \frac{1}{3}$	$1/6$
794	$X_{10} \subset \mathbb{P}(1,2,2,5)$	$5 \times \frac{1}{2}$	$1/2$

What is this theorem about? If a general K3 surface X, A has $g = 3$ then the sections of A embed X (modulo -2 curves) as a hypersurface $X_4 \subset \mathbb{P}^3$. The generality assumption in this case is exactly the same as asking that a genus 3 curve be nonhyperelliptic: indeed, X, A is general in this sense if and only if the general element of the linear system of A is a nonhyperelliptic curve of genus 3. If $g > 3$ then again with some generality hypothesis A already embeds X in an ordinary projective space. One could view these $g \geq 3$ cases as satisfactory. The theorem studies the other cases, those for which A does *not* have enough sections to embed X, assuming in addition that X cannot be constructed by

Table 4. K3 surfaces with $g = -1$ and no Gorenstein projections

No.	K3	Basket $\mathcal{B}$	Degree A^2	cod
1	$X_{24,30} \subset \mathbb{P}(8,9,10,12,15)$	$\frac{1}{2},\frac{1}{3},\frac{1}{4},\frac{1}{5}(2,3),\frac{1}{9}$	1/180	2
4	$X_{50} \subset \mathbb{P}(7,8,10,25)$	$\frac{1}{2},\frac{1}{5}(2,3),\frac{1}{7}(3,4),\frac{1}{8}$	1/280	1
5	$X_{36} \subset \mathbb{P}(7,8,9,12)$	$\frac{1}{3},\frac{1}{4},\frac{1}{7}(2,5),\frac{1}{8}$	1/168	1
9	$X_{18,30} \subset \mathbb{P}(6,8,9,10,15)$	$2\times\frac{1}{2},2\times\frac{1}{3},\frac{1}{5},\frac{1}{8}$	1/120	2
12	$X \subset \mathbb{P}(6,6,7,8,9,10,11)$	$2\times\frac{1}{2},2\times\frac{1}{3},\frac{1}{6},\frac{1}{7}$	1/42	4
18	$X_{40} \subset \mathbb{P}(5,7,8,20)$	$\frac{1}{4},2\times\frac{1}{5}(2,3),\frac{1}{7}$	1/140	1
21	$X_{66} \subset \mathbb{P}(5,6,22,33)$	$\frac{1}{2},\frac{1}{3},\frac{1}{5}(2,3),\frac{1}{11}(5,6)$	1/330	1
22	$X_{38} \subset \mathbb{P}(5,6,8,19)$	$\frac{1}{2},\frac{1}{5},\frac{1}{6},\frac{1}{8}(3,5)$	1/120	1
24	$X_{27} \subset \mathbb{P}(5,6,7,9)$	$\frac{1}{3},\frac{1}{5},\frac{1}{6},\frac{1}{7}(2,5)$	1/70	1
26	$X_{16,\dots,20} \subset \mathbb{P}(5,6,7,8,9,10)$	$\frac{1}{2},\frac{1}{3},\frac{1}{5},\frac{1}{5}(2,3),\frac{1}{7}$	1/42	3
33	$X \subset \mathbb{P}(5,6,6,7,8,9,10)$	$\frac{1}{2},\frac{1}{3},\frac{1}{5}(2,3),2\times\frac{1}{6}$	1/30	4
42	$X_{14,\dots,18} \subset \mathbb{P}(5,5,6,7,8,9)$	$\frac{1}{5},2\times\frac{1}{5}(2,3),\frac{1}{6}$	1/30	3
48	$X_{34} \subset \mathbb{P}(4,6,7,17)$	$2\times\frac{1}{2},\frac{1}{4},\frac{1}{6},\frac{1}{7}(2,5)$	1/84	1
50	$X_{16,18} \subset \mathbb{P}(4,6,7,8,9)$	$2\times\frac{1}{2},\frac{1}{3},2\times\frac{1}{4},\frac{1}{7}$	1/42	2
59	$X_{54} \subset \mathbb{P}(4,5,18,27)$	$\frac{1}{2},\frac{1}{4},\frac{1}{5}(2,3),\frac{1}{9}(4,5)$	1/180	1
61	$X_{32} \subset \mathbb{P}(4,5,7,16)$	$2\times\frac{1}{4},\frac{1}{5},\frac{1}{7}(2,5)$	1/70	1
64	$X_{30} \subset \mathbb{P}(4,5,6,15)$	$2\times\frac{1}{2},\frac{1}{3},\frac{1}{4},2\times\frac{1}{5}$	1/60	1
68	$X_{14,16} \subset \mathbb{P}(4,5,6,7,8)$	$\frac{1}{2},2\times\frac{1}{4},\frac{1}{5}(2,3),\frac{1}{6}$	1/30	2
130	$X_{12,14} \subset \mathbb{P}(4,4,5,6,7)$	$2\times\frac{1}{2},3\times\frac{1}{4},\frac{1}{5}$	1/20	2
162	$X_{24} \subset \mathbb{P}(3,6,7,8)$	$\frac{1}{2},4\times\frac{1}{3},\frac{1}{7}$	1/42	1
167	$X_{48} \subset \mathbb{P}(3,5,16,24)$	$2\times\frac{1}{3},\frac{1}{5},\frac{1}{8}(3,5)$	1/120	1
170	$X_{21} \subset \mathbb{P}(3,5,6,7)$	$3\times\frac{1}{3},\frac{1}{5}(2,3),\frac{1}{6}$	1/30	1
184	$X_{42} \subset \mathbb{P}(3,4,14,21)$	$\frac{1}{2},2\times\frac{1}{3},\frac{1}{4},\frac{1}{7}(3,4)$	1/84	1
191	$X_{24} \subset \mathbb{P}(3,4,5,12)$	$2\times\frac{1}{3},2\times\frac{1}{4},\frac{1}{5}(2,3)$	1/30	1
200	$X_{18} \subset \mathbb{P}(3,4,5,6)$	$\frac{1}{2},3\times\frac{1}{3},\frac{1}{4},\frac{1}{5}$	1/20	1
269	$X_{10,12} \subset \mathbb{P}(3,4,4,5,6)$	$\frac{1}{2},2\times\frac{1}{3},3\times\frac{1}{4}$	1/12	2
420	$X_{15} \subset \mathbb{P}(3,3,4,5)$	$5\times\frac{1}{3},\frac{1}{4}$	1/12	1
544	$X_{30} \subset \mathbb{P}(2,6,7,15)$	$5\times\frac{1}{2},\frac{1}{3},\frac{1}{7}$	1/42	1
551	$X_{42} \subset \mathbb{P}(2,5,14,21)$	$3\times\frac{1}{2},\frac{1}{5},\frac{1}{7}(2,5)$	1/70	1
555	$X_{26} \subset \mathbb{P}(2,5,6,13)$	$4\times\frac{1}{2},\frac{1}{5}(2,3),\frac{1}{6}$	1/30	1
580	$X_{22} \subset \mathbb{P}(2,4,5,11)$	$5\times\frac{1}{2},\frac{1}{4},\frac{1}{5}$	1/20	1
608	$X_{30} \subset \mathbb{P}(2,3,10,15)$	$3\times\frac{1}{2},2\times\frac{1}{3},\frac{1}{5}(2,3)$	1/30	1
627	$X_{18} \subset \mathbb{P}(2,3,4,9)$	$4\times\frac{1}{2},2\times\frac{1}{3},\frac{1}{4}$	1/12	1
820	$X_{12} \subset \mathbb{P}(2,3,3,4)$	$3\times\frac{1}{2},4\times\frac{1}{3}$	1/6	1
1611	$X_{14} \subset \mathbb{P}(2,2,3,7)$	$7\times\frac{1}{2},\frac{1}{3}$	1/6	1
2577	$X_{6,6} \subset \mathbb{P}(2,2,2,3,3)$	$9\times\frac{1}{2}$	1/2	2

unprojection from some other K3 surface. To avoid the question of generality, the theorem is phrased in terms of Hilbert series, as explained in Section 3. The (families of) K3 surfaces listed in the theorem each comprise an open subset of their Hilbert scheme, so X itself will lie in these families if it is sufficiently general.

Let me be clear about the proof of this theorem: one simply reads the result from the datagraphs of K3 surfaces for each $g \leq 2$ — I use a computer routine, but it could be done by hand. It could be viewed as fortunate that the result does not involve high codimension K3 surfaces that I would not be able to analyse, although the two codimension 4 surfaces in Table 4 — Altınok's two most difficult examples — are already mysterious. In Section 6, I indicate how the K3 datagraph is used in this proof: it could even be used to help to find a construction of these two surfaces by locating useful projections.

3   Varieties, Graded Rings and Hilbert Series

I outline briefly the geometric background from [ABR]. Let R be a finitely generated graded ring. We usually assume that $R_0 = \mathbb{C}$ and that $R = \bigoplus_{n \geq 0} R_n$ with $R_n \cdot R_m \subset R_{n+m}$ and each R_n a finite dimensional vector space over $\mathbb{C}$. The *Hilbert series of* R is the formal power series

$$P_R(t) = \sum_{n=0}^{\infty} (\dim R_n)\, t^n.$$

Let X, A be a polarised variety, that is a projective variety X together with an ample Weil divisor A. (In practice both will be required to satisfy additional hypotheses.) It is common to take some very ample multiple nA of A and consider X in its embedding by the sections $H^0(nA)$ of nA. The classical example is when X is a nonhyperelliptic curve and A is its canonical class. Then A is already very ample and the embedding is the canonical embedding of X. Even if X is hyperelliptic, one can take $2A$ or $3A$ to embed it.

It is often better not to restrict to a single multiple of A, but to take them all at once. To do this, one embeds X by all generators of the finitely generated ring

$$R(X, A) = \bigoplus_{n \geq 0} H^0(nA)$$

in some *weighted* projective space. Then $X = \operatorname{Proj} R(X, A)$ and A is the natural $\mathcal{O}_X(1)$. Again there is a classical example, the embedding of a hyperelliptic curve $(y^2 = f(x)) \subset \mathbb{P}(1, g+1, 1)$ by all multiples of $A = 2p$ where $p \in X$ is a Weierstraß point. The weights of the projective space are the same as the degrees of the generators of $R(X, A)$ (to fix up the homogeneity of the embedding). In this case, the two 1s are the degrees of the functions defining the pencil $g_2^1 = |A|$ while the $g + 1$ is the degree of the first section of a multiple of A which is not a power of the g_2^1.

So there is a general setup as follows:

polarised variety		graded ring		Hilbert series
X, A	$\longleftrightarrow$	$R(X, A)$	$\longrightarrow$	$P_R(t)$

where the arrows are natural translations. In an attempt to construct varieties we will work from right to left. The obvious problem is that there is no translation in that direction. Nonetheless, we will prescribe some invariants — ones that appear in the Riemann–Roch formula — of varieties X, A and use Riemann–Roch to compute $h^0(nA) = \dim H^0(nA)$ for each n. We package all these values in a

Hilbert series $P(t)$. (It is true that we have to concern ourselves with the higher cohomology, but often we work with varieties having good vanishing results for cohomology.) The challenge is then to construct X, A from the Hilbert series. This is not an exact science, but many tricks are known, most of which can be applied mechanically to a datagraph.

4 Computer Generated Lists and Datagraphs

The point of datagraphs comes from the experience of building lists in algebraic geometry. Here I explain what datagraphs are and give a simplistic example. In Sections 5 and 6 I will work through the main example of this paper and show how they work in practice.

4.1 Datagraphs

To build a list of polarised geometric objects one first classifies the possible invariants that appear in a Riemann–Roch formula associated to the geometry. Each such piece of input generates a Hilbert series. The second stage, and the main problem, is to realise this as the Hilbert series of a suitable geometric object or even just show that such an object exists.

One thinks of the first part of the problem as being an easy combinatorial step (even when it is not) and the second part as being a difficult geometrical step. In the case of K3 surfaces, although we may have a plausible-looking Hilbert series, we have no general reason why there should be a surface X, A whose coordinate ring has this Hilbert series. And even if there is such an X, it may be difficult for us to prove that or to construct it.

Rather than regarding a database as being a list, we take it to be a directed combinatorial graph, the *datagraph*. It comprises a collection of vertices and a collection of directed edges, or arrows, between them. The basic skeleton of such a graph will be built at the combinatorial stage. Its vertices are in bijection with Hilbert series, while the directed edges correspond to Gorenstein projections between the corresponding K3 surfaces if they exist. The work required to prove existence, or deduce any other property of K3 surfaces does not then happen surface by surface, but for the most part inherits information from other surfaces by pulling it back along directed edges.

4.2 A Naive Example of the Method

Suppose I present you with a set V of n numbered points in the plane, no three of which are collinear, and ask for a list of the combinatorial graphs having the points of V as vertices and straight lines as edges. This is an active problem, not one which any fixed database you might already own will answer directly.

It is certainly easy to write down a list of all possible graphs L_n on n numbered vertices (having at most one edge between any two vertices). This is a raw pre-database. Some items on it may be realised as planar graphs and some may not, and it may be hard to decide which.

The graph with no edges and also graphs having exactly one edge can all be realised on vertices V. So we attempt an induction on the number of edges. We check whether a graph in L_k with k edges can be realised by removing an edge, answering the question for the resulting graph on $k - 1$ edges and then trying to redraw the edge. So there are two steps. The first is to identify in L_{k-1} those graphs which arise by removal of a single edge from a graph in L_k. This is part of the combinatorial step. The second is an algorithm which checks whether or not a new edge can be added to a given graph. This is the geometrical step.

The combinatorial step can be solved right at the beginning once and for all. Rather than taking the initial database of graphs to be a list of graphs $\cup L_n$, take it to be a *directed* graph G_c. The vertices of G_c are the graphs. There is a separate edge from any graph to each of the graphs which result by removing an edge. If you can implement an algorithm that checks whether a particular proposed edge can be added to a planar graph with result another planar graph, then you can set it working on G_c starting from the unique terminal vertex, the graph with no edges, creating the full subgraph G of G_c of all planar graphs, as required. The graph G is then a *datagraph*.

This datagraph certainly solves the initial problem — it lists all planar straight-line graphs on vertices V. And it already solves plenty of other problems. It is interesting to find maximal planar graphs, and in this context they are simply the vertices of G which have no edges directed into them. (For K3 surfaces, the interesting question is the complementary one of finding minimal vertices: this is Theorem 1.) Moreover, there are plenty of modifications to the question for which G_c is already useful. If I change the position of one of the n points, you only need to repeat the second geometrical step to G_c with the new points as basic data. Or maybe I relax the condition that edges must be straight lines, in which case one needs to re-implement the edge-adding algorithm before running it through G_c.

Presumably this is not a terribly effective approach to databases of graphs: such databases are usually large so the graph G_c would be unwieldy. But this approach is very effective for lists occurring in birational geometry: every feature of the process just described has a substantial counterpart there.

5   Building a Datagraph of K3 Surfaces

We will construct a datagraph of pairs X, A where X is a K3 surface and A is an ample Weil divisor (which we assume generates the analytic local class group at each singular point of X). The edges of the datagraph will be Gorenstein projections from the A_n singularities of X to other K3 surfaces.

5.1   The Combinatorial Step

Each K3 surface X, A is characterised by two pieces of data. The first is an integer $g \geq -1$, called the *genus*, equal to $h^0(A) - 1$ as in Theorem 1. The second is a *basket* $\mathcal{B}$ of singularities. A basket is a sequence of the polarised A_{r-1} singularities of X, $\frac{1}{r}(a, r - a)$ denoted by a pair of coprime integers $[r, a]$ with $1 \leq a \leq r/2$ as in Section 2.

The Riemann–Roch formula for K3 surfaces then computes $h^0(nA)$ for every $n \geq 1$. For any pair $(g, \mathcal{B})$, we collect these computations in a Hilbert series $P(t) = \sum_{n=0}^{\infty} h^0(nA)t^n$:

$$P(t) = \frac{1+t}{1-t} + \frac{t(1+t)}{2(1-t)^3}A^2 - \sum_{[r,a] \in \mathcal{B}} \frac{1}{2r(1-t^r)} \sum_{i=1}^{r-1} \overline{bi}(r - \overline{bi})t^i \qquad (1)$$

where b is the reciprocal of a modulo r and $\overline{c}$ denotes the smallest residue of c modulo r. This is completed by the formula

$$A^2 = 2g - 2 + \sum_{[r,a] \in \mathcal{B}} \frac{b(r - b)}{r}. \qquad (2)$$

Both formulas are worked out in Altınok's paper [A2].

There is one standard condition coming from the geometry of K3 surfaces which the basket of any K3 surface must satisfy:

$$\sum_{[r,a] \in \mathcal{B}} (r - 1) \leq 19.$$

The point is that resolving the A_n singularities produces a nonsingular K3 surface. Such a surface has at most 20 linearly independent divisors lying on it. One is the hyperplane section A. Each A_n singularity contributes a further $n - 1$ curves all independent of one another and of A. Imposing only this condition (and K3 experts will be aware of others), it is easy to check that there are exactly 6640 possible baskets that any K3 surface might have. So the collection of all possible combinatorial input is the infinite collection of pairs $(g, \mathcal{B})$ where $g \geq -1$ is an integer and $\mathcal{B}$ is one of the 6640 possible baskets. Since A is ample, it must be true that $A^2 > 0$ which is expressed by equation (2) as a condition on the pair $(g, \mathcal{B})$. One checks that this condition reduces the number of allowable baskets as follows:

genus	number of baskets
−1	4293
0	6491
1	6639
≥ 2	6640

The simple tricks one uses to try to construct a variety from a Hilbert series are far more effective when working with low genus so restricting our attention to pint-sized g, less that 2 say, is the better part of valour for now.

The combinatorial solution G_c is a directed graph as follows. First fix a small genus g. The graph G_c will have vertices in bijection with the collection of all baskets $\mathcal{B}$ for which $A^2 > 0$. For each pair $(g, \mathcal{B})$ we compute the Hilbert series $h_{\mathcal{B}}$. These power series have a natural order: $h_{\mathcal{B}_1} < h_{\mathcal{B}_2}$ if and only if the leading coefficient of $h_{\mathcal{B}_2} - h_{\mathcal{B}_1}$ is positive. We number the vertices $v_1, v_2 \ldots$ of G_c in this order.

We now define a surgery operation on baskets. Let $[r, a]$ be any singularity in a basket $\mathcal{B}$ and let $r = a + b$. Make a new basket by replacing $[r, a]$ with the two singularities $[a, b']$ and $[b, a']$, where b' is $b \mod a$ and a' is $a \mod b$. The notation is a bit unhelpful here: we must interpret each of these new singularities correctly. If $r = 2$ we simply delete $[r, a]$. If $a = 1$ then we do not include $[a, b']$ at all. If $b' > a/2$ then write $[a, a - b']$ rather than $[a, b']$; similarly for $[b, a']$.

There is an edge from vertex v_i to v_j if and only if $\mathcal{B}_j$ can be constructed from $\mathcal{B}_i$ by a single such basket surgery operation. Gorenstein projection is decreasing with respect to the order on Hilbert series so one only needs to search for images among v_j with $j < i$.

5.2   Geometrising the Vertices

At this stage we have a directed graph G_c which is the skeleton of our datagraph. Its vertices correspond to plausible Hilbert series for the graded homogeneous coordinate rings of polarised K3 surfaces, although we have only checked two simple conditions on the series so we must expect that sometimes no such surface actually exists. Its edges correspond to surgery operations on baskets that will be realised by Gorenstein projections if the corresponding surfaces exist. Now we begin to turn this into a datagraph of K3 surfaces.

Suppose given a Hilbert series $P(t) = \sum_{n \geq 0} h^0(A) t^n$. This power series can be written as a rational function

$$P(t) = \frac{p(t)}{(1 - t^{d_1}) \cdots (1 - t^{d_r})} \tag{3}$$

where $p(t)$ is a polynomial in t and $d_1, \ldots, d_r$ are positive integers. As described in [R3] Section 3.6, if X is embedded in $\mathbb{P}(d_1, \ldots, d_r)$ with hyperplane section A and Hilbert series $P(t)$ then a minimal free resolution of the graded ring $R(X, A)$ determines a rational expression for $P(t)$ in the form of equation (3) with Hilbert numerator

$$p(t) = 1 - \sum t^{e_j} + \cdots \pm t^k \tag{4}$$

where the integers e_j are the weights of the equations of X and k is the adjunction number of X. (Unfortunately the syzygies between the equations will contribute positive terms in t, some of which might well mask some of the t^{e_j}. But that is not the immediate problem.) How does one find this expression in advance?

In fact it is rather easy to generate expressions like (3). For example, if $g = -1$ and the basket $\mathcal{B}$ comprises $[2, 1]$, $[8, 3]$, $[11, 5]$ then one sees that

$$P(t) = \frac{t^{30} - 1}{(1 - t^5)(1 - t^6)(1 - t^8)(1 - t^{11})}. \tag{5}$$

One way is to write down the rational form of equation (1) immediately and manipulate that, giving a rational expression for $P(t)$: one constructs the unique

minimal expression by cancelling common factors from numerator and demoni-
nator. To understand $P(t)$ as a variety in weighted projective space we want to
see the form of equation (3). There are many such expressions. To make one, we
simply need to ensure that there are enough factors $1-t^d$ in the denominator to
absorb all the cyclotomic factors in the minimal denominator. When the target
variety X, A is a complete intersection, there is a unique minimal choice of such
d_i; in codimension 3 or more complete intersections are rare so we must expect
things to be more complicated.

The $1-t^d$ factors in the denominator are there to account for the periodic
contributions arising from the singularities so the index of each singularity must
divide the d occuring in some factor. For X_{30} this gives a first guess for a denominator of $(1-t^8)(1-t^{11})$. The ring $R(X, A)$ is 3-dimensional so we need to
include another factor before the numerator becomes a polynomial. The factor
$1-t$ is harmless since it is a factor any $1-t^d$. We compute then that

$$(1-t)(1-t^8)(1-t^{11})P(t) = t^{20} - t^{19} + t^{15} - t^{13} + t^{10} - t^7 + t^5 - t + 1.$$

Reading this Hilbert numerator, we see that there is an equation in degree 1.
There is no point having a generator and an equation in degree 1 so the factor
$1-t$ isn't quite right. We increase this power of t. Repeating the same reasoning
reveals $1-t^5$ as the first sensible choice. Multiplying out again shows that a
generator is needed in degree 6 and we are done.

We seek a projective variety whose homogeneous coordinate ring has the
Hilbert series (5). Guided by the rule of thumb (4) we consider $X_{30} \subset \mathbb{P}(5, 6, 8, 11)$.
Any such variety has the right Hilbert series and as in the introduction one checks
that if the equation defining X_{30} is general, then X_{30} is indeed a K3 surface with
singularities as encoded by $\mathcal{B}$. Success!

5.3   Improving the Datagraph: Unprojection

By this stage we have a datagraph with candidates for K3 surfaces at its vertices
and we would like to determine which really exist and which do not. They are
not all as clear-cut as X_{30} above. In codimension higher than 3 we cannot simply
write down equations and calculate. We will use the edges of the datagraph to
guide an alternative construction process called *unprojection*.

Consider a special case of the main theorem of [PR]:

Theorem 2. *If $Y \subset \mathbb{P} = \mathbb{P}(d_1, \ldots, d_n)$ is a quasi-smooth K3 surface and one
of the coordinate lines $D = \mathbb{P}(a, b) \subset \mathbb{P}$ lies on Y then there is a K3 surface
$X \subset \mathbb{P}(d_1, \ldots, d_n, r)$ containing a singularity p of index $r = a + b$ and type
$\frac{1}{r}(a, b)$. Moreover, Y is the Type I Gorenstein projection of X from p.*

One views this theorem as an implicit construction of X from the pair $Y \supset D$,
an *unprojection*. Most K3 surfaces admit a Type I projection, especially when
$g \neq -1$, so this approach promises some success. But Theorem 2 has a condition
that needs to be checked. It is easy to find $D = \mathbb{P}(a, b)$ as a coordinate line,
but we must also find an element of the family Y which contains this line. This

is a monomial condition on the equations of Y so is easy to check if we know their degrees. But there is tension: too many conditions and the surface Y may acquire bad singularities and not be K3.

With this in mind, suppose we have a pair of candidate K3 surfaces with a Gorenstein projection between them $X \dashrightarrow Y$ and suppose inductively that we have a description of Y embedded in some weighted projective space. Now if X actually exists, and if the projection is Type I from a singularity $\frac{1}{r}(a, b) \in X$ then there is a curve $D \subset Y$ as in Theorem 2 and the weights of X are same as those of Y together with a new variable s of weight r. The projection is then simply the elimination of the variable s from the homogeneous coordinate ring of X and a, b are also weights of Y. (The surfaces have one trick up their sleeve. It can happen that some generators of Y are expressed in terms of the others once s is included so that not all the weights of Y are visible on X. In practice this only happens in codimension 1. The projection from $\frac{1}{11}(5, 6) \in X_{30}$ is of this tricky Type I.)

So immediately we can tell whether there is any hope of realising X as a Type I unprojection of Y: the confirmed weights of Y should include a, b and the Hilbert series of X must be compatible with the weights of Y together with $a + b$, in the sense that an expression of the form (3) must be possible with these weights. When these conditions are satisfied we say that the projection $X \dashrightarrow Y$ is a *numerical Type I* projection. We also record the predicted weights of X. The task of proving that suitable Y exists remains — it is essentially a Gröbner basis problem, so something symbolic which could also be run through the datagraph — but already we expect to have a better collection of candidates. The experience of an analogous calculation in [BR] is that suitable Y often exist, but may do so in several distinct ways.

With datagraphs in hand, it is easy to write code to carry out this analysis. For example, one sees that

Proposition 1. *If $g = 1$ then there are unique weights for each of the 6639 possible K3 baskets so that each of the 20721 projections are either known Type I or numerical Type I projections with the single exception:*

$$\mathbb{P}(1, 1, 3, 5) \supset X_{10} \dashrightarrow X_{12} \subset \mathbb{P}(1, 1, 4, 6),$$

one of two projections to the $g = 1$ surface of Theorem 1.

In fact, all but six of the Type I projections are simply the elimination of a single variable. Apart from the exceptional case of the proposition, there are five other projections from K3s in codimension 1, which are necessarily the slightly tricky Type I. And let me emphasise again: there is no prima facie reason why any of these 6639 pieces of data should be realised by K3 surfaces, even though I expect most of them will be.

6   The Magma Implementation

Datagraphs for K3 surfaces are implemented in prototype on Magma, see [K3] or [Ma] for information. A version that can be tailored to other geometric situations will be made in the future. Here we build the datagraph for K3 surfaces of genus −1 from scratch and then find the 36 surfaces of Table 4. The Magma commentary describes the steps discussed above. The > sign is the Magma prompt: the line following that is input, the rest is output. I leave you to work out the Magma syntax for yourself.

```
> G := K3BuildDatabaseGraph(-1);
  There are 4293 baskets with genus -1 ...     Time 108s
  Computing RR for baskets ...                 Time 173s
  Sorting the Hilbert series ...               Time 229s
  Building a graph from the numerical data ... Time 496s
  Analysing vertices of the graph ...          Time 830s
  Datagraph complete.
> K3PrintDatabase(G);
Database graph of 4293 K3 surfaces and 13078 projections
> #[ X : X in K3Surfaces(G) | #K3Projections(X) eq 0 ],
36
```

And that's it! The K3 surfaces that have no projection are exactly those on the datagraph with no arrows pointing away from them. And the number of these is exactly what the last line of input requested: remove the # to get the sequence of them all. Precisely, the result has identified Hilbert series of polarised K3 surfaces which cannot have a Gorenstein projection.

References

K3.   Lists of K3 surfaces are available from
       www.maths.warwick.ac.uk/~gavinb/k3graph.html and k3db.html
A.   Altınok, S. (1998): Graded rings corresponding to polarised K3 surfaces and
       ℚ-Fano 3-folds. Thesis, Univ. of Warwick 93 + vii pp.
A2.   S. Altınok, Hilbert series and applications to graded rings, to appear in International journal of Mathematics and Mathematical science
ABR.   S. Altınok, G. Brown and M. Reid, Fano 3-folds, K3 surfaces and graded rings, to appear in Contemp Math (SISTAG 2001 Proceedings, ed Berrick et al.)
B.   A. Beauville, Complex algebraic surfaces. Second edition. LMS Student Texts, 34. CUP, 1996. x+132 pp
BC.   S-M. Belcastro, Picard lattices of families of K3 surfaces, Comm. Algebra 30 (2002), no. 1, 61–82
BR.   G. Brown and M. Reid, Extreme K3 surfaces with high unprojections, in preparation
CPR.   A. Corti, A. Pukhlikov and M. Reid, Birationally rigid Fano hypersurfaces, in Explicit birational geometry of 3-folds, A. Corti and M. Reid (eds.), CUP 2000, 175–258
D.   A. Durfee, Fifteen characterisations of Du Val singularities, Enseign. Math. (2) 25 (1979), no. 1-2, 131–163
Fl.   A.R. Iano-Fletcher, Working with weighted complete intersections, in Explicit birational geometry of 3-folds, CUP 2000, pp. 101–173

JK.   J. Johnson and J. Kollár, Fano hypersurfaces in weighted projective 4-spaces, Experiment. Math. 10 (2001), no. 1, 151–158

KS.   M. Kreuze and H. Skarke, Lists of K3 and Calabi-Yau on the web at hep.itp.tuwien.ac.at/~kreuzer/CY/

Ma.   Magma (John Cannon's computer algebra system): W. Bosma, J. Cannon and C. Playoust, The Magma algebra system I: The user language, J. Symb. Comp. **24** (1997) 235–265.
See also www.maths.usyd.edu.au:8000/u/magma

PR.   S. Papadakis and M. Reid, Kustin–Miller unprojection without complexes, J. algebraic geometry (to appear), preprint math.AG/0011094, 15 pp

R1.   M. Reid, Surfaces with $p_g = 0$, $K^2 = 1$. J. Fac. Sci. Univ. Tokyo Sect. IA Math. 25 (1978), no. 1, 75–92

R2.   M. Reid, Chapters on algebraic surfaces. Complex algebraic geometry (Park City, UT, 1993), 3–159, IAS/Park City Math. Ser., 3, Amer. Math. Soc., Providence, RI, 1997

R3.   M. Reid, Graded rings and birational geometry, in Proc. of algebraic geometry symposium (Kinosaki, Oct 2000), K. Ohno (Ed.), 1–72, available from www.maths.warwick.ac.uk/~miles/3folds

SZ.   I. Shimada and D-Q. Zhang, Classification of extremal elliptic $K3$ surfaces and fundamental groups of open $K3$ surfaces, Nagoya Math. J. 161 (2001), 23–54

Sz.   B. Szendrői, Calabi-Yau threefolds with a curve of singularities and counterexamples to the Torelli problem. Internat. J. Math. 11 (2000), no. 3, 449–459.

Resultants and Neighborhoods of a Polynomial

Valentina Marotta

Department of Mathematics and Computer Science
Viale A. Doria, 6 - 95125, Catania, Italy
marotta@dmi.unict.it *

Abstract. In this paper the concept of neighborhood of a polynomial is analyzed. This concept is spreading into Scientific Computation where data are often uncertain, thus they have a limited accuracy. In this context we give a new approach based on the idea of using resultant in order to know the common factors between an empirical polynomial and a generic polynomial in its neighborhood. Moreover given a polynomial, the Square Free property for the polynomials in its neighborhood is investigated.

1   Introduction

In Computer Algebra and Symbolic Computation we deal with exact data, but in many real life situations data come either from physical measurements or observations or they are the output of numerical computations. Moreover computers work with Floating Points, i.e. with finite arithmetic, so a real number is represented approximately because of the limited number of available digits. For these reasons it is growing up the necessity of combining methods in computer algebra and numerical analysis. This led to a new branch of the classic polynomial algebra, the *numerical polynomial algebra* [2,11], whose aim is to accommodate the presence of inaccurate data and inexact computations. Since a polynomial is not exact, we consider a family of polynomials called *neighborhood*. For instance, given a polynomial $p(x) = 1.23x^3 + 5.78x^2 + 12x + 11.98$, if we suppose that the last decimal digit is off by one, then the coefficient 1.23 does not mean a precise rational number but rather any real number from a sufficiently small neighborhood, i.e. a real number belonging to the interval $[1.22, 1.24]$. Therefore computations with polynomial neighborhoods involve some notions of interval analysis [1].

The paper is organized as follows. In Section 2 the basic definition of polynomial neighborhood as in [9] is introduced, while the notion of pseudozero, as in [9,11], is introduced in Section 3. In Section 4 the problem of finding the nearest polynomial with a given zero is briefly discussed. In Section 5 we give a new definition for the neighborhood imposing stronger conditions by means of the resultant. In Section 6 the Square Free property for the polynomials in the neighborhood is the object of our interest. Finally in Section 7 it is shown the geometric meaning of the problems previously discussed.

---

* Supported by University of Naples "Federico II" and INTAS grant n.99-1222.

F. Winkler and U. Langer (Eds.): SNSC 2001, LNCS 2630, pp. 225–239, 2003.

Many definitions can be given for multivariate polynomials and the results in the paper can be extended in the future to multivariate polynomials case. The author has used two computer algebra packages: MAPLE 6 [3] and MATHEMATICA 4.1 [12]. The use of MATHEMATICA 4.1 is essential because this package works with interval arithmetic.

2   Polynomial Neighborhoods

Let $\mathbb{C}$ be the field of complex numbers and let $p(x) = p(x_1, \ldots, x_s)$ in $\mathbb{C}[x_1, \ldots, x_s]$ be a multivariate polynomial in the variables $x_1, \ldots, x_s$ with $\deg(p) = n$. Let $x^j = x_1^{j_1} \cdots x_s^{j_s}$ with $j = (j_1, \ldots, j_s) \in \mathbb{N}_0^s$. The polynomial $p(x)$ can be written as

$$p(x) = \sum_{j=0,\ldots,k} a_j x^j$$

with $a_j \in \mathbb{C}$, $k = \binom{n+s}{n}$ and $\sum_{h=1,\ldots,s} j_h \leq n$ for all j.

The *tolerance* e associated with $p(x)$ is a non negative vector $e = (e_k, \ldots, e_0)$ such that

$$e_j \in \mathbb{R} \quad e_j \geq 0 \text{ for } j = 0, \ldots, k$$

Let $p(x)$ be a polynomial with at least one coefficient having limited accuracy, i.e. a $p(x)$ such that there exists $j \in \{0, \ldots, k\}$ with $e_j > 0$. $p(x)$ will be called an *empirical* polynomial.

Let us give some different definitions for the polynomial neighborhood. We start introducing the definition given by Stetter as in [9].

Definition 1. *The* **neighborhood** $\mathcal{N}(p, e)$ *of an empirical polynomial p with tolerance e is the family of polynomials $\tilde{p} \in \mathbb{C}[x_1, \ldots, x_s]$, $\tilde{p} = \sum \tilde{a}_j x^j$, having vector of coefficients $\tilde{a} = (\tilde{a}_k, \ldots, \tilde{a}_0) \in \mathbb{R}^{k+1}$, such that*

$$\tilde{a}_j = a_j \text{ if } e_j = 0, \text{ i.e. } a_j \text{ is exact}$$

$$\left\| \left(\ldots, \frac{|\tilde{a}_j - a_j|}{e_j}, \ldots \right) \right\|^* \leq 1 \tag{1}$$

with $\| \cdot \|^*$ a generic norm over $\mathbb{R}^{k+1}$.

For instance if the polynomial $p(x) \in \mathbb{R}[x]$ and choosing $\|(\ldots, b_j, \ldots)\|^* = \max_j |b_j|$ as norm, the (1) becomes

$$|\tilde{a}_j - a_j| \leq e_j \text{ for } j = 0, \ldots, k \tag{2}$$

This means that the exact value $\tilde{a}_j \in [a_j - e_j, a_j + e_j]$, i.e. $\tilde{a}_j = a_j + \delta_j$ such that $|\delta_j| \leq e_j$. Each δ_j, for $j = 0, \ldots, k$, is called *perturbation* and the generic polynomial $\tilde{p}(x)$ in the neighborhood is called *perturbed polynomial*.

Another definition of polynomial neighborhood is the following as in [5].

Definition 2. *Let* $p(x) = \sum a_j x^j \in \mathbb{C}[x_1, \ldots, x_s]$ *be an empirical polynomial. The neighborhood* $\mathcal{N}(p)$ *of* p *is the family of polynomials* $\tilde{p}$ *in* $\mathbb{C}[x_1, \ldots, x_s]$, $\tilde{p} = \sum \tilde{a}_j x^j$ *such that* $\|p - \tilde{p}\| \leq \epsilon$, *with* $\epsilon \geq 0$, $\epsilon \in \mathbb{R}$, *and with* $\| \cdot \|$ *a generic coefficient vector norm.*

A special case of the Definition 1 is the following definition as in [8].

Definition 3. *Let* $p(x_1, \ldots, x_s) \in \mathbb{C}[x_1, \ldots, x_s]$ *be a multivariate polynomial with tolerance* e. *Let* $\prec$ *be a term ordering on* $PP(x_1, \ldots, x_s) = \{x_1^{t_1} \cdots x_s^{t_s} : (t_1, \ldots, t_s) \in \mathbb{N}_0^s\}$. *We say that a polynomial* $\tilde{p} \in \mathbb{C}[x_1, \ldots, x_s]$ *is a lower-order perturbed of* p *if* $\tilde{p} \in \mathcal{N}(p, e)$ *and* $HTerm_{\prec}(\tilde{p} - p)$ *is less than* $HTerm_{\prec}(p)$.

In the sequel of the paper we will use the notion of neighborhood as in Definition 1 using the maximum norm and somewhere we will compare the results for different definitions.

3   Pseudozero

We are interested in finding meaningful solutions for algebraic problems with polynomial data of limited accuracy. The treatment of data with limited accuracy is based on the *backward error principle*. An approximate result is meaningful if it is the exact result for a problem within a tolerance neighborhood of the specified problem. Whenever we have an empirical polynomial the notion of zero must be replaced by the notion of "pseudozero".

Definition 4. *A value* $z \in \mathbb{C}^s$ *is a pseudozero of an empirical polynomial* p *with tolerance* e *if it is a zero of some polynomial* $\tilde{p} \in \mathcal{N}(p, e)$.

Definition 5. *A pseudozero domain of an empirical polynomial* p *with tolerance* e *is the set of all pseudozeros*

$$\mathcal{Z}(p, e) = \{z \in \mathbb{C}^s : \exists \tilde{p} \in \mathcal{N}(p, e) : \tilde{p}(z) = 0\}$$

For further details about pseudozeros see [9,10,11].

3.1   Univariate Case

Our aim is studying the behaviour of the pseudozero domain of a polynomial, i.e. where the zeros lie. A simple case occurs when we work in $\mathbb{R}[x]$ and we choose the maximum norm (obtaining (2)). Let $p(x) = \sum_{j=0}^{n} a_j x^j \in \mathbb{R}[x]$ be an empirical polynomial and let $e = (e_n, \ldots, e_0)$ be its tolerance vector. A perturbed polynomial in the neighborhood is $\tilde{p}(x) = \sum_{j=0}^{n}(a_j + \delta_j)x^j$; if we define the polynomial $p_\Delta(x) = \sum_{j=0}^{n} \delta_j x^j$, then $\tilde{p}(x) = p(x) + p_\Delta(x)$. It holds

$$\forall z : \tilde{p}(z) = 0 \Rightarrow p(z) + p_\Delta(z) = 0 \Rightarrow p(z) = -p_\Delta(z) \tag{3}$$

and then $|p_\Delta(z)| = |\sum_{j=0}^{n} \delta_j z^j| \leq \sum_{j=0}^{n} |\delta_j||z^j| \leq \sum_{j=0}^{n} e_j |z|^j$, i.e.

$$-\sum_{j=0}^{n} e_j |z|^j \leq p_\Delta(z) \leq \sum_{j=0}^{n} e_j |z|^j \tag{4}$$

We can distinguish 3 cases:

1. $z = 0$: trivially $a_0 + \delta_0 = 0$ and then $-e_0 \leq a_0 \leq e_0$;
2. $z > 0$: the inequalities (4) become $\sum_{j=0}^{n}(-e_j)z^j \leq p_\Delta(z) \leq \sum_{j=0}^{n} e_j z^j$, and from (3) z has to satisfy the following inequalities:

$$\sum_{j=0}^{n}(a_j + e_j)z^j \geq 0 \qquad \sum_{j=0}^{n}(a_j - e_j)z^j \leq 0 \qquad (5)$$

3. $z < 0$: from (3) and (4) z has to satisfy the following inequalities:

$$\sum_{h=0}^{n}(a_h + (-1)^h e_h)z^h \geq 0 \qquad \sum_{h=0}^{n}(a_h + (-1)^{h+1} e_h)z^h \leq 0 \qquad (6)$$

In this way we can find where all pseudozeros lie; the pseudozeros are distributed into disjoint components named *pseudozero components*.

Example 1. Let $p(x) = 1.1x^2 + 2.3x + 1$ be an empirical polynomial and let $e = (10^{-2}, 10^{-2}, 10^{-1})$ be the tolerance vector. We look for the pseudozero components. From equations (5) and (6), we have to solve these two systems:

$$\begin{cases} 1.11x^2 + 2.31x + 1.1 \geq 0 \\ 1.09x^2 + 2.29x + 0.9 \leq 0 \end{cases} \qquad \begin{cases} 1.11x^2 + 2.29x + 1.1 \geq 0 \\ 1.09x^2 + 2.31x + 0.9 \leq 0 \end{cases}$$

By solving the former system we obtain the components $[-1.57, -1.34]$ and $[-0.73, -0.52]$; since $z > 0$ by hypothesis (case 2) we can say that there are not positive pseudozeros; from the latter system we obtain $[-1.60, -1.30]$ and $[-0.76, -0.51]$, that are the ranges where all pseudozeros lie.

There is a different way of checking where all pseudozeros lie. It is sufficient to consider a perturbed polynomial $\tilde{p}(x) = (1.1+\delta_2)x^2 + (2.3+\delta_1)x + 1 + \delta_0$ in the neighborhood. We substitute the interval $[-e_j, e_j]$ to each δ_j and we solve the equation $\tilde{p}(z) = 0$ in the variable z by using interval arithmetic. If we perform all these computations with MATHEMATICA 4.1, then we obtain the pseudozero components $[-1.60, -1.30]$ and $[-0.78, -0.49]$, i.e. components larger than the previous ones, that we have obtained by using formulas (5) and (6). It follows that the computation based on these formulas is more careful than the computation involving intervals (i.e. interval polynomials), which is affected by a *bad* interval arithmetic.

Roughly speaking, it seems that small perturbations in the coefficients of the polynomial give small variations of the roots, but it is not always true.

Example 2. Let $p(x) = (x - 1)(x - 2) \cdots (x - 20) = x^{20} - 210x^{19} + \cdots + 20!$. This is a well-known polynomial in numerical analysis, named the *Wilkinson's polynomial*. If we change the coefficient -210 with $-210 + 10^{-7}$, then the roots $1, 2, \ldots, 20$ of the polynomial change dramatically (in fact there are a lot of complex roots).

A polynomial as in Example 2 is said to be *sensitive*, which means that the relative error in the result is much greater than the relative error in the data, i.e. it is *ill-conditioned* (or similarly *ill-posed*). Pseudozeros searching analysis can be very useful for investigating more accurately this class of cases and for checking the sensitivity of a polynomial.

4   Nearest Polynomial with a Given Zero

Given an empirical polynomial with tolerance vector e and an approximate zero we want to investigate how we can check that it is a *pseudozero*, i.e. an exact zero of a polynomial that we can not distinguish from the specified one because of the indetermination in its data.

Assigned an empirical polynomial $p(x) \in \mathbb{R}[x]$, $p(x) = \sum_{j=0}^{n} a_j x^j$, with $\deg(p) = n$, and a value $z \in \mathbb{R}$, we look for a polynomial $\tilde{p}(x) = \sum_{j=0}^{n} \tilde{a}_j x^j$ in $\mathcal{N}(p, e)$ such that $\tilde{p}(z) = 0$. We can find more than one of such polynomials, but there is only one of these polynomials whenever we look for the *nearest* one, i.e. the polynomial with the smallest perturbations from the original coefficients.

The problem of finding the nearest polynomial with a given zero can be formulated in two different manners depending on the chosen definition for the polynomial neighborhood. The former case occurs when the distance between the two polynomials $p(x)$ and $\tilde{p}(x)$ is measured by using a norm as in Definition 2. In this case the concept of *nearest* depends on the chosen norm. The problem is the following: we want to compute

$\delta_j \in \mathbb{R}$, $j = 0, \ldots, k$, such that $\|p - \tilde{p}\|$ is minimized for a given norm $\| \cdot \|$

If we choose the Euclidean norm, i.e. $\|p - \tilde{p}\|_2 = (\sum (a_j - \tilde{a}_j)^2)^{\frac{1}{2}}$, then it is straightforward to find the δ_j's such that $\tilde{p}(z) = 0$ and $\|p - \tilde{p}\|_2$ is minimized. In fact we have the following theorem

Theorem 1. *Let $p(x)$ be an empirical polynomial with $\deg(p) = n$. Let $\| \cdot \|_2$ be the Euclidean norm and let $z \in \mathbb{R}$. $\|p - \tilde{p}\|_2$ is minimized whenever the δ_j's satisfy the following conditions:*

$$\delta_j = \frac{-z^j p(z)}{z^{2n} + z^{2n-2} + \ldots + z^2 + 1} \quad j = 0, \ldots, n \tag{7}$$

Proof. The formula can be obtained minimizing a quadratic form as in [6].

If $p(x) \in \mathbb{C}[x]$ and $z \in \mathbb{C}$, then the formula for the δ_j's is the following

$$\delta_j = \frac{-\bar{z}^j p(z)}{\sum_{i=0}^{n} (z\bar{z})^i} \quad j = 0, \ldots, n \tag{8}$$

where $\bar{z}$ is the conjugate of z; a complete proof of formula (8) can be found in [4]. There are similar results using $\| \cdot \|_\infty$ [5].

The latter case is obtained using Definition 1 for the polynomial neighborhood. Whenever we impose that the perturbed polynomial $\tilde{p}(x) = \sum_{j=0}^{n}(a_j + \delta_j)x^j$ has a zero z, we locate a hyperplane in the space $(\delta_0, \ldots, \delta_n)$:

$$\delta_n z^n + \ldots + \delta_1 z + \delta_0 + p(z) = 0 \tag{9}$$

that is a linear manifold of codimension 1 in the space $(\delta_0, \ldots, \delta_n)$. The value for the δ_j's such that the distance between the origin $(0, \ldots, 0)$ and the hyperplane is minimum can be computed as follows. The straight line perpendicular to the hyperplane and crossing the origin has parametric equation

$$\begin{cases} \delta_n = z^n t \\ \vdots \\ \delta_1 = zt \\ \delta_0 = t \end{cases} \tag{10}$$

Computing the intersection between the hyperplane and the straight line, after simple manipulations, we obtain

$$t = \frac{-p(z)}{z^{2n} + z^{2n-2} + \ldots + z^2 + 1}$$

and then

$$\delta_j = \frac{-z^j p(z)}{z^{2n} + z^{2n-2} + \ldots + z^2 + 1} \quad j = 0, \ldots, n \tag{11}$$

Example 3. Given the polynomial $p(x) = 1.1x^2 + 2.3x + 1$ and its tolerance $e = (10^{-2}, 10^{-2}, 10^{-1})$ we look for the nearest polynomial in $\mathcal{N}(p, e)$ having $z = -1.5$ as a zero. Using formula (9) we obtain the hyperplane $2.25\delta_2 - 1.5\delta_1 + \delta_0 + 0.025 = 0$. The straight line as in formula (10) becomes

$$\begin{cases} \delta_2 = 2.25t \\ \delta_1 = -1.5t \\ \delta_0 = t \end{cases} \tag{12}$$

Using the method previously described we find the point $(\delta_2 = -0.0068, \delta_1 = 0.0045, \delta_0 = -0.003)$ belonging to the hyperplane and having minimum distance $d = 0.00866$ from the origin. Then the polynomial $\bar{p}(x) = 1.0932x^2 + 2.3045x + 0.997$ is the nearest polynomial in $\mathcal{N}(p, e)$ having $z = -1.5$ as a zero.

Remark 1. Actually the formula as in (11) is correct only if every coefficient of the empirical polynomial has a nonzero tolerance. For instance if we know by hypothesis that the leading coefficient of the empirical polynomial is exact, the equation (9) becomes $\delta_{n-1}\alpha^{n-1} + \ldots + \delta_1\alpha + \delta_0 + p(\alpha) = 0$, then, after computations as above, we find the following formula for the perturbations:

$$\delta_j = \frac{-\alpha^j p(\alpha)}{\sum_{k=0}^{n-1} \alpha^{2k}} \quad j = 0, \ldots, n-1$$

In a more general case, given a tolerance vector $e = (e_n, \ldots, e_0)$, let us denote by $\Omega(p) = \{j \in \{0, \ldots, n\} \mid e_j > 0\}$. Now it is simple to prove that the formula for the perturbations can be written as:

$$\delta_j = \frac{-\alpha^j p(\alpha)}{\sum_{k \in \Omega(p)} \alpha^{2k}} \quad j \in \Omega(p) \tag{13}$$

5   New Properties by Using Resultants

Sometimes beside the acceptable variations for the coefficients of the empirical polynomial $p(x)$, we know (often experimentally) more properties of the exact polynomial. For instance, we can know a zero of the exact polynomial we are looking for. So we construct $p(x)$, such that it has the same zero but its coefficients have limited accuracy. Similarly we can know more than one zero or a common factor between $p(x)$ and the 'generic $\tilde{p}(x)$. In these cases we use the notion of resultant.

The resultant of a polynomial $p(x) = \sum_{j=0}^{n} a_j x^j$ and a generic perturbed polynomial in the neighborhood $\tilde{p}(x) = \sum_{j=0}^{n} (a_j + \delta_j) x^j$ is computed. By definition the resultant is the determinant of the Sylvester matrix of the two polynomials; in our case we work with the following $(n + n) \times (n + n)$ matrix:

$$S(p, \tilde{p}) = \begin{pmatrix} a_n & \cdots & a_0 & 0 & \cdots & 0 \\ 0 & a_n & \cdots & a_0 & 0 & \cdots \\ \vdots & \ddots & \ddots & \cdots & \ddots & \cdots \\ 0 & \cdots & 0 & a_n & \cdots & a_0 \\ a_n + \delta_n & \cdots & a_0 + \delta_0 & 0 & \cdots & 0 \\ 0 & a_n + \delta_n & \cdots & a_0 + \delta_0 & 0 & \cdots \\ \vdots & \ddots & \ddots & \cdots & \ddots & \cdots \\ 0 & \cdots & 0 & a_n + \delta_n & \cdots & a_0 + \delta_0 \end{pmatrix}$$

and $\text{Res}(p, \tilde{p}) = \det(S(p, \tilde{p}))$.

It is well known in linear algebra that the determinant of the matrix does not change (and the resultant too), when we substitute the $(n + 1)$-th row with the difference between the $(n + 1)$-th row and the first row, the $(n + 2)$-th row with the difference between the $(n + 2)$-th row and the second row and so on until the $2n$-th row that we substitute with the difference between the $2n$-th row and the n-th row. We obtain a new matrix:

$$S_1(p, \tilde{p}) = \begin{pmatrix} a_n & \cdots & a_0 & 0 & \cdots & 0 \\ 0 & a_n & \cdots & a_0 & 0 & \cdots \\ \vdots & \ddots & \ddots & \cdots & \ddots & \cdots \\ 0 & \cdots & 0 & a_n & \cdots & a_0 \\ \delta_n & \cdots & \delta_0 & 0 & \cdots & 0 \\ 0 & \delta_n & \cdots & \delta_0 & 0 & \cdots \\ \vdots & \ddots & \ddots & \cdots & \ddots & \cdots \\ 0 & \cdots & 0 & \delta_n & \cdots & \delta_0 \end{pmatrix}$$

Since $\det(S(p,\tilde{p})) = \det(S_1(p,\tilde{p}))$, we can write the resultant between $p(x)$ and $\tilde{p}(x)$ as the resultant between $p(x)$ and a new polynomial $p_\Delta(x) = \sum_{j=0}^{n} \delta_j x^j$.

Next theorems on resultants are well known in the literature and they are very useful for our purposes. They hold in the general case of polynomials with coefficients in a U.F.D. (Unique Factorization Domain).

Theorem 2. *Let R be a U.F.D., let $A(x) = \sum_{i=0}^{n} a_i x^i$, $B(x) = \sum_{j=0}^{m} b_j x^j$ be primitive polynomials in $R[x]$, then*

$$\mathrm{Res}(A, B) = 0 \Leftrightarrow \mathrm{GCD}(A, B) \neq 1$$

By Theorem 2 $p(x)$ and $\tilde{p}(x)$ have a common factor if and only if $\mathrm{Res}(p, \tilde{p}) = 0$. The resultant between $\tilde{p}$ and p is a polynomial in the indeterminates $\delta_0, \ldots, \delta_n$ hence $\mathrm{Res}(\tilde{p}, p) = 0$ is equivalent to give some conditions on the δ_j's.

We can give the following two new definitions.

Definition 6. *Let $p(x) \in \mathbb{R}[x]$ be an empirical polynomial and let e be a vector of tolerance. The polynomial $\tilde{p}(x) \in \mathbb{R}[x]$ is a k common-factor perturbed of $p(x)$ if $\tilde{p}(x) \in \mathcal{N}(p, e)$ and it has at least a common factor of degree k with $p(x)$.*

Definition 7. *Let $p(x) \in \mathbb{R}[x]$ be an empirical polynomial and let e be a vector of tolerance. A k common-factor neighborhood $\mathcal{N}_k(p, e)$ of a polynomial $p(x)$ is the set of all k common-factor perturbeds.*

Let us see two very important theorems about resultants [7]:

Theorem 3. *Let R be a U.F.D. and let $A(x) = \sum_{i=0}^{n} a_i x^i$ and $B(x) = \sum_{j=0}^{m} b_j x^j$ be primitive polynomials in $R[x]$. Let $D = \mathrm{GCD}(A, B)$. If the rank of the Sylvester matrix is $m + n - k$ then $\deg(D) = k$.*

Theorem 4. *Let R be a U.F.D. and let $A(x) = \sum_{i=0}^{n} a_i x^i$ and $B(x) = \sum_{j=0}^{m} b_j x^j$ be primitive polynomials in $R[x]$. Let $S_d(A, B)$ be the matrix obtained by the resultant matrix $S(A, B)$ when we put it in triangular echelon form by using row transformations only. The first nonzero row from the bottom of the matrix $S_d(A, B)$ gives the coefficients of a GCD of $A(x)$ and $B(x)$.*

Once the resultant matrix is given, then we can put it in triangular echelon form by using row transformations only. By using the Laidacker's result, the entry in the $2n$-th row and column is the resultant of the two polynomials. If the first nonzero row from the bottom is the $(2n - h)$-th, then the entries of such row are the coefficients of the polynomial $q(x) = \mathrm{GCD}(p, \tilde{p})$, having degree h. For example, if $\mathrm{Res}(p, \tilde{p}) = 0$, then the polynomials have, at least, a common linear factor. $\mathrm{Res}(p, \tilde{p})$ is a homogeneous polynomial of degree $2n$ in the δ_j's and a_j's and a homogeneous polynomial of degree n as polynomial in the only δ_j's.

Remark 2. A trivial case $\mathrm{Res}(p, \tilde{p}) = 0$ occurs when the coefficients of the polynomials are the same up to a nonzero constant. This case occurs when we have the following choice for the perturbations:

$$\delta_n \in [-e_n, e_n], \quad \delta_j = a_j \frac{\delta_n}{a_n} \quad j = 0, \ldots, n-1 \tag{14}$$

where $a_n \neq 0$, since $n = \deg(p)$. In order to guarantee that for any choice of δ_n the corresponding values of each δ_j belong to their intervals, we have to compute

$$\tau = \min_{j=0,\ldots,n-1} \left\{ \left| \frac{e_j a_n}{a_j} \right| \right\} \quad \text{with} \quad a_j \neq 0$$

Now if $\delta_n \in [-\tau, +\tau] \cap [-e_n, e_n]$ by applying (14) we can find all polynomials in the neighborhood, whose coefficients are proportional to the coefficients of $p(x)$.

In the general case we have the following: if $p(x)$ and $\tilde{p}(x)$ have at least a k degree common factor, with $1 \leq k \leq n$, then all entries from the $(2n - k + 1)$-th row to the $2n$-th row of the resultant matrix $S_d(p, \tilde{p})$ must vanish. Furthermore, denoted by $S_d[2n - k, 2n - k]$ the entry in the $(2n - k)$-th row and $(2n - k)$-th column of the matrix $S_d(p, \tilde{p})$, if $S_d[2n - k, 2n - k]$ is different from zero, then $p(x)$ and $\tilde{p}(x)$ have exactly a k degree common factor and $S_d[2n - k, 2n - k]$ is its leading coefficient. We find a system of $\frac{k(k+1)}{2}$ homogeneous equations in the δ_j's. We observe that, by construction of the matrix S_d, if the entries in the $(2n - k + 1)$-th row vanish then all entries from the $(2n - k)$-th row to the $2n$-th row vanish too; then the system of $\frac{k(k+1)}{2}$ equations in the δ_j's is equivalent to a system of only k equations. If $k = n$ then the only solution is given by (14), i.e. $p(x)$ and $\tilde{p}(x)$ have proportional coefficients. Otherwise we find the conditions on the δ_j's in order to satisfy that $\tilde{p}(x)$ is a k common-factor perturbed of $p(x)$.

5.1   An Example: Polynomials of Degree Two

Let $p(x) = a_2 x^2 + a_1 x + a_0$ be a generic polynomial of degree two and let $\tilde{p}(x) = (a_2 + \delta_2)x^2 + (a_1 + \delta_1)x + (a_0 + \delta_0)$ be a perturbed polynomial in $\mathcal{N}(p, e)$, for a given tolerance e. Their Sylvester matrix in triangular form (using the Fraction Free Gaussian Elimination) is

$$\begin{pmatrix} a_2 & a_1 & a_0 & 0 \\ 0 & a_2^2 & a_1 a_2 & a_0 a_2 \\ 0 & 0 & \delta_1 a_2^2 - a_1 a_2 \delta_2 & \delta_0 a_2^2 - a_2 a_0 \delta_2 \\ 0 & 0 & 0 & R \end{pmatrix}$$

By Laidacker's theorem $R = \text{Res}(p, \tilde{p}) = -a_2^2 \delta_0^2 - a_0 a_2 \delta_1^2 + 2 a_2 a_0 \delta_0 \delta_2 + a_1 a_2 \delta_0 \delta_1 + a_0 a_1 \delta_1 \delta_2 - a_0^2 \delta_2^2 - a_1^2 \delta_0 \delta_2$. If we impose the condition $\text{Res}(p, \tilde{p}) = 0$ then we have to satisfy

$$(a_1 \delta_2 - a_2 \delta_1)(a_0 \delta_1 - a_1 \delta_0) - (a_2 \delta_0 - a_0 \delta_2)^2 = 0 \tag{15}$$

Moreover if $\delta_1 a_2^2 - a_1 a_2 \delta_2 \neq 0$, i.e. $\delta_1 a_2 - a_1 \delta_2 \neq 0$ since $a_2 \neq 0$ because $\deg(p) = 2$, then p and $\tilde{p}$ have exactly the following common linear factor

$$(\delta_1 a_2^2 - a_1 a_2 \delta_2)x + \delta_0 a_2^2 - a_2 a_0 \delta_2 \tag{16}$$

If $\tilde{p}(x)$ is a lower-order perturbed as in Definition 3, with $\deg(\tilde{p} - p) = 1$, then we impose $e_2 = 0$, and in (15) and (16) we have to substitute the value $\delta_2 = 0$.

Example 4. We compute the resultant between the empirical polynomial $p(x) = 1.1x^2 + 2.3x + 1$ and a generic polynomial in its neighborhood $\tilde{p}(x) = (1.1 + \delta_2)x^2 + (2.3 + \delta_1)x + (1 + \delta_0)$. We choose the tolerance vector $e = (10^{-2}, 10^{-2}, 10^{-1})$ and we assume to work with 3 digits after the point. $\text{Res}(p, \tilde{p}) = (2.3\delta_2 - 1.1\delta_1)(\delta_1 - 2.3\delta_0) - (1.1\delta_0 - \delta_2)^2 = 0$ in order to have at least a common factor. This condition gives the following choice of the perturbations: $\delta_1 \in [-e_1, e_1], \delta_2 \in [-e_2, e_2], \delta_0 = \frac{23}{22}\delta_1 - \frac{309}{242}\delta_2 \pm \frac{1}{242}\sqrt{89}(-23\delta_2 + 11\delta_1)$. For instance if we choose $\delta_0 = -0.025, \delta_1 = -0.002, \delta_2 = 0.01$, then we find the polynomial $1.11x^2 + 2.298x + 0.975$ in the neighborhood having the common root -1.474 with the initial polynomial.

Remark 3. In the paper the notion of polynomial neighborhood is introduced using the standard basis $\{1, x, x^2, \ldots, x^n\}$ for polynomials, but other bases can be used, e.g. Bernstein polynomials, Lagrangian polynomials, etc. Unfortunately some problems can be found in order to define the resultant with respect to a different basis. Since there is a transformation matrix from the standard basis to a new basis, then a procedure to define a corresponding resultant can be given.

6   Square Free Property

Let $p(x)$ be a univariate polynomial in $\mathbb{R}[x]$. $p'(x)$ will denote the usual derivative of $p(x)$ with respect to x.

Definition 8. *A polynomial $p(x)$ is square free if* $\text{GCD}(p, p') = 1$, *i.e. it has no factors with multiplicity greater than one.*

For instance, square free property is important in order to use the Sturm's algorithm for finding the zeros of a polynomial in a given real interval (a, b). Given an empirical polynomial $p(x) = \sum_{j=0}^{n} a_j x^j \in \mathbb{R}[x]$, we suppose that it is square free (respectively not square free). Our aim is to investigate if this property holds for any perturbed polynomial in the neighborhood of $p(x)$. We compute the resultant between $p(x)$ and its derivative $p'(x)$. If $p(x)$ is square free then, by Theorem 2, $\text{Res}(p, p') \neq 0$. Similarly given $\tilde{p}(x)$ we compute $\text{Res}(\tilde{p}, \tilde{p}')$. $\tilde{p}(x)$ is square free if $\text{Res}(\tilde{p}, \tilde{p}') \neq 0$.

Theorem 5. *Let $p(x) = \sum_{j=0}^{n} a_j x^j \in \mathbb{R}[x]$ be an empirical polynomial with $\deg(p) = n$ and let e be its tolerance vector. Let $\tilde{p}(x) = \sum_{j=0}^{n}(a_j + \delta_j)x^j$ be a generic perturbed in the neighborhood of $p(x)$. We have*

$$\text{Res}(\tilde{p}, \tilde{p}') = \text{Res}(p, p') + \phi(\delta_0, \ldots, \delta_n) \tag{17}$$

where $\phi(\delta_0, \ldots, \delta_n)$ is a polynomial in the indeterminate δ_j's of degree less or equal than $2n - 1$.

Proof. By definition, the resultant $\mathrm{Res}(\tilde{p}, \tilde{p}')$ is the determinant of the following $(2n - 1) \times (2n - 1)$ matrix

$$\begin{pmatrix} a_n + \delta_n & \cdots & a_1 + \delta_1 & a_0 + \delta_0 & 0 & \cdots & 0 \\ 0 & a_n + \delta_n & \cdots & a_1 + \delta_1 & a_0 + \delta_0 & 0 & \cdots \\ \vdots & \ddots & \cdots & \ddots & \cdots & \ddots & \cdots \\ 0 & \cdots & 0 & a_n + \delta_n & \cdots & a_1 + \delta_1 & a_0 + \delta_0 \\ n(a_n + \delta_n) & \cdots & a_1 + \delta_1 & 0 & 0 & \cdots & 0 \\ 0 & n(a_n + \delta_n) & \cdots & a_1 + \delta_1 & 0 & 0 & \cdots \\ \vdots & \ddots & \ddots & \cdots & \ddots & \cdots & \vdots \\ 0 & \cdots & 0 & 0 & n(a_n + \delta_n) & \cdots & a_1 + \delta_1 \end{pmatrix}$$

Applying a well known property of the matrices to the first row, and iterating this process, we obtain

$$\mathrm{Res}(\tilde{p}, \tilde{p}') = \det \begin{pmatrix} a_n & \cdots & a_1 & a_0 & 0 & \cdots & 0 \\ 0 & a_n & \cdots & a_1 & a_0 & 0 & \cdots \\ \vdots & \ddots & \cdots & \ddots & \cdots & \ddots & \cdots \\ 0 & \cdots & 0 & a_n & \cdots & a_1 & a_0 \\ na_n & \cdots & a_1 & 0 & 0 & \cdots & 0 \\ 0 & na_n & \cdots & a_1 & 0 & 0 & \cdots \\ \vdots & \ddots & \ddots & \cdots & \ddots & \cdots & \vdots \\ 0 & \cdots & 0 & 0 & na_n & \cdots & a_1 \end{pmatrix} + \cdots$$

i.e. $\mathrm{Res}(\tilde{p}, \tilde{p}') = \mathrm{Res}(p, p') + r$, where r is the sum of the other $2^{2n-1} - 1$ determinants and it is a polynomial function of the δ_j's.

Since by hypothesis $|\delta_j| \leq e_j$ for $j = 0, \ldots, n$, then we can identify each δ_j with an interval $\delta_j = [-e_j, e_j]$. By using interval arithmetic let $[\rho, \eta]$ be the interval obtained by substituting $\delta_j = [-e_j, e_j]$ as interval in the formula of ϕ. If we look at $\mathrm{Res}(p, p')$ as a degenerate interval $[\mathrm{Res}(p, p'), \mathrm{Res}(p, p')]$, then

$$\mathrm{Res}(\tilde{p}, \tilde{p}') = [\mathrm{Res}(p, p') + \rho, \mathrm{Res}(p, p') + \eta]$$

By working with intervals, whenever we ask $\mathrm{Res}(\tilde{p}, \tilde{p}') \neq 0$, it must hold that $0 \notin [\mathrm{Res}(p, p') + \rho, \mathrm{Res}(p, p') + \eta]$, i.e. we have to guarantee that either

$$\begin{cases} \mathrm{Res}(p, p') + \rho > 0 \\ \mathrm{Res}(p, p') + \eta > 0 \end{cases} \quad \text{or} \quad \begin{cases} \mathrm{Res}(p, p') + \rho < 0 \\ \mathrm{Res}(p, p') + \eta < 0 \end{cases} \tag{18}$$

On the other hand if we want that $\tilde{p}(x)$ is not square free, we must impose

$$\begin{cases} \rho + \mathrm{Res}(p, p') < 0 \\ \eta + \mathrm{Res}(p, p') > 0 \end{cases} \tag{19}$$

so $0 \in [\mathrm{Res}(p, p') + \rho, \mathrm{Res}(p, p') + \eta]$. If we know that $p(x)$ is square free and $\mathrm{Res}(p, p') > 0$ (respectively $\mathrm{Res}(p, p') < 0$), then we can evaluate (17) in $\delta_j = 0$ for all $j = 0, \ldots, n$. We find

$$\mathrm{Res}(\tilde{p}, \tilde{p}')|_{\{\delta_0 = 0, \ldots, \delta_n = 0\}} = \mathrm{Res}(p, p') + \phi(\delta_0, \ldots, \delta_n)|_{\{\delta_0 = 0, \ldots, \delta_n = 0\}} = \mathrm{Res}(p, p') > 0.$$

By using Bolzano's theorem, there is a neighborhood of the point $(0, \ldots, 0)$ in the space of $(\delta_0, \ldots, \delta_n)$ where all perturbed polynomials in the neighborhood are square free.

Let us see some numerical examples performed with MATHEMATICA 4.1.

Example 5. Let $p(x) = 1.1x^2 + 2.3x + 1$ be an empirical polynomial with vector of tolerance $e = (10^{-2}, 10^{-2}, 10^{-1})$. $p(x)$ is square free, in fact $\mathrm{Res}(p, p') = -0.979 \neq 0$. A perturbed polynomial in $\mathcal{N}(p, e)$ is $\tilde{p}(x) = (1.1 + \delta_2)x^2 + (2.3 + \delta_1)x + (1 + \delta_0)$, with $|\delta_2| \leq 10^{-2}$, $|\delta_1| \leq 10^{-2}$ and $|\delta_0| \leq 10^{-1}$. By working with the δ_j's as intervals we compute $\mathrm{Res}(\tilde{p}, \tilde{p}') = [-1.852, -0.087]$, and then we can assert that every polynomial in the neighborhood of $p(x)$ is square free.

Example 6. Let $q(x) = 0.03x^2 - 0.4x + 1.4$ be an empirical polynomial with vector of tolerance $e = (10^{-2}, 10^{-1}, 10^{-1})$. Let us work with four digits after the point. $q(x)$ is square free because $\mathrm{Res}(q, q') = 0.0002$, but for a perturbed $\tilde{q}(x) = (0.03 + \delta_2)x^2 + (-0.4 + \delta_1)x + 1.4 + \delta_0 \in \mathcal{N}(q, e)$ we have $\mathrm{Res}(\tilde{q}, \tilde{q}') = [-0.016, 0.016]$. So there exist non-square free polynomials in the neighborhood; for instance if we choose $\delta_2 = 0.01$, $\delta_1 = -0.08$ and $\delta_0 = 0.04$ we obtain the polynomial $\bar{p} = 0.04x^2 - 0.48x + 1.44 = (0.2x - 1.2)^2$, i.e. a non-square free polynomial.

6.1   Nearest Non-square Free Polynomial

Given an empirical polynomial $p(x) = \sum_{j=0}^{n} a_j x^j$ and its tolerance vector e, we suppose that $p(x)$ is square free. We are interested in computing the minimal values $(\bar{\delta}_n, \ldots, \bar{\delta}_0)$ for the perturbations δ_j's with $|\bar{\delta}_j| \leq e_j$, for $j = 0, \ldots, n$, such that the polynomial $\bar{p}(x) = \sum_{j=0}^{n} (a_j + \bar{\delta}_j)x^j \in \mathcal{N}(p, e)$ is not square free, i.e. $\mathrm{Res}(\bar{p}, \bar{p}') = 0$. Let $\tilde{p}(x) = \sum_{j=0}^{n} (a_j + \delta_j)x^j$ be a generic perturbed polynomial in the neighborhood. The equation $\mathrm{Res}(\tilde{p}, \tilde{p}') = 0$ gives an hypersurface in the space $(\delta_0, \ldots, \delta_n)$. In order to find the minimum values for the perturbations we have to minimize the distance between the origin $(0, \ldots, 0)$ and a generic point $(\delta_0, \ldots, \delta_n)$ belonging to the hypersurface, then we have to minimize the value $\delta_0^2 + \ldots + \delta_n^2$ under the constraint $\mathrm{Res}(\tilde{p}, \tilde{p}') = 0$. A well-known method to determine the minima and maxima of a function under given constraints is the *Lagrange Multipliers Method*. In our case, we have the function $\mathcal{D} = \delta_0^2 + \ldots + \delta_n^2$ to minimize and the constraint $\mathrm{Res}(\tilde{p}, \tilde{p}') = 0$. We construct a new function in the indeterminates $\delta_0, \ldots, \delta_n$ with a new indeterminate λ:

$$F(\delta_0, \ldots, \delta_n, \lambda) = \delta_0^2 + \ldots + \delta_n^2 + \lambda \, \mathrm{Res}(\tilde{p}, \tilde{p}')$$

and we compute the solution of the following system:

$$
\begin{cases}
\dfrac{\partial F(\delta_0,\ldots,\delta_n,\lambda)}{\partial \delta_0} = 0 \\
\;\vdots \\
\dfrac{\partial F(\delta_0,\ldots,\delta_n,\lambda)}{\partial \delta_n} = 0 \\
\dfrac{\partial F(\delta_0,\ldots,\delta_n,\lambda)}{\partial \lambda} = 0
\end{cases}
\tag{20}
$$

Example 7. For the polynomial $q(x) = 0.03x^2 - 0.4x + 1.4$ as in Example 6, using the Lagrange Multipliers Method, we find the following assignment for the perturbations: $\delta_2 = -0.0014$, $\delta_1 = -0.0002$, $\delta_0 = -0.00003$. The correspondent polynomial $\bar{q}(x) = 0.286x^2 - 0.4002x + 1.39997 = (0.1691x - 1.183203)^2$ is not square free.

Let us give a final example [1] where we use all concepts till now discussed.

Example 8. Let $p(x) = x^4 - 2.83088x^3 + 0.00347x^2 + 5.66176x - 4.00694$ be an empirical polynomial and let $e = (0, 10^{-5}, 10^{-5}, 10^{-5}, 10^{-5})$ be its tolerance vector. We look for the zeros of $p(x)$. Using the method as described in Sect. 3.1, we find the interval $[-1.4142168, -1.4142104]$, that gives the value -1.41421 as a good approximation of a zero of $p(x)$. For the other three zeros of the polynomial the situation becomes more complicated, in fact we can find three distinct real zeros or a real zero and a complex conjugate pair. In this case it is not possible to attribute individual pseudozero sets to each of the three zeros, but it must be considered as a single cluster.

Computing the resultant between a generic perturbed $\tilde{p} \in \mathcal{N}(p, e)$ and its derivative $\tilde{p}'$, we obtain $\mathrm{Res}(\tilde{p}, \tilde{p}') = [-18.8098, 18.8098]$ and $\mathrm{Res}(p, p') = 9.96352 \times 10^{-16}$. This means that $p(x)$ is square free by an algebraic point of view, but "numerically" $p(x)$ is not square free because $\mathrm{Res}(p, p')$ is zero with respect to the precision we are working to. For instance the polynomial $\tilde{p}_1(x) = (x - \sqrt{2})^3(x + \sqrt{2})$ belongs to $\mathcal{N}(p, e)$ and it is not square free. This is why we cannot assign individual pseudozero sets to the other three zeros of $p(x)$.

7 Geometric Interpretation

In this section we analyze the geometric meaning of the concept of polynomial neighborhood and we focus *geometrically* the conditions discussed in the previous sections.

Let $p(x) = \sum_{j=0}^{n} a_j x^j$ be an empirical polynomial, let e be its tolerance vector and let $\tilde{p}(x) = \sum_{j=0}^{n}(a_j + \delta_j)x^j \in \mathcal{N}(p, e)$ be a perturbed polynomial. We have two possibilities.

(i). Let us consider the affine (n+1)-dimensional space over $\mathbb{R}$ given by the (n+1)-tuples $(\tilde{a}_0, \ldots, \tilde{a}_n)$. This means that we identify the initial polynomial $p(x)$ with the point $(a_0, \ldots, a_n)$ and the neighborhood with the polytope having points $(a_0 + \delta_0, \ldots, a_n + \delta_n)$, with $|\delta_j| \le e_j$ for all $j = 0, \ldots, n$.

---

[1] Found in [11].

(ii). Let us consider the affine (n+1)-dimensional space over $\mathbb{R}$ given by the (n+1)-tuples $(\delta_0, \ldots, \delta_n)$. This means that we identify the initial polynomial $p(x)$ with the point $(0, \ldots, 0)$ and the neighborhood with the polytope having points $(\delta_0, \ldots, \delta_n)$, with $|\delta_j| \le e_j$ for all $j = 0, \ldots, n$.

We obtain case (ii) from case (i) by the linear isomorphism of affine spaces defined by $T_j = t_j - a_j$ for $j = 0, \ldots, n$, where t_j is a coordinate system in the space as in (i) and T_j is the coordinate system of the space as in (ii). Let denote by Δ^{n+1} the space of (n+1)-tuples $(\delta_0, \ldots, \delta_n)$. In our case we prefer the space Δ^{n+1} because, as in Sections 5 and 6, we find algebraic conditions on the δ_j's.

In Section 5 if we impose $\text{Res}(p, \tilde{p}) = 0$, then we require more conditions on the δ_j's. If $\deg(p) = n$ then the resultant is a homogeneous polynomial of degree n in the δ_j's. It represents a hypersurface in the space Δ^{n+1}. The points where the polytope intersects this hypersurface are in one to one correspondence with the perturbed polynomials having at least a linear common factor with $p(x)$. Similarly when we ask for a k common-factor neighborhood, with $1 \le k \le n$, we have k homogeneous equations in the δ_j's. All common points between these hypersurfaces and the polytope are the $(n+1)$-tuple $(\delta_0, \ldots, \delta_n)$'s, that are in one to one correspondence with the $\tilde{p}(x)$ and have at least a common factor of degree k with $p(x)$.

Finally, in Section 6 we study the case when the polynomial $p(x)$ is square free. If the polynomial $\tilde{p}$ has no multiple factors, then we must have $\text{Res}(\tilde{p}, \tilde{p}') \ne 0$. If $\deg(p) = n$ then $\text{Res}(\tilde{p}, \tilde{p}')$ is a polynomial of degree $2n - 1$ in the δ_j's. The condition $\text{Res}(\tilde{p}, \tilde{p}') = 0$ locates a hypersurface in the space Δ^{n+1}. If the hypersurface intersects the polytope, then we can find the values of δ_j's, with $|\delta_j| \le e_j$, such that $\text{Res}(\tilde{p}, \tilde{p}') = 0$, i.e. such that the polynomial $\tilde{p}$ is not square free.

Acknowledgement

I would like to thank Prof. Giuseppa Carrà Ferro (University of Catania) and Prof. Hans J. Stetter (Tech. Univ. Vienna) for their useful comments.

References

1. Alefeld, G., Claudio, D. (1998): The basic properties of interval arithmetic, its software realization and some applications. Computer and Structures, **67**, 3–8
2. Emiris, I. Z. (1998): Symbolic-numeric algebra for polynomials. In: Encyclopedia of Computer Science and Technology, A. Kent and J.G. Williams, editors, vol. 39, 261–281, Marcel Dekker, New York
3. Geddes, K.O., Monogan, M.B., Heal, K.M., Labahn, G., Vorkoetter, S.M., McCarron, S. (2000): Maple 6, Programming Guide. Waterloo Maple Inc., Canada
4. Hitz, M.A., Kaltofen, E. (1998): Efficient algorithms for computing the nearest polynomial with constrained root. In: Proceed. ISSAC'98 (Ed. O.Gloor), 236–243
5. Hitz, M.A., Kaltofen, E. (1999): Efficient algorithms for computing the nearest polynomial with a real root and related problems. In: Proceed. ISSAC'99 (Ed. S.Dooley), 205–212

6. Karmarkar, N., Lakshman, Y.N. (1996): Approximate polynomial greatest common divisors and nearest singular polynomials. In: Proceed. ISSAC'96, 35–39
7. Laidacker, M.A. (1969): Another theorem relating Sylvester's matrix and the greatest common divisors. Mathematics Magazine **42**, 126–128
8. Jia, R. (1996): Perturbation of polynomial ideals. Advances in Applied Mathematics **17**, 308–336
9. Stetter, H.J. (1999): Polynomial with coefficients of limited accuracy. In: Computer Algebra in Scientific Computing (Eds. V.G.Ganzha, E.W.Mayr, E.V.Vorozhtsov), Springer, 409–430
10. Stetter, H.J. (1999): Nearest polynomial with a given zero, and similar problems. SIGSAM Bull. **33** n. 4, 2–4
11. Stetter, H.J. (2000): Numerical polynomial algebra: concepts and algorithms. In: ATCM 2000, Proceed. 5th Asian Technology Conf. in Math. (Eds.: W.-Ch. Yang, S.-Ch. Chu, J.-Ch. Chuan), 22–36, ATCM Inc. USA
12. Wolfram, S. (1999): The Mathematica Book. 4th ed., Wolfram Media, Cambridge University Press

Multi-variate Polynomials and Newton-Puiseux Expansions

Frédéric Beringer and Françoise Richard-Jung

LMC-IMAG, Tour Irma 51, rue des Mathématiques 38041 Grenoble cedex (France)
{Frederic.Beringer,Francoise.Jung}@imag.fr

Abstract. The classical Newton polygon method for the local resolution of algebraic equations in one variable can be extended to the case of multi-variate equations of the type $f(y) = 0$, $f \in \mathbb{C}[x_1, \ldots, x_N][y]$. For this purpose we will use a generalization of the Newton polygon - the Newton polyhedron - and a generalization of Puiseux series for several variables.

1 Introduction

Starting with the local analysis of non linear differential equations[2], we were induced to study equations $f(y) = 0$, where f is a polynomial in y, whose coefficients involve logarithms and exponentials[4].To solve such equations, we proposed to consider each *independent* transcendental function as a new variable, and so we were brought to solve equations where the coefficients of f are polynomials (or formal Laurent series) in several variables.

For this, we have used the well-known Newton method in one variable in an iterative way. For example, let $f \in \mathbb{C}[[u,v]][y]$, we can apply the process considering $f \in \mathbb{C}((u))((v))[y]$ and obtain solutions in $\mathbb{C}((u^{\frac{1}{m}}))((v^{\frac{1}{q}}))$, $m, q \in \mathbb{N}^*$, or considering $f \in \mathbb{C}((v))((u))[y]$ and obtain solutions in $\mathbb{C}((v^{\frac{1}{m}}))((u^{\frac{1}{q}}))$ (for some other m and q). In both cases, the ramifications are bounded, but in the first case, the series solutions can involve arbitrary large negative powers of $u^{\frac{1}{m}}$, and of $v^{\frac{1}{q}}$ in the second. This leads to some difficulties to define a notion of convergence for such objects.

In this paper, we propose an other approach to solve polynomial equations in N variables ($f(y) = 0$, with $f \in \mathbb{C}[x_1, \ldots, x_N][y]$), which gives a construction of series expansions with exponents in a cone. This construction is based on the Newton polyhedron of f and the algorithm exposed in [5] and has been implemented in *Maple*. The major improvment lies in the fact that the algorithm we will present provides a full set of solutions without the choose of an irrational vector. Furthermore, the cones in which lie the series solutions are known from the beginning on. Finally, we will explain how this construction can be used to reduce the non linearity of differential equations $f(t, y, y') = 0$ with $f \in \mathbb{C}[[t]][y, y']$.

F. Winkler and U. Langer (Eds.): SNSC 2001, LNCS 2630, pp. 240–254, 2003.

2   Regular Case for Multi-variate Equations

In this section, we will extend the notion of *Regular case* for equations in one variable [4] to multi-variate equations:

$$f(y) = 0 \text{ with } f \in \mathbb{C}[[x_1, \ldots, x_N]][y]. \tag{1}$$

We note $\mathbf{x} = (x_1, \ldots, x_N)$, $\mathbb{C}[[\mathbf{x}]] = \mathbb{C}[[x_1, \ldots, x_N]]$ and we define the expression $f \bmod \mathbf{x}$ by $f \bmod \mathcal{I}$ where $\mathcal{I}$ is the ideal generated in the ring $\mathbb{C}[[\mathbf{x}]][y]$ by $x_1, \ldots, x_N$. In the same way, for all $n \geq 1$, we define $f \bmod \mathbf{x}^n$ by $f \bmod \mathcal{I}^n$.

With these definitions we will say that the equation (1) satisfies the regular case if there exists $y_0 \in \mathbb{C}$ such that:

$$\begin{cases} f(y_0) = 0 & \bmod \mathbf{x} \\ f'(y_0) \neq 0 & \bmod \mathbf{x}. \end{cases}$$

Then we can define: $\forall n \in \mathbb{N}^*$, $\quad y_n = y_{n-1} - \frac{f(y_{n-1})}{f'(y_{n-1})} \bmod \mathbf{x}^{2^n}$. By induction on n, it can be proved that: $\forall n \in \mathbb{N}$, $\quad f(y_n) = 0 \bmod \mathbf{x}^{2^n}$.

Conclusion: if $\hat{y}$ is the element of $\mathbb{C}[[\mathbf{x}]]$ such that $\hat{y} = y_n \bmod \mathbf{x}^{2^n}$, $\forall n \in \mathbb{N}$, then $f(\hat{y}) = 0$.

This result, also known as lifting theorem [14] is interesting for two reasons: first, it proves the existence of a solution of the equation $f(y) = 0$ as a formal power series in the variables $x_1, \ldots, x_N$ and second, it produces a practical algorithm for computing the coefficients of the series in an efficient way. Of course, there is no reason for having exactly d solutions of this form, if d is the degree of the polynomial equation. So, in order to obtain a complete set of solutions, we have to look for more general solutions and to introduce a generalization of both the Newton polygon and the Puiseux series.

3   A First Resolution Algorithm

The first resolution algorithm appearing in the literature using a polytope in $\mathbb{R}^{N+1}$ is due to John McDonald in [5]. In order to understand our algorithm's improvment, we first summarize this work. In this section, f is supposed to be polynomial in all the variables, $f \in \mathbb{C}[\mathbf{x}][y]$ and we will use the following notations:

$$f(x_1, \ldots, x_N, y) = \sum_{k=0}^{d} \bar{a}_k y^k, \text{ where } \bar{a}_d \neq 0, \bar{a}_k = \sum_{\mathbf{K} \in \mathbb{N}^N} a_{\mathbf{K}}^k \mathbf{x}^{\mathbf{K}}, a_{\mathbf{K}}^k \in \mathbb{C}$$

$$\text{and} \quad \mathbf{x}^{\mathbf{K}} = x_1^{k_1} \ldots x_N^{k_N}, \text{ if } \mathbf{K} = (k_1, \ldots, k_N).$$

The ring of fractional power series in the variables $x_1, \ldots, x_N$ with exponents in the cone σ will be denoted by $\mathbb{C}[[\mathcal{S}_\sigma]]$, and more generally, the ring of fractional power series with exponents in some translate of σ by $\mathbb{C}((\mathcal{S}_\sigma))$.

3.1   A Total Order on $\mathbb{Q}^N$

The first important idea in [5] is to put a total order on $\mathbb{Q}^N$, by fixing an irrational vector $w \in \mathbb{R}^N$. If $\alpha \in \mathbb{Q}^N$, we define the order of α

$$\bigcirc_w(\alpha) = \langle w, \alpha \rangle = \sum_{i=1}^{N} w_i \alpha_i.$$

This order induces a total order on the monomials $\mathbf{x}^\alpha$, $\alpha \in \mathbb{Q}^N$, and induces the definition of the order of a non null series $\xi = \sum_{\alpha \in \mathbb{Q}^N} c_\alpha \mathbf{x}^\alpha$:

$$\bigcirc_w(\xi) = \inf_{c_\alpha \neq 0} \bigcirc_w(\alpha) \text{ and } \bigcirc_w(0) = -\infty.$$

3.2   The Newton Polytope

The algorithm that we will describe in the next paragraph is based on $\mathcal{P}(f)$, the Newton polytope of f, which is the convex hull of its support (in $\mathbb{R}^{N+1}$).

We will use the vocabulary defined in [5] and add the definition of some terms which will be usefull in the description of the resolution algorithm.

- The characteristic polynomial, Φ_e, of an edge, e, will be defined as in [2].
- The barrier cone associated to a monotone coherent edge path is by definition the cone: $\mathcal{C}_E = \sum_{i=1}^r \mathcal{C}_{e_i}$, where $\mathcal{C}_{e_i}$ denotes the barrier cone of the edge e_i.
- An admissible edge of the Newton polytope is said simple if its characteristic polynomial has only simple roots. A monotone edge path of the Newton polytope is said simple if the edges which compose it are all simple.

3.3   The Newton Algorithm

In fact the principle of this algorithm is very close to the one exposed in [3] for the case of the algebraic equations in two variables. The common idea of these algorithms is to look for solutions under the form $y = c\mathbf{x}^\alpha + \tilde{y}$ where α is the opposite of the slope, $\mathbf{s}_e = (\frac{p_1}{q_1}, \dots \frac{p_N}{q_N})$, of an admissible edge e of the Newton polytope of f and c a root of the associated characteristic polynomial $\Phi_e(X^q)$ where $q = lcm(q_i)$. We consider then the following change of unknown $\tilde{f}(\mathbf{x}, \tilde{y}) = f(\mathbf{x}, c\mathbf{x}^\alpha + \tilde{y})$, and search a series $\tilde{y}$ solution of $\tilde{f}(\mathbf{x}, \tilde{y}) = 0$ such that $\bigcirc_w(\alpha) < \bigcirc_w(\tilde{y})$.

The first result of [5] is to say that this process can be iterated infinitly if the given irrational vector $w \in \mathcal{C}_e^*$.

By this way it is possible to associate to each admissible edge e of $\mathcal{P}(f)$ and to each irrational vector $w \in \mathcal{C}_e^*$ a strongly convex polyhedral rational cone $\mathcal{C}_{w,e}$ such that $w \in \mathcal{C}_{w,e}^*$ and l_e solutions (counting multiplicity) $\phi \in \mathbb{C}((\mathcal{S}_{\mathcal{C}_{w,e}}))$ such that $\bigcirc_w(\phi) = -\mathbf{s}_e$.

Then it is shown in [5] that for each monotone coherent edge path $E = \{e_1, \dots, e_r\}$ and for each irrational vector $w \in \cap_{i=1}^r \mathcal{C}_{e_i}^*$, we can find a strongly convex polyhedral rational cone $\mathcal{C}_w$ such that $w \in \mathcal{C}_w^*$ and d solutions (counting multiplicity) in $\mathbb{C}((\mathcal{S}_{\mathcal{C}_w}))$.

Furthermore, if $\mathcal{P}(f)$ possesses a simple coherent edge path, $\mathcal{C}_w = \sum_{i=1}^r \mathcal{C}_{e_i}$ is suitable in the result above.

Example. Let us consider the following equation[1]:

$$f(\mathbf{x}, y) = y^2 + 2x_1 y^2 + 2x_1 y + x_1^2 y^2 + 2x_1^2 y + x_1^2 + x_2^2. \tag{2}$$

The associated Newton polytope is represented on the figure 1 below. It has four coherent edge paths $E_1 = \{e_1\}$, $E_2 = \{e_2\}$, $E_3 = \{e_3\}$, $E_4 = \{e_4\}$.

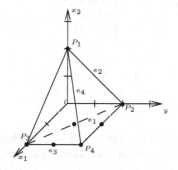
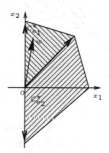

Fig. 1. Newton polytope of f.   **Fig. 2.** Choice of w.

Suppose that we are interested in the computation of the series solutions associated to the path E_2. We have then to choose a vector w in the dual of the barrier cone of the admissible edge e_2, $w \in \mathcal{C}_{e_2}^*$. The associated characteristic polynomial is $\Phi_{e_2}(X) = X^2 + 1$. This polynomial has two simple roots $\alpha_1 = i$ and $\alpha_2 = -i$. So we can define two begining of solutions $\psi_1 = ix_2$ and $\psi_2 = -ix_2$ ($\mathbf{s}_{e_2} = (0, -1)$, $q = 1$) and compute the solutions with the algorithm presented in the previous section

$$\begin{cases} \phi_1 = ix_2 - x_1 - ix_1 x_2 + x_1^2 + ix_1^2 x_2 + \dots \\ \phi_2 = -ix_2 - x_1 + ix_1 x_2 + x_1^2 - ix_1^2 x_2 + \dots \end{cases}$$

Furthermore, since the edge e_2 is simple, we can make precise that $\phi_i \in x_2 \mathbb{C}[[\mathcal{S}_{\mathcal{C}_{e_2}}]]$ with $\mathcal{C}_{e_2} = Pos((1, 0), (1, -1))$.

Let us now have a look at the edge e_1. Let us choose an irrational vector w in the dual of the barrier cone of e_1. We can choose it as on the figure 2.

The characteristic polynomial associated to e_1 is $\Phi_{e_1}(X) = 1 + 2X + X^2 = (1 + X)^2$. This polynomial has -1 as root of multiplicity 2. This path is not simple. So we define $\psi_1 = -x_1$ ($\mathbf{s}_{e_1} = (-1, 0)$, $q = 1$) and compute

$$f_1(x_1, x_2, y) = f(x_1, x_2, \psi_1 + y) = y^2 - 2x_1^2 y + 2x_1 y^2 + x_1^4 - 2x_1^3 y + x_1^2 y^2 + x_2^2$$

The Newton polytope associated to f_1 (fig. 3) has again four coherent edge paths: two composed of an unique simple edge $e_{1,2}$ and $e_{1,4}$, and two others composed of an edge, $e_{1,1}$ and $e_{1,3}$, whose characteristic polynomial has a root of multiplicity two. The barrier cones are represented on figure 4.

---

[1] Of course, we know a "closed-form" of the solutions: $y = \frac{\pm ix_2 - x_1}{1 + x_1}$

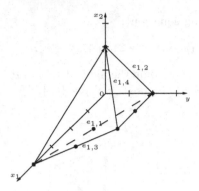

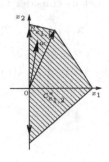

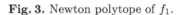

Fig. 4. Barrier cones of
the edges $e_{1,1}$, and $e_{1,2}$.

Fig. 3. Newton polytope of f_1.

Since $w \in \mathcal{C}^*_{e_{1,1}}$, we continue the process with the edge $e_{1,1}$. This edge is not a simple edge, so we will again have a multiple root for the associated characteristic polynomial and we are not yet able to define the cone of the series solution at this step. We can remark that the normal cone of the edge $e_{1,1}$ has shrunk and it will continue if we carry on the algorithm. At the opposite, the normal cone of the edge $e_{1,2}$ is increasing. So there will be a step n_0 for which the chosen vector w will belong to $\mathcal{C}^*_{e_{n_0,2}}$. This time, we will continue the process with the edge $e_{n_0,2}$ which is simple. The (unique at this step) beginning of solution will split into two simple solutions whose supports are included in a translate of the cone $\mathcal{C}_{e_{n_0,2}}$.

We see through this example that the choice of w can strongly influence the number of steps we will have to compute until we are able to separate the two solutions. In this example, the more the vector w is close to the x_2-axes, the more steps we have to compute in the algorithm.

Suppose for example that $n_0 = 3$: we obtain the set of solutions

$$\begin{cases} \varphi_1 = -x_1 + x_1^2 - x_1^3 + ix_2 + x_1^4 - ix_1x_2 + \dots \\ \varphi_2 = -x_1 + x_1^2 - x_1^3 - ix_2 + x_1^4 + ix_1x_2 + \dots \end{cases}$$

In fact, since $\mathcal{C}_{e_{n_0,2}} \subset \mathcal{C}_{e_2}$ and f has at most two solutions in $\mathbb{C}((\mathcal{S}_{\mathcal{C}_{e_2}}))$, $\{\varphi_1, \varphi_2\} = \{\phi_1, \phi_2\}$.

For example, $\varphi_1 = \phi_1$, only the order of the terms depends on the order on $\mathbb{Q}^N$, that is to say on the choice of the vector w.

The main drawback of this version of the Newton algorithm lies in the fact that we have an infinite possibility to choose the vector w, and in the impossibility of determining a priori if the solutions obtained with two vectors w_1 and w_2 are equal or not.

We will now see that it is possible to improve all points and particularly to determine the cones of the series solution from the begining, even in the case of non simple edges. For that we will use some theoretical results due to Perez ([7]) that we will expose in the next section.

4   Theoretical Results on the Cones of the Solutions

Up to now we consider a strongly convex rational polyhedral cone ρ of dimension N and a non constant polynomial f (of degree d) with coefficients in $\mathbb{C}((\mathcal{S}_\rho))$,

$$f(y) = 0 \text{ with } f \in \mathbb{C}((\mathcal{S}_\rho))[y]. \tag{3}$$

4.1   Newton ρ-Polyhedron

The Newton ρ-polyhedron of a non null series $\phi \in \mathbb{C}((\mathcal{S}_\rho))$ is the Minkowski sum of the convex hull of its exponents and the cone ρ.

The Newton ρ-polyhedron of a polynom $f \in \mathbb{C}((\mathcal{S}_\rho))[y]$, is the Minkowski sum of the convex hull of its exponents and the cone $\rho \times \{0\}$.

The Newton ρ-polyhedron of ϕ (resp. f) is noted $\mathcal{N}_\rho(\phi)$ (resp. $\mathcal{N}_\rho(f)$).

When $\rho = (\mathbb{R}^+)^N$, we will just call it Newton polyhedron of ϕ (resp. f) and simply denote it by $\mathcal{N}(\phi)$ (resp. $\mathcal{N}(f)$).

From an algorithmic point of view, the **construction** of the Newton ρ-polyhedron of a polynom f is based on the fact that it is a polyhedral set in $\mathbb{R}^{N+1}$. This result can be found in [8] in the case $\rho = (\mathbb{R}^+)^N$ and extended in the general case. So it can be represented by a matrix $\mathbf{A}$ with $N+1$ columns and a vector $\mathbf{d}$ such that $z \in \mathcal{N}_\rho(f) \Longleftrightarrow \mathbf{A}z \geq \mathbf{d}$. In the particular case of the Newton ρ-polyhedron, we will isolate the last coordinate of z, and separate $\mathbf{A}$ in a matrix $\mathbf{M}$ (the N first columns) and a vector $\mathbf{b}$ (the last column) so that:

$$\mathcal{N}_\rho(f) = \left\{ (\mathbf{x}, y) \in \mathbb{R}^N \times \mathbb{R} \mid \mathbf{M}\mathbf{x} + y\mathbf{b} \geq \mathbf{d} \quad (3) \right\}.$$

Each line of the inequation (3) $M_{i1}x_1 + \ldots + M_{iN}x_N + b_i y \geq d_i$ defines a supporting hyperplane of a facet from the Newton ρ-polyhedron, with integer coefficients M_{ij}, b_i, d_i.

4.2   Newton Algorithm

The algorithm described briefly in the above section can be transposed in this context and gives the following result:

Theorem 1. *[7] For all irrational vector $w \in \rho^*$, there exists a strongly convex rational cone $\mathcal{C}_w$ of dimension N such that $w \in \mathcal{C}_w^* \subset \rho^*$ and f admits d solutions in $\mathbb{C}((\mathcal{S}_{\mathcal{C}_w}))$.*

More interesting for our purpose is the relation between the cones of the series solutions and the fan of the discriminant of the polynomial f.

4.3   Results on the Cones of Series Solutions

We consider a primitive polynomial of degree $d \geq 1$ and the equation

$$f(\mathbf{x}, y) = \sum_{k=0}^{d} \overline{a}_k y^k = 0 \text{ with } \overline{a}_k \in \mathbb{C}\{\{\mathcal{S}_\rho\}\}. \tag{4}$$

For the <u>discriminant</u>, we use the definition of [10]. If $\mathcal{A}$ is an integral domain, and P in $\mathcal{A}[y]$, the discriminant of P is: $\Delta_y(P) = Resultant_y(P, P')$.

A great deal of information concerning the Newton polytope of the discriminant is given in [11]. Here we will use the following precise result:

Theorem 2. *[7] For each cone σ of dimension N of the fan associated to the Newton ρ-polyhedron $\mathcal{N}_\rho(\Delta_y(f))$, f can be decomposed in the ring $\mathbb{C}\{\{\mathcal{S}_{-\sigma^*}\}\}$.*

4.4   Example

Consider the equation: $f(\mathbf{x}, y) = x_2 y^2 + y^2 + x_2^2 y + x_1^2 y + x_1^2 x_2 y - x_1^2 x_2$. Its Newton polyhedron is graphically represented on figure 5

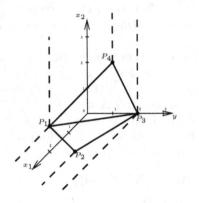
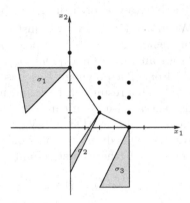

Fig. 5.   Fig. 6.

and in terms of linear inequalities:
$$\underbrace{\begin{pmatrix} 3 & 2 \\ 1 & 2 \\ 1 & 0 \\ 1 & 0 \\ 0 & 1 \\ 0 & 1 \\ 0 & 0 \\ 0 & 0 \end{pmatrix}}_{M} \begin{pmatrix} x_1 \\ x_2 \end{pmatrix} + y \underbrace{\begin{pmatrix} 4 \\ 2 \\ 0 \\ 2 \\ 0 \\ 1 \\ 1 \\ -1 \end{pmatrix}}_{b} \geq \underbrace{\begin{pmatrix} 8 \\ 4 \\ 0 \\ 2 \\ 0 \\ 1 \\ 0 \\ -2 \end{pmatrix}}_{d} \tag{5}$$

Moreover
$$\Delta_y(f) = -(x_2 + 1)(x_2^4 + 2x_1^2 x_2^3 + 6x_1^2 x_2^2 + x_1^4 x_2^2 + 2x_1^4 x_2 + 4x_1^2 x_2 + x_1^4).$$

The Newton polyhedron of this *series* and the associated fan is represented on the figure 6. In this case, there are three cones of dimension 2 in the fan, the cones σ_1, σ_2 and σ_3 represented on figure 6. The theorem 2 says us that we can find $d = 2$ series solution with exponents in a translate of each cone $-\sigma_1^*$, $-\sigma_2^*$ and $-\sigma_3^*$. The full set of solutions we are looking for will be composed of three cones and two series solutions with support in a translate of each cone.

4.5   Full Set of Solutions

By definition, the full set of solutions of the equation (4) is the union, when σ is any cone of dimension N of the fan of the ρ-polyhedron $\mathcal{N}_\rho(\Delta_y(f))$, of the sets of the d solutions computed in $\mathbb{C}((\mathcal{S}_{-\sigma^*}))$. We will now justify the term of "full set" of solutions.[2] For this purpose, we state the following lemma:

Lemma 1 *Let $\sigma \subset \mathbb{R}^N$ be a rational cone. The following propositions are equivalent:*

- *(a) There exists an irrational vector, w, in σ.*
- *(b) $dim(\sigma) = N$.*

This lemma permits to demonstrate the theorem:

Theorem 3. *Let $w \in \rho^*$ be an irrational vector. Denote by $\{\phi_i\}$ the set of d solutions existing in $\mathbb{C}((\mathcal{S}_{\mathcal{C}_w}))$ (Th 1).*
*   Then, for $i = 1 \ldots d$, ϕ_i appears in the full set of solutions.*

Proof: Since the fan $\Sigma\left(\mathcal{N}_\rho(\Delta_y(f))\right) = -\rho^*$, there exists a rational cone $\sigma \in \Sigma\left(\mathcal{N}_\rho(\Delta_y(f))\right)$ such that $w \in -\sigma$. By the previous lemma, this cone is of dimension N, and we denote by $\{\varphi_i\}$ the set of d solutions in $\mathbb{C}((\mathcal{S}_{-\sigma^*}))$ (Th. 2). By definition, the solutions φ_i belong to the full set of solutions.
    The fact that $w \in (-\sigma) \cap \mathcal{C}_w^*$ implies with the previous lemma that $dim((-\sigma) \cap \mathcal{C}_w^*) = N$. That means that the cone $\tau = ((-\sigma) \cap \mathcal{C}_w^*)^* = (-\sigma)^* + \mathcal{C}_w$ is strongly convex ([9]). But $\{\phi_i\} \subset \mathbb{C}((\mathcal{S}_\tau))$ and $\{\varphi_i\} \subset \mathbb{C}((\mathcal{S}_\tau))$ and then we can conclude that $\{\phi_i\} = \{\varphi_i\}$.   □

5   Computation of the Full Set of Solutions

In this section, we consider an equation $f(\mathbf{x}, y) = 0$ where $f \in \mathcal{C}[[\mathcal{S}_\rho^\nu]][y]$, where $\nu \in \mathbb{N}^*$ and $\mathcal{S}_\rho^\nu = \rho \cap \frac{1}{\nu}\mathbb{Z}^N$.

5.1   Paths Associated to a Cone of the Fan of $\mathcal{N}_\rho(\Delta_y(f))$

Let us consider σ an N dimensional cone of the fan of $\mathcal{N}_\rho(\Delta_y(f))$. Let us take w an irrational vector in $-\sigma$. We know ([7]) that this vector defines a monotone coherent edge path E in $\mathcal{N}_\rho(f)$, which permits to compute d solutions of the equation with exponents in a translate of a cone $\mathcal{C}_w$.
    We give now a criterion which associates to each cone σ a set of monotone coherent edge paths capable of furnishing solutions in $\mathbb{C}((\mathcal{S}_{-\sigma^*}))$.
    The monotone coherent edge path E with barrier cone $\mathcal{C}_E$ is associated to the cone $\sigma \in \Sigma\left(\mathcal{N}_\rho(\Delta_y(f))\right)$ if:

$$dim((\mathcal{C}_E + (-\sigma)^*)^*) = dim(\mathcal{C}_E^* \cap (-\sigma)) = N$$

---

[2] Notice that this term does not have the same meaning as in [5]

Theorem 4. *A path E with barrier cone $\mathcal{C}_E$, associated to the cone σ of the fan $\Sigma\left(\mathcal{N}_\rho(\Delta_y(f))\right)$, will permit to compute d solutions of the considered equation in $\mathbb{C}((\mathcal{S}_{-\sigma^*}))$.*

Proof: Let w be an irrational vector in $(\mathcal{C}_E + (-\sigma)^*)^*$. Since $\mathcal{C}_E + (-\sigma)^*$ is a strongly convex rational polyhedral cone, containing ρ, we have that $f \in \mathbb{C}((\mathcal{S}_{\mathcal{C}_E + (-\sigma)^*}))$. Then we apply theorem 1, $\exists \mathcal{C}_w$ such that $(-\sigma)^* + \mathcal{C}_E \subset \mathcal{C}_w$ with d solutions of the equation in $\mathbb{C}((\mathcal{S}_{\mathcal{C}_w}))$. Since $-\sigma^* \subset \mathcal{C}_w$ we conclude with theorem 1 that these d solutions computed in $\mathbb{C}((\mathcal{S}_{\mathcal{C}_w}))$ are the d solutions existing in $\mathbb{C}((\mathcal{S}_{-\sigma^*}))$. $\square$

In the next section we will detail the resolution process for an edge. We will use the same notations as in section 5.

5.2   Solutions Associated to an Edge

The main idea is to bring the equation to the regular case through iterative changes of unknown and changes of variables. After what we can easily find, as we have seen it in section 2, the associated series solution.

1. Consider an **admissible edge** e of the Newton ρ-polyhedron, $\mathcal{N}_\rho(f)$. By definition of an admissible edge, the line supporting the edge intersects the null hyperplane: let B be this intersection point, $B = (\beta, 0)$ with $\beta = (\beta_1, \ldots, \beta_N)$. So the edge is supported by the line of equation: $\mathbf{x} = \beta + y\mathbf{s}_e$. Besides this line is also the intersection of supporting hyperplanes of the Newton ρ-polyhedron. So we can extract from the matrix and vectors defining $\mathcal{N}_\rho(f)$ a full rank matrix $\mathbf{M}_e$, and two vectors $\mathbf{b}_e$ and $\mathbf{d}_e \in \mathbb{R}^N$ such that a point $P = (\mathbf{K}, h)$ belongs to the edge if and only if $\mathbf{M}_e\mathbf{K} + h\mathbf{b}_e = \mathbf{d}_e$.
   The fact that B is the intersection with the null hyperplane can be expressed by $\mathbf{M}_e\beta = \mathbf{d}_e$ and the barrier cone $\mathcal{C}_e$ of the edge is the set of points $\mathbf{K} \in \mathbb{R}^N$ such that $\mathbf{M}_e\mathbf{K} \geq 0$.
2. Let us calculate the associated **characteristic polynomial** Φ_e.
3. Let c be a **(non null) root** of $\Phi(X^q)$ in $\mathbb{C}$, of multiplicity r. We put: $\Phi(X^q) = (X - c)^r \psi(X)$, $\psi \in \mathbb{C}[X]$ and $\psi(c) \neq 0$.
4. We make a **change of unknown**, putting $y = \mathbf{x}^{-\mathbf{s}_e}(c + \tilde{y})$ and we calculate $\tilde{f}(\tilde{y}) = \mathbf{x}^{-\beta} f(\mathbf{x}^{-\mathbf{s}_e}(c + \tilde{y}))$. In this transformation, a monomial $a_{\mathbf{K}}^h \mathbf{x}^{\mathbf{K}} y^h$ becomes

$$\sum_{m=0}^{h} \binom{h}{m} a_{\mathbf{K}}^h c^m \mathbf{x}^{\mathbf{K} - \beta - h\mathbf{s}_e} \tilde{y}^{h-m}.$$

Geometrically, this means that the point $P = (\mathbf{K}, h)$ gives rise to $h+1$ points of coordinates $(\mathbf{K} - \beta - h\mathbf{s}_e, h - m)$. In the following, we will denote by $P' = (\mathbf{K}', h')$ an arbitrary point obtained from P by the above computation. If P belongs to the edge e, P' lies on the y-axis. Else, $\mathbf{M}_e\mathbf{K} + h\mathbf{b}_e \geq \mathbf{d}_e$ implies that $\mathbf{M}_e\mathbf{K}' \geq 0$.
This means that $\tilde{f} \in \mathbb{C}[[\mathcal{S}_{\mathcal{C}_e}^{\nu q}]][y]$. Moreover, for all w irrational vector in $\mathcal{C}_e^*$, the terms of null order with respect to w are coming from

$$\sum_{(\mathbf{K}, h) \in e \cap \mathbb{S}(f)} a_{\mathbf{K}}^h (c + \tilde{y})^h = (c + \tilde{y})^j \Phi((c + \tilde{y})^q) = (c + \tilde{y})^j \tilde{y}^r \psi(c + \tilde{y}).$$

Then we can write $\tilde{f} = \bar{b}_0 + \bar{b}_1\tilde{y} + \ldots + \bar{b}_d\tilde{y}^d$, with $\bigcirc_w(\bar{b}_i) > 0$ for $i = 0, \ldots, r-1$ and $\bigcirc_w(\bar{b}_r) = 0$.

So two cases are to be considered:

- c is a simple root: we do a **change of variables**, introducing new variables $\tilde{x}_1, \ldots, \tilde{x}_N$ such that $\tilde{f} \in \mathbb{C}[[\tilde{x}_1, \ldots, \tilde{x}_N]][\tilde{y}]$.
  This transformation can be built in two steps:

  - first we introduce the intermediate variables $t_i = x_i^{\frac{1}{\nu q}}$; a monomial $\mathbf{x}^{\mathbf{K}'}$ becomes $\mathbf{t}^{q\mathbf{K}'}$, so that the exponents are integers.
  - next we calculate the Hermite normal form of $\mathbf{M}_e^t$, that is to say an upper triangular matrix $\mathbf{H}$ and a unimodular matrix $\mathbf{U}$ such that:

$$\begin{cases} \mathbf{H} = \mathbf{U}^t\mathbf{M}_e^t \\ \mathbf{H}_{ii} \geq 0 \quad \forall\, 1 \leq i \leq N \\ \mathbf{H}_{ij} < 0 \quad \forall\, i < j. \end{cases}$$

  Now we ask the monomial $\mathbf{t}^{\mathbf{K}}$ to become $\tilde{\mathbf{x}}^{\tilde{\mathbf{K}}}$, where $\mathbf{K} = \mathbf{U}\tilde{\mathbf{K}}$. For that, we have to put $\tilde{x}_i = t_1^{a_{1i}} \ldots t_N^{a_{Ni}}$, where $(a_{1i}, \ldots, a_{Ni})$ is the ith column of $\mathbf{U}$.
  Finally, we can see easily that $\mathbf{M}_e\mathbf{K} \geq 0$ implies $\mathbf{H}^t\tilde{\mathbf{K}} \geq 0$, and then that all coordinates of $\tilde{\mathbf{K}}$ are positive integers.

  Conclusion: the vectors of exponents $\tilde{\mathbf{K}}$ are in $\mathbb{N}^N$.
  Then $\tilde{f} \in \mathbb{C}[[\tilde{\mathbf{x}}]][\tilde{y}]$. Moreover $\tilde{y}_0 = 0$ satisfies the hypothesis of the **regular case** for $\tilde{f}$:

  - $\tilde{f}(\tilde{y}_0) = 0 \mod \tilde{\mathbf{x}}$
  - $\tilde{f}'(\tilde{y}_0) \neq 0 \mod \tilde{\mathbf{x}}$.

  So we can compute a solution $\tilde{y} \in \mathbb{C}[[\tilde{\mathbf{x}}]]$ of the equation $\tilde{f}(\tilde{\mathbf{x}}, \tilde{y}) = 0$, which gives rise to a solution $y = \mathbf{x}^{-\mathbf{s}_e}(c + \tilde{y}) \in \mathbf{x}^{-\mathbf{s}_e}\mathbb{C}[[\mathcal{S}_{\mathcal{C}_e}^{\nu q}]]$ of $f(\mathbf{x}, y) = 0$.
  Remark: the matrix M_e is not unique (up to the order of the lines), so the matrix U is not unique. The different choices lead to different orderings of the terms of the solution $\tilde{y}$, but do not induce the computation of different solutions. This part of the algorithm could have been replaced by a choice of an irrational vector w in $\mathcal{C}_e$ and the developpement of the solution with respect of $\bigcirc_w$. In the same manner, the choice of w is not unique and leads to different orderings of the terms of the solution.

- if c is a root of multiplicity $r > 1$: we go to step 1 above, with $\tilde{f}$ in place of f, taking into account only the part of the Newton $\mathcal{C}_e$-polyhedron of $\tilde{f}$ located between the hyperplanes $\tilde{y} = 0$ and $\tilde{y} = r$.

5.3 Description of the Algorithm

Even if the theoretical algorithm has been presented in the frame of polynomials with power series coefficients, the effective implementation deals withs multi-variate polynomials. So we consider $f(y) = 0$ with $f \in \mathbb{C}[x_1, \ldots, x_N, y]$.

1. First of all, we perform the squarefree factorization of f:

$$f = \prod_{i=1}^{h} f_i^{\alpha_i} \text{ with } f_i \in \mathbb{C}[x_1, \ldots, x_N][y] \text{ and } \alpha_i \in \mathbb{N}^*$$

We will solve separately each simplified equation $f_i = 0$ with f_i squarefree. Each root of the equation $f_i = 0$ will then be a root of multiplicity α_i for the equation $f = 0$. From now on, we suppose that f is **squarefree**.

2. We calculate the discriminant $\Delta_y(f)$ of f with respect to the variable y and construct its Newton polyhedron.

3. We construct the Newton polyhedron of f.

4. To each N-dimensional cone σ of the fan of $\mathcal{N}(\Delta_y(f))$, we associate a (nonempty) set of coherent monotone edge paths. If this set contains a simple path, we will consider this one else we take one path in the set. For the considered path $E = \{e_1, \ldots, e_r\}$, we build the set of solutions in the following way: for each edge in the path, we iterate the transformations as described in the above paragraph. Simple roots (if there exists any) give rise to solutions in $\mathbb{C}((\mathcal{S}_{-\sigma^*}))$. We stop the process when the set of solutions associated to the path is complete (that is to say, contains a number of solutions equal to the length of the considered edge). The fact that f is squarefree assures that this situation will be reached after a finite number of steps.

The algorithms presented here can be summarized in the following way:

algorithm *Newton-Puiseux(f, n)*
input:
 $-$ $f \in \mathbb{C}[\mathbf{x}][y]$
 $-$ n, an integer for the number of terms in the solutions expansions
output:
 $-$ A set of cones and associated truncated generalized Puiseux series.
Begin
    Squarefree factorization(f)
    $d := Degree(f, y)$
    $\rho := (\mathbb{R}^+)^N$
    $\nu := 1$
    $Sol := \mathbf{Compute\text{-}NP}(f, n, \rho, \nu, d)$
    Return(Sol)
End.

The procedure **Compute-NP** computes the solutions expansions of f by considering only the part of the Newton-ρ-polyhedron located between the two hyperplanes of equation $y = 0$ and $y = r$. This procedure will be used recursively.

procedure *Compute-NP(f, n, ρ, ν, r)*
input:
 $-$ $f \in \mathbb{C}[\mathcal{S}_\rho^\nu][y]$
 $-$ n, an integer for the number of terms in the solutions expansions
 $-$ ρ, the set of vectors $\{u_1, \ldots, u_p\}$ defining the cone

– ν, an integer defining the ramification of the expansions solutions
– r, an integer defining the equation of the hyperplane $y = r$
output:
– A set of cones and associated truncated generalized Puiseux series.
Begin

$Sol := \emptyset$

$\Delta := Discriminant(f, y)$

$\mathcal{N}_\rho(\Delta) := Newton\ \rho\text{-}polyhedron\ of\ \Delta$

$\mathcal{N}_\rho(f) := Newton\ \rho\text{-}polyhedron\ of\ f\ between\ the$
$hyperplanes\ y = 0\ and\ y = r$

$E(\mathcal{N}_\rho(f)) := Set\ of\ monotone\ coherent\ paths\ of\ \mathcal{N}_\rho(f)$

$\Sigma(\mathcal{N}_\rho(\Delta)) := Associated\ fan\ of\ \mathcal{N}_\rho(\Delta)$

For each $\sigma \in \Sigma(\mathcal{N}_\rho(\Delta))$ *of dimension N* **do**

   $E(\sigma) := Paths\ of\ E(\mathcal{N}_\rho(f))\ associated\ to\ the\ cone\ \sigma$

   Choose a path E in $E(\sigma)$ *(simple if possible)*

   For each $e \in E$ **do**

      $Sol_e := $ **Compute-Edge-Solution**(f, n, ν, e)

      $Sol := Sol$ **union** Sol_e

Return(Sol)

End.

The procedure **Compute-Edge-Solution** computes the solutions associated to an edge by iterating the transformations described in the section 7.2.

procedure *Compute-Edge-Solution*(f, n, ν, e)
input:
– $f \in \mathbb{C}[\mathcal{S}_\rho^\nu][y]$
– n, an integer for the number of terms in the solutions expansions
– ν, an integer defining the ramification of the expansions solutions
– e, the set of the points of the edge
output:
– A set of truncated generalized Puiseux series.
Begin

$Sol_e := \emptyset$

$C_e := Barrier\ cone(e)$

$\mathbf{s}_e := Slope(e) := (p_1/q_1, \ldots, p_N/q_N)$

$q := lcm(q_i)$

$\Phi_e := Characteristic\ polynomial(e)$

$C := Set\ of\ non\ null\ roots\ of(\Phi_e)$

For each $c \in C$ **do**

   Compute $\tilde{f}(\tilde{y})$

   If $(c\ simple)$ **then**

      $Sol_c := Regular\ case(\tilde{f}, n, e, \nu, q)$

   Else

      $r := Multiplicity(c)$

      $Sol_c := $ **Compute-NP**$(\tilde{f}, n, C_e, \nu q, r)$

   $Sol_e := Sol_e$ **union** Sol_c

Return(Sol_e)

End.

The algorithms of the resolution process are implemented in Maple. For the elementary functions on the cones they use the package $\mathcal{C}$onvex developped by Mattias Franz, Universität Konstanz, Germany.

5.4   Example

We now run the algorithm on the equation 2 from the section 3.3.

1. f is squarefree.
2. $\Delta(f) = 4x_2^2(1 + x_1)^4$. The Newton polyhedron of this expression and the associated fan are represented on the figure 7 below. The associated fan is in fact composed of one unique cone $\sigma = Pos\big((-1,0),(0,-1)\big)$.
3. The Newton polyhedron of f is given by the figure 8.
4. To the cone σ, there is two monotone coherent edge paths associated: $E_1 = \{[P_1P_2]\}$, $E_2 = \{[P_3P_2]\}$. In the case of this example, the path E_1 is simple and the other is not. To compute the series solutions associated to the cone σ we use the simple path and obtain then the following full set of solutions: the series ϕ_1 and ϕ_2, elements of $\mathbb{C}((\mathcal{S}_{-\sigma^*}))$, computed in section 3.3.

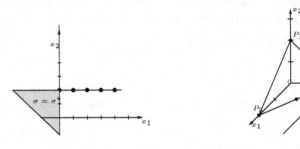

Fig. 7.	**Fig. 8.**

6   Link with Differential Equations

This work is linked with the study of differential equations in the following way: it permits to transform an implicit differential equation of order one into an explicit one. Indeed, consider $f(t,y,y') = 0$, $f \in \mathbb{C}[[t]][y,y']$. It is possible to apply directly a Newton algorithm as explained in [2], but it is also possible to run the above algorithm on $f(t,y,p)$. By this way, we obtain after a finite number of steps, $p = \dfrac{dy}{dt}$ as a series in t and y, with exponents in a cone, or more precisely

$$p = t^{-s_1(1)}y^{-s_1(2)}\big(c_1 + t^{-s_2(1)}y^{-s_2(2)}(c_2 + \ldots + \tilde{p})\big)$$
$$\text{where } \tilde{p} \in \mathbb{C}[[\tilde{t},\tilde{y}]], \text{ (with finitely many } c_i,\ s_i).$$

$$(6)$$

The variables are linked by the following relations:

$$\begin{cases} \tilde{y}^q = t^{\alpha_1} y^{\beta_1} \\ \tilde{t}^q = t^{\alpha_2} y^{\beta_2} \end{cases} \Longleftrightarrow \begin{cases} y = \tilde{t}^{q\alpha_1'} \tilde{y}^{q\beta_1'} \\ t = \tilde{t}^{q\alpha_2'} \tilde{y}^{q\beta_2'} \end{cases},$$

with q, α_i, β_i, α_i', $\beta_i' \in \mathbb{Z}$. Let us differentiate the previous expressions

$$\begin{cases} q\tilde{y}^{q-1} d\tilde{y} = \alpha_1 t^{\alpha_1-1} y^{\beta_1} dt + \beta_1 t^{\alpha_1} y^{\beta_1-1} dy \\ q\tilde{t}^{q-1} d\tilde{t} = \alpha_2 t^{\alpha_2-1} y^{\beta_2} dt + \beta_2 t^{\alpha_2} y^{\beta_2-1} dy \end{cases}$$

and finally

$$\left(\alpha_2 t^{\alpha_2-1} y^{\beta_2} + \beta_2 t^{\alpha_2} y^{\beta_2-1} \frac{dy}{dt} \right) \tilde{y}^{q-1} \frac{d\tilde{y}}{d\tilde{t}} = \tilde{t}^{q-1} \left(\alpha_1 t^{\alpha_1-1} y^{\beta_1} + \beta_1 t^{\alpha_1} y^{\beta_1-1} \frac{dy}{dt} \right)$$

Since $p = \dfrac{dy}{dt}$ can be expressed as a series in $\tilde{y}$ and $\tilde{t}$ with the relation (6) we can observe that this relation is in fact of the form

$$a(\tilde{t}, \tilde{y}) \frac{d\tilde{y}}{d\tilde{t}} + b(\tilde{t}, \tilde{y}) = 0 \text{ with } a, b \in \mathbb{C}[[\tilde{t}, \tilde{y}]]$$

which are the type of differential equations which have been the subject of many investigations. Some results on Newton algorithms and Puiseux series solutions can be found in [13].

7   Conclusion

From the computer algebra point of view, the local resolution of multi-variate equations is achieved. Of course, it will be interesting to complete the algebraic study presented here, by an analytic study of the solutions (definition of the domain of convergence of the series expansions). But we are convinced that the rings $\mathbb{C}((\mathcal{S}_{-\sigma^*}))$ are the good way of generalizing the Puiseux series in several variables, and that this approach for solving multi-variate polynomial equations is very promising and will have many fruitfull developments (local study of differential equations, of systems of multi-variate algebraic equations).

References

1. Brieskorn, E., Knörrer, H. (1986): Plane algebraic curves. Birkhäuser Verlag
2. Della-Dora, J., Richard-Jung, F. (1997): About the newton polygon algorithm for non linear ordinary differential equations. Proceedings of the International symposium on symbolic and algebraic computation
3. Walker, R. J. (1950): Algebraic curves. Dover edition
4. Beringer, F., Jung, F. (1998) Solving "Generalized Algebraic Equations". Proceedings of the International symposium on symbolic and algebraic computation
5. McDonald, J. (1995): Fiber polytopes and fractional power series. Journal of Pure and applied Alebra, **104**, 213–233

6. Ewald, G. Combinatorial Convexity and Algebraic Geometry. Graduate Texts in Mathematics, Springer

7. González Pérez, P.D. (2000): Singularités quasi-ordinaires toriques et polyèdre de Newton du discriminant. Canad. J. Math., **52**, 348–368

8. Alonso, M.E. Luengo, I., Raimondo, M. (1989): An algorithm on quasi-ordinary polynomials. Applied Algebra, Algebraic Algorithms and Error-Correcting Codes, Lecture Notes in Computer Science., **357**, 59–73

9. Oda, T. (1988): Convex bodies and algebraic geometry: an introduction to the theory of toric varieties Annals of Math. Studies Springer-Verlag., **131**.

10. von zur Gathen, J., Gerhard, J. (1999): Modern Computer Algebra. Cambridge University Press

11. Gelfand, M. Kapranov, M. M., Zelevinsky, A. V. (1994): Discriminants, Resultants and multidimensional determinants. Birkhauser

12. Aroca Bisquert, F. (2000): Metodos algebraicos en ecuaciones differenciales ordinaries en el campo complejo. Thesis, Universidad de Valladolid

13. Cano, J. (1992): An extension of the Newton-Puiseux polygon construction to gives solutions of Pfaffian forms. Preprint, Universidad de Valladolid

14. Geddes, K. O. Czapor, S. R. Labahn, G. (1992): Algorithms for computer algebra. Kluwer Academic Publishers

Wavelets with Scale Dependent Properties

Peter Paule[1], Otmar Scherzer[2], and Armin Schoisswohl[3]

[1] Research Institute for Symbolic Computation, Johannes Kepler University,
A–4040 Linz, Austria,
Peter.Paule@risc.uni-linz.ac.at
[2] Applied Mathematics, Department of Computer Science, University Innsbruck,
A–6020 Innsbruck, Austria,
otmar.scherzer@uibk.ac.at
[3] GE Medical Systems Kretz Ultrasound,
A–4871 Zipf, Austria,
armin.schoisswohl@med.ge.com

Abstract. In this paper we revisite the constitutive equations for coefficients of orthonormal wavelets. We construct wavelets that satisfy
alternatives to the vanishing moments conditions, giving orthonormal
basis functions with scale dependent properties. Wavelets with scale dependent properties are applied for the compression of an oscillatory one-
dimensional signal.

1 Introduction

In this paper we construct filter coefficients of real, compactly supported, orthonormal wavelets with scale dependent properties.

Daubechies' construction of wavelets is based on the existence of a *scaling
function* ϕ, such that for $m \in \mathbf{Z}$ the functions $\phi_{m,k} := 2^{-m/2}\phi(2^{-m}x - k)$,
$k \in \mathbf{Z}$, are orthonormal with respect to $L^2(\mathbf{R})$. Moreover, ϕ is chosen in such a
way that

$$V_m := \overline{\text{span}\{\phi_{m,k}, \ k \in \mathbf{Z}\}}, \qquad m \in \mathbf{Z}$$

form a multiresolution analysis on $L^2(\mathbf{R})$, i.e.,

$$V_m \subset V_{m-1}, \qquad m \in \mathbf{Z},$$

with

$$\bigcap_{m \in \mathbf{Z}} V_m = \{0\} \text{ and } \overline{\bigcup_{m \in \mathbf{Z}} V_m} = L^2(\mathbf{R}).$$

The wavelet spaces W_m are the orthogonal complements of V_m in V_{m-1}, i.e.,

$$W_m := V_m^{\perp} \cap V_{m-1}.$$

The *mother wavelet* ψ is chosen such that $\psi_{m,k} := 2^{-m/2}\psi(2^{-m}x - k)$, $k \in \mathbf{Z}$,
form an orthonormal basis of W_m. Since $\phi = \phi_{0,0} \in V_0 \subset V_{-1}$, the scaling
function ϕ must satisfy the *dilation equation*

$$\phi(x) = \sum_{k \in \mathbf{Z}} h_k \phi(2x - k), \qquad (1)$$

F. Winkler and U. Langer (Eds.): SNSC 2001, LNCS 2630, pp. 255–265, 2003.

where the sequence $\{h_k\}$ is known as the *filter sequence* of the wavelet ψ. The filter coefficients have to satisfy certain conditions in order to guarantee that the scaling function ϕ and ψ satisfy various desired properties. These properties of wavelets and scaling functions will be reviewed in Section 2. Daubechies' construction principle guarantees that the properties imposed on the mother wavelet ψ carry over to any $\psi_{m,k}$.

However, in practical applications (like in digital signal and image processing) one actually utilizes only a finite number of scales m. In data compression with wavelets of medical image data we experienced that five scales are sufficient to achieve high compression ratios.

This experience stimulated our work to design wavelets that have additional properties on an a–priori prescribed number of scales. As we show in Section 3, Daubechies' construction principle leaves enough freedom to design such wavelets. In Section 4 we present some examples of scale dependent wavelets. Moreover, we compare compression using Daubechies' wavelets with compression using scale dependent wavelets, which are designed to include a-priori information of the underlying data. Finally we put our work in relation to recent work on rational spline wavelets [8,9,10].

2   A Review on Daubechies' Wavelets

Following Daubechies [1,2] the construction of orthonormal wavelet functions is reduced to the design of the corresponding filter sequence $\{h_k\}$ in (1). Moreover, one assumes that the mother wavelet ψ satisfies

$$\psi(x) = \sum_{k \in \mathbf{Z}} (-1)^k h_{1-k} \phi(2x - k) . \tag{2}$$

In particular this choice guarantees that the wavelet ψ and the scaling function ϕ are orthogonal.

In orthogonal wavelet theory due to Daubechies the desired properties on the scaling function and wavelets are:

1. For fixed integer $N \geq 1$ the scaling function ϕ has support in the interval $[1 - N, N]$. This in particular holds when the filter coefficients satisfy

$$h_k = 0, \qquad (k < 1 - N \text{ or } k > N). \tag{3}$$

2. The existence of a scaling function ϕ satisfying (1) requires that

$$\sum_{k \in \mathbf{Z}} h_k = 2 . \tag{4}$$

3. In order to impose orthonormality of the integer translates of the scaling function ϕ, i.e., $\int_{\mathbf{R}} \phi(x - l)\phi(x)dx = \delta_{0,l}$, the filter coefficients $\{h_k\}$ have to satisfy

$$\sum_{k \in \mathbf{Z}} h_k h_{k-2l} = 2\delta_{0,l}, \qquad (l = 0, \ldots, N - 1). \tag{5}$$

4. The wavelet ψ is postulated to have N vanishing moments, i.e.,

$$\int_{\mathbf{R}} x^l \psi(x)dx = 0, \qquad (l = 0, \ldots, N-1) \tag{6}$$

which requires the filter sequence to satisfy

$$\sum_{k \in \mathbf{Z}} (-1)^k h_{1-k} k^l = 0, \qquad (l = 0, \ldots, N-1). \tag{7}$$

3   Wavelets with Scale Dependent Properties

In this section we are particularity interested in constructing wavelets that satisfy alternatives to the vanishing moments condition (7). Our motivation is to have more flexibility in adapting wavelets to practical needs.

All along this paper we restrict our attention to filter coefficients that satisfy the following general conditions:

1. the wavelet and the scaling function are compactly supported,
2. the scaling function is orthogonal to its integer translates on every scale.

To satisfy these properties we assume that the filter coefficients satisfy (3)–(5).

To derive alternatives to the vanishing moments condition (7) we redirect our attention to the connection between (6) and (7).

To this end we consider families $\{s_j : j \in I\}$ of functions on $\mathbf{R}$, which are othogonal to all wavelets on a fixed scale m (I being a suitable set of indices). In other words, we want that for all j in I and all integers k,

$$\int_{\mathbf{R}} s_j(x)\psi_{m,k}dx = 0. \tag{8}$$

In order to achieve this goal we assume for all j in I the existence of functions $\{t_{j,k} : k \in I\}$ on $\mathbf{R}$ such that

$$s_j(x+y) = \sum_{k \in I} s_k(x)t_{j,k}(y). \tag{9}$$

The following computation will show that property (9) together with the conditions

$$\sum_{k \in \mathbf{Z}} (-1)^k h_{1-k} \, s_j(2^{m-1}k) = 0, \qquad (\text{for all } j \in I) \tag{10}$$

are sufficient to guarantee the orthogonality (8).

Namely, the left side of (8) equals

$$2^{-m/2} \int_{\mathbf{R}} s_j(x)\psi(2^{-m}x - k)dx$$

$$= 2^{(m-2)/2} \sum_{l \in \mathbf{Z}} (-1)^l h_{1-l} \int_{\mathbf{R}} \phi(t)s_j\left(2^{m-1}(t + 2k + l)\right) dt,$$

where we applied (2) and then substituted $x \to 2^{m-1}(t+2k+l)$. Next, invoking (9) with $x = 2^{m-1}l$ and $y = 2^{m-1}(t+2k)$ results in

$$
2^{(m-2)/2} \sum_{i \in I} \left(\sum_{l \in \mathbf{Z}} (-1)^l h_{1-l} s_i(2^{m-1}l) \right) \times
$$

$$
\int_{\mathbf{R}} \phi(t) \, t_{j,i} \left(2^{m-1}(t+2k) \right) dt = 0, \tag{11}
$$

which vanishes according to (10).

We conclude this section by making some obvious choices for $\{s_j\}$ and $\{t_{j,k}\}$; corresponding examples for computations of filter coefficients $\{h_k\}$ are presented in the next section. Alternative choices that guarantee (9) will be discussed in a forthcoming paper.

1. Sheffer Relations. If we restrict ourselves to polynomial sequences, there is the classical theory of *Sheffer sequences* $\{s_j : j \geq 0\}$ satisfying the Sheffer identity

$$
s_j(x+y) = \sum_{k=0}^{j} \binom{j}{k} s_k(x) p_{j-k}(y), \qquad (j \geq 0) \tag{12}
$$

where $\{p_j : j \geq 0\}$ is called an *associated* sequence. For further information see, for instance, Roman's book [3] which is devoted to Rota's view of umbral calculus.

In our context (12) is obtained from (9) by choosing $t_{j,k}(x) = \binom{j}{k} p_{j-k}(x)$. It is important to note that in order to be able to compute the filter coefficients $\{h_k\}$, it is necesssary to restrict the index set I to a *finite* subset of $\mathbf{N}$, for instance, to $\{0, 1, \ldots, N-1\}$.

For the particular choice

$$
s_j(x) = x^j \quad \text{and} \quad t_{j,k}(x) = \binom{j}{k} s_{j-k}(x)
$$

the relations (9) and (12) are nothing but the binomial theorem. Additionally, taking $I = \{0, 1, \ldots, N-1\}$ the conditions (10) turn into Daubechies' situation of (7), i.e., orthogonality holds on any scale m.

Finally we remark that special types of Sheffer sequences, namely *Appell* sequences, appear in other recent work [4, Remark 7] on wavelets, but in a different context of analysing orthonormal systems of multiwavelets.

2. Exponential Relations. The corresponding setting in full generality is as follows. Let $\{\omega_j : j \in I\}$ be a sequence of complex parameters. For all j, k in I define

$$
s_j(x) = q^{\omega_j x} \quad \text{and} \quad t_{j,k}(x) = \delta_{j,k} \, s_j(x),
$$

with $\delta_{j,k}$ being the Kronecker function and q being a fixed nonzero complex number. Then (9) turns into

$$
q^{\omega_j(x+y)} = q^{\omega_j x} q^{\omega_j y};
$$

additionally, (10) becomes

$$\sum_{k\in\mathbf{Z}}(-1)^k h_{1-k}\, q^{2^{m-1}\omega_j k} = 0 \qquad (j\in I).\tag{13}$$

In the following examples section we examine two special cases over the reals: q-*wavelets* where for all j in I we set $\omega_j = 1$ and q to a fixed positive real number; and sin-*wavelets* where we set $q = \exp(i)$, the complex exponential, and where we split (13) into two groups of equations over the reals by taking the real *and* the imaginery part of

$$\exp(i\,2^{m-1}\omega_j\, k) = \cos(2^{m-1}\omega_j\, k) + i\,\sin(2^{m-1}\omega_j\, k).$$

4   Examples

In this section we present several examples of wavelets with scale dependent properties. The resulting systems for $\{h_k\}$ are considered as algebraic equations which can be solved by the combined symbolic/numerical approach described in [5].

4.1 q-Wavelets

Here we derive wavelet filter coefficients of orthonormal, compactly supported wavelet functions $\psi_{m,k}$ that are orthogonal to $s_j(x) = q^x$ on scales $m = 0, -1, \ldots,$ $-(N-2)$. Since the s_j are independent of j this means, we can set $I = \{0\}$. For $N > 1$ the equations for the wavelet filter coefficients are

$$\sum_{k\in\mathbf{Z}} h_k = 2\,,$$
$$\sum_{k\in\mathbf{Z}} h_k h_{k-2l} = 2\delta_{0,l}\,, \qquad (l = 0,\ldots,N-1)\tag{14}$$
$$\sum_{k\in\mathbf{Z}}(-1)^k h_{1-k} q^{2^{-(\overline{m}+1)}k} = 0\,, \qquad (\overline{m} = 0,\ldots,N-2)\,.$$

For $N = 2$ the solution $h = (h_{-1}, h_0, h_1, h_2)$ of this system is

$$\left(h_{-1}, h_0, h_1, h_2\right) = \left(\tfrac{q^{3/2}-q-\sqrt{q}Q}{q(-1+q)},\ \tfrac{q-1-Q}{-1+q},\ \tfrac{q^2-q^{3/2}+\sqrt{q}Q}{q(-1+q)},\ \tfrac{Q}{-1+q}\right)\tag{15}$$

where

$$Q \equiv Q_{\pm} = \tfrac{q^2+q-q^{3/2}-\sqrt{q}}{2(q+1)}$$
$$\pm\tfrac{\sqrt{q^4-5\,q^3+2\,q^{7/2}+4\,q^{5/2}-5\,q^2+2\,q^{3/2}+q}}{2(q+1)}.$$

In Figure 1 we have plotted the scaling functions and the wavelets according to the filter coefficients (15). From such plottings one can see that for $q \to 1$ the scaling function and the wavelet converge to Daubechies' scaling function and wavelet, respectively. The proof of this observation is immediate since applying $q\,D_q$, where D_q is derivation with respect to q, $N - 1$ times to the last equation in (14) and then setting $q = 1$ results in (7).

Moreover, from (15) it follows that for $q \to 0$ and $q \to \infty$ the coefficients of the q-wavelets according to Q_+ approach the coefficients of the Haar wavelet, which confirms the plots in Figure 1.

In [7, Section 3] all cofficients h_k, $k \in \mathbf{Z}$ satisfying (3) – (5) (with $N = 2$) are calculated. It can be shown that the family of solutions can be parametrized by a single real parameter. Thus the q-wavelets for $N = 2$ form a subset of the coefficients obtained by Daubechies.

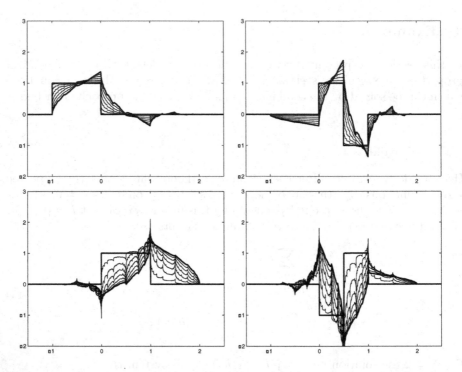

Fig. 1. q–scaling functions (left) and q–wavelets (right) for values of $q \in (1, \infty)$: Solutions for Q_+ (first line), and for Q_- (second line). Note that the plots show clearly the smooth interpolation between Daubechies' wavelet with two vanishing moments ($q \to 1+$) and the Haar wavelet ($q \to \infty$) in both cases. The solutions for $q \in (0, 1)$ are just the mirrored/antimirrored versions of the solutions for $1/q$.

4.2   Variations of Daubechies' Wavelets

In this subsection we investigate wavelet filter coefficients of orthonormal, compactly supported wavelet functions $\psi_{m,k}$ that have $K \geq 1$ vanishing moments and which, in addition, are orthogonal to $s_j(x) = q^x$ on scales $m = 0, -1, \ldots,$ $-(N - K - 1)$. As in the previous section we can again set $I = \{0\}$ since s_j is independent of j. Thus the equations for the filter coefficients are

$$\sum_{k \in \mathbf{Z}} h_k = 2,$$

$$\sum_{k \in \mathbf{Z}} h_k h_{k-2l} = 2\delta_{0,l}, \quad (l = 0, \ldots, N - 1),$$

$$\sum_{k \in \mathbf{Z}} (-1)^k h_{1-k} q^{2^{-(\overline{m}+1)} k} = 0, \quad (\overline{m} = 0, \ldots, N - K - 1),$$

$$\sum_{k \in \mathbf{Z}} (-1)^k h_{1-k} k^l = 0, \quad (l = 0, \ldots, K - 1). \tag{16}$$

In Figure 2 we have plotted the associated scaling function and wavelet for $q \in \{2, 4, 16, 32\}$, $K = 2$, and $N = 3$.

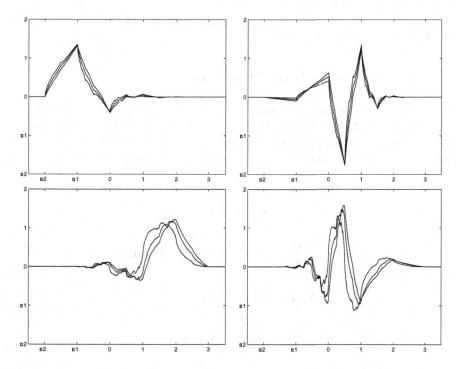

Fig. 2. Variation of Daubechies' Wavelets for $q \in \{2, 4, 16, 32\}$, $K = 2$, and $N = 3$. Scaling functions (left) and wavelets (right) for the two different solutions.

4.3   sin-Wavelets

Wavelet filter coefficients of orthonormal, compactly supported wavelet functions $\psi_{m,k}$ that are orthogonal on scale $m = 0$ to $\sin(2^{m-1}\omega_j x)$ and $\cos(2^{m-1}\omega_j x)$ for $j \in I = \{0, \ldots, K - 1\}$; here $\omega_j \neq 0$ is a sequence of real numbers.

Setting $N = 2K + 1$ the equations for such filter coefficients are

$$\sum_{k \in \mathbf{Z}} h_k = 2\,,$$

$$\sum_{k \in \mathbf{Z}} h_k h_{k-2l} = 2\delta_{0,l}, \quad (l = 0, \ldots, N - 1)\,,$$

$$\sum_{k \in \mathbf{Z}} (-1)^k h_{1-k} \sin(\omega_j k/2) = 0, \quad (j = 0, \ldots, K - 1)\,, \tag{17}$$

$$\sum_{k \in \mathbf{Z}} (-1)^k h_{1-k} \cos(\omega_j k/2) = 0, \quad (j = 0, \ldots, K - 1)\,.$$

Note that the last two sets of equations correspond to taking real and imaginary parts as explained in Section 3.

In Figure 3 we have plotted the scaling function and wavelet for $N = 3$, $K = 1$, and $\omega_0 \in \{\pi/2, \pi/4, \pi/16\}$.

5   Compression with sin-Wavelets

In this section we compare the compression of an oscillatory signal with Daubechies wavelets and sin-wavelets. The a-priori information that we are using is that the signal contains a ground frequency $\omega_0 = \frac{\pi}{4}$. The equations for the coefficients of the sin-wavelets with 6 coefficients and ground frequency $\frac{\pi}{4}$ read as follows

$$\sum_{k=-2}^{3} h_k = 2\,,$$

$$\sum_{k=-2}^{3} h_k h_{k-2l} = 2\delta_{0,l}\,, \quad (l = 0, \ldots, 2)\,,$$

$$\sum_{k=-2}^{3} (-1)^k h_{1-k} \sin(\pi k/8) = 0\,,$$

$$\sum_{k=-2}^{3} (-1)^k h_{1-k} \cos(\pi k/8) = 0\,.$$

In the following numerical example we used sin-wavelets and Daubechies' Wavelets with 3 vanishing moments to compress an oscillatory signal. The compression was performed by considering just two scales and putting the smallest 90% of the wavelet coefficients to zero (cf. Figure 4). From the reconstruction we find that both, the error in the maximum norm (ME), and the mean squares error (SE) are significantly smaller. In this particular situation the ME,SE errors occuring

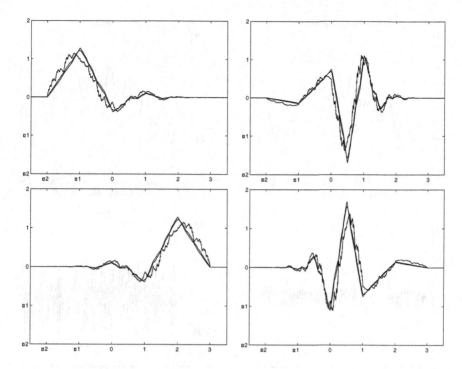

Fig. 3. sin-scaling functions (left) and wavelets (right) for $N = 3$, $K = 1$, and $\omega_0 \in \{\pi/2, \pi/4, \pi/16\}$. here the two different solutions (first and second line) are just mirrored/antimirrored versions.

by compression with the sin-wavelet are 17%, 51%, respectively, of the errors occuring by compressing using Daubechies wavelet. The improvement could be achieved by taking into account a-priori information on the ground frequency in the signal. It is of course no general tendency that sin-wavelets outperform Daubechies wavelets. In many practical applications, especially when no a-priori information is available, still Daubechies wavelets might perform better.

6 Discussion and Related Work

We have constructed orthogonal families of wavelet-functions with scale dependent properties. The constructed wavelet families and functions involve additional parameters: the parameter q for q-wavelets, the sequence of frequencies in the sin wavelets.

Striking is the obvious visual relationship between q-wavelets and fractional splines [8,9,10]: one may think of fractional splines of degree α as functions of the form

$$s^{\alpha}(x) = \sum_{k \in \mathbf{Z}} a_k (x - x_k)_+^{\alpha},$$

264   Peter Paule, Otmar Scherzer, and Armin Schoisswohl

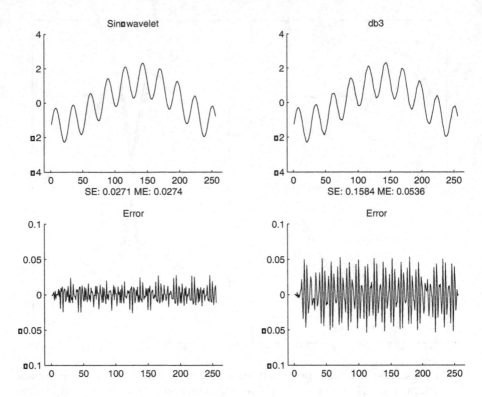

Fig. 4. Comparison of a compression with Daubechies' wavelet with 3 vanishing moments and the sin-wavelet ($N = 3, \omega_0 = \pi/4$). The ME,SE errors for the sin-wavelet are 17%, 51%, respectively, of the errors for the Daubechies wavelet

where x_k are the knots of the spline and $(\cdot)_+^\alpha$ denotes the one–sided power function. As becomes evident from the impressive graphics and the mathematical analysis in [10] (even more impressive graphics can be found at `http://bigwww.epfl.ch/art`) rational splines "interpolate" the common B-splines of integer order. In particular they interpolate between the Haar scaling function and B-splines of any order. The q-Wavelets in our paper reveal a similar interpolation behavior, where of course due to our construction we can only expect "interpolation" between the wavelets. A comparison with the work of Blu and Unser [8,9,10] immediately gives rise to the following open question. Smoothness of the fractional splines is equivalent by the (fractional) power α. We expect that also for q-wavelets smoothness can be directly linked to the parameter q. This question is also related to the existence of the scaling and wavelet functions.

Acknowledgement

The work of O.S. has been partly suported by the Austrian Science Foundation (FWF), SFB-F1310 and Y123-INF. The work of P.P. has been partly supported by the Austrian Science Foundation (FWF) SFB-F1305. The authors thank Georg Regensburger (University Innsbruck) for providing the compression example with sin-wavelets.

References

1. I. Daubechies, "Orthonormal bases of compactly supported wavelets," *Commun. Pure Appl. Math.*, vol. 41, no. 7, pp. 901–996, 1988.
2. I. Daubechies, *Ten Lectures on Wavelets*, vol. 61 of *CBMS-NSF Regional Conference Series in Applied Mathematics*, SIAM, Philadelphia, 1992.
3. S. Roman, *The Umbral Calculus*, Academic Press, Orlando, 1984.
4. J. Lebrun, "High order balanced multiwavelets: Theory, factorization and design," *IEEE Trans. Signal Processing*, vol. 49,pp. 1918-1930, 2001.
5. F. Chyzak, P. Paule, O. Scherzer, A. Schoisswohl, and B. Zimmermann, "The construction of orthonormal wavelets using symbolic methods and a matrix analytical approach for wavelets on the interval" *Experimental Mathematics*, vol. 10, pp. 67–86, 2001.
6. J.R. Williams and K. Amaratunga, "Introduction to wavelets in engineering," *Int. J. Numer. Methods Eng.*, vol. 37, no. 14, pp. 2365–2388, 1994.
7. I. Daubechies, "Orthonormal bases of compactly supported wavelets. II: Variations on a theme," *SIAM J. Math. Anal.*, vol. 24, no. 2, pp. 499–519, 1993.
8. M. Unser, "Splines-a perfect fit for signal and image processing," *IEEE Signal Processing Magazine*, vol. 16, pp. 22–38, 1999.
9. T. Blu and M. Unser, "The fractional spline wavelet transform: definition and implementation," *preprint*, 1999.
10. M. Unser and T. Blu, "Fractional splines and wavelets," *SIAM Rev.*, vol. 42, pp. 43–67, 2000.

Accurate Numerical Fourier Transform in d-Dimensions

Normand Beaudoin* and Steven S. Beauchemin**

The University of Western Ontario, London, Canada N6A 5B7
{normand,beau}@csd.uwo.ca

Abstract. The classical method of numerically computing Fourier transforms of digitized functions in one or in d-dimensions is the so-called *Discrete Fourier Transform* (*DFT*) efficiently implemented as *Fast Fourier Transform* (*FFT*) algorithms. In many cases, the *DFT* is not an adequate approximation of the continuous Fourier transform. Because the *DFT* is periodic, spectrum aliasing may occur. The method presented in this contribution provides accurate approximations of the continuous Fourier transform with similar time complexity. The assumption of signal periodicity is no longer posed and allows to compute numerical Fourier transforms in a broader domain of frequency than the usual half-period of the *DFT*. The aliasing introduced by periodicity can be reduced to a negligible level even with a relatively low number of sampled data points. In addition, this method yields accurate numerical derivatives of any order and polynomial splines of any odd order with their optimum boundary conditions. The numerical error on results is easily estimated. The method is developed in one and in d-dimensions and numerical examples are presented.

1 Introduction

The ubiquitous Fourier transform and its numerical counterpart, the Discrete Fourier Transform (*DFT*), in one or many dimensions, are used in many fields, such as mathematics (linear systems, random processes, probability, boundary-value problems), physics (quantum mechanics, optics, acoustics, astronomy), chemistry (spectroscopy, crystallography), and engineering (telecommunications, signal processing, image processing, computer vision, multidimensional signal processing) [1,2,3,4,5].

Although it is usual to consider the *DFT* as a mathematical tool with its own properties, it certainly makes sense to conceptualize it as the discrete version of the analytical Fourier transform and as an approximation of the latter [1]. In

---

* Normand Beaudoin is a post-doctoral fellow at the Department of Computer Science, Middlesex College, University of Western Ontario, London, Canada, N6A 5B7. Phone: (519) 661-2111.

** Steven S. Beauchemin is with the Department of Computer Science, Middlesex College, University of Western Ontario, London, Canada, N6A 5B7. Phone: (519) 661-2111.

F. Winkler and U. Langer (Eds.): SNSC 2001, LNCS 2630, pp. 266–278, 2003.
© Springer-Verlag Berlin Heidelberg 2003

this regard, the DFT, usually computed via a fast Fourier transform (FFT) algorithm, must be used with caution since it is not a correct approximation in all cases [6,7,8,9]. First, the DFT is periodical and it is only on one half of a period that it constitutes an approximation of the Fourier transform. Second, the sampling rate of the function to be submitted to the DFT is a critical issue. Without sampling the time[1] function at a sufficiently high rate, a phenomenon known as aliasing may become intolerable and spoil the accuracy of the DFT as an approximation of the Fourier transform. It could be thought that if the Nyquist criterion is fulfilled, everything should come out fine. The Nyquist criterion states that the sampling rate must be at least twice the highest frequency of the initial function [1,2,10]. However, in applied science, a function may be defined between 0 and T only. Hence, the highest frequency of such a time-limited function is infinite. Consequently, the DFT produces aliasing[2]. One could argue that, even though the highest frequency is infinite, it is possible to sufficiently increase the number of sampled data points such that the error of the DFT becomes as small as one desires. However, the required number of data points could be huge. As an example, for the function $h(t) = e^{-50t}$, $t \in [0,1]$, the error on $DFT\{h\}$, around $f = 64$, decreases roughly as $N^{-1/3}$. Hence, one must increase N by a factor of 1000 to decrease the error by a factor of 10.

In some cases where the result of the DFT is used qualitatively a high accuracy is not absolutely mandatory. But in some applications, such as in deconvolution where a division is performed in the frequency domain, a slight error in the denominator function, particularly when it is close to zero, can seriously distort the result [11].

However, one may increase the accuracy of the numerical Fourier transform when the number of sampled data points is limited. This can be implemented through the assumption that the function from which the sampled data points are extracted and its derivatives are continuous. The sampling process, performed through the so-called Dirac comb [1], in a sense, isolates each data point and considers them as independent from each other. The function and its derivatives are no longer continuous. By re-establishing the continuity between the sampled data points, a method that yields a highly accurate numerical Fourier transform can be devised.

2   Theory in d-Dimension

Let $\boldsymbol{t} = (t_1, t_2 \ldots t_d) \in \mathbb{R}^d$ and $\boldsymbol{f} = (f_1, f_2 \ldots f_d) \in \mathbb{R}^d$, $d \in \mathbb{N}^*$. $\mathbb{R}$ is the set of real numbers, $\mathbb{N}$ the set of nonnegative integers and $\mathbb{N}^* = \mathbb{N} \backslash \{0\}$. Let us define

---

[1] Without loss of generality, the reciprocal variables *time* (t) and *frequency* (f) are used throughout this article.

[2] The usual method to avoid aliasing is to filter out the high frequency components thus modifying the original signal.

a d-dimensional Heaviside's function:

$$\chi : \mathbb{R}^d \to \mathbb{R}, \quad \chi(t) = \prod_{i=1}^{d} \chi(t_i) \tag{1}$$

in which χ and χ are Heaviside's functions in d-dimensions and in one dimension respectively. Let us define two d-dimensional rectangular functions such as:

$$R(t) = \chi(t - 0^-)\chi(-t + T^+) \quad \text{and} \quad S(t) = \chi(t)\chi(-t + T) \tag{2}$$

with $0^- = (0^-, 0^-, \ldots 0^-)$ and $T^+ = (T_1^+, T_2^+, \ldots, T_d^+)$, $T_\alpha \in \mathbb{R}$, $T_\alpha > 0$, $\forall \alpha$, and in which:

$$0^- = \lim_{\varepsilon \to 0}(0 - \varepsilon), \quad T_\alpha^+ = \lim_{\varepsilon \to 0}(T_\alpha + \varepsilon), \quad \varepsilon \in \mathbb{R}, \varepsilon > 0. \tag{3}$$

Let $g : \mathbb{R}^d \to (\mathbb{R} \text{ or } \mathbb{C})$, ($\mathbb{C}$ is the field of complex numbers) be a continuous function that admits directional derivatives of any order in any directions for all t such that $S(t) \neq 0$. We now define the following function:

$$h(t) = R(t)g(t). \tag{4}$$

This way of defining $h(t)$ ascertain that the function is continuous and derivable between 0 and T_α and at 0 and T_α for all values of α.

We adopt the following definition for the Fourier transform:

$$\mathcal{F}\{h(t)\} = \int_{\mathbb{R}^d} h(t)e^{-i2\pi f \cdot t} dt. \tag{5}$$

By expanding the inner product and reorganizing the terms, (5) becomes:

$$\mathcal{F}\{h(t)\}$$

$$= \int_{-\infty}^{\infty} \cdots \int_{-\infty}^{\infty} \left[\int_{-\infty}^{\infty} h(t_1, \ldots, t_d)e^{-i2\pi f_1 \cdot t_1} dt_1\right] e^{-i2\pi f_2 \cdot t_2} dt_2 \cdots e^{-i2\pi f_d \cdot t_d} dt_d. \tag{6}$$

It is a known fact, evident from (6), that a d-dimensional Fourier transform of a function can be performed by d successive one dimensional Fourier transforms. Consequently, in the next section we develop the theory in one dimension. In that case, generic non-indexed variables as t, f, T... that stand for any indexed variable of a particular dimension of the d-dimensional space are used.

3   Theory in One Dimension

In virtue of the properties of the differentiation of Heaviside's and Dirac-delta functions (δ) [2,12], the nth derivative of h with respect to t is:

$$h^{(n)}(t) = \chi(t - 0^-)\chi(-t + T^+)g^{(n)}(t) + D_n(t) \tag{7}$$

in which $D_n(t)$ is defined as:

$$D_n(t) =$$

$$
\begin{cases}
0 & \text{if} \quad n = 0 \\[2ex]
\displaystyle\sum_{m=0}^{n-1} \left\{ g^{(m)}(0^-)\,\delta^{n-m-1}(t-0^-) - g^{(m)}(T^+)\,\delta^{n-m-1}(t-T^+) \right\} & \text{if} \quad n \in \mathbb{N}^*.
\end{cases}
$$

$$(8)$$

Equation (7) and (8) express the fact that the nth derivative of h with respect to t is the ordinary n^{th} derivative of the function h strictly inside the rectangular box where it is continuous and differentiable, in addition to the nth derivative of h in the regions where it is discontinuous.

According to our definition of the Fourier transform, we have:

$$\mathcal{F}\left\{ h^{(n)}(t) \right\} = \int_{-\infty}^{\infty} h^{(n)}(t)\, e^{-i2\pi ft}\, dt \; . \tag{9}$$

We can expand the integral in (9) into parts to form:

$$\mathcal{F}\left\{ h^{(n)}(t) \right\}$$

$$= \int_{-\infty}^{0} h^{(n)}(t)\, e^{-i2\pi ft}\, dt + \int_{0}^{T} h^{(n)}(t)\, e^{-i2\pi ft}\, dt + \int_{T}^{\infty} h^{(n)}(t)\, e^{-i2\pi ft}\, dt \; . \tag{10}$$

The sum of the first and last integrals of the right hand side of (10) clearly is $\mathcal{F}\{D_n(t)\}$. Hence, (10) becomes:

$$\mathcal{F}\left\{ h^{(n)}(t) \right\} = \int_{0}^{T} h^{(n)}(t)\, e^{-i2\pi ft}\, dt + \mathcal{F}\{D_n(t)\} \; . \tag{11}$$

By separating the interval $[0, T]$ into N equal $\Delta t = T/N$ subintervals, (11) can be rewritten as:

$$\mathcal{F}\left\{ h^{(n)}(t) \right\} = \sum_{j=0}^{N-1} \left\{ \int_{j\Delta t}^{(j+1)\Delta t} h^{(n)}(t)\, e^{-i2\pi ft}\, dt \right\} + \mathcal{F}\{D_n(t)\} \; , \quad j \in \mathbb{N}. \tag{12}$$

Since $h^{(n)}$ is continuous and differentiable between 0 and T, it can be approximated for $t \in [j\Delta t, (j+1)\,\Delta t]$, for each $j \in [0, N-1]$, by a Taylor expansion:

$$h^{(n)}(t) = \sum_{p=0}^{\infty} \frac{h_j^{(p+n)}\,(t - j\Delta t)^p}{p!} \; , \quad p \in \mathbb{N} \tag{13}$$

where $h_j^{(m)}$ is the mth derivative of h at $t = j\Delta t$. Merging (12) and (13) yields:

$$\mathcal{F}\left\{h^{(n)}(t)\right\}$$

$$= \sum_{j=0}^{N-1}\left\{\int_{j\Delta t}^{(j+1)\Delta t}\left(\sum_{p=0}^{\infty}\frac{h_j^{(p+n)}(t-j\Delta t)^p}{p!}\right)e^{-i2\pi ft}dt\right\} + \mathcal{F}\left\{D_n(t)\right\}, \quad j \in \mathbb{N}.$$

$$(14)$$

With the substitution $\tau = t - j\Delta t$ and an adequate permutation of the integral and sums on j and p, (14) becomes:

$$\mathcal{F}\left\{h^{(n)}(t)\right\}$$

$$= \sum_{p=0}^{\infty}\left\{\left(\int_0^{\Delta t}\frac{\tau^p e^{-i2\pi f\tau}}{p!}d\tau\right)\left(\sum_{j=0}^{N-1}h_j^{(p+n)}e^{-i2\pi fj\Delta t}\right)\right\} + \mathcal{F}\left\{D_n(t)\right\}.$$

$$(15)$$

To numerically compute the Fourier transform of h, we must evaluate it for some discrete values of f. Let $f = k\Delta f = k/T$, $k \in \mathbb{N}$ be these discrete variables. In addition, let us define H_k as the discrete version of $\mathcal{F}\left\{h^{(n)}(t)\right\}$. The integral in (15) depends only on the variable f (or k) and on the parameters p and Δt and can be evaluated analytically, whether f is continuous or discrete, once and for all, for each value of p as:

$$I_p = \frac{1}{p!}\int_0^{\Delta t}\tau^p e^{-i2\pi f\tau}d\tau. \tag{16}$$

Since the integral in the definition of I_p is always finite and, in the context of the Gamma function [13], $p! = \pm\infty$ when p is a negative integer, then $I_p = 0$ for $p < 0$.

The summation on j in (15), when $f = k\Delta f = k/T$, is the discrete Fourier transform of the sequence $h_j^{(p+n)}$, $j \in [0, N-1] \subset \mathbb{N}$ [1]. We denote it as $F_{p+n,k}$. Since $\Delta t = T/N$ and $f = k/T$, we have:

$$F_{p+n,k} = \sum_{j=0}^{N-1}h_j^{(p+n)}e^{-i2\pi\frac{kj}{N}}. \tag{17}$$

One should note that although we wrote I_p and $F_{p+n,k}$ instead of $I_p(f \text{ or } k)$ and $F_{p+n,k}(f \text{ or } k)$, these functions always depend on f or k.

Substituting (16) and (17) in (15), we obtain the following result:

$$\mathcal{F}\left\{h^{(n)}(t)\right\} = \sum_{p=0}^{\infty}I_p F_{p+n,k} + \mathcal{F}\left\{D_n(t)\right\}. \tag{18}$$

When $n = 0$, (18) becomes:

$$H_k = \sum_{p=0}^{\infty}I_p F_{p,k}. \tag{19}$$

Now, integrating by parts the right hand side of (9) yields:

$$\mathcal{F}\left\{h^{(n+1)}\right\} = i2\pi f \mathcal{F}\left\{h^{(n)}\right\} . \tag{20}$$

Defining $b_n = i2\pi f \mathcal{F}\left\{D_n\right\} - \mathcal{F}\left\{D_{n+1}\right\}$, combining (18) and (20) and reorganizing the terms yields:

$$-i2\pi f I_0 F_{n,k} + \sum_{p=1}^{\infty} \left(I_{(p-1)} - i2\pi f I_p\right) F_{p+n,k} = b_n . \tag{21}$$

With the definition $J_\alpha = I_{\alpha-1} - i2\pi f I_\alpha$, (21) becomes:

$$J_0 F_{n,k} + \sum_{p=1}^{\infty} J_p F_{p+n,k} = b_n . \tag{22}$$

Given the definition of g and h, we have $g^{(n)}\left(0^-\right) = g^{(n)}\left(0\right) = h^{(n)}\left(0\right)$ and $g^{(n)}\left(T^+\right) = g^{(n)}\left(T\right) = h^{(n)}\left(T\right)$. Using these facts in addition to the properties of Fourier transforms and those of Dirac delta functions [12], one easily observes that expanding b_n results in the simple following form:

$$b_n = h^{(n)}\left(T\right) e^{-i2\pi fT} - h^{(n)}\left(0\right) . \tag{23}$$

In the discrete case, where $f = k/T$, (23) takes the following simple and significant form:

$$b_n = h^{(n)}\left(T\right) - h^{(n)}\left(0\right) = h_N^{(n)} - h_0^{(n)} . \tag{24}$$

Up to this point, all equations are rigorously exact since p tends towards infinity. However, in practical situations we introduce an approximation by limiting the range on p. Let us define $\theta \in \mathbb{N}$, the truncating parameter, which, for reasons discussed later, is always chosen as an odd integer. We refer to it as the order of the system.

Let us expand (22) for each value of $n \in [0, \theta - 1] \subset \mathbb{N}$. This generates a system of θ different equations and, for each of these, we let p range from 1 to $\theta - n - 1$. This gives the following system, which is written in matrix form:

$$\begin{bmatrix} J_0 & J_1 & \cdots & J_{\theta-1} \\ 0 & J_0 & \cdots & J_{\theta-2} \\ \vdots & \vdots & \ddots & \vdots \\ 0 & 0 & \cdots & J_0 \end{bmatrix} \begin{bmatrix} F_{0,k} \\ F_{1,k} \\ \vdots \\ F_{\theta-1,k} \end{bmatrix} \simeq \begin{bmatrix} b_0 \\ b_1 \\ \vdots \\ b_{\theta-1} \end{bmatrix} \tag{25}$$

or, more compactly as:

$$\mathsf{M}_a \mathsf{F}_a \simeq \mathsf{B} . \tag{26}$$

Note that the matrix M_a is completely known since each of its terms depends only on f. The general expression for the elements of M_a is:

$$\left(\mathsf{M}_a\right)_{\mu\nu} = I_{\mu-\nu-1} - i2\pi f I_{\nu-\mu} = J_{\nu-\mu} . \tag{27}$$

Matrix B is unknown. If it were known, we could evaluate F_a from (26):

$$F_a \simeq M_a^{-1} B . \qquad (28)$$

However, for $f = \kappa N \Delta f$, $\kappa \in \mathbb{N}$, the solutions would strictly diverge. Indeed, for these particular values of f, $\det(M_a) = 0$. However, for values of f around $N\Delta f/2$, the approximation (28) is quite accurate. We take advantage of this fact to compute B.

The first element of F_a, which is $(F_a)_1 = F_{0,k}$, is the DFT of the sequence h_j. It is completely determined for each value of k. It is not the same situation for the other elements of F_a which are still unknown. Furthermore, the elements of matrix M_a^{-1} are given for each value of f. We can then extract the following from (28):

$$F_{0,k} \simeq \left(row_1 M_a^{-1}\right) B = \sum_{\mu=1}^{\theta} \left(M_a^{-1}\right)_{1,\mu} (B)_\mu . \qquad (29)$$

Let us now define Ω , an interval of θ values of k, centered at $N/2$:

$$\Omega = \left[\frac{N}{2} - \left(\frac{\theta-1}{2}\right) \quad , \quad \frac{N}{2} + \left(\frac{\theta-1}{2}\right)\right] = [k_1, k_2, \ldots, k_\theta] \subset \mathbb{N} . \qquad (30)$$

Let us expand (29) for each value of $k \in \Omega$. (It is understood that in practical cases, for each instance of f in each term, one has to replace it by $k\Delta f$.) Doing so yields the following system of linear equations:

$$\begin{bmatrix} F_{0,k_1} \\ F_{0,k_2} \\ \vdots \\ F_{0,k_\theta} \end{bmatrix} \simeq \begin{bmatrix} \left(\left(M_a^{-1}\right)_{1,1} |_{k_1}\right) & \left(\left(M_a^{-1}\right)_{1,2} |_{k_1}\right) & \cdots & \left(\left(M_a^{-1}\right)_{1,\theta} |_{k_1}\right) \\ \left(\left(M_a^{-1}\right)_{1,1} |_{k_2}\right) & \left(\left(M_a^{-1}\right)_{1,2} |_{k_2}\right) & \cdots & \left(\left(M_a^{-1}\right)_{1,\theta} |_{k_2}\right) \\ \vdots & \vdots & \ddots & \vdots \\ \left(\left(M_a^{-1}\right)_{1,1} |_{k_\theta}\right) & \left(\left(M_a^{-1}\right)_{1,2} |_{k_\theta}\right) & \cdots & \left(\left(M_a^{-1}\right)_{1,\theta} |_{k_\theta}\right) \end{bmatrix} \begin{bmatrix} b_0 \\ b_1 \\ \vdots \\ b_{\theta-1} \end{bmatrix} . $$

$$\qquad (31)$$

Note that $\left(M_a^{-1}\right)_{1,\alpha} |_{k_\beta}$ is compact notation for $\left(M_a^{-1}\right)_{1,\alpha}$ evaluated at $k = k_\beta$ and $f = k_\beta \Delta f$. Let us express (31) in a more compact form as $F_c \simeq WB$, from which we directly deduce the following matrix equation:

$$B \simeq W^{-1} F_c . \qquad (32)$$

Equation (32) completely determines B from $F_{0,k}$ (the discrete Fourier transform of the digitized function h). According to (24), the knowledge of B specifies b_m ($m \in [0, \theta-1]$). They are the boundary conditions of the system.

Although B is completely determined, (28) cannot be used, for reasons mentioned earlier, to evaluate F_a. Considering the first element of F_a as known (it is the DFT of h), we again expand (22), but in a slightly different manner than we did to obtain (25). We once more expand it for each value of $n \in [0, \theta-1] \subset \mathbb{N}$. This generates a system of θ equations, and for each of these we let p range, this time, from 1 to $\theta - n$. Thereafter, the terms are reorganized to obtain the following system, which is written again in matrix form:

$$
\begin{bmatrix}
J_1 & J_2 & \cdots & J_\theta \\
J_0 & J_1 & \cdots & J_{\theta-1} \\
\vdots & \vdots & \ddots & \vdots \\
0 & 0 & \cdots & J_1
\end{bmatrix}
\begin{bmatrix}
F_{1,k} \\
F_{2,k} \\
\vdots \\
F_{\theta,k}
\end{bmatrix}
\simeq
\begin{bmatrix}
b_0 \\
b_1 \\
\vdots \\
b_{\theta-1}
\end{bmatrix}
+
\begin{bmatrix}
-J_0 F_{0,k} \\
0 \\
\vdots \\
0
\end{bmatrix}
\tag{33}
$$

or, more compactly as:

$$
\mathsf{M}_b \mathsf{F}_b \simeq \mathsf{B} + \mathsf{C} .
\tag{34}
$$

The general expression for elements of M_b is thus:

$$
(\mathsf{M}_b)_{\mu\nu} = I_{\mu-\nu} - i 2 \pi f I_{\nu-\mu+1} = J_{\nu-\mu+1} .
\tag{35}
$$

Let us now write (34) as:

$$
\mathsf{F}_b \simeq \mathsf{M}_b^{-1} (\mathsf{B} + \mathsf{C}) .
\tag{36}
$$

The advantage of (28) over (36) is that it allows, through (32), to compute B. However, it shows singularities at $f = \kappa N \Delta f$, $\kappa \in \mathbb{N}$ which prevent us to compute F_a. Conversely, the advantage of (36) is that it exhibits a higher order than (28) and, provided that θ is odd, F_b is computable for values of $f = k \Delta f$. One should note that, with (36), undetermined values appear for $f = \kappa N \Delta f$, but they can always be solved by Hospital's rule for any odd order. For even orders, there are singularities at $f = (\kappa + 1/2) N \Delta f$ that cannot be removed. This is an imperative reason to choose odd values for θ.

With the knowledge of F_b from (36), the terms of (19), for $p \in [0, \theta]$, are completely determined. Thus, the truncated version of (19) can be written as:

$$
H_k \simeq \sum_{p=0}^{\theta} I_p F_{p,k} .
\tag{37}
$$

Let us define a one-row matrix as $\mathsf{I}_\theta = [I_1 \ I_2 \ \cdots \ I_\theta]$, and write (37) as follows:

$$
H_k \simeq I_0 F_{0,k} + \mathsf{I}_\theta \mathsf{F}_b .
\tag{38}
$$

With (38), we approximate the Fourier transform (or the inverse Fourier transform) of a digitized function in one dimension. The digitized Fourier transform calculated with (38) is not band-limited (as with the DFT which is periodical). Equation (38) remains valid and accurate as an approximation of the analytical Fourier transform for positive or negative values of k such that $|k| > N/2$ or even for $|k| > N - 1$. (This last property and the following, briefly mentioned in the rest of this section, have been discussed and illustrated with examples in [14].)

A close examination of (31), (33) and (38) reveals that the computation of only one FFT is required. The other terms form a correcting operation to be applied once on each of the N values of the FFT. The time complexity of the correcting operation is $O(N)$ and the time complexity of the FFT is $O(N \log N)$. Hence, the time complexity of the entire algorithm is $O(N \log N)$ when θ is kept constant. The time complexity relatively to θ, the order of the system, is $O\left(\theta^2\right)$, but, as long as $\theta \leq \theta_{opt}$, the errors on computed results decrease exponentially with the increase of the order θ. Hence, as long as one can afford to increase θ, the trade-off is strongly beneficial.

Equation (38) contains the symbolic form of F_b which can be used as is to form a single symbolic formula without having its terms evaluated numerically. On the other hand, if, for instance, (36) is used to numerically compute each term of F_b for values of k from 0 to $N-1$, it produces θ different sequences of numbers which are actually accurate approximations of the DFT of the derivatives $h_j^{(p)}$, for values of $p \in [1, \theta]$. Thus, applying the inverse DFT operation to each of these sequences generates the corresponding sequences $h_j^{(p)}$ that are very accurate numerical derivatives of the initial function $h_j^{(0)}$ of all orders from 1 to θ. This implies that one can numerically compute very accurately the derivatives of any order of a digitized function or signal.

Derivatives calculated in that way are continuous in-between and at each data point. Thus, we obtain spline polynomials of any odd order, with their corresponding properties, merely with DFT (FFT). Furthermore, since (32) is used to compute B, these spline interpolation polynomials, whatever their order, always exhibit optimal boundary conditions, that is to say, accurate end derivatives for each order [15]. Such accurate high-order spline interpolation polynomials allow integrals between any limit to be accurately computed.

Let R_θ be any result (Fourier transform, derivative or integral) obtained with an arbitrary order θ. As long as $\theta + 2 \leq \theta_{opt}$, the error on R_θ (noted E_θ) can be fairly estimated since $R_{\theta+2}$ relatively to R_θ can be considered almost as the exact result. To do so, one can use the following relation: $E_\theta = \mathcal{O}\left(R_{\theta+2} - R_\theta\right)$, where $\mathcal{O}$ is any operator one can define to meet specific needs.

4   Back in d-Dimensions

In the previous section we have obtained a highly accurate method to compute the Fourier transform in one dimension. According to (6), this method can be applied sequentially to compute an accurate d-dimensional Fourier transform. In this multidimensional case, for each $\alpha \in \{1, 2, \cdots, d\}$ we have $t_\alpha \in [0, T_\alpha]$. This interval is separated into N_α equal $\Delta t_\alpha = T_\alpha/N_\alpha$ parts, and $f_\alpha = k_\alpha \Delta f_\alpha = k_\alpha/T_\alpha$. As with the ordinary DFT (FFT), the order in which the dimensions are treated is irrelevant. The number of times (38) has to be applied to compute a d-dimensional Fourier transform is:

$$PQ \quad \text{where} \quad P = \prod_{\alpha=1}^{d} N_\alpha \quad \text{and} \quad Q = \sum_{\beta=1}^{d} \frac{1}{N_\beta} \, . \tag{39}$$

The time complexity is then $O\left(P \log P\right)$. Let us put $N_\alpha = a_\alpha N$, $\forall \alpha$, a_α being constants. It is easy to show that the time complexity is $O\left(N^d \log N\right)$ which is the same as for the DFT in d-dimensions.

5   Example in 2 Dimensions

In this section, an example in two dimensions is used to illustrate the algorithm. The choice of such an example is not obvious. That is to say, the function should not be a trivial one; it must be difficult enough for the computation of the Fourier transform to be numerically demanding. On the other hand, for purpose of comparison and accuracy testing, the Fourier transform of the function must be analytically known. The chosen initial function for our example is then the following complex function:

$$h\left(t_1, t_2\right)$$

$$= \left(\cos\left(9t_1\right)\cos\left(11t_1 + 17t_2\right)e^{-2.5t_1}, e^{-2(t_1+t_2)} + e^{\left[-100(t_1-0.5)^2 - 50(t_2-0.5)^2\right]}\right)$$

$$T_1 = T_2 = 1.$$

$$(40)$$

The real part is a combination of damped oscillations that are slanted in virtue of the damping. The imaginary part is a non-symmetrical Gaussian peak purposely slanted by an exponential to avoid error cancellation by symmetry. For both variables, the function is discontinuous at 0 and at T_α, $\forall \alpha$. Figures 1.a and 1.b show, respectively, the modulus of (40) and of its analytical Fourier transform for $N_1 = N_2 = 128$. The formula of the analytical Fourier transform of this two dimensional function is not shown here since it requires several pages of text.

Figure 2.a shows the DFT of h. The expected periodical behavior is evident. Figure 2.b shows the modulus of the error of the DFT relatively to the analytical

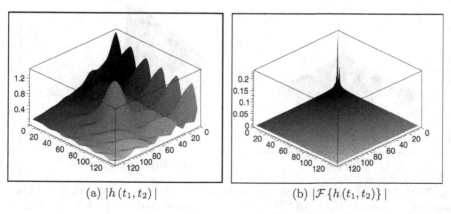

(a) $|h\left(t_1, t_2\right)|$ (b) $|\mathcal{F}\left\{h\left(t_1, t_2\right)\right\}|$

Fig. 1. Modulus of h and of $\mathcal{F}\left\{h\right\}$.

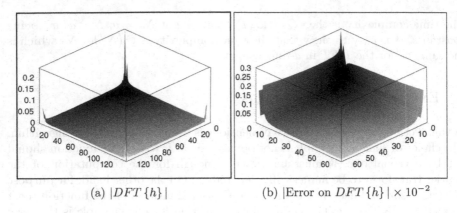

(a) $|DFT\{h\}|$ (b) $|$Error on $DFT\{h\}| \times 10^{-2}$

Fig. 2. Modulus of $DFT\{h\}$ and of error on $DFT\{h\}$ $(\times 10^{-2})$. The error is computed on the first 64×64 data points only. Maximum error $= 0.3 \times 10^{-2}$.

Fourier transform. To be fair, this error must indeed be computed on the first $(N_1/2) \times (N_2/2)$ data points only, since the DFT is periodical.

Figure 3.a shows the numerical Fourier transform of h computed with (38) for $\theta = 13$. It is clearly seen that this approximation behaves as the analytical Fourier transform and is not periodical. Figure 3.b shows the modulus of the error of this approximation relatively to the analytical Fourier transform, computed, this time, on the full range of the $N_1 \times N_2$ data points. For comparison, one should note the vertical scale factor in Fig. 2.b and 3.b.

This same equation (38), used for Fig. 3, is used again, on the same function (40) with different values of $N = N_1 = N_2$ and θ. The averages of the moduli of the error on the numerical Fourier transform given by (38) relatively to the

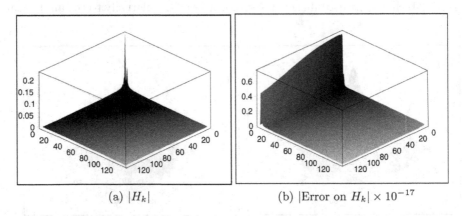

(a) $|H_k|$ (b) $|$Error on $H_k| \times 10^{-17}$

Fig. 3. Modulus of the numerical Fourier transform given by (38), with $\theta = 13$, and modulus of error on H_k $(\times 10^{-17})$. The error is computed on the full range of the 128×128 data points. Maximum error $= 0.7 \times 10^{-17}$.

exact analytical Fourier transform are shown in Table 1 in addition to the results obtained with the DFT.

We observe that for small values of N (actually for $N = 8$), increasing θ does not improve accuracy. It means that the optimum value of θ called θ_{opt} is already reached. If N is slightly increased, θ_{opt} grows rapidly and the smallest possible error decreases dramatically. For $N = 8$, $\theta_{opt} < 3$. For $N = 16$, $\theta_{opt} < 5$. For $N = 32$, $\theta_{opt} = 5$. For $N \geq 64$, $\theta_{opt} \geq 13$. We also note that, except for small values of N, the error from (38) is always much smaller than the error from the DFT and, for any θ, it decreases more rapidly with the increase of N than does the error from the DFT with the same increase in N.

Table 1. Average of the modulus of the error on the numerical Fourier transform of the function given by (40) and computed with (38) for different values of N and θ. The last line is the average of the modulus of the error of the DFT of the same function.

θ	N				
	8	16	32	64	128
1	1×10^{-2}	1×10^{-3}	2×10^{-4}	2×10^{-5}	3×10^{-6}
3	3×10^{-1}	1×10^{-3}	9×10^{-6}	3×10^{-7}	1×10^{-8}
5		1×10^{-2}	8×10^{-7}	6×10^{-9}	5×10^{-11}
7			4×10^{-6}	1×10^{-10}	3×10^{-13}
9				3×10^{-12}	2×10^{-15}
11				8×10^{-14}	9×10^{-18}
13				2×10^{-15}	8×10^{-20}
DFT	3×10^{-2}	9×10^{-3}	3×10^{-3}	8×10^{-4}	2×10^{-4}

6   Conclusion

An analytical function contains an infinite quantity of information [16]. It is possible, in principle, at least when the Fourier integral is analytically obtainable, to compute its Fourier transform exactly. In all other cases we have to resort to numerical techniques and depart from analytical forms. The classical sampling of a function uses the Dirac comb which rejects an infinite quantity of information to make the digitized function manageable on a finite computer. The usual tool to numerically compute the Fourier transform of a digitized function in d-dimensions is the DFT, efficiently implemented as the FFT algorithm. However, in many cases the DFT is not an adequate numerical approximation of the Fourier transform.

On the one hand, there is the perfect analytical Fourier transform that is, most of the time, not practical and, on the other hand, we have the very efficient FFT which computes an approximation of the Fourier transform without any attempt to reduce the unavoidable ravages of the Dirac comb. The method presented in this contribution is in-between these two extremes; its position being adjusted by the value of θ, the order of the system. The method provides accurate approximations of the continuous Fourier transform, is no longer periodical and computes the numerical Fourier transform in a broader frequency domain than the half-period of the DFT. The aliasing can be reduced to a negligible level even with a relatively low number of sampled data points. The method gives accurate numerical partial derivatives of any order and the polynonial splines of any odd order with their optimal boundary conditions. The error can be easily computed by comparing the results of two successive odd orders. The time complexity, relatively to the number of sampled data points, is the same as for the FFT. The time complexity, relatively to θ (independent of the time complexity relatively to N) is $O\left(\theta^2\right)$, while the accuracy increases exponentially with θ. Hence, the numerical accuracy increases much more rapidly than the computational cost of the proposed method.

References

1. Brigham, E. O. (1974): *The Fast Fourier Transform*. Prentice-Hall
2. Morrison, N. (1994): *Introduction to Fourier Analysis*. Wiley-Interscience
3. Marchildon, L. (2000): *Mécanique Quantique*. De Boeck Université
4. Kittel, C. (1983): *Physique de l'État Solide*. Dunod Université
5. Shapiro, L. G., Stockman G. C. (2001): *Computer Vision*. Prentice Hall
6. Schutte J. (1981): *New Fast Fourier Transform Algorithm for Linear System Analysis Applied in Molecular Beam Relaxation Spectroscopy*. Rev. Sci. Instrum., 52(3), 400-404
7. Makinen S. (1982): *New Algorithm for the Calculation of the Fourier Transform of Discrete Signals*. Rev. Sci. Instrum., 53(5), 627-630
8. Sorella S., Ghosh S. K. (1984): *Improved Method for the Discrete Fast Fourier Transform*. Rev. Sci. Instrum., 55(8), 1348-1352
9. Froeyen M., Hellemans L. (1985): *Improved Algorithm for the Discrete Fourier Transform*. Rev. Sci. Instrum., 56(12), 2325-2327
10. Gaskill, J. D. (1974): *Linear Systems, Fourier Transform, and Optics*. Prentice-Hall
11. Beaudoin, N. (1999): *PhD. Thesis*. Université du Québec à Trois-Rivières. Canada
12. Butkov, E. (1968): *Mathematical Physics*. Addison-Wesley Reading
13. Kreyszig, E. (1983): *Advanced Engineering Mathematics*. Wiley
14. Beaudoin N. (1998): *A High-accuracy Mathematical and Numerical Method for Fourier Transform, Integral, Derivative, and Polynomial Splines of any Order*. Canadian Journal of Physics, 76(9), 659-677
15. Gerald C. F., Wheatley P. O. (1989): *Applied Numerical Analysis*. Addison-Wesley Reading
16. Ventsel H. (1973): *Théorie des Probabilités*. Editions Mir

Exact Real Computation in Computer Algebra

Gábor Bodnár, Barbara Kaltenbacher, Petru Pau, and Josef Schicho

SFB Numerical and Symbolic Computation
Johannes Kepler University, A-4040 Linz, Austria

Abstract. Exact real computation allows many of the advantages of numerical computation (e.g. high performance) to be accessed also in symbolic computation, providing validated results. In this paper we present our approach to build a transparent and easy to use connection between the two worlds, using this paradigm. The main discussed topics are: representation of exact real objects, operations on exact real matrices, polynomial greatest common divisor and root computation. Some of these problems are ill-posed; we use regularization methods to solve them.

1 Introduction

Many algorithms in computer algebra rely on exactness. The usual way to perform exact computation is to work with the rationals or with algebraic numbers. The latter are traditionally represented by the minimal polynomial and either an isolating interval or a sequence of signs in the real case (see [10]). Sometimes it can be a problem that these subfields are not closed under transcendental functions. Moreover, computations with algebraic numbers tend to be very expensive (see [17]).

The method of exact real computation seems to be a reasonable alternative: it provides mathematically consistent representation of elements of metric spaces. That is, the objects behave exactly like the represented mathematical entities. Recall that this does not hold, for instance, in the case of floating point numbers, which is one of the reasons why it is often so difficult to work with this representation in computer algebra.

Several ways to represent real numbers have already been invented; we can mention here infinite (lazy) strings of digits ([9], with golden ratio notation, [13], [8]) and functional representation, where the reals are given as functions that produce approximations on demand (see [5]). Those representations realize only very basic algorithms. On the other hand, there has been some recent research on numerical algorithms in algorithmic computer algebra, e.g. [18, 6, 16]. What lacks is an approach to exact real computation integrating up-to-date symbolic and numerical algorithms.

In this paper we present a new approach to exact real computation in a generic setup, where the represented elements come from metric spaces, and there exist numerical algorithms that can compute arbitrarily close approximates which are efficiently representable on the computer. The type specifications were

F. Winkler and U. Langer (Eds.): SNSC 2001, LNCS 2630, pp. 279–292, 2003.

designed to provide a transparent interface to the represented mathematical entities, while we endeavored to make it as easy as possible to define functions, like algebraic operations, on exact real objects. We used the paradigm of object oriented programming, obtaining a template parametrized by the class of underlying metric spaces.

In the type specifications we strictly separated the layer of the interface to the exact real objects from the layer of the interface to the numerical routines that compute approximates for the represented mathematical entity. This way the computations of approximates are separated from the computations related to error propagation; therefore an implementation of an exact real type is independent of the underlying numerical library. In this paper we focus on the representation of exact real objects together with the analysis of error propagation in a few selected computations. We do not discuss anything under the interface to the numerical algorithms.

For many problems the main task is to control error propagation. Often this information can be found in the literature of numerical analysis (see e.g. [11]). However, there are problems which lead to fundamental difficulties in numerical computation, because their solution does not depend continuously on the input data. They are called *ill-posed*, and some of them are in the center of our attention: pseudoinverse of rank-deficient matrices, polynomial greatest common divisor, multivariate polynomial factorization and curve and surface parameterization. Strictly speaking, these problems cannot be solved within the exact real paradigm (see [1]). As we definitely need a substitute for them, we apply the common technique of *regularization*, in which we replace the ill-posed problem by a nearby continuous one, and require the distance to the exact problem to be an additional input parameter. In this paper we apply Tikhonov regularization (see [2] and [7]) to the computation of pseudoinverses and GCD's.

The paper is organized as follows: In Sect. 2 we give the base type specification for exact real objects and provide some examples for the computable real numbers. Section 3 presents the analysis of the error propagation in some nontrivial operations on real matrices. In Sect. 4, we introduce polynomials with exact real coefficients with the most important operations.

We want to express our gratitude to R. Corless, S. Ratschan and H.-J. Stetter for inspiring remarks. This work has been supported by the Austrian science fund (FWF) in the frame of the special research area "numerical and symbolic scientific computing" (SFB013, projects F1303, F1308).

2 Type Specification for Exact Real Objects

Our final goal is to make powerful numerical methods readily available in symbolic computation, that is, to establish reliable and transparent connection between numeric and symbolic computation using the exact real paradigm.

In this paper we stay on higher abstraction levels treating the actual implementation of the underlying numerical algorithms as black boxes with which we communicate via standard interfaces. It is not our intention to look behind these interfaces and to discuss efficiency, stability, etc., of the possible implementations. However, as it will be shown in Sect. 3 and Sect. 4, as soon as we have numerical algorithms whose specifications fulfill the requirements of the interface, the reliability of the computations can be ensured by controlled error propagation, which can be sustained even in some ill-posed problems.

On the other hand, we cannot escape from choosing an appropriate representation for the approximates, which are produced on user request to provide information on the mathematical object in question. The approximating elements must be finitely representable (e.g. matrices of multi-precision floating point numbers), and they, together with the objects which they approximate, have to come from some metric space, so that the notion of distance or accuracy makes sense. This is perfectly conform with the theory of interval computation (see e.g. [14]).

We use the following notation: an abstract mathematical object is denoted by a standard italic character (e.g. $x \in \mathbf{R}$, say $x := \sqrt{2}$), a numerical computer representations by a typewriter character (e.g. $\mathtt{x} := 1.41$), which can be naturally considered as an element of the space of x (i.e. $1.41 \in \mathbf{R}$), and an exact real object for x is denoted by a calligraphic character $\mathcal{X}$.

Definition 1. *Let us fix some metric space M. We call a pair $(\mathtt{x}, \mathtt{e}) \in M \times \mathbf{R}_{\geq 0}$ an* approximate *for $x \in M$, if $|x - \mathtt{x}| \leq \mathtt{e}$. From a given list of approximates for x, $[(\mathtt{x}_1, \mathtt{e}_1), \ldots, (\mathtt{x}_n, \mathtt{e}_n)]$, $(\mathtt{x}_i, \mathtt{e}_i)$ the* most accurate *one if $\mathtt{e}_i \leq \mathtt{e}_j$ for all $j = 1, \ldots, n$ and i is the minimal index with this property.*

At a given step of a computation, we can consider the list L_x of approximates for an object x that were computed so far, using temporal order. The most accurate approximate for x at the given point of the computation is then the most accurate approximate for x with respect to L_x.

2.1   Base Type of Exact Real Objects

The other main goal is to achieve transparency. For this, we have to provide a suitable abstract type specification with a fixed communication interface, basic and higher level constructors.

Definition 2. *An exact real object $\mathcal{X}$ that represents an element x from some metric space M with metric d, is defined by:*

Data members:

- *Name: $\mathtt{knownApprox}$. At any point of the lifetime of the object it stores the most accurate approximate, of the form $(\mathtt{x}, \mathtt{e})$, $\mathtt{e} \in \mathbf{R}_{\geq 0}$, which was computed for x till then.*

Methods:

- *Name: $\mathtt{epsint}$. Input: a positive real value $\mathtt{e}$. Output: an approximating element $\mathtt{x}''$, such that $d(\mathtt{x}'', x) \leq \mathtt{e}$. Side effect: if the accuracy of $\mathtt{x}''$ is $\mathtt{e}''(\leq \mathtt{e})$ and $\mathtt{e}'' < \mathtt{e}'$, where $(\mathtt{x}', \mathtt{e}')$ is the $\mathtt{knownApprox}$ member of $\mathcal{X}$ at the time point of the procedure call, the $\mathtt{knownApprox}$ member of $\mathcal{X}$ becomes $(\mathtt{x}'', \mathtt{e}'')$.*

- *Name:* **refine**. *Input: none. Output: an approximate* (x, e) *of* x, *such that* $e < e'/2$ *if* $e' > 0$, *where* (x', e') *is the* **knownApprox** *member of* $\mathcal{X}$ *at the time point of the procedure call, or* $e = e'$ *if* $e' = 0$. *Side effect: the* **knownApprox** *member of* $\mathcal{X}$ *becomes* (x, e).
- *Constructor. Input: an initial approximate* (x, e) *for* x *and, if* $e > 0$, *at least one procedure which matches one of the following specifications:*
    - *Input: a positive real value* e. *Output: an approximate* (x', e') *for* x, *such that* $e' \le e$. *(Referred as:* **approx***)*
    - *Input: an approximate* (x, e) *of* x. *Output: an approximate* (x', e') *of* x, *such that* $e' < e$ *if* $e > 0$, *or* $e' = e$ *if* $e = 0$. *(Referred as:* **iterate***)*

Remark 1. (i) The definition provides a template, with parameter M (d is given by M, but we do not denote it, e.g. by d_M, to simplify notation). (ii) In the specification of the iterative procedures we prescribe a lower bound on the improvement of the accuracy of approximation in order to ensure the termination of algorithms that need to produce approximating elements with given accuracies. (iii) When **epsint** is called with $e \ge e'$, x' is returned without any computation. (iv) The input initial approximate for the constructor has to be provided by the user. This is a trivial task if an **approx** procedure is available, however, it provides indispensable information when only an **iterate** procedure is given.

2.2 An Example for Real Numbers

In order to demonstrate how the above framework can be filled, we present a down-to-earth example in detail. We write the procedures in the programming language of the computer algebra system Maple, which also has a powerful numeric subsystem to exploit. This language does not support object oriented programming natively, thus class definitions and construction of objects has to be carried out implicitly. In particular, the methods need to be pointed explicitly to the object in context, which will be their first argument by default.

The simplest mathematical entities that offer themselves naturally to be represented in the exact real paradigm are the computable real numbers. The information defining the class is collected in the table **exactReal**, named after the class.

The constructor of the class just instantiates an object with the provided data members, from which **approx** and **iterate** are actually (references to) procedures. The **epsint** and **refine** methods provide the intended behavior relying on the **approx** and/or **iterate** members of the object in context.

```
exactReal := table([
"constructor" = proc(ka::list(float),ap,it)
 if ka[2]<0 then ERROR("Invalid approximate!"); fi;
 if ka[2]>0 and not (type(ap,procedure) or type(it,procedure)) then
  ERROR("At least one procedure must be given!");fi;
 RETURN(table(["approx"=ap, ''iterate"=it, ''knownApprox"=ka])); end,
"epsint" = proc(x::table, e::nonneg) local ap,ae;
 ap:=x["knownApprox"]; if ap[2]=0 then RETURN(ap[1]); fi; ae:=e;
```

```
while ap[2]>e do
 if type(x["approx"],procedure) then ap:=x["approx"](ae);ae:=ae/10;
 else ap:=x["iterate"](ap); fi; od;
x["knownApprox"]:=ap; RETURN(ap[1]); end,
"refine" = proc(x::table) local ap,e,ae;
ap:=x["knownApprox"];
if ap[2]=0 then RETURN(ap[1]); fi; e:=ap[2]/10; ae:=e;
while ap[2]>e do
 if type(x["iterate"],procedure) then ap:=x["iterate"](ap);
 else ap:=x["approx"](ae); ae:=ae/10; fi;od;
x["knownApprox"]:=ap; RETURN(ap[1]); end]);
```

Now we can define, for instance, an exact real object for π, using the routine of Maple for numerical expansion, and we can use the queries of the class on it:

```
knownApprox := [3.14, 0.1];
approx := proc(eps::nonneg) [evalf(Pi,max(2,2-ilog10(ops))),eps] end;
exPi := exactReal["constructor"](knownApprox, approx, 0);
exactReal["epsint"](exPi, 1e-10);
                        3.14159265359
```

2.3 Higher Level Constructors

Higher level constructors are basically the procedures that implement functions on exact real objects. Let $f : M \to M'$ be a continuous function, M, M' being metric spaces; in particular let d' be the metric on M'. We assume that the numerical algorithm F, which implements f, for any input $\mathbf{x} \in M$, returns a pair $(\mathbf{x}', \mathbf{e}') \in M' \times \mathbf{R}_{\geq 0}$, such that $d'(\mathbf{x}', f(\mathbf{x})) \leq \mathbf{e}'$.

Then a higher level constructor $\mathcal{F}$ that implements f, takes an exact real object $\mathcal{X}$ as argument, which represents a fixed argument x of f, and builds an exact real object $\mathcal{X}'$ for $x' := f(x)$, using the basic constructor for the exact real type for M'. Note that $\mathcal{F}$ has to generate at least one procedure and an initial approximate (according to definition 2), and that in both of these tasks it has to rely on F, the underlying numerical algorithm.

The most important task in the implementation of $\mathcal{F}$ is to reliably encode error propagation. The inaccuracy of the approximates of x' originates not only in that F generates numerical error, but also in the fact that the input of F (i.e. an approximating element $\mathbf{x}$, obtained from $\mathcal{X}$ as knownApprox or by epsint or refine) is itself inaccurate.

The *forward error computation* determines for a given approximate $(\mathbf{x}, \mathbf{e})$ of a fixed element $x \in M$, an upper bound for $d'(f(x), f(\mathbf{x}))$. It is performed by $\mathcal{F}$ when it creates an initial approximate for $\mathcal{X}'$, and also by the iterate procedure of $\mathcal{X}'$, which is generated by $\mathcal{F}$.

The *backward error computation*, for fixed $x \in M$ and given positive real value $\mathbf{e}'$, determines $\mathbf{e}$ such that for all approximates $(\mathbf{x}, \mathbf{e})$ of x, $d'(f(x), f(\mathbf{x})) \leq \mathbf{e}'$. It is performed in the approx procedure of $\mathcal{X}'$, which is generated by $\mathcal{F}$.

More advanced examples of forward and backward error computation are presented in the consecutive sections. Here we give only a simple higher level constructor; the addition of real numbers as exact real objects. We insert **addReals** as a new method into the class definition of **exactReal**.

In this constructor we provide only one procedure with the input and output specification of **epsint**. The backward error computation is very simple: in case of exact addition we can just take approximating elements for the operands with $e/2$ accuracy. However, in general we also have to count with rounding error, thus, by default, we distribute the allowed error equally between the inaccuracies of the approximating elements and the rounding error of the addition.

The forward error is just the sum of inaccuracies of the approximates which we added to produce the initial approximation. This is the straightforward application of interval arithmetic (see [14]). Again, the rounding errors of the computations increase the inaccuracy of the initial approximating element of the sum.

```
"addReals" = proc(x1::table,x2::table)
local ka,approx,dx1,dx2;
 approx := proc(eps::nonneg) local ap,e,en;
 e:=divRD(eps,3); en:=e;
 ap:=numAdd(exactReal["epsint"](x1,e),exactReal["epsint"](x2,e));
 # if numAdd generates too much error we distribute eps unequally
 while ap[2]>en do
  e:=divRD(e,2); en:=subRD(subRD(eps,e),e);
  ap:=numAdd(exactReal["epsint"](x1,e),exactReal["epsint"](x2,e));
 od; RETURN([ap[1],eps]); end;
dx1:=x1["knownApprox"]; dx2:=x2["knownApprox"];
ka:=numAdd(dx1[1],dx2[1]);
RETURN(exactReal["constructor"]([ka[1],addRU(ka[2],dx1[2],dx2[2])],
 approx,0)); end
```

Here **divRD** and **subRD** perform numerical division and subtraction respectively, introducing only rounding down errors, and **addRU** is numerical addition with only rounding up error. A simple procedure for numeric addition, which works well as long as the mantissas do not exceed the maximal allowed lengths in Maple, can be defined as:

```
numAdd := proc(x1::float,x2::float)
 RETURN([evalf(x1+x2, 1+max(length(op(1,x1))+op(2,x1),
 length(op(1,x2))+op(2,x2))-min(op(2,x1),op(2,x2))),0]); end;
```

As an additional remark to this section, let us note that often it is natural to generate iterative approximation procedures in the higher level constructors, applying only forward error computation, which can rely on firm interval analysis techniques. However, in critical situations one cannot always ensure sufficiently large improvement of the accuracy of the approximation in an iteration step because the intervals may have the tendency of blowing up. To prevent looping, one can encode heuristics in the **epsint** and **refine** procedures, that change from **iterate** to **approx** (if it is available) when a critical situation is predicted.

3   Real Linear Algebra

In this section we present examples of the essential part of defining new exact real types, i.e. the formulas of forward and backward error computation for nontrivial operations on multidimensional exact real objects. We describe matrix inversion and the computation of the Moore-Penrose generalized matrix inverse (pseudoinverse) for matrices over the computable real numbers.

The straightforward way of describing multidimensional entities of real numbers would be to build them up from elementary exact real objects. However, in this case all the standard methods of linear algebra should be re-implemented to work with them, and the resulting exact real entries of vectors and matrices would be deeply cascaded, which would lead far from our prescribed goals.

In our approach, any multidimensional real entity is treated as an exact real object in its corresponding metric space. In this section the metric is induced by the 2-norm by default.

The usual arithmetic operations on real vectors and matrices can be easily described by means of forward and backward error analysis, analogously to the case of real numbers, thus we skip them here.

3.1   Inverse of a Nonsingular Matrix

Let $A \in \mathbf{R}^{n \times n}$ be a nonsingular matrix with known approximate (A, d). We have

$$\|A^{-1} - \mathsf{A}^{-1}\| = \|A^{-1}A(A^{-1} - \mathsf{A}^{-1})\mathsf{A}\mathsf{A}^{-1}\| < \tag{1}$$
$$\leq \|A^{-1}\|\|\mathsf{A} - A\|\|\mathsf{A}^{-1}\| \leq \mathsf{d}\|A^{-1}\|\|\mathsf{A}^{-1}\|.$$

This yields an upper bound for the norm of the inverse: $\|A^{-1}\| \leq \|\mathsf{A}^{-1}\|/(1 - \mathsf{d}\|\mathsf{A}^{-1}\|)$ (note that $\mathsf{d}\|\mathsf{A}^{-1}\| < 1$ if d is small enough). Assuming that (A, d) fulfills this condition; for the forward error we get

$$\|A^{-1} - \mathsf{A}^{-1}\| \leq \frac{\mathsf{d}\|\mathsf{A}^{-1}\|^2}{1 - \mathsf{d}\|\mathsf{A}^{-1}\|}.$$

Let now a positive real number e be given; the goal is to find d_e such that for all approximate $(\mathsf{A}_\mathsf{e}, \mathsf{d}_\mathsf{e})$ of A, we have $\|A^{-1} - \mathsf{A}_\mathsf{e}^{-1}\| \leq \mathsf{e}$. We apply (1) for A and A_e, with the fact $\|\mathsf{A}_\mathsf{e}^{-1}\| \leq \|A^{-1}\| + \mathsf{e}$, also using the upper bound for $\|A^{-1}\|$. We get

$$\|A^{-1} - \mathsf{A}_\mathsf{e}^{-1}\| \leq \mathsf{d}_\mathsf{e} \frac{\|\mathsf{A}^{-1}\|}{1 - \mathsf{d}\|\mathsf{A}^{-1}\|} \left(\frac{\|\mathsf{A}^{-1}\|}{1 - \mathsf{d}\|\mathsf{A}^{-1}\|} + \mathsf{e} \right).$$

We have to choose d_e small enough to get the right hand side smaller than e; for instance, we may set

$$\mathsf{d}_\mathsf{e} := \mathsf{e} / \left(\frac{\|\mathsf{A}^{-1}\|}{1 - \mathsf{d}\|\mathsf{A}^{-1}\|} \left(\frac{\|\mathsf{A}^{-1}\|}{1 - \mathsf{d}\|\mathsf{A}^{-1}\|} + \mathsf{e} \right) \right) = \frac{\mathsf{e}(1 - \mathsf{d}\|\mathsf{A}^{-1}\|)^2}{\|\mathsf{A}^{-1}\|^2 + \mathsf{e}\|\mathsf{A}^{-1}\| - \mathsf{e}\mathsf{d}\|\mathsf{A}^{-1}\|^2}.$$

These error bounds are valid only if the inverse computation on a nonsingular numeric matrix is exact, otherwise e has to be distributed between the approximation error and the rounding errors.

The computations of this subsection are valid under any consistent norm.

3.2   Pseudoinverse

Given $A \in \mathbf{R}^{m \times n}$, the *pseudoinverse* of A is the unique matrix $A^{\dagger} \in \mathbf{R}^{n \times m}$, which, for any $b \in \mathbf{R}^m$, provides $x := A^{\dagger} b$ in $\mathbf{R}^n$ that minimizes $\|Ax - b\|$ with the smallest $\|x\|$. In other terms, our main interest is to find the "best" solution for a linear equation system in a numerically stable way. If A is nonsingular, the pseudoinverse coincides with the inverse. For the extensive study of the topic we refer to [3] and [20].

In the general case, pseudoinverse computation is an ill-posed problem; in order to solve it we use a special technique, known as *Tikhonov regularization*.

We solve the following optimization problem:

$$\text{for } \alpha > 0 \text{ find } x \text{ such that } \|Ax - b\|^2 + \alpha \|x\|^2 \text{ is minimal.} \tag{2}$$

For all $\alpha > 0$ problem (2) is continuous in A, b, and the solution can be expressed as $A^{\alpha} b$, where, as $\alpha \to 0$, A^{α} converges to $A^{\dagger}$.

We find the optimum of the quadratic function in x, at the zero of the derivative $(2\langle A\dot{x}, Ax - b \rangle + 2\alpha \langle \dot{x}, x \rangle = 2\langle \dot{x}, A^T(Ax - b) + \alpha x \rangle)$ which vanishes if $A^T(Ax - b) + \alpha x = 0$, or in other terms at

$$x = (A^T A + \alpha I^{n \times n})^{-1} A^T b,$$

where $I^{n \times n}$ is just the identity in $\mathbf{R}^{n \times n}$. Because the solution of problem (2) approximates $A^{\dagger} b$ as α goes to zero, we have $\lim\limits_{\alpha \to 0} A^{\alpha} = A^{\dagger}$, where $A^{\alpha} := (A^T A + \alpha I^{n \times n})^{-1} A^T$. We also remark that $A^T A + \alpha I^{n \times n} \in \mathbf{R}^{n \times n}$ is an invertible matrix.

Let A_0 approximate A, and let σ be the minimum of the positive singular values of A; assume that we have knowledge of this value. We want to determine how good does A_0^{α} approximate $A^{\dagger}$. We have

$$\|A^{\dagger} - A_0^{\alpha}\| \leq \|A^{\dagger} - A^{\alpha}\| + \|A^{\alpha} - A_0^{\alpha}\|, \tag{3}$$

where the second summand can be bounded by standard error bounds computed for the basic operations on matrices. In the following, we determine an upper bound for the first summand.

Let $A = U \Delta V$ be a singular value decomposition, where $U \in \mathbf{R}^{m \times m}, V \in \mathbf{R}^{n \times n}$ are orthogonal matrices, and $\Delta \in \mathbf{R}^{m \times n}$ is a diagonal matrix, with elements $(\sigma_1, \ldots, \sigma_i, \ldots, 0)$. Using this decomposition of A, we have

$$A^{\alpha} = (V^T \Delta^2 V + \alpha I^{n \times n})^{-1} V^T \Delta U^T = (V^T(\Delta^2 + \alpha I^{n \times n})V)^{-1} V^T \Delta U^T =$$
$$= V^T(\Delta^2 + \alpha I^{n \times n})^{-1} \Delta U^T.$$

Thus, by the convergence of A^{α} to $A^{\dagger}$, we have $A^{\dagger} = V^T \Delta^{\dagger} U^T$, and since we use 2-norm we can write

$$\|A^{\dagger} - A^{\alpha}\| = \|V^T(\Delta^{\dagger} - (\Delta^2 + \alpha I^{n \times n})^{-1} \Delta)U^T\| =$$
$$= \|\Delta^{\dagger} - (\Delta^2 + \alpha I^{n \times n})^{-1} \Delta\|.$$

We have that
$$(\Delta^2 + \alpha I^{n\times n})^{-1}\Delta = \mathrm{diag}\left(\frac{\sigma_i}{\sigma_i^2 + \alpha}\right).$$

On the other hand, we have $\Delta^\dagger = \mathrm{diag}(\sigma_i^\dagger)$, where

$$\sigma_i^\dagger = \begin{cases} 1/\sigma_i & \text{if } \sigma_i \neq 0 \\ 0 & \text{otherwise.} \end{cases}$$

Therefore we get

$$\Delta^\dagger - (\Delta^2 + \alpha I^{n\times n})^{-1}\Delta = \mathrm{diag}\left(\sigma_i^\dagger - \frac{\sigma_i}{\sigma_i^2 + \alpha}\right) = \mathrm{diag}(\delta_i),$$

where

$$\delta_i = \begin{cases} \frac{\alpha}{\sigma_i(\sigma_i^2 + \alpha)} & \text{if } \sigma_i \neq 0 \\ 0 & \text{otherwise.} \end{cases}$$

Finally, we get

$$\|A^\dagger - A^\alpha\| = \|\mathrm{diag}(\delta_i)\| = \max_{\sigma_i > 0} \frac{\alpha}{\sigma_i(\sigma_i^2 + \alpha)} = \frac{\alpha}{\sigma(\sigma^2 + \alpha)}.$$

The first step in the forward and backward error computation is to find a lower bound s for σ. Such an s is called regularization information for the problem, and it usually depends on A. This information, from our point of view, has to come from outside, that is, the user has to provide it. If the given s is too big, the exact real matrix of the solution may represent a different matrix than $A^\dagger$.

Let (A, d) be an approximate for A, and let s be a positive lower bound for σ. Let e_d denote the forward error of the computation $(A^T A + \alpha' I^{n\times n})^{-1}A^T$ (which can be determined from the forward errors of the basic matrix operations). The forward error of the pseudoinverse computation is then

$$\frac{\alpha'}{s(s^2 + \alpha')} + e_d.$$

We would like to determine a close to optimal value for α', which minimizes the forward error. Under the assumptions $\alpha' \leq \|A\|^2$ and $d \ll s \ll 1$, by expanding e_d, we get a formula of the following form $\alpha'/(s^3 + s\alpha') + d/\alpha' + c$, where c is constant in α'. After simplifying the denominator of this expression, we have to minimize $\alpha'/s^3 + d/\alpha'$. We get that a good choice is $\alpha' := \sqrt{s^3 d}$.

In the backward error computation, for given positive e, we distribute the error between the parts $\|A^\dagger - A^\alpha\|$ and $\|A^\alpha - A_0^\alpha\|$ in (3) equally. Then, we have to determine a suitably small α', so that we have

$$\frac{\alpha'}{s(s^2 + \alpha')} \leq \frac{e}{2}.$$

For instance, if e s < 2, we may choose $\alpha' = e\, s^3/(2 - e\, s)$.

Finally, we set up an exact real matrix $\mathcal{A}^{\alpha'}$ for $(A^T A + \alpha' I^{n \times n})^{-1} A^T$ by means of the basic matrix operations, and ask for an $e/2$ approximation: $A^{\alpha'}_{e/2}$. According to inequality 3, by the choice of α', $(A^{\alpha'}_{e/2}, e)$ is an approximate for $A^\dagger$, provided that the regularization information is correct.

4   Real Polynomials

In this section, we describe some basic algorithms for univariate polynomials with real coefficients. From our point of view, the polynomials of degree at most d are just the exact objects in the metric space $\mathcal{P}_d := \mathbf{R}^{d+1}$. For many operation on polynomials, we also need the metric space $\mathcal{M}_d := \mathbf{R}^d$ of monic polynomials of degree d. We have some obvious inclusion maps, and a partial map *normation*: $\mathcal{P}_d \rightarrow \mathcal{M}_d$ which is defined iff the formal leading coefficient is not zero.

It is rather straightforward to develop algorithms for addition, multiplication, composition, evaluation, division by monics and derivative. In fact, all these functions could be implemented using exact real arithmetic for the coefficients; however, we a get better performance if we implement these operations from the scratch, using the scheme described in Sect. 2 with more accurate forward and backward error. For instance, the forward error of evaluation predicted by the mean value theorem of calculus $f(b) - f(a) = (b - a)f'(\xi)$ is much smaller than the the accumulation of the forward errors arising by Horner's scheme.

In the following subsections, we discuss some problems that cannot be solved solely by arithmetic.

4.1   GCD and Squarefree Factorization

The GCD operation is a map from $\mathcal{M}_m \times \mathcal{M}_n$ to $\bigcup_{i=0}^{\min(m,n)} \mathcal{M}_i$. Clearly, the function is discontinuous, because the degree is discrete and cannot depend continuously on the input. Thus, the problem of GCD computation is ill-posed. To solve it, we use a regularization technique similar to the one for computing the pseudo-inverse in Sect. 3, and which was proposed in [6].

Let $P \in \mathcal{M}_m, Q \in \mathcal{M}_n$ be monic polynomials, $m \leq n$. The Sylvester matrix $S = S(P, Q)$ is an $(m+n) \times (m+n)$ matrix with entries from the set of coefficients of P and Q (we refer to [21] for a more precise definition).

Let $V := \mathbf{R}^{m+n}$ be endowed with the ∞-norm. The function $sv : \mathbf{R}^{N \times N} \rightarrow \mathbf{R}^N$ computing the ordered N-tuple of singular values is continuous; in fact we have $\|sv(A) - sv(B)\| \leq \|A - B\|$, which gives a very easy formula for the forward and backward error (see [11]). Suppose that we have a lower bound s for the smallest nonzero singular value of the Sylvester matrix S. Then we can compute the rank of S by computing an $s/2$-estimate of $sv(S)$: the singular values with approximating element less than $s/2$ are zero, and the singular values with approximating element greater than $s/2$ must be nonzero. If the rank is $m + n - r$, then r is the degree of the GCD (see [21]).

But if we know the degree of the GCD, then we can compute the GCD by solving a system of linear equations. If R is the GCD, and $\deg(R) = k$, then there is a unique pair of polynomials $U \in \mathcal{P}_{n-k-1}$, $V \in \mathcal{P}_{m-k-1}$, such that $UP + VQ = R$. The polynomial $UP + VQ$ has formal degree $m + n - k - 1$, and we know the coefficients of $x^{m+n-k-1}$, $x^{m+n-k-2}$, ..., x^k (namely $0, \ldots, 0, 1$). This gives a system $(*)$ of $m + n - 2k$ linear equations in equally many unknowns. The system is nonsingular, so we could solve it exactly by the methods described in Sect. 3. However, we prefer to solve it via the pseudo-inverse, i.e. treating it as a least squares problem. The reason is the following.

The regularization information s has to come from the user. If the smallest nonzero singular value was estimated too low, then s is still a lower bound, and the computed result is still exact (the effect giving too small bounds just leads to a worse performance, because the singular values must be computed more accurate than necessary). If the smallest nonzero singular value was estimated too high, then the rank of S is larger than expected, and the degree of the GCD is smaller than expected. The system $(*)$ becomes singular, and the exact algorithm for resolving it does not terminate. On the other hand, if we treat $(*)$ as a least squares problem, then we get a solution anyway. The computed GCD R' is still a linear combination of P and Q but it is not an exact divisor of P and Q. However, the remainder of P or Q modulo R' is small when the the given s is small.

Squarefree factorization is another ill-posed problem, but we can reduce it to GCD computation. There are several ways to do this for polynomials over the rational numbers (or over any other computable field). Not all these reductions work also in the exact real case; we have to be careful that we never estimate the degree of an intermediate result too high, whereas an error in the other direction is not so disturbing (note that wrong degree estimates are unavoidable in the case that the regularization information was incorrect). Algorithm 1 works because it has this monotonicity property (see also [1]).

Algorithm 1 Squarefree(P)

---

  if $\deg(P) = 1$ **then**
    return $((P, 1))$;
  else
    $P' :=$ derivative of p divided by $\deg(P)$;
    $Q := \mathrm{GCD}(P, P')$;
    $L :=$ Squarefree(Q);
    for all (F, e) in L **do**
      $P :=$ quotient(P, F^{e+1});
      in L, replace (F, e) by $(F, e + 1)$;
    end for
    append $(P, 1)$ to L;
    return L
  end if

If the given regularization information (used in the GCD computation) was correct, then the algorithm returns a list $((F_1, e_1), \ldots, (F_r, e_r))$, such that the

F_i are squarefree and relatively prime, and $F_1^{e_1} \ldots F_r^{e_r} = P$. Otherwise, the F_i are still squarefree and relatively prime, but $\tilde{F}_1^{e_1} \ldots \tilde{F}_r^{e_r}$ is not exactly equal to P (but close if the estimated lower bound s is small). In this case, the sum of the degrees of the F_i' is smaller than the number of distinct complex roots of P (in other words, we have made the mistake of identifying two roots which are closely together).

4.2   Roots Computation

Let $\mathbf{C}$ be the space of complex numbers, endowed with the Euclidean norm. Any polynomial of degree n has n complex roots, counted with multiplicity; but there is no continuous map $\mathcal{M}_n \to \mathbf{C}^n$, and the reason is that there is no continuous way of selecting one particular root. To avoid this difficulty, we introduce the space $\mathcal{S}_n$ as the set of orbits under the permutation group S_n permuting the coordinates of $\mathbf{C}^n$. The metric is defined by $d([x], [y]) := \min_{\sigma \in S_n} d_0(x, \sigma(y))$, where d_0 is the metric induced by the ∞-norm in $\mathbf{C}^n$. The roots map $rt :$ $\mathcal{M}_n \to \mathcal{S}_n$ is continuous, and we want to compute it in the sense of exact real computation. It turns out that it is more practical to create this exact real object by an `iterate` function rather than by an `approx` function, because the a posteriori error bound is typically much smaller than the a priori error bound of a root, i.e. we can find a more realistic accuracy when we already know the approximating value.

For the moment, let us assume that the input polynomial P is squarefree. Then the approximating polynomial P is also squarefree, if the accuracy e is sufficiently small. We compute its roots by a numerical algorithm. For each computed root $\mathbf{z_i}$ (which is in general not an exact root of P), we expect a nearby root of the exact polynomial P. If $P'(\mathbf{z_i}) \neq 0$, then there is a root with distance at most $\frac{|P(\mathbf{z_i})|}{|nP'(\mathbf{z_i})|}$ (see [19]). This does not necessarily imply that we have information about every root of P: it could be that a root of P lies in two of the disks with midpoint $\mathbf{z_i}$ and radius $\mathbf{r_i} \geq \frac{nP(\mathbf{z_i})}{P'(\mathbf{z_i})}$, and there is another root of P which is outside of all disks. But for small e, the radii become small, and the disks will be disjoint. Thus, we have computed an approximate $((\mathbf{z_1}, \ldots, \mathbf{z_n}), \max_{i=1}^n \mathbf{r_i})$ of $rt(P)$.

Now, let us drop the assumption that P is squarefree. Then we can compute its squarefree decomposition with an arbitrary regularization parameter s > 0 (we take s $= 10^{-5}$ by default). If the regularization information is correct, i.e. if s is a lower bound for all positive singular values in all Sylvester matrices occurring in algorithm 1, then we get the correct decomposition $P = F_1^{e_1} \ldots F_r^{e_r}$. Otherwise, we get the squarefree decomposition of a nearby polynomial P_1, and we know that the number of distinct roots of P_1 is smaller than the number of distinct roots of P. By decreasing s, we can make the distance $\|P - P_1\|$ arbitrarily small.

Now we compute the roots of the $F_1, \ldots, F_r$ by the above algorithm for the squarefree case. Let $\mathbf{z_{i,j}}$ be a computed root of F_i. We expect a cluster of e_i roots, counted with multiplicity, near $\mathbf{z_{i,j}}$. There are several methods for cluster validation, see [15, 12, 1, 4]; below we do the validation by a simple formula.

In the case $P^{e_i}(\mathbf{z_{i,j}}) \neq 0$, the distance can be estimated by the formula below. This inequation is true for small regularization parameter: if our decomposition is exact, then it is certainly true, otherwise we have $P_1^{e_i}(\mathbf{z_{i,j}}) \neq 0$ and this implies the claim whenever $\|P - P_1\|$ is sufficiently small. Here is the precise statement.

Lemma 1. *Let P be a monic polynomial of degree n. Let $z \in \mathbf{C}$ be such that $P^{(k)}(z) \neq 0$. Then there are at least k roots, counted with multiplicity, in the disk around z with radius*

$$M_{P,z,k} := 2n \max_{j=0}^{k-1} \sqrt[k-j]{\frac{kk!|P^{(j)}(z)|}{j!|P^{(k)}(z)|}}.$$

Proof. Without loss of generality, we assume $z = 0$. Let $p_0, \ldots, p_{n-1}, p_n = 1$ be the coefficients of P. We assume also that $p_0 \neq 0$; the case $p_0 = 0$ follows by continuity.

Let $y_1, \ldots, y_n$ be the roots of the reverse polynomial $\bar{P}(x) := x^n P(1/x)$, ordered by absolute value, largest first. Assume, indirectly, that there are no k roots of P in the circle of radius $M := M_{P,0,k}$. Then $|y_i| < \frac{1}{M}$ for $i = k, \ldots, n$. For $i = 0, \ldots, k$, we define A_i as the sum of all products of i roots from $y_1, \ldots, y_{k-1}$, and B_i as the sum of all products of i roots from $y_k, \ldots, y_n$ $(A_0 = B_0 = 1)$, and $C_i := \sum_{j=0}^{i} \left|\frac{p_j}{p_0}\right| \left(\frac{2n}{M}\right)^{i-j}$. The B_i can be estimated by

$$B_i < \binom{n-k+1}{i} \frac{1}{M^i} \leq \left(\frac{n}{M}\right)^l.$$

We claim that $|A_i| \leq C_i$ for $i = 0, \ldots, k - 1$. From Vieta's formula, we obtain

$$\left|\frac{p_i}{p_0}\right| = \left|\sum_{j=0}^{i} A_{i-j} B_j\right|$$

and by induction on i, we get

$$B_i \leq \left|\frac{p_i}{p_0}\right| + \sum_{j=0}^{i-1} |A_j B_{i-j}| \leq \left|\frac{p_i}{p_0}\right| + \sum_{j=0}^{i-1} R_j \left(\frac{n}{M}\right)^{i-j} =$$

$$= \left|\frac{p_i}{p_0}\right| + \sum_{j=0}^{i-1} \left|\frac{p_j}{p_0}\right| (2^{i-j} - 1) \left(\frac{n}{M}\right)^{i-j} \leq R_i,$$

which proves the claim. Now we get

$$\left|\frac{p_k}{p_0}\right| \leq \sum_{i=0}^{k-1} R_i \left(\frac{n}{M}\right)^{k-i} = \sum_{i=0}^{k-1} \left|\frac{p_i}{p_0}\right| (2^{k-i} - 1) \left(\frac{n}{M}\right)^{k-i} < \sum_{i=0}^{k-1} \left|\frac{p_i}{p_0}\right| \left(\frac{2n}{M}\right)^{k-i}.$$

By the choice of M, we have $\left(\frac{2n}{M}\right)^{k-i} \leq \left|\frac{p_0}{kp_i}\right|$ for all $i = 0, \ldots, k - 1$. So we get $\left|\frac{p_k}{p_0}\right| < \left|\frac{p_k}{p_0}\right|$, which is a contradiction.

As the regularization parameter and the accuracy tend to zero, the radii $M_{P,z_{i,j},e_i}$ become arbitrary small. Eventually, one of two things will happen. Either the disks will become disjoint. In this case, we have computed an approximate of $rt(P)$. Or the rank of the Sylvester matrix increases in one of the GCD computations required in the squarefree computation, which means that one of the clusters splits into several. The second can only happen a finite number of times, so that we will always come to an approximate of $rt(P)$.

References

[1] Aberth, O. (1998): Precise numerical methods using C++. Academic Press Inc., San Diego, CA
[2] Arsenin, V., and Tikhonov, A. N. (1977): Solutions of ill-posed problems. Wiley
[3] Ben-Israel, A., and Greville, T. N. E. (1974): Generalized Inverses: Theory and Applications. Wiley
[4] Bini, D., and Fiorentino, G. (2000): Design, analysis, and implementation of a multiprecision polynomial rootfinder. Num. Alg. 23, 127–173
[5] Boehm, H., and Cartwright, R. (1990): Exact real arithmetic, formulating real numbers as functions. In: Research Topics in Functional Programming, Turner D. (ed.) Addison-Wesley, 43–64
[6] Corless, R., Gianni, P., Trager, B., and Watt, S. (1995): The singular value decomposition for polynomial systems. In: Proc. ISSAC'95, ACM Press, 195–207
[7] Engl, H. (1999): Regularization methods for solving inverse problems. In: Proc. ICIAM 99
[8] Escardó, M. H. (2000): Exact numerical computation. In: ISSAC'00, ACM Press, tutorial.
[9] Gianantonio, P. (1993): Real number computability and domain theory. In: Proc. Math. Found. Comp. Sci. '93, Springer, 413–422
[10] Gonzalez-Vega, L., Rouillier, F., Roy, M.-F., and Trujillo, G. (2000): Symbolic recipes for real solutions. Algorith. Comp. Math. 4, 121–167
[11] Higham, N. J. (1993): Accuracy and stability of numerical algorithms. SIAM
[12] Hribernig, V., and Stetter, H.-J. (1997) Detection and validation of clusters of polynomial zeroes. Journal of Symbolic Computation 24, 667–682
[13] Ménissier-Morain, V. (2001): Arbitrary precision real arithmetic: design and algorithms. Preprint
[14] Moore, R. E. (1966): Interval Anaysis. Prentice-Hall, Englewood Cliffs, N.J.
[15] Neumaier, A. (1988): An existence test for root clusters and multiple roots. ZAMM 68, 257–259
[16] Noda, M. T., and Sasaki, T. (1989): Approximate square-free decomposition and rootfinding for ill-conditioned algebraic equations. J. Inf. Proc. 12, 159–168
[17] Roy, M.-F., and Szpirglas, A. (1990): Complexity of computation of real algebraic numbers. Journal of Symbolic Computation 10, 39–51
[18] Stetter, H.-J. (2000): Condition analysis of overdetermined algebraic problems. In: Proc. CASC 2000, 345–365.
[19] Strzeboński, A. (1997): Computing in the field of complex algebraic numbers. Journal of Symbolic Computation 24, 647–656
[20] Tran, Q.-N. (1996): A Hybrid Symbolic-Numerical Approach in Computer Aided Geometric Design (CAGD) and Visualization. PhD thesis, Johannes Kepler University, RISC-Linz
[21] Winkler, F. (1996): Polynomial algorithms in computer algebra. Springer

Symbolic Methods
for the Element Preconditioning Technique [*]

Ulrich Langer[1], Stefan Reitzinger[1], and Josef Schicho[2]

[1] Institute of Computational Mathematics, University of Linz, Austria,
{ulanger,reitz}@numa.uni-linz.ac.at
[2] Research Institute of Symbolic Computation, University of Linz, Austria,
Josef.Schicho@risc.uni-linz.ac.at

Abstract. The method of element preconditioning requires the construction of an M-matrix which is as close as possible to a given symmetric positive definite matrix in the spectral sense. In this paper we give a symbolic solution of the arising optimization problem for various subclasses. This improves the performance of the resulting algorithm considerably.

1 Introduction

The condition number of the stiffness matrix K_h arising from the finite element discretization of second-order elliptic boundary value problems typically behaves like $O(h^{-2})$ as h tends to zero, where h denotes the usual discretization parameter characterizing a regular finite element discretization. This means practically, that the stiffness matrix has a large condition number on a fine grid. That is the reason why the classical iterative methods exhibit slow convergence This drawback can be avoided by multigrid methods (see, e.g. [6]) or multigrid preconditioned iterative methods [8, 9].

In contrast to the geometric multigrid method (MGM) that is based on a hierarchy of finer and finer meshes, the algebraic multigrid (AMG) method needs only single grid information, usually the matrix and the right-hand side of the system that is to be solved. In AMG, the hierarchy of coarser and coarser representation of the fine grid problem must be generated algebraically.

It is well known that an AMG method works well if K_h is an M-matrix (see Section 2.2 for the definition of M-matrices), but this is hardly the case in many real life applications. Thus, if K_h is not an M-matrix then it is desirable to derive an AMG preconditioner for K_h from a nearby, spectrally equivalent M-matrix B_h that is sometimes called regularizer [9]. In [5], we construct such an M-matrix regularizer. The main idea is to localize the problem: For each element, we compute an M-matrix which is as close as possible (in the spectral sense) to

[*] This work has been supported by the Austrian Science Fund under the grant 'Fonds zur Förderung der wissenschaftlichen Forschung (FWF)' - under the grant SFB F013 'Numerical and Symbolic Scientific Computing' and under the project P 14953 'Robust Algebraic Multigrid Methods and their Parallelization'.

F. Winkler and U. Langer (Eds.): SNSC 2001, LNCS 2630, pp. 293–308, 2003.

the element stiffness matrix. Then we assemble these M-matrices by the help of the element connectivity relations and get our regularizator B_h from which we afterwards derive the AMG-preconditioner (see Subsection 2.2 for details).

In [5], we solve these optimization problems numerically, by a sequential quadratic programming algorithm using a Quasi-Newton update formula for estimating the Hessian of the objective function. Unfortunately, this subtask turns out to be a bottleneck, because in some practically important cases we have to solve a large number of such optimization problems. Moreover, the numerical solution does not always get close to the global optimum, because there are several local optima, and on some of them the objective function is not even differentiable.

In this paper, we solve various cases of these optimization problems symbolically. For various element matrices involving only one symbolic parameter, we can find a closed form solution in terms of polynomials. A similar formula is given for general 2×2 matrices, but here there are several closed forms, and some inequalities need to be checked in order to determine which one has to taken. For general 3×3 matrices, a similar closed form would be theoretically possible, except that we need also square roots and – in one case – roots of higher degree polynomials (see Section 3). But such a closed formula would be too large to be useful, so we prefer to give a "formula" consisting of a program with arithmetic or square root (and in one case higher order root) assignments and **if then else** branches, but no loops. This is done in Section 3. Using these formula, we can compute the optimal preconditioner faster and more accurately (see Section 4).

2   Problem Formulation

In this section we explain the idea of the element preconditioning technique proposed in [5], and isolate the most crucial subproblem, namely the problem of the construction of M-matrices which are as close as possible to the finite element stiffness matrices in some spectral sense.

2.1   Finite Element Discretization

For simplicity, let us consider the weak formulation

$$\text{Find } u \in \mathbb{V} : \quad \int_\Omega (\nabla^t u \nabla v + \sigma u v)\, dx = \int_\Omega f v\, dx \quad \forall v \in \mathbb{V} \qquad (1)$$

of the potential equation $-\Delta u + \sigma u = f$ in Ω under homogeneous Neumann conditions on the boundary $\partial\Omega$ of Ω as some model problem for explaining the element preconditioning technique. The computational domain $\Omega \subset \mathbb{R}^d$ (with $d = 2, 3$ the spatial dimension) is bounded with a sufficiently smooth boundary $\partial\Omega$. The test space $\mathbb{V}$ coincides with the Sobolev space $H^1(\Omega)$, $\sigma \geq 0$ denotes a real parameter and $f \in L_2(\Omega)$ is a given right-hand side. In order to ensure solvability of the boundary value problem (1) in the case $\sigma = 0$, we assume that f is L_2-orthogonal to all constants.

A finite element (FE) discretization of the computational domain Ω results in some FE-mesh $\mathcal{T}_h = \{\overline{\Omega}^r : r \in \tau_h\}$ (see Figure 1 for an FE-discretization of the unit square by rectangular and triangular elements) such that

$$\overline{\Omega} = \bigcup_{r \in \tau_h} \overline{\Omega}^r,$$

with the index set τ_h of all finite elements, the set of all nodes $\{x_i : i \in \overline{\omega}_h\}$, the index set $\overline{\omega}_h$ of all nodes and the typical mesh size h. The FE-basis $\Phi =$

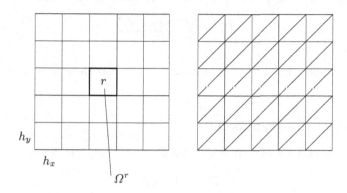

Fig. 1. Finite element discretizations of the unit square.

$\{\varphi^{[j]}(x), j \in \overline{\omega}_h\}$ spans the FE-space $\mathbb{V}_h = \mathrm{span}\{\Phi\} \subset \mathbb{V}$ and changes (1) into the FE-scheme

$$\text{Find } u_h \in \mathbb{V}_h : \quad \int_\Omega (\nabla^t u_h \nabla v_h + \sigma u_h v_h) \, dx = \int_\Omega f v_h \, dx \quad \forall v_h \in \mathbb{V}_h \qquad (2)$$

that is equivalent to the linear system of FE-equations

$$\text{Find } \underline{u}_h \in \mathbb{R}^{N_h} : \quad K_h \underline{u}_h = \underline{f}_h \quad \text{in } \mathbb{R}^{N_h} \qquad (3)$$

for defining the coefficients $u_h^{[i]}$ in the FE-ansatz

$$u_h(x) = \sum_{i \in \overline{\omega}_h} u_h^{[i]} \cdot \varphi^{[i]}(x), \qquad (4)$$

where $K_h \in \mathbb{R}^{N_h \times N_h}$ denotes the stiffness matrix, $\underline{f}_h \in \mathbb{R}^{N_h}$ denotes the load vector, and $\underline{u}_h = (u_h^{[i]}) \in \mathbb{R}^{N_h}$ is the solution vector of nodal unknowns $u_h^{[i]}$. We assume that the FE-basis functions are of Lagrangian type such that $\varphi^{[j]}(x_i) = \delta_{ij}$, where δ_{ij} denotes the Kronecker delta. Since the FE-bases is assumed to have local support the corresponding integrals are evaluated locally on each element, i.e.,

$$\int_\Omega (\nabla^t u_h \nabla v_h + \sigma u_h \, v_h) \, dx = \sum_{r \in \tau_h} \int_{\Omega^r} (\nabla^t u_h \nabla v_h + \sigma u_h \, v_h) \, dx.$$

By using the FE-ansatz (4) we get, on each element, an integral of the form

$$(K_h^r)_{ji} = \int_{\Omega^r} (\nabla^t \varphi^{[i]} \nabla \varphi^{[j]} + \sigma \varphi^{[i]} \varphi^{[j]}) \, dx$$

for calculating the coefficients $(K_h^r)_{ji}$ of the element stiffness matrix K_h^r. In the case of n_r bases functions on element $\overline{\Omega}^r$ ($r \in \tau_h$), we arrive at an $K_h^r \in \mathbb{R}^{n_r \times n_r}$ that is symmetric and positive (semi)definite. Then the global stiffness matrix $K_h \in \mathbb{R}^{N_h \times N_h}$ will be assembled from the local element stiffness matrix by the usual assembling procedure that can be represented in the form

$$K_h = \sum_{r \in \tau_h} C_r^t K_h^r C_r, \tag{5}$$

where the matrices $C_r \in \mathbb{R}^{n_r \times N_h}$ denote the so-called element connectivity matrices.

2.2   Element Preconditioning

The condition number of some regular matrix $A \in \mathbb{R}^{n \times n}$ is defined by

$$\kappa(A) = \|A\| \cdot \|A^{-1}\|,$$

where $\| \cdot \|$ is an appropriate matrix norm. For an SPD matrix A, the so-called spectral condition number

$$\kappa(A) = \frac{\lambda_{\max}(A)}{\lambda_{\min}(A)}$$

is based on the spectral norm $\|A\| = \sqrt{\lambda_{\max}(A^t A)} = \lambda_{\max}(A)$, where $\lambda_{\max}(A)$ and $\lambda_{\min}(A)$ denotes the maximal and minimal eigenvalue of the matrix A, respectively. The matrix A^t denotes the transpose of A.

The SPD matrices $A, B \in \mathbb{R}^{n \times n}$ are called spectrally equivalent if

$$\exists c_1, c_2 \in \mathbb{R}^+ : c_1 \cdot \langle Bu, u \rangle \leq \langle Au, u \rangle \leq c_2 \cdot \langle Bu, u \rangle \qquad \forall u \in \mathbb{R}^n \tag{6}$$

which is briefly denoted by $c_1 \cdot B \leq A \leq c_2 \cdot B$. Therein $\langle \cdot, \cdot \rangle$ denotes the Euclidean inner product. In addition, $\langle \cdot, \cdot \rangle_A$ denotes the A-energy inner product corresponding to the SPD matrix A, i.e., $\langle \cdot, \cdot \rangle_A = \langle A \cdot, \cdot \rangle$. Obviously the best possible constants c_1 and c_2 are given by the solution of the generalized eigenvalue problem

$$Au = \lambda Bu$$

with $c_1 = \lambda_{\min}(B^{-1/2} A B^{-1/2})$ and $c_2 = \lambda_{\max}(B^{-1/2} A B^{-1/2})$.

An important subclass of regular matrices are M-matrices. For SPD matrices, this class is defined as follows:

$$M_n = \left\{ A \in \mathbb{R}^{n \times n} : a_{ii} > 0, \, a_{ij} \leq 0, i \neq j, \sum_{j=1}^{n} a_{ij} \geq 0 \right\}.$$

A crucial point in the FE-simulation is the solution of the linear, symmetric and positive semidefinite (positive definite in the case $\sigma > 0$) system of equations (3). It is well known that an AMG method works well if $K_h \in M_{N_h}$, but this is hardly the case in many real life applications. Thus, it is desirable to derive an AMG preconditioner for K_h from a nearby, spectrally equivalent matrix B_h that is sometimes called regularizator [9]. Especially, in our case we need an M-matrix for applying the standard AMG efficiently. Consequently, if we are able to construct such a symmetric, positive definite regularizator B_h in the class of the M-matrices, then we can derive a good preconditioner C_h for K_h by applying a symmetric AMG cycle to B_h instead of K_h. Finally, the constructed preconditioner C_h is used in a preconditioned conjugate gradient solver. This approach was presented in [5].

In order to be able to construct an M-matrix B_h efficiently we have to localize the problem. The basic idea is the content of the following lemma.

Lemma 1. *Let the stiffness matrix $K_h \in \mathbb{R}^{N_h \times N_h}$ be SPD, and K_h be assembled from SPD element matrices $K_h^r \in \mathbb{R}^{n_r \times n_r}$, $r \in \tau_h$, i.e., K_h can be represented in the form (5). Further, let us suppose that, for all $r \in \tau_h$, there are SPD matrices $B_h^r \in M_{n_r}$ such that the spectral equivalence inequalities*

$$c_1^r \cdot B_h^r \le K_h^r \le c_2^r \cdot B_h^r \qquad \forall r \in \tau_h \tag{7}$$

hold, with h-independent, positive spectral equivalence constants c_1^r and c_2^r. Then the matrix

$$B_h = \sum_{r \in \tau_h} C_r^t B_h^r C_r \tag{8}$$

is spectrally equivalent to the stiffness matrix K_h, i.e.,

$$c_1 \cdot B_h \le K_h \le c_2 \cdot B_h, \tag{9}$$

with the spectral equivalence constants

$$c_1 = \min_{r \in \tau_h}\{c_1^r\} \quad and \quad c_2 = \max_{r \in \tau_h}\{c_2^r\}.$$

Additionally, the matrix B_h is SPD and belongs to the class M_{N_h}.

Proof. see [5].

Our particular interest in this paper consists in the construction of such an SPD M-matrices B_h^r that are as close as possible to K_h^r in the spectral sense. Thus, Lemma 1 provides the theoretical background for Algorithm 1. This algorithm returns the best SPD matrix $B_h \in M_{N_h}$ in the sense of the localized problem.

Remark 1.

1. We note that in the 2D case $n_r = 3$ and $n_r = 4$ correspond to linear and bilinear elements, respectively. Similarly, linear and trilinear elements for the 3D case are represented by $n_r = 4$ and $n_r = 8$, respectively.

Algorithm 1 GeneralSpectralMatrix ()

---

for all $r \in \tau_h$ **do**
   Get the element matrix $K_h^r \in \mathbb{R}^{n_r \times n_r}$
   if $K_h^r \notin M_{n_r}$ **then**
      Calculate B_h^r from the restricted minimization problem

$$\frac{\lambda_{max}((B_h^r)^{-1/2} K_h^r (B_h^r)^{-1/2})}{\lambda_{min}((B_h^r)^{-1/2} K_h^r (B_h^r)^{-1/2})} \rightarrow \min$$

      subject to $B_h^r \in M_{n_r}$ and B_h^r is SPD
   else
      Set $B_h^r = K_h^r$
   end if
   Assemble B_h^r
   Assemble K_h^r
end for

---

2. In the case of symmetric positive semidefinite element matrices K_h^r the technique applies again. Now the generalized spectral condition number has to be minimized, i.e., the eigenvalue $\lambda_{min}((B_h^r)^{-1/2} K_h^r (B_h^r)^{-1/2})$ is replaced by $\lambda_{min} = \min\{\lambda((B_h^r)^{-1/2} K_h^r (B_h^r)^{-1/2}), \lambda \neq 0\}$.
3. For our special case $\sigma = 0$, the element stiffness matrices are symmetric positive semidefinite. But such K_h^r can be transformed to an SPD one by eliminating the last row and column (kernel elimination), see [5].

For the rest of the paper we skip the indices of the element stiffness matrices, i.e. $K = K_h^r$ whenever no ambiguities can occur. In the following we give three typical examples from the FE-discretization where the M-matrix property is lost.

Example 1. We study the case of anisotropic rectangular and triangular elements (see Figure 2) where we can establish the dependency of the condition number on the anisotropic parameter q explicitly. Further we always use the variational form (1) for our examples.

1. Let us consider the case of an anisotropic rectangular element with bilinear FE-functions and set $\sigma = 0$. The element stiffness matrix has the form

$$K = \frac{1}{6q} \begin{pmatrix} 2 + 2q^2 & 1 - 2q^2 & -2 + q^2 & -1 - q^2 \\ 1 - 2q^2 & 2 + 2q^2 & -1 - q^2 & -2 + q^2 \\ -2 + q^2 & -1 - q^2 & 2 + 2q^2 & 1 - 2q^2 \\ -1 - q^2 & -2 + q^2 & 1 - 2q^2 & 2 + 2q^2 \end{pmatrix}. \tag{10}$$

After eliminating the last row and column (i.e. kernel elimination, see Remark 1.3), the element stiffness matrix $\tilde{K} \notin M_3$ for $0 < q < \sqrt{1/2}$.
2. The second example is due to the triangle with $\sigma = 0$. In this case the element stiffness matrix has the form

$$K = \frac{1}{4q} \begin{pmatrix} 1 + q^2 & 1 - q^2 & -2 \\ 1 - q^2 & 1 + q^2 & -2 \\ -2 & -2 & 4 \end{pmatrix}. \tag{11}$$

Again after a proper reduction to the SPD case $\tilde{K} \notin M_2$ for $0 < q < 1$.

3. As a third example we consider our model bilinearform with $\sigma = 1$ and calculate the element stiffness matrix on the triangle, where some mass lumping is used, i.e.,

$$K = \frac{1}{4q} \begin{pmatrix} 1 + 2q^2 & 1 - q^2 & -2 \\ 1 - q^2 & 1 + 2q^2 & -2 \\ -2 & -2 & 4 + q^2 \end{pmatrix}. \tag{12}$$

Again, $K \notin M_3$ for $0 < q < 1$.

Fig. 2. Thin finite element structures.

In the case of higher-order ansatz functions that are very important in practical application, we can not expect to get M-matrices at all.

2.3   An Optimization Problem

The critical step in Algorithm 1 is the solution of the restricted minimization problem: Given an SPD matrix K, find an SPD M-matrix B such that the condition number of $B^{-1/2}KB^{-1/2}$ is as small as possible. In this situation, we say that B is the closest M-matrix to K in the spectral sense.

We need to solve many instances of the problem, in general, one for each element. If the dimension of the matrices is equal to n, then we have an optimization problem with $\frac{(n+1)(n-2)}{2}$ variables (for the coefficients of B, up to scalar multiplication) depending on the same number of parameters (for the coefficients of K, up to scalar multiplication). The number n is relatively small, i.e. $n \leq 30$; in typical applications (i.e. low-order finite elements), we have $n = 2, 3, 4$ (see Example 1).

The search space is a convex polyhedron in projective space, which is independent of the parameters. We can reformulate the problem in such a way that the objective function is independent of the parameters (of course, this makes the search space vary).

We write $K = A^t A$ by the Cholesky decomposition. Then the inverse of the matrix $B^{-1/2}KB^{-1/2}$ is similar to

$$C := AB^{-1/2}(B^{1/2}K^{-1}B^{1/2})B^{1/2}A^{-1} = (A^t)^{-1}BA^{-1}.$$

Finding the closest M-Matrix B is equivalent to finding an SPD matrix C such that $A^t C A$ is an M-Matrix (equal to B), which has smallest possible condition number.

In the reformulation, the search space is described by linear inequalities in the coefficients of C. More precisely, the inequalities are of the type

$$\langle a_i, a_j \rangle_C \leq 0 \text{ for } 1 \leq i < j \leq n+1,$$

where $a_1, \ldots, a_n$ are the columns of A, and $a_{n+1} := -a_1 - \cdots - a_n$, and $\langle u, v \rangle_C = \langle u, Cv \rangle$ is again the scalar product defined by the SPD matrix C. This follows from the fact that $\langle a_i, a_j \rangle_C$ is the (i,j)-th entry of the matrix $A^t C A$ for $i, j \leq n$, and that $\langle a_i, -a_{n+1} \rangle_C$ is the i-th row sum. The search space depends on the vectors $a_1, \ldots, a_{n+1}$ in a symmetric way. The vectors $a_1, \ldots, a_{n+1}$ fulfill the symmetric condition $a_1 + \cdots + a_{n+1} = 0$ and the restriction that each n of them are linearly independent.

In [5], the restricted optimization problem is solved by a sequential quadratic programming algorithm using a Quasi-Newton update formula for estimating the Hessian of the objective. An additional difficulty is the fact that the objective function is not everywhere differentiable: The gradient does not exist for matrices with multiple maximal or minimal eigenvalue. This is a subset of measure zero, but unfortunately it contains the optimum in some cases, as experiments have shown.

3   A Symbolic Solution of the Optimization Problem

As the restricted problem is strictly algebraic, we can approach it by general quantifier elimination methods such as the method of Gröbner bases [1, 2] or the method of cylindrical algebraic decomposition [3, 7, 4]. For $n = 2$, this indeed gives a formula for the solution (see Remark 2 below). For $n = 3$ or higher, the number of variables is too large for such an approach to work. It is therefore necessary to exploit the specific structure of the problem.

3.1   Cases Where the Objective Function Is Differentiable

In the space of all SPD matrices modulo scalar multiplication, the objective function has only one local minimum, namely the identity matrix I. If the optimum is assumed in the interior of the search space $\Sigma(a_1, \ldots, a_{n+1})$, then this optimum must be equal to this local optimum. Clearly, this happens only if the given matrix K is already an M-matrix.

In the other case, the optimum is assumed on the boundary. We like to distinguish cases specifying on which face the boundary is assumed. In order to do this in a convenient way, we introduce some terminology.

Let C be a point on the boundary of $\Sigma(a_1, \ldots, a_{n+1})$ (or Σ for short). Then there is a unique face F such that C is contained in the interior of of F, where the interior of a face is defined as the face minus the union of all subfaces. We call this face $F(C)$. The linear subspace carrying $F(C)$ is denoted by $E(C)$. It is defined by equalities $\langle a_i, a_j \rangle_C = 0$ for a set of index pairs (i, j), $i \neq j$; this set is denoted by $\pi(C)$. If we replace each defining equation by its corresponding inequality $\langle a_i, a_j \rangle_C \leq 0$, then we get a convex cone, which we denote by $U(C)$.

Example 2. Assume that C lies on the interior of the maximal face (also called facet) defined by vectors a_1, a_2. Then we have

- $E(C) = \{C \mid \langle a_1, a_2 \rangle_C = 0\}$;
- $F(C) = E(C) \cap \Sigma$;
- $U(C) = \{C \mid \langle a_1, a_2 \rangle_C \leq 0\}$;
- $\pi(C) = \{(1, 2)\}$.

In order to find the optimum in the case where it lies in the interior of a face F, we restrict the objective function to the corresponding linear subspace E and study the local minima.

Assume that E is a hyperplane, defined by $\langle a_i, a_j \rangle_C = 0$, where $\langle a_i, a_j \rangle > 0$. (If $\langle a_i, a_j \rangle \leq 0$, then the unity matrix is the global optimum of the half space.) Let $\overline{a_i}, \overline{a_j}$ be the normalizations of a_i, a_j to unit length. We set $s := \overline{a_i} + \overline{a_j}$ and $d := \overline{a_i} - \overline{a_j}$. For any matrix C in E, we have $\langle s, s \rangle_C = \langle d, d \rangle_C$, and therefore

$$\text{cond}(C) = \frac{\max_{\|u\|=1} \langle u, u \rangle_C}{\min_{\|v\|=1} \langle v, v \rangle_C} \geq \frac{\langle d, d \rangle_C / \|d\|^2}{\langle s, s \rangle_C / \|s\|^2} = \frac{\|s\|^2}{\|d\|^2}.$$

Equality is assumed iff s and d are eigenvectors to the eigenvalues $\frac{\|s\|^2}{\|d\|^2}\lambda$ and λ, respectively, for some scaling factor λ, and all other eigenvalues lie between these two values. Since s and d are orthogonal, this is indeed possible. The set of all matrices satisfying these conditions is denoted by $\Pi(a_i, a_j)$.

Remark 2. In the case $n = 2$, $\Pi(a_i, a_j)$ has exactly one element (up to scaling). This element is easy to compute. Moreover, the hyperplanes are the only possible faces, because any matrix in a lower-dimensional subspace is already singular. Carrying out the computation, one obtains the following formula for the closest M-matrix to $K = \begin{pmatrix} a & b \\ b & c \end{pmatrix}$:

1. If $b > 0$ then $B = \begin{pmatrix} a & 0 \\ 0 & c \end{pmatrix}$.

2. If $a + b < 0$, then $B = \begin{pmatrix} a & -a \\ -a & 2a + 2b + c \end{pmatrix}$.

3. If $b + c < 0$, then $B = \begin{pmatrix} a + 2b + 2c & -c \\ -c & c \end{pmatrix}$.

4. Otherwise, K is already an M-matrix, and $B = K$.

The same result can also be obtained by the general method of Gröbner bases mentioned above.

The following lemma is useful because it allows to reduce other cases to the hyperplane case.

Lemma 2. *Let C_0 be a boundary point of Σ. Assume that the objective function* cond, *restricted to $U(C_0)$, has a local minimum at C_0. Assume that the maximal and the minimal eigenvalue of C_0 are simple. Then C_0 is a global minimum. Moreover, there exist two vectors $a, b \in \mathbb{R}^n$, such that*

(a) $E(C_0)$ is contained in the set $\{C \mid \langle a, b \rangle_C = 0\}$.
(b) $U(C_0)$ is contained in the set $\{C \mid \langle a, b \rangle_C \leq 0\}$.
(c) $\cos(\angle(a, b)) = \frac{\text{cond}(C_0) - 1}{\text{cond}(C_0) + 1}$.

We call the pair of vectors (a, b) the "guard" of C_0.

Proof. Let u and v be the normalized eigenvectors to the eigenvalues λ_{max} and λ_{min} of C_0. The gradient of the cond function at C_0 is equal to (see [5])

$$(\text{grad cond})_{C_0} = \left(\frac{\lambda_{min} u_i u_j - \lambda_{max} v_i v_j}{\lambda_{min}^2} \right)_{i,j}.$$

Since C_0 is a local minimum in $E(C_0)$, the gradient must be orthogonal to $E(C_0)$, or equivalently

$$\lambda_{min} \langle u, u \rangle_C - \lambda_{max} \langle v, v \rangle_C = 0$$

for all $C \in E(C_0)$. With

$$a := \sqrt{\lambda_{min}} u + \sqrt{\lambda_{max}} v, \quad b := -\sqrt{\lambda_{min}} u + \sqrt{\lambda_{max}} v,$$

we obtain (a).

Since C_0 is also a local minimum in $U(C_0)$, the gradient must have nonnegative scalar product with all vectors from C_0 into $U(C_0)$. Equivalently, we can write

$$\lambda_{min} \langle u, u \rangle_C - \lambda_{max} \langle v, v \rangle_C \geq 0$$

for all $C \in U(C_0)$, and (b) follows.

Since u and v are orthogonal, the equation (c) follows by a straightforward computation. Finally, for all $C \in U(C_0)$ we have

$$\text{cond}(C) \geq \frac{\langle u, u \rangle_C}{\langle v, v \rangle_C} \geq \frac{\lambda_{max}}{\lambda_{min}} \tag{13}$$

by the above equation. For C_0, equality holds in (13) and hence the minimum is global.

Let π be a set of pairs of indices, and let F, E and U be the corresponding face, linear subspace, and convex cone. If the optimum lies in the interior of F, then the guard must satisfy the conditions (a) and (b) of Lemma 2. Therefore, we call any pair (a, b) of vectors satisfying (b), (c), and $\langle a, b \rangle > 0$ – this is a consequence of (c) in Lemma 2 – a "possible guard" for π.

The set of all possible guards may be infinite. (We consider two guards as different if they do not arise one from the other by scaling or by switching the vectors. Note that we can scale both factors independently with positive factors, and the whole pair with -1.) For instance, let $\pi := \{(1,2),(1,3)\}$, and assume that $\langle a_1, a_2 \rangle > 0$ and $\langle a_1, a_3 \rangle > 0$. Then $(a_1, \lambda a_2 + \mu a_3)$ is a possible guard for any $\lambda > 0, \mu > 0$.

Let (a, b) a possible guard. Then we can show that $\mathrm{cond}(C) \geq \frac{1 + \cos(\measuredangle(a,b))}{1 - \cos(\measuredangle(a,b))}$ for any $C \in U$, as in the proof for the hyperplane case above. Therefore each possible guard gives a lower bound for the objective function. If (a, b) are guards for the optimum C_0, then this lower bound is assumed for C_0. If (a', b') is another guard with $\measuredangle(a', b') > \measuredangle(a, b)$, then (a', b') gives a lower bound which cannot be achieved inside U, because it is smaller than the condition number of the global optimum. Therefore, we only need to consider the possible guards enclosing the smallest possible angle.

For the above example, we will see later that the set of all possible guards is equal to $\{(a_1, \lambda a_2 + \mu a_3) \mid \lambda \geq 0, \mu \geq 0, \lambda \text{ or } \mu > 0\}$. If the orthogonal projection b of a_1 into the plane spanned by a_2, a_3 is a positive linear combination of a_1, a_2, then (a_1, b) is the positive guard enclosing the smallest possible angle.

Algorithm 2 OptimizeByGuards (n=3)

---

   for all sets π of pairs of indices **do**
      compute the possible guard (a, b) enclosing the smallest possible angle
      if $\Pi(a, b) \cap \Sigma \neq \emptyset$ **then**
         return a matrix in $\Pi(a, b) \cap \Sigma$
         exit
      end if
   end for
   return failure

In the case $n = 3$, we can compute the possible guard enclosing the smallest possible angle by sorting out cases, as shown below. For each guard (a, b), the set $\Pi(a, b)$ of possible optima is a line segment. It can be represented by its two deliminating points, which can be computed easily. Intersecting a line segment with a polyhedron given by linear inequalities is again easy: We only need to evaluate the left hand side at the two deliminating points, check the signs, and compute the linear combination giving a zero value if the two signs are different. Using Algorithm 2, we can compute the global optimum assuming that its largest and smallest eigenvalue are simple, i.e. that the objective function is differentiable at this point.

In the following case by case analysis, we make the following global assumptions: The indices i, j, k, l are pairwise different. We assume that (u, v) is a possible guard. The vectors u, v are expressed in the basis a_i, a_j, a_k: $u = u_1 a_i + u_2 a_j + u_3 a_k$, $v = v_1 a_i + v_2 a_j + v_3 a_k$. We define positive semi-definite matrices $C_1 := a_i a_i^t$, $C_2 := a_j a_j^t$, $C_3 := a_k a_k^t$, $C_4 := (a_i - a_j)(a_i - a_j)^t$, $C_5 := (a_j - a_k)(a_j - a_k)^t$, $C_6 := (a_i - a_k)(a_i - a_k)^t$. Note that if one of these matrices C_r lies in E, (resp. U), then it must also satisfy $\langle u, v \rangle_{C_r} = (\text{resp. } \leq) \, 0$, by condition (a) and (b) of Lemma 2 and by continuity.

Pairs: $\pi = \{(i,j)\}$. Obviously, the only possible guard is (a_i, a_j).

3-chains: $\pi = \{(i,j),(j,k)\}$. Since C_1, C_2, C_3, C_6 are in E, we get

$$u_1 v_1 = u_2 v_2 = u_3 v_3 = (u_1 - u_3)(v_1 - v_3) = 0.$$

Up to scaling and switching, the general solution is

$$u_1 = u_3 = v_2 = 0, u_2 = 1, v_1 = \lambda, v_3 = \mu.$$

Since C_4, C_5 are in U, we get $\lambda \geq 0$, $\mu \geq 0$. The smallest possible angle is enclosed for choosing v to be the projection of $u = a_1$ to the plane spanned by a_2, a_3, if this projection is a positive linear combination, or by choosing $v = a_2$ or a_3 otherwise.

Double pairs: $\pi = \{(i,j),(k,l)\}$. Since C_1, C_2, C_5, C_6 are in E, we get

$$u_1 v_1 = u_2 v_2 = (u_1 - u_3)(v_1 - v_3) = (u_2 - u_3)(v_2 = v_3) = 0.$$

Up to scaling and switching, we get the following three solutions:

$$u = (1,0,0), v = \pm(0,1,0);$$
$$u = (1,0,1), v = \pm(0,1,1);$$
$$u = (1,0,0), v = \pm(1,1,1).$$

Since C_3, C_4 are in U, we can discard the second solution. Therefore, we have precisely two possible pairs, namely (a_i, a_j) and (a_k, a_l).

Triangles: $\pi = \{(i,j),(j,k),(i,k)\}$. Since C_1, C_2, C_3 are in E, we get

$$u_1 v_1 = u_2 v_2 = u_3 v_3 = 0.$$

All solutions have already appeared as possible pairs of a 3-chain. Therefore, the set of possible pairs is equal to the union of the sets of possible pairs of the three 3-chains contained in π.

4-chains: $\pi = \{(i,j),(j,k),(k,l)\}$. Since C_1, C_2, C_6 are in E, we get

$$u_1 v_1 = u_2 v_2 = (u_1 - u_3)(v_1 - v_3) = 0.$$

The are three parameterized solutions:

$$u = (1,0,1), v = (0,\lambda,\mu);$$
$$u = (0,1,0), v = (\lambda,0,\mu);$$
$$u = (0,0,1), v = (\lambda,\mu,\lambda).$$

The second and the third have already appeared as possible guards for the two 3-chains contained in π. The first is new. Since C_3, C_4, C_6 are in U, we get the additional restrictions $\lambda \leq 0$, $\mu \geq 0$. The smallest possible angle is enclosed for choosing v to be the projection of $u = a_1 + a_2 = -a_3 - a_4$ to the plane spanned by a_2, a_3, if this projection is a linear combination with coefficients $\lambda \leq 0$, and $\mu \geq 0$.

Other: It is not possible to have an a_i which is orthogonal to all a_j, $j \neq i$ – these other a_j form a basis of $\mathbb{R}^3$ and a_i, being different from zero, cannot be orthogonal to the whole $\mathbb{R}^3$. We also cannot have a 4-cycle, because this leads to the contradiction

$$\langle a_i + a_k, a_i + a_k \rangle_C = -\langle a_i + a_k, a_j + a_l \rangle_C = 0.$$

Therefore there are no other cases.

Remark 3. For $n \geq 4$, a similar case by case analysis is possible, and we can determine the possible guards enclosing a minimal angle. It is less obvious to determine whether $\Pi(u, v) \cap \Sigma \neq \emptyset$ in this case. But if $\Pi(u, v)$ has a non-empty intersection with the border of Σ, then there is a superset of index pairs such that an optimum can be found on the corresponding smaller linear subspace. Thus, it suffices to do this check assuming that the $\Pi(u, v)$ has empty intersection with the border, and this is easy: Take a single element and check whether it is contained in Σ.

For $n = 3$, it is easier to do the full completeness check, because in this way we can omit testing possible pairs that are also possible pairs of a subset of pairs of indices.

3.2   Cases where the Objective Function Is Not Differentiable

If there is no optimum C with largest and smallest eigenvalue being simple, then Lemma 2 does not help in finding the optimum. Our strategy is to restrict the objective function so that it becomes differentiable again. We could give a complete symbolic solution in the case $n = 3$.

Let D be the set of all SPD 3×3 matrices with at most two different eigenvalues. Naive dimension counting would suggest that $\dim(D) = 5$, because we have one algebraic condition, namely the discriminant of the characteristic polynomial has to vanish. But we have two degrees of freedom to choose the eigenvalues, and two degrees of freedom to choose the eigenvector for the simple eigenvalue – the two-dimensional eigenspace for the double eigenvalue is uniquely determined as the plane orthogonal to the simple eigenvector. Thus, we get $\dim(D) = 4$.

Let C be a matrix in D. Let λ be the unique eigenvalue which is at least double. Then $C - \lambda I$ has rank at most 1. Therefore we can write C as $\lambda I \pm xx^t$ with $x \in \mathbb{R}^3$. This representation is unique and gives a parameterization of D by 4 parameters. As we check local minima on linear subspaces, we get additional restrictions on the parameters, which are linear in λ and quadratic in the three coordinates of x. The objective function is given by $\frac{\lambda + x^t x}{\lambda}$ or $\frac{\lambda}{\lambda - x^t x}$, depending on the sign of the rank 1 summand.

By Lemma 2, we only need to compute local minima in the cases where there is no local minimum on U with distinct eigenvalues (in other words, if U has a local minimum that can be found by guards then we can cross out this case even if this local minimum does not lie in Σ). Especially, we do not need to consider pairs and 3-chains. It remains to compute the local minima in the cases of double pairs, triangles, and 4-chains. Here is a case by case analysis.

Double pairs: The search space is given by two equations in λ, x_1, x_2, x_3. The linear variable λ can be eliminated, so that the search space is actually a conic in the projective plane (factoring out scalar multiplication as usual). Using well-known algorithms for curve parameterization, see e.g. [10], we can parameterize the conic, thereby reducing the search space to a projective line. The objective function transforms to a rational function of degree 4. There are 6 (maybe complex) stationary points, which are candidates for being local minima.

Triangles and 4-chains: The three equations in λ, x_1, x_2, x_3 can be solved symbolically by standard methods. There are four solutions, all of them real; three of them are singular because the λ-component vanishes. So, the search space restricts to a single point.

4   Comparison

In this section the efficiency and robustness of the symbolic approach to the element preconditioning technique is shown, by comparing with the numerical approach used in [5]. All numerical studies are done on an SGI Octane 300MHz.

First of all we calculate the best possible M-matrices of Example 1 and compare it to the numerical solution. For these cases the matrices which depend on a single parameter $0 < q < 1$ can be calculated once in a Maple implementation, and this leads to the following results:

Example 3. Now we give the results of Example 1. Only the transformed matrix is presented.

1. First of all the anisotropic rectangle with $\sigma = 0$ is considered. The best possible M-matrix is given by

$$\tilde{B} = \begin{pmatrix} 1 + 4q^2 & 0 & -1 - q^2 \\ 0 & 1 + 4q^2 & -q^2 \\ -1 - q^2 & -q^2 & 1 + 4q^2 \end{pmatrix}$$

with a condition number of $\kappa = 3/(1 + 4q^2)$ for $0 < q < \sqrt{1/2}$.

2. Next the anisotropic triangle with $\sigma = 0$ was already explicitly solved in Remark 2 and leads to the matrix

$$\tilde{B} = \begin{pmatrix} 1 + q^2 & 0 \\ 0 & 1 + q^2 \end{pmatrix}$$

with condition number $\kappa = 1/q^2$.

3. Finally the anisotropic triangle with $\sigma = 1$ yields

$$B = \begin{pmatrix} q^2 + 2 & 0 & -2 \\ 0 & q^2 + 2 & -2 \\ -2 & -2 & q^2 + 4 \end{pmatrix}$$

and the condition number behaves like $\kappa = \frac{q^2 + 2}{3q^2}$.

For such types of elements where the element stiffness matrix depends only on one parameter the solution can be done in advanced. However, for more general matrices the number of parameters is too large to be efficiently solved with the Maple implementation. Therefore we make a C++ implementation in order to get a fast solution of the optimization problem. Note that no initial guess is necessary for the calculation and we always get the optimal solution. This is in contrast to a numerical approach where a different initial guess might lead to a different result, see Table 1.

Table 1. Different initial guess for the numerical case.

q^2	$\kappa((B_h^r)^{-1/2} K_h^r (B_h^r)^{-1/2})$	$\kappa((B_h^r)^{-1/2} K_h^r (B_h^r)^{-1/2})$
1.0	1.00	1.00
0.9	2.35	1.23
0.5	6.50	4.00
0.1	150.5	100

Finally we compare the CPU-time of the numerical approach and the symbolic approach. These studies are done on the triangle with $\sigma = 1$ and on the rectangle with $\sigma = 0$ (see Example 1). In Table 2 the results are presented. The C++ implementation is much faster than the numerical approach (by the factor 10) and has the advantage that no initial guess has to be used. We emphasize that in general the optimization problem must be solved for each finite element separately. In such cases, our symbolic approach can obviously save a lot of computer time on fine meshes with several hundred thousands or millions of elements.

Table 2. Comparison of numerical and symbolic approaches.

name	numerical (sec)	symbolic (sec)
triangle	0.0052	0.00047
rectangle	0.0060	0.00048

References

[1] B. Buchberger, *An algorithm for finding a basis for the residue class ring of a zero-dimensional polynomial ideal*, Ph.D. thesis, Universitat Innsbruck, Institut fur Mathematik, 1965, German.

[2] ———, *Gröbner bases: An algorithmic method in polynomial ideal theory*, Recent Trends in Multidimensional Systems Theory (N. K. Bose, ed.), D. Riedel Publ. Comp., 1985.

[3] G. E. Collins, *Quantifier elimination for the elementary theory of real closed fields by cylindrical algebraic decomposition*, Lecture Notes In Computer Science, Springer, 1975, Vol. 33, pp. 134–183.

[4] G. E. Collins and H. Hong, *Partial cylindrical algebraic decomposition for quantifier elimination*, J. Symb. Comp. **12** (1991), no. 3, 299–328.

[5] G. Haase, U. Langer, S. Reitzinger, and J. Schöberl, *Algebraic multigrid methods based on element preconditioning*, International Journal of Computer Mathematics **78** (2001), no. 4, 575–598.

[6] W. Hackbusch, *Multigrid methods and application*, Springer Verlag, Berlin, Heidelberg, New York, 1985.

[7] H. Hong, *Improvements in cad–based quantifier elimination*, Ph.D. thesis, The Ohio State University, 1990.

[8] M. Jung and U. Langer, *Applications of multilevel methods to practical problems*, Surveys Math. Indust. **1** (1991), 217–257.

[9] M. Jung, U. Langer, A. Meyer, W. Queck, and M. Schneider, *Multigrid preconditioners and their application*, Proceedings of the 3rd GDR Multigrid Seminar held at Biesenthal, Karl-Weierstraß-Institut für Mathematik, May 1989, pp. 11–52.

[10] F. Winkler, *Polynomial algorithms in computer algebra*, Springer, 1996.

Solving Symbolic and Numerical Problems in the Theory of Shells with *MATHEMATICA*®

Ryszard A. Walentyński

The Department of Building Structures Theory, The Faculty of Civil Engineering,
The Silesian University of Technology, ul. Akademicka 5, PL44-101 Gliwice, Poland,
rwal@kateko.bud.polsl.gliwice.pl

1 Introduction

The theory of shells describes the behaviour (displacement, deformation and stress analysis) of thin bodies (thin walled structures) defined in the neighbourhood of a curved surface in the 3D space. Most of contemporary theories of shells use differential geometry as a mathematical tools and tensor analysis for notations. Examples are [1, 2, 3].

Tensor notation has a lot of advantages, first of all it makes it possible to denote general problems with elegant and short expressions. Therefore it is very suitable for lecturing purposes (from the point of view of the teacher) and writing scientific essays.

However, it seems to be to be very abstract for most of engineers and boring for students. Moreover, it is prone to error especially when summations are done by hand. Therefore most of the publications considering tensor problems are at least suspected to contain errors in the indices. Thus, all of them need careful scrutiny. Despite of verification the error can be repeated. Some theories presumed some a'priori simplification of the expressions, especially in constitutive relations. They make final results shorter but valid under certain circumstances.

The paper will try to show how these problems can be avoided using computer assisted symbolic calculations. ***MathTensor***™ [1], an external package of the computer algebra system *MATHEMATICA*® [2], can be used to solve symbolic tensor problems. The package was written by Parker and Christensen [4].

First the paper presents selections from the *MATHEMATICA*® notebooks[3] illustrating solution of some problems of the theory of shells. The aim is to show that it is possible to proceed from very general equations to ones ready for numerical computations and then solve them within one system. Each step requires specific approach. The translation of the task into the language of *MATHEMATICA*® and ***MathTensor***™ and the information which can be received will be discussed.

The obtained symbolic differential equations that should be approximated numerically are very complex. Moreover due to the essential difference in the shear and bending stiffness of shells the task is usually numerically ill-conditioned. In a lot of cases it is a boundary layer problem unstable in the Lyapunov sense. Therefore numerical approaches like Finite Differences or Finite Element methods may fail in some cases.

---

[1] ***MathTensor***™ is a registered trademark of MathTensor, Inc.

[2] *MATHEMATICA*® is a registered trademark of Wolfram Research, Inc.

[3] Full notebooks can be requested by email.

F. Winkler and U. Langer (Eds.): SNSC 2001, LNCS 2630, pp. 309–346, 2003.
© Springer-Verlag Berlin Heidelberg 2003

There are several engineering two step approaches, represented for example by Bielak [2], which are based on the assumption that the membrane state dominates in shells in most of the domain except boundary layers. It leads to the reduction of the order of differential operator of the problem. It results in better stability but also a loss of generality. Membrane state cannot satisfy all boundary conditions. They are satisfied locally using so called bending theory in the limited boundary layer in the second step. The approach fails in cases when bending of shells cannot be neglected. These problems were investigated in [5].

The Refined Least Squares method (RLS) has been applied to this problem and it has been found a new approach to the solution of the tasks of long cylindrical shells. It has been discovered experimentally that the process can be divided into two steps somehow similar to the [2] approach.

The first step of this approach consists in computing the base solution (approximation of the 10th order differential operator) satisfying only the 4 essential boundary conditions. This solution is feasible for most of the problem domain except the boundary layer. Other conditions are neglected. They are adjusted in the second step where only the boundary layer is considered.

This approach results in better convergence and stability. First of all it is not connected with reduction of the order of the differential operator. Therefore moments and transverse forces are not neglected what makes the method much more general than ones based on the membrane state.

It is hoped that the ideas presented will be impressive, particularly for researchers and engineers who deals with other problems of the Mechanics of Continuum and Mathematical Physics which involve tensor analysis and boundary layer tasks. First of all I tried to present how tools of the computer algebra can be applied effectively and what can be done additionally to moderate simplification process and exploit its symbolic possibilities more deeply, effectively and how to obtain reliable results. The least squares method functional has been appended with terms responsible for boundary conditions. It resulted first of all with simpler algorithm, which has been implemented in MATHEMATICA®. The very important matter is the discussed numerical occurrence that the boundary value problem with 10th order differential operator can be approximated with consideration of only 4 boundary conditions. It can be applied for other problems with a boundary layer phenomenon, which are present for many engineering and physical tasks.

The considered symbolic problems are based on the theory of shells by Bielak [2]. However, discussed ideas can be applied for another theory of shells. It is assumed that the reader is familiar with the notations rules used in the MATHEMATICA® and **MathTensor**™ language. Only the crucial aspects are discussed and some input and output information has been intentionally omitted. The system output and formulas derived with MATHEMATICA® has been prepared for publication. Despite editorial scrutiny it seems to be more sensible and effortless to use the presented commands and ideas to receive own output than apply directly the final formula.

Notations are explained in the text, some of them: for tensor problems are collected in the end.

2 Symbolic Problems – Implementations of *MathTensor*™

Presented are some selections from three *MATHEMATICA* sessions[4]. The crucial points of the process of the derivation of differential equations are discussed. Starting with the very general relations in tensor notation form, the final equations in terms of displacements are obtained, which are ready for numerical computations.

2.1 Derivation of Differential Equations from Tensor Equations of Equilibrium

Three steps in the process can be distinguished, each of them is different.

Step 1 – Changing Covariant Derivatives in Equations to (Ordinary) Partial Ones.
The first equation of shell equilibrium takes the following form:

$$N^{ij}{}_{;i} - b_i{}^j\, Q^i + P^j = 0. \tag{1}$$

The left hand side of the equation can be denoted with the following *MathTensor*™ definition.

```
In[1]:= CD[n[ui,uj],li] - q[ui] b[li,uj] + p[uj]
Out[1]= N^{ij}{}_{;i} + P^j - (b_i{}^j) (Q^i)
```

MathTensor™ was written for *MATHEMATICA* 2. Therefore all symbols in it are denoted with Latin letters. To receive Greek letters the output form have to redefined. For example the symbols of the affine connection **AffineG** have got the standard output form G. This redefines it to Γ.

```
In[2]:= Format[AffineG[a__]] := PrettyForm[Γ, a]
```

The same can be done for Kronecker delta **Kdelta** to be expressed with δ or even indices to be represented with small Greek letters.

That command changes the covariant derivatives to ordinary partial ones. Dummy indices are automatically recognised and denoted with $p, q, r \ldots$ letters.

```
In[3]:= CDtoOD[%]
Out[3]= (Γ^j{}_{pq}) (N^{qp}) + (Γ^p{}_{pq}) (N^{qj}) + N^{pj}{}_{,p} + P^j - (b_p{}^j) (Q^p)
```

The space is metric so the affine connection can be expressed in terms of a metric tensor. It is done with the **AffineToMetric** function. *MathTensor*™ function **Tsimplify** will simplify the expression taking into account all symmetries of the tensors involved.

```
In[4]:= Tsimplify[AffineToMetric[%]]
```
$$Out[4]= \tfrac{1}{2}\,(g^{pj})\,(N^{qr})\,(g_{pq,r}) + \tfrac{1}{2}\,(g^{pq})\,(N^{rj})\,(g_{pq,r}) +$$
$$\tfrac{1}{2}\,(g^{pj})\,(N^{qr})\,(g_{pr,q}) - \tfrac{1}{2}\,(g^{pj})\,(N^{qr})\,(g_{qr,p}) +$$
$$N^{pj}{}_{,p} + P^j - (b_p{}^j)\,(Q^p)$$

---

[4] The section collects and extends topics which were presented in [8], [9], [10] and [13]

Both *MATHEMATICA*ᵃ and ***MathTensor***™ have advanced algebraic simplification tools, but sometimes the use of automatic function does not result in the simplest form. Using more complex instruments better results can be obtained. The set of commands below carry out the following operations. It is easy to find that the expression above contain 3 terms with N^{qr}. Therefore using the **Collect** terms can be collected in the expression with respect to it and then simplify it term by term with **Simplify**. It is done with the so called mapping **Map** function in the infix form **/@**. The result is saved for further use.

In[5]:= **r1 = Simplify/@Collect[%, n[u2, u3]]**

It results in the following formula, which contains only "ordinary" partial derivatives.

$$N^{pj}{}_{,p} + \frac{1}{2} g^{pq} g_{pq,r} N^{rj} + \frac{1}{2} g^{pj} \left(g_{pq,r} + g_{pr,q} - g_{qr,p} \right) N^{qr} - b_p{}^j Q^p + P^j = 0. \quad (2)$$

Step 2 – Summations. The next step consists in the summation of the expression. The problem of the shell is two dimensional and the system should be informed about it before starting the summation process.

In[6]:= **Dimension = 2;**

The first equation of equilibrium in tensor notation produces two partial differential equations. The metric tensor on the reference surface is equal to the first differential form $g_{ij} = a_{ij}$.

In[7]:= **R1[uj_] = MakeSum[r1]**

Presentation of the result which contains 21 terms is omitted here.

Step 3 – Transformation to *MATHEMATICA*ᵃ Differential Equations. In the end the equation can be transformed into an "ordinary" partial differential equation. First tensor the components have to be represented as the "normal" functions of two variables.

In[8]:= **n[1, 1] := n11[x, y];**
       n[1, 2] := n12[x, y];
       n[2, 1] := n21[x, y];
       n[2, 2] := n22[x, y]; (*and so on*)

Next ordinary differentiation has to be turned on.

In[9]:= **On[EvaluateODFlag]**

The first differential equation takes the form:

In[10]:= **eqn[1] = R1[1]**

$Out[10]=$ p1 [x, y] – bm11 [x, y] q1 [x, y] – +

      bm12 [x, y] q2 [x, y] + n21$^{(0,1)}$ [x, y] +

      $\dfrac{1}{2}$ n22 [x, y] (ag12 [x, y] a22$^{(0,1)}$ [x, y] + ag11 [x, y]

          (2 a12$^{(0,1)}$ [x, y] – a22$^{(1,0)}$ [x, y])) +

      $\dfrac{1}{2}$ n12 [x, y] (ag11 [x, y] a11$^{(0,1)}$ [x, y] +

          ag12 [x, y] a22$^{(1,0)}$ [x, y]) +

      $\dfrac{1}{2}$ n11 [x, y] (2 ag11 [x, y] a11$^{(1,0)}$ [x, y] –

          ag12 [x, y] (a11$^{(0,1)}$ [x, y] – 4 a12$^{(1,0)}$ [x, y]) +

          ag22 [x, y] a22$^{(1,0)}$ [x, y]) +

      $\dfrac{1}{2}$ n21 [x, y] (2 ag11 [x, y] a11$^{(0,1)}$ [x, y] +

          ag22 [x, y] a22$^{(0,1)}$ [x, y] + ag12 [x, y]

          (2 a12$^{(0,1)}$ [x, y] + a22$^{(1,0)}$ [x, y])) + n11$^{(1,0)}$ [x, y]

We will return to this equation later, in point 2.4.

2.2 Nonlinear Equations of the Second Order Theory

The approach can be easily extended to nonlinear problems, this is a simple example.
The third equation of shell equilibrium in tensor notation is:

$$M^{ij}_{;i} - Q^j = 0. \tag{3}$$

In the second order theory the additional moments caused by the stretching (tensile) forces acting on the normal displacements are taken into consideration.

$$M^{ij} \rightarrow M^{ij} + N^{ij} w^3. \tag{4}$$

It is implemented with the following function:

$In[11]:=$ **m[ui_, uj_] := m1 [ui, uj] + n[ui, uj] w3**

The moment tensor now has the following form:

$In[12]:=$ **m[ui, uj]**
$Out[12]=$ M^{ij} + w^3 (N^{ij})

The left hand side (lhs) of the third equation is:

$In[13]:=$ **CD[m[ui, uj], li] – q[uj]**
$Out[13]=$ M$^{ij}_{;i}$ + w^3 (N$^{ij}_{;i}$) + (w$^3_{;i}$) (N^{ij}) – Q^j

Using a similar approach as that presented in the previous section the following result is obtained:

$$M^{pj}_{,p} - \frac{1}{2} M^{pq} g^{rj} \left(g_{pq,r} - g_{pr,q} - g_{qr,p} \right) + \frac{1}{2} M^{pj} g^{qr} g_{qr,p} - Q^j +$$
$$+ N^{pj} w^3_{,p} + \frac{1}{2} w^3 \left(g^{pq} N^{rj} g_{pq,r} + g^{pj} N^{qr} \left(g_{pq,r} + g_{pr,q} - g_{qr,p} \right) + N^{pj}_{,p} \right) = 0. \tag{5}$$

Nonlinear terms are written in the second line.

2.3 Receiving and Simplification of the Constitutive Relations

One of the task of the theory of shells is to reduce the three dimensional problem of the theory of elasticity into a two dimensional one[5]. The analysis of stresses is reduced to the analysis of internal forces: stretching (tensile) forces and moments. The respective tensors are computed from the following integrals:

$$N^{ij} = \int_{-h}^{h} \sqrt{\frac{g}{a}} \left(\delta_r{}^j - z\, b_r{}^j \right) \tau^{ri}\, dz, \tag{6}$$

$$M^{ij} = \int_{-h}^{h} \sqrt{\frac{g}{a}} \left(\delta_r{}^j - z\, b_r{}^j \right) \tau^{ri}\, z\, dz. \tag{7}$$

The square root in these formulas is the following function:

$$Z = \sqrt{\frac{g}{a}} = 1 - 2\,H\,z + K\,z^2. \tag{8}$$

$In[1] := \mathbf{Z := 1 - 2\,H\,z + K\,z^2;}$

The stress tensor for isotropic material can be derived from the formula:

$$\tau^{ij} = E \left(\frac{\nu}{1-\nu^2}\, g^{ij}\, g^{pq} + \frac{1}{1+\nu}\, g^{ip}\, g^{jq} \right) \overset{*}{\gamma}_{pq}. \tag{9}$$

$In[2] := \mathbf{tau[ui\_, uj\_] := \left(\dfrac{\nu\, e}{1-\nu^2}\, Metricg[ui, uj]\, Metricg[u1, u2] +}$

$\dfrac{e}{1+\nu}\, \mathbf{Metricg[ui, u1]\, Metricg[uj, u2] \right)\, gammastar[11, 12]}$

A strain tensor in 3D space can be expressed in terms of the strain tensor of the reference surface.

$$\overset{*}{\gamma}_{ij} = \gamma_{ij} - 2\,z\,\rho_{ij} + z^2\,\vartheta_{ij}. \tag{10}$$

$In[3] := \mathbf{gammastar[li\_, lj\_] :=}$

$\mathbf{gamma[li, lj] - 2\,z\,rho[li, lj] + z^2\, theta[li, lj]}$

Contravariant components of the metric tensor 3-D shell can be computed from the formula which is a function of the variable z measured along the normal to the reference surface.

$$g^{ij} = \frac{a^{ij}\left(1 - K\,z^2\right) - 2\left(2\,H\,a^{ij} - b^{ij}\right)\left(1 - H\,z\right)z}{Z^2}. \tag{11}$$

$In[4] := \mathbf{Metricg[ui\_, uj\_] :=}$

$\dfrac{\mathbf{a[ui, uj]\left(1 - K\,z^2\right) - 2\,(2\,H\,a[ui, uj] - b[ui, uj])\,(1 - H\,z)\,z}}{\mathbf{z^2}}$

The integrals (6) and (7) appear to be simple but after substituting into them functions (8), (9), (10) and (11), they become a bit more complicated but using possibilities of the

---

[5] The consideration is made in a new *MATHEMATICA* session

system they can be computed with arbitrary precision with respect to the thickness $2h$ expansion. It can be done with the function:

```
In[5]:= n[ui_,uj_][k_] :=
           Simplify/@Tsimplify[
             AbsorbKdelta[
               integ/@Expand[
                 Normal[
                   Series[
                     Z (Kdelta[13,uj] - z b[13,uj])
                       tau[ui,u3],{z,0,k}]]]]]
```

This complex multi-function does the following: the considered integrand –
(**Kdelta[13,uj]** - **z b[13,uj]**) **tau[ui,u3]**
is expanded into a power series with **Series** and **Normal**. The result is algebraically expanded with **Expand** and then integrated term by term (using the mapping procedure **/@**) with a predefined function of integration **integ**:

$$In[6]:= \textbf{integ[x\_]} := \int_{-h}^{h} \textbf{xdz}$$

This approach is necessary as *MATHEMATICA* is a program and only a program. If the argument of the function is complicated it takes a lot of time to deal with it. Therefore it is usually sensible to divide the task into a set of simpler problems. This approach speeds up computations. Mapping is a very useful tool in this process.

After integration the Kronecker delta is absorbed with **AbsorbKdelta** and simplified with the already mentioned functions **Tsimplify** and **Simplify**.

The parameter **k** defines the precision of the computation, for engineering purposes it is sufficient for the calculations to be done with a precision to the third power of the shell thickness. Then the parameter **k** in the function should take value 2.

```
In[7]:= noriginal[ui_,uj_] = n[ui,uj][2]
```

The result can be used directly in further computations but is rather long. It contains 24 terms, so its presentation is omitted here. Nevertheless using a set of the *MATHEMATICA* and *MathTensor*™ tools it can be presented in a shorter form. Among them we can find: **Dum[%]** for finding pairs of dummy indices, **Expand[%]** for the expression expansion, **Absorb[%,a]** and **AbsorbKdelta** for lowering and raising indices, **Canonicalize** for finding the canonical form of tensor expression and the already mentioned **Tsimplify** and **Simplify** for tensor and algebraic simplification.

Automatic simplification does not necessarily give the simplest form. The system can be helped. An example of enforcing the required behaviour is given below. The terms are grouped and each group is simplified.

```
In[8]:= Simplify[%[[1]] + %[[6]]] +
          Simplify[%[[2]] + %[[4]]] +
          Simplify[%[[3]] + %[[5]]]
```

Replacement is a very useful tool for controlling the simplification process. The well known identities:

$$b_{pq} b_{jp} \equiv 2 H b_q{}^j - K a_q{}^j,$$
$$b_p{}^i b_{jp} \equiv 2 H b^{ij} - K a^{ij}, \tag{12}$$

$$\gamma^{pj} \equiv \gamma^{pq} \delta_q{}^j \tag{13}$$

are applied with the following functions, respectively:

```
In[9]:= %/.{b[11, 12] b[uj, u1] →
          2 H b[12, uj] - K a[12, uj],
        b[11, ui] b[uj, u1] →
          2 H b[ui, uj] - K a[ui, uj]}
```

```
In[10]:= %/.gamma[u1, uj] →
          gamma[u1, u2] Kdelta[12, uj]
```

MATHEMATICA® simplifications tools do not always find the simplest form. Replacement can be very useful in such cases, for example:

```
In[11]:= %/.aa_ v + bb_ v → (aa + bb) v
```

```
In[12]:= %/.aa_ b[11, 12] + bb_ b[11, 12] →
          (aa + bb) b[11, 12]
```

```
In[13]:= %//.{  aa_
               ------- → -  aa
               -1 + v²      1 - v²  ,
        aa_ (-1 + v) → -aa (1 - v),
        aa_ (- 1 + h² K) → -aa (1 - h² K)}
```

At the end the following result is obtained:

$$
\begin{aligned}
N^{ij} = {} & \frac{2 E h \left(1 - h^2 K\right) \left(v\, a^{ij}\, \gamma_p{}^p + (1 - v)\, \gamma^{ij}\right)}{1 - v^2} + \\[4pt]
& + \frac{4 E h^3\, \gamma^{pq} \left(v\, b_{pq} \left(H\, a^{ij} + b^{ij}\right) + (1 - v)\, b_p{}^i \left(b_q{}^j + H\, \delta_q{}^j\right)\right)}{3 \left(1 - v^2\right)} + \\[4pt]
& - \frac{8 E h^3 \left(v\, a^{ij}\, b_{pq}\, \rho^{pq} + (1 - v)\, b_p{}^i\, \rho^{pj}\right)}{3 \left(1 - v^2\right)} + \\[4pt]
& + \frac{4 E h^3 \left(2 H \delta_q{}^j - b_q{}^j\right) \left(v\, a^{qi}\, \rho_p{}^p + (1 - v)\, \rho^{qi}\right)}{3 \left(1 - v^2\right)} + \\[4pt]
& + \frac{2 E h^3 \left(v\, a^{ij}\, \vartheta_p{}^p + (1 - v)\, \vartheta^{ij}\right)}{3 \left(1 - v^2\right)} + O\!\left(h^5\right).
\end{aligned}
\tag{14}
$$

This is a formula for the tensor of moments. It is received by a similar procedure.

$$M^{ij} = \frac{4\,E\,h^3\,\left(v\,a^{ij}\,b_{pq}\,\gamma^{pq} + (1-v)\,b_p{}^i\,\gamma^{pj}\right)}{3\,(1-v^2)} +$$
$$-\frac{2\,E\,h^3\,\left(2\,H\,\delta_q{}^j - b_q{}^j\right)\left(v\,a^{qi}\,\gamma_p{}^p + (1-v)\,\gamma^{qi}\right)}{3\,(1-v^2)} + \tag{15}$$
$$-\frac{4\,E\,h^3\,\left(v\,a^{ij}\,\rho_p{}^p + (1-v)\,\rho^{ij}\right)}{3\,(1-v^2)} + O\left(h^5\right).$$

The next step is to check if simplified results obtained satisfy the last equation of equilibrium, which has the following form and should be satisfied as an identity.

$$\varepsilon_{pq}\left(N^{pq} - b_r{}^p\,M^{qr}\right) = 0. \tag{16}$$

The check is carried out by the following function, which shows that it is satisfied. Here an totally antisymmetric object ε_{pq} is denoted with. **EpsDown[** $l1$ **,** $l2$ **]**. The function contain a lot of simplification tools like tensor simplification **Tsimplify**, canonicalization **Canonicalize**, absorbtion **Absorb** and **AbsorbKdelta** and expansion **Expand**. Moreover an identity:

$$b_{pq}\,b_r{}^p \equiv 2\,H\,b_{qr} - K\,a_{qr} \tag{17}$$

is applied with **b[** $l1$ **,** $l2$ **] b[** $l3$ **, u1]** $\rightarrow$ **2 H b[** $l2$ **,** $l3$ **] - K a[** $l2$ **,** $l3$ **], a**.

```
In[14]:= Tsimplify[
            Absorb[
              Canonicalize[
                Absorb[
                  AbsorbKdelta[
                    Expand[
                      EpsDown[l1, l2]
                        (nfinal[u1, u2]-
                            b[l3, u1] mfinal[u3, u2])]
                    ], a
                  ]
                ]/.b[l1, l2] b[l3, u1] →
                    2 H b[l2, l3] - K a[l2, l3], a
              ]
            ] == 0
Out[14]= True
```

2.4 Description of an Arbitrary Shell with *MathTensor*™

The problems presented in the previous two sections are general consideration. The next step is to carry out the more detailed calculations for a concrete shell, which is done in a new *MATHEMATICA* session.

Geometrical Description of the Reference Surface. A surface in 3D space can be parameterised with two variables x^i. This is an example of parameterisation of the catenoide shown in figure 1.

$$In[1]:= \mathbf{r} := \left\{ \mathbf{Cos[x[2]]} \sqrt{\mathbf{s_o}^2 + \mathbf{x[1]}^2}, \right.$$
$$\left. \mathbf{Sin[x[2]]} \sqrt{\mathbf{s_o}^2 + \mathbf{x[1]}^2}, \mathbf{s_o} \mathbf{ArcSinh}\left[\frac{\mathbf{x[1]}}{\mathbf{s_o}}\right] \right\};$$

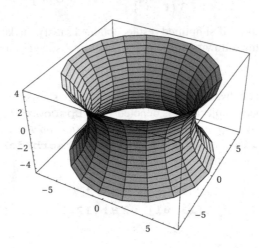

Fig. 1. Catenoide

The components of the covariant curvilinear basis are computed as a derivative of the vector $\mathbf{r}$ with respect to the parameter x^i.

$$\mathbf{r}_i = \partial_{x^i} \mathbf{r}. \tag{18}$$

The condition **NegIntegerQ[i]** restricts the definition to covariant components.

$$In[2]:= \mathbf{ri[i\_]}/; \mathbf{NegIntegerQ[i]} := \mathbf{ri[i]} = \partial_{\mathbf{x[-i]}}\mathbf{r}$$

Geometrical Properties. The first differential form of the reference surface is defined by the following scalar product:

$$a_{ij} = \mathbf{r}_i \cdot \mathbf{r}_j. \tag{19}$$

$$In[3]:= \mathbf{a[i\_,j\_]}/; \mathbf{NegIntegerQ[i]\&\&NegIntegerQ[j]} :=$$
$$\mathbf{a[i,j]} = \mathbf{Simplify[ri[i].ri[j]]}$$

It is sensible to collect these coefficients into the matrix.

$$In[4]:= \mathbf{aLowerMatrix} := \mathbf{aLowerMatrix} =$$
$$\mathbf{Table[a[-i,-j], \{i,2\}, \{j,2\}]}$$

This is a definition of the determinant *a* of this matrix.

In[5]:= **Deta := Deta =**
          Simplify[Det[Table[a[-i,-j],{i,2},{j,2}]]]

The third vector of the curvilinear basis can now be computed, it is normal to the mid-surface. It is obtained from the formula:

$$r_3 = \frac{r_1 \times r_2}{\sqrt{a}}. \tag{20}$$

In[6]:= **ri[-3] := ri[-3] = Simplify[$\dfrac{\text{ri[-1] x ri[-2]}}{\text{PowerExpand}\left[\sqrt{\text{Deta}}\right]}$]**

The normal vector is necessary to calculate the second and the third differential form. In this example this computations are omitted because the process is very similar to the first differential form derivation.

Tensor a_{ij} is a metric tensor on the reference surface so its contravariant components can be computed from the inversion of **aLowerMatrix**.

In[7]:= **aUpperMatrix := aUpperMatrix =**
          Simplify[Inverse[aLowerMatrix]]

Contravariant coefficients a^{ij} are elements of this matrix and can be computed from the following definition. Here the condition **PosIntegerQ[i]** restricts the definition to the contravariant components.

In[8]:= **a[i_, j_]/; PosIntegerQ[i]&&PosIntegerQ[j] :=**
          a[i,j] = aUpperMatrix[[i,j]]

Kinematical Relations. The displacement and rotation vectors can be decomposed in the covariant basis:

$$w = w^k r_k + w^3 r_3, \tag{21}$$

$$d = d^k r_k + d^3 r_3. \tag{22}$$

The last term in the rotation vector is negligible for linear problems. Attention is focused by limiting further consideration to geometrically linear theory.

In[9]:= **w := w = ww[1] ri[-1] + ww[2] ri[-2] + ww[3] ri[-3]**

In[10]:= **d := d = dd[1] ri[-1] + dd[2] ri[-2] (* + dd[3] [ri[-3]] *)**

It has to be emphasised that **MakeSum[]** cannot be used in those definitions because it results in an error. It is probably not a bug in the package but is caused by further definitions which express displacements with their physical components. Nevertheless it is a good example of the need to be critical of the results obtained with computer assistance, they need careful scrutiny.

Derivatives of the displacement and rotation vectors are objects of valence one.

$$w_i = \partial_{x^i} w, \tag{23}$$

$$d_i = \partial_{x^i} d. \tag{24}$$

$In[11]:= \text{wi}[i\_]/; \text{NegIntegerQ}[i] := \text{wi}[i] = \text{Simplify}[\partial_{x[-i]}w]$

$In[12]:= \text{di}[i\_]/; \text{NegIntegerQ}[i] := \text{di}[i] = \text{Simplify}[\partial_{x[-i]}d]$

The physical components of the displacement and rotation components can be computed from the following formulae:

$$w_i = w^i \sqrt{a_{ii}}, \tag{25}$$

$$d_i = d^i \sqrt{a_{ii}}. \tag{26}$$

Thus, the following definition is made:

$In[13]:= \text{ww}[i\_] := \text{ww}[i] = \dfrac{w_i[x[1],x[2]]}{\text{PowerExpand}\left[\sqrt{a[-i,-i]}\right]}$

$In[14]:= \text{dd}[i\_] := \text{dd}[i] = \dfrac{d_i[x[1],x[2]]}{\text{PowerExpand}\left[\sqrt{a[-i,-i]}\right]}$

Strains. The generalised formula for the first strain tensor (containing terms responsible for temperature distortions) is:

$$\gamma_{ij} = \frac{1}{2}\left(r_i \cdot w_j + r_j \cdot w_i\right) - \epsilon\, a_{ij} + \text{(nonlinear terms)}. \tag{27}$$

It is denoted by:

$In[15]:= \text{gamma}[i\_,j\_]/; \text{NegIntegerQ}[i]\&\&\text{NegIntegerQ}[j] :=$
  $\text{gamma}[i,j] =$
  $\dfrac{1}{2}(\text{ri}[i].\text{wi}[j] + \text{ri}[j].\text{wi}[i]) - \epsilon[x[1],x[2]]\,a[i,j]$

The other two strain tensors of the reference surface can be computed with the similar definitions.

Internal Forces and Equations in Terms of Displacements. The formula for the moments is presented in point 2.3. Having already computed the kinematical relations and strain tensors internal forces can be expressed in terms of displacements. Here is an example of one of the component of moment tensor for the considered parameterisation of the catenoidal shell.

$In[16]:= \text{mRef}[1,1]$

$Out[16]= -\dfrac{2\,e\,h^3\,s_0\,\epsilon[x,y]}{3\,(-1+v)\,(x^2+s_0^2)} + \dfrac{2\,e\,h^3\,\kappa[x,y]}{3-3\,v} - \dfrac{2\,e\,h^3\,x\,v\,d_1[x,y]}{3\,(-1+v^2)\,(x^2+s_0^2)} +$

$\dfrac{4\,e\,h^3\,s_0^2\,w_3[x,y]}{3\,(-1+v^2)\,(x^2+s_0^2)^2} + \dfrac{2\,e\,h^3\,v\,d_2^{(0,1)}[x,y]}{(3-3\,v^2)\,\sqrt{x^2+s_0^2}} +$

$\dfrac{2\,e\,h^3\,d_1^{(1,0)}[x,y]}{3-3\,v^2} + \dfrac{4\,e\,h^3\,s_0\,w_1^{(1,0)}[x,y]}{3\,(-1+v)\,(1+v)\,(x^2+s_0^2)}$

The internal forces are substituted into the differential equations, obtained in point 2.1, to receive them in terms of displacements. They are long, for example the first one for the catenoidal shell contains 17 terms so it will not be presented here.

It can now be stated that it is possible to proceed from the very general equations to very specific ones which are ready for numerical computations. The next section will deal with these problems.

3 Numerical Tasks – Boundary Value Problems Using the Refined Least Squares Method

3.1 Basic Features

The least squares method is a well–known, [15], meshless way of finding an approximate solution to a boundary value problem. The classical approach consists in minimising the functional based on algebraic, differential or integral equations, or on a system of equations with a set of independent functions which satisfy boundary conditions.

The refined least squares (RLS) method uses the following approach. The minimized functional is supplemented with the terms responsible for boundary conditions. The approximating functions do not have to satisfy the boundary or initial conditions but they must be linearly independent. The basic features of the RLS method are described in [6] and [7].

Described below is an example of applying the RLS method to the tasks of computing a short and long cylindrical shell, which are specific boundary layer problems. Another example is presented in [12].

The set of equations presented below, functions in boundary conditions have been developed with *MATHEMATICA* and the **MathTensor**™ package in a way presented in the previous section.

3.2 Problem Description

Physical Description. Let us consider two steel (Young modulus $E = 2 \cdot 10^8$, Poisson ratio $v = \frac{3}{10}$) cylindrical shells of length $l = 4$ m (short) and $l = 60$ m (long), thickness $2h = 10$ mm, and cylinder radius $s_o = 2$ m. It is subjected to the periodical, with regard to the direction of parallel, load normal to the reference surface $p_3 = P_3 \cos{(ny)}$, $n = 5$, $P_3 = 1$ kN/m² normal to the cylinder mid–surface, tangent load components $p_1 = 0$, $p_2 = 0$.

The shapes of the undeformed reference surfaces are presented in the figures 2 and 3.

The variable $y \in \langle 0, 2\pi \rangle$ is measured along parallel of the cylinder, and $x \in \langle x_a, x_b \rangle$ along the meridian. The cylinder is fixed on both edges $x_a = -\frac{l}{2}$ and $x_b = \frac{l}{2}$.

Due to the axial symmetry of the cylinder the two–dimensional problem can be reduced to one–dimensional task by Fourier expansion.

From the engineering point of view the problem can be considered as a static of a steel pipe loaded with a fifth component of the wind.

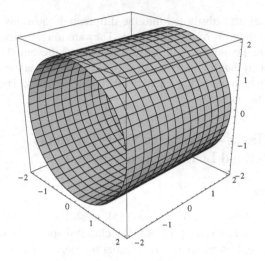

Fig. 2. Short shell

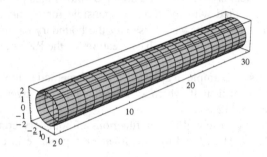

Fig. 3. Long shell, from the plane of symmetry $x = 0$ to the fixed end $x = 30$

Equations. The problem is described by a system of five second–order ordinary differential equations obtained within *MATHEMATICA* notebooks discussed in the previous section. The following notation is used:

$\mathcal{D}_1(x) \cos(y)$— meridian rotation component,
$\mathcal{D}_2(x) \sin(y)$— parallel rotation component,
$\mathcal{W}_1(x) \cos(y)$— meridian displacement component,
$\mathcal{W}_2(x) \sin(y)$— parallel displacement component,
$\mathcal{W}_3(x) \cos(y)$— normal displacement component.

The system of equations represents the equilibrium state of the shell. The right hand sides (rhs) of the equations are equal to zero, and below are their left hand sides (lhs):

$$
e_1 := -\frac{E\,n^2\,\mathcal{D}_1(x)\,h^3}{3\,s_o^3\,(1+v)} - \frac{2\,E\,\mathcal{D}_1''(x)\,h^3}{3\,s_o\,(1-v^2)} - \frac{E\,(h^2 + 3\,s_o^2)\,n^2\,\mathcal{W}_1(x)\,h}{3\,s_o^4\,(1+v)} +
$$
$$
+ \frac{2\,E\,\mathcal{W}_1''(x)\,h}{1-v^2} + \frac{n\,E\,\mathcal{W}_2'(x)\,h}{s_o\,(1-v)} - \frac{2\,E\,v\,\mathcal{W}_3'(x)\,h}{s_o\,(1-v^2)} + P_1(x),
$$

(28a)

$$e_2 := \frac{5\,h\,E\,\mathcal{D}_1'(x)}{6\,(1+v)} + \frac{h\,n\,E\,\left(4\,h^2 + 5\,(1-v)\,s_o^2\right)\mathcal{D}_2(x)}{6\,s_o^3\,\left(1-v^2\right)} +$$

$$+ \frac{2\,h\,E\,v\,\mathcal{W}_1'(x)}{s_o\,\left(1-v^2\right)} + \frac{h\,n\,E\,\left(4\,h^2 + (17-5\,v)\,s_o^2\right)\mathcal{W}_2(x)}{6\,s_o^4\,\left(1-v^2\right)} + \tag{28b}$$

$$- \frac{E\,h\,\left(4\,h^2 + \left(5\,(1-v)\,n^2 + 12\right)s_o^2\right)\mathcal{W}_3(x)}{6\,s_o^4\,\left(1-v^2\right)} + \frac{5\,h\,E\,\mathcal{W}_3''(x)}{6\,(1+v)} + P_3(x),$$

$$e_3 := -\frac{E\,\left(2\,h^2\,n^2 + 5\,s_o^2\right)\mathcal{D}_1(x)\,h}{6\,s_o^2\,(1+v)} + \frac{2\,E\,\mathcal{D}_1''(x)\,h^3}{3\,\left(1-v^2\right)} + \frac{n\,E\,\mathcal{D}_2'(x)\,h^3}{3\,s_o\,(1-v)} +$$

$$- \frac{E\,n^2\,\mathcal{W}_1(x)\,h^3}{3\,s_o^3\,(1+v)} - \frac{2\,E\,\mathcal{W}_1''(x)\,h^3}{3\,s_o\,\left(1-v^2\right)} - \frac{5\,E\,\mathcal{W}_3(x)\,h}{6\,(1+v)}, \tag{28c}$$

$$e_4 := -\frac{E\,\left(4\,h^2\,n^2 + 5\,(1-v)\,s_o^2\right)\mathcal{D}_2(x)\,h}{6\,s_o^4\,\left(1-v^2\right)} - \frac{E\,\mathcal{D}_2''(x)\,h^3}{3\,s_o^2\,(1+v)} + \frac{E\,n\,\mathcal{W}_1'(x)\,h}{s_o^2\,(1-v)} +$$

$$+ \frac{E\,\mathcal{W}_2''(x)\,h}{s_o\,(1+v)} - \frac{E\,\left(4\,h^2\,n^2 + \left(12\,n^2 - 5\,v + 5\right)s_o^2\right)\mathcal{W}_2(x)\,h}{6\,s_o^5\,\left(1-v^2\right)} + \tag{28d}$$

$$+ \frac{n\,E\,\left(4\,h^2 + (17-5\,v)\,s_o^2\right)\mathcal{W}_3(x)\,h}{6\,s_o^5\,\left(1-v^2\right)} + P_2(x),$$

$$e_5 := -\frac{E\,n\,\mathcal{D}_1'(x)\,h^3}{3\,s_o^2\,(1-v)} - \frac{E\,\left(4\,h^2\,n^2 + 5\,(1-v)\,s_o^2\right)\mathcal{D}_2(x)\,h}{6\,s_o^3\,\left(1-v^2\right)} + \frac{E\,\mathcal{D}_2''(x)\,h^3}{3\,s_o\,(1+v)} +$$

$$- \frac{E\,\left(4\,h^2\,n^2 + 5\,(1-v)\,s_o^2\right)\mathcal{W}_2(x)\,h}{6\,s_o^4\,\left(1-v^2\right)} - \frac{E\,\mathcal{W}_2''(x)\,h^3}{3\,s_o^2\,(1+v)} + \tag{28e}$$

$$+ \frac{n\,E\,\left(4\,h^2 + 5\,(1-v)\,s_o^2\right)\mathcal{W}_3(x)\,h}{6\,s_o^4\,\left(1-v^2\right)}.$$

Boundary Conditions. It is obvious that the system of 5 second–order differential equations system (28) requires 10 boundary conditions. Two types of boundary conditions can be distinguished: essential conditions and boundary layer conditions.

The **essential conditions** (lhs) for our problem are:

$$b_1 := \mathcal{W}_1(x_a),$$
$$b_2 := \mathcal{W}_2(x_a),$$
$$b_3 := \mathcal{W}_1(x_b), \tag{29}$$
$$b_4 := \mathcal{W}_2(x_b).$$

Their (rhs) are equal to zero. These boundary conditions are the same as used in the membrane approach to the problem. Here it needs to be stated that the membrane approach is not correct for the considered task as the cylindrical shell is significantly bent by this type of load so the moments and transverse forces cannot be neglected.

The **boundary–layer conditions** (lhs) for our problem are:

$$
\begin{aligned}
b_5 &:= \mathcal{D}_1\left(x_a\right), \\
b_6 &:= \mathcal{D}_2\left(x_a\right), \\
b_7 &:= \mathcal{W}_3\left(x_a\right), \\
b_8 &:= \mathcal{D}_1\left(x_b\right), \\
b_9 &:= \mathcal{D}_2\left(x_b\right), \\
b_{10} &:= \mathcal{W}_3\left(x_b\right).
\end{aligned}
\tag{30}
$$

3.3   Method Description

For the considered system (28) a–e of the equations and boundary conditions (29) and (30) the following functional can be built:

$$
\mathcal{F} = \int_{x_a}^{x_b} \left(\sum_{n=1}^{5} \left(\alpha_n\, e_n\right)^2 \right) dx + \sum_{k=1}^{10} \left(\beta_k\, b_k\right)^2,
\tag{31}
$$

where α_n and β_k are scale factors or scale functions.

The RLS method consists in minimising of this functional. If the solution is exact the value of the functional is zero, otherwise it is positive. Minimisation is done by the Ritz method. According to it the approximation of $f\,(x)$ can be predicted in a form of linear combination of m independent functions $u_i\,(x)$.

$$
f\,(x) = \sum_{i=1}^{m} C_i\, u_i\,(x).
\tag{32}
$$

Substituting (32) into (31) for each $f\,(x)$ (it stands here for $\mathcal{D}_1\,(x)$, $\mathcal{D}_2\,(x)$, $\mathcal{W}_1\,(x)$, $\mathcal{W}_2\,(x)$ and $\mathcal{W}_3\,(x)$, respectively) one can compute the derivative of $\mathcal{F}$ with respect to C_i. Minimisation of the functional is equivalent to the condition:

$$
\frac{\partial \mathcal{F}}{\partial C_i} = 0.
\tag{33}
$$

Doing it for each unknown C_i a system of algebraic equations is obtained. For a linear problem it is linear one $\left(A_{ij}\, C_j = B_i\right)$ with a symmetrical and positive definite matrix A_{ij}.

The formulae for developing the terms of the system and the *MATHEMATICA* implementation are presented below.

3.4   Method Implementation

Each unknown function of the system of differential equations (28) can be approximated with a linear combination of monic Chebyshev polynomials. As the considered problem is symmetrical with regard to the plane $x = 0$, the functions $\mathcal{W}_1\,(x)$ and $\mathcal{D}_1\,(x)$

are antisymmetrical and $W_2(x)$, $W_3(x)$ and $D_2(x)$ are symmetrical. The system can be informed about it with the following definitions, for example:

$$In[1] := W_1[x\_] :=$$

$$\overset{\text{PolyDegree}}{\underset{i=0}{\sum}} c[5\,i]\; \text{MonicChebyshevT}\left[2i+1,\; \frac{2\,(x-bx)}{ax-bx}-1\right];$$

$$In[2] := W_2[x\_] :=$$

$$\overset{\text{PolyDegree}}{\underset{i=0}{\sum}} c[5\,i+1]\; \text{MonicChebyshevT}\left[2i,\; \frac{2\,(x-bx)}{ax-bx}-1\right];$$

where **ax** stands for x_a, **bx** for x_b, **c[k]** for C_k and:

$$In[3] := \text{MonicChebyshevT}[n\_, x\_] := \frac{\text{ChebyshevT}[n, x]}{2^{n-1}}$$

It is not difficult to notice that the real domain $\langle x_a, x_b \rangle$ is transformed to the interval $\langle -1, 1 \rangle$.

Each differential equation is scaled and saved into the variable, for example:

$$In[4] := \text{DifferentialEquation}[1] :=$$

$$\text{DifferentialEquation}[1] = \frac{r1}{h};$$

According to this approximation the scaled differential equation $\alpha_k\, e_k$ and boundary condition $\beta_n\, b_k$ can be rewritten in the following form:

$$\alpha_k\, e_k(x) = f_k(x) + \sum_{i=0}^{5(p+1)-1} C_i\, m_{ik}(x), \tag{34}$$

$$\beta_k\, b_k = g_k + \sum_{i=0}^{5(p+1)-1} C_i\, n_{ik}. \tag{35}$$

where p is an assumed polynomial degree.

Free terms f_k are extracted from (34) with the following functions:

$$In[5] := \text{DiffEqnFreeTerm}[k\_] := \text{DiffEqnFreeTerm}[k] =$$

$$\text{Expand}[-\text{DifferentialEquation}[k]/.c[\_] \to 0];$$

The functions $m_{ik}(x)$ can also be found with the very similar procedure. Standard *MATHEMATICA* functions could be used but this one is faster.

$$In[6] := \text{DiffEqnCoefficient}[i\_, k\_] := \text{DiffEqnCoefficient}[i, k] =$$

$$\text{Expand}[\text{DifferentialEquation}[k] +$$

$$\text{DiffEqnFreeTerm}[k]/.c[i] \to 1/.c[\_] \to 0]$$

Similarly each scaled boundary condition for the task is saved into the variable. Here the full set of *MATHEMATICA* input for the edge x_a can be seen, the other 5 are similar:

$$In[7] := \text{BoundaryCondition}[1] := \text{BoundaryCondition}[1] = W_1[ax]\; e;$$

In[8]:= **BoundaryCondition[2] := BoundaryCondition[2] = $\mathcal{W}_2$[ax] e;**

In[9]:= **BoundaryCondition[5] := BoundaryCondition[5] = $\mathcal{D}_1$[ax] e;**

In[10]:= **BoundaryCondition[6] := BoundaryCondition[6] = $\mathcal{D}_2$[ax] e;**

In[11]:= **BoundaryCondition[7] :=**
        BoundaryCondition[7] = $\mathcal{W}_3$[ax] e;

The following commands are applied to make the extractions of $n_{ik}(x)$ and g_k from the boundary conditions (35).

In[12]:= **BoundCondFreeTerm[i_] := BoundCondFreeTerm[i] =**
        Expand[-BoundaryCondition[i]/.c[_] → 0];

In[13]:= **BoundCondCoefficient[k_, i_] :=**
        BoundCondCoefficient[k, i] =
          Expand[BoundaryCondition[i] +
            BoundCondFreeTerm[i]/.c[k] → 1/.c[_] → 0]

As $x \in \langle x_a, x_b \rangle$, coefficients of the symmetric matrix can be computed with the following formula:

$$A_{ij} = \int_{x_a}^{x_b} \left(\sum_{k=1}^{5} m_{ik}(x)\, m_{jk}(x) \right) dx + \sum_{k=1}^{10} n_{ik}\, n_{jk}. \qquad (36)$$

It is done with:

In[14]:= **MatrixCoefficient[i_, j_] :=**
        MatrixCoefficient[i, j] = MatrixCoefficient[j, i] =
          Expand$\Big[$integ1$\Big[$Expand$\Big[$

$$\sum_{k=1}^{5}$$ **DiffEqnCoefficient[i, k] DiffEqnCoefficient[j, k]$\Big]\Big]$ +**

          Expand$\Big[$

$$\sum_{k=1}^{10}$$ **BoundCondCoefficient[i, k]**

        BoundCondCoefficient[j, k]$\Big]\Big]$

where **integ1** is a function for computing a definite integral. There are predefined formulas for integrating monomials and dealing with sums, it makes calculations faster.

In[15]:= **integ1[z_Plus] := xxinteg1/@z;**

In[16]:= **integ1[z_] := xxinteg1[z];**

In[17]:= **xxinteg1[z_] := integx[z, ax, bx];**

In[18]:= **integx[a_ x^n_., ax_, bx_] := $\dfrac{a\,(-ax^{1+n} + bx^{1+n})}{1+n}$ /; FreeQ[a, x];**

$In[19]:=$ **integx$[x^{n\_}, ax\_, bx\_] :=$** $\dfrac{bx^{1+n} - ax^{1+n}}{1+n}$ **;**

$In[20]:=$ **integx$[a\_, ax\_, bx\_] :=$ a (bx - ax) /; FreeQ[a, x];**

$In[21]:=$ **integx$[z\_, ax\_, bx\_] :=$** $\displaystyle\int_{ax}^{bx}$ **zdx;**

System matrix is built from the matrix coefficients. The matrix is symmetrical so only the lower triangle should be saved.

$In[22]:=$ **SystemMatrix := SystemMatrix =**
   Table[Table[
    MatrixCoefficient[j, i],
     {i, 0, j}], {j, 0, NumberOfEqns}]

where, for our task:

$In[23]:=$ **NumberOfEqns = 5 (PolyDegree + 1) - 1**

Analogously elements of the free vector can be obtained:

$$B_i = \int_{x_a}^{x_b} \left(\sum_{k=1}^{5} m_{ik}(x)\, f_k \right) dx + \sum_{k=1}^{10} n_{ik}\, g_k. \tag{37}$$

It is done with:

$In[24]:=$ **FreeVecCoefficient$[i\_] :=$**
   FreeVecCoefficient[i] =
   Expand$\Big[$integ1$\Big[$Expand$\Big[$
    $\displaystyle\sum_{k=1}^{5}$**DiffEqnCoefficient[i, k] DiffEqnFreeTerm[k]$\Big]\Big]$**
   +Expand$\Big[$
    $\displaystyle\sum_{k=1}^{10}$**BoundCondCoefficient[i, k]**
    BoundCondFreeTerm[k]$\Big]\Big]$

$In[25]:=$ **FreeVector := FreeVector =**
   Table[FreeVecCoefficient[i],
    {i, 0, NumberOfEqns}]

As the matrix A_{ij} is positive definite and symmetrical the Cholesky-Banachiewicz method is used for the solution of the linear system of equations. This is the only part of the procedure done numerically. The results of the approximation are functions.

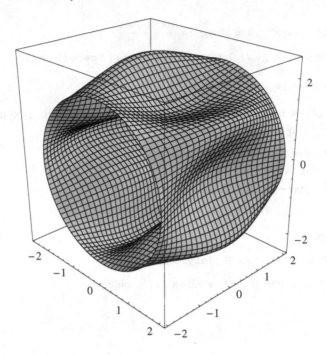

Fig. 4. Short shell: deformation (exaggeration 5000 times)

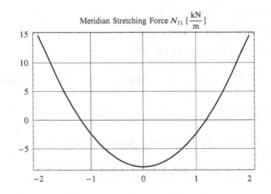

Fig. 5. Short shell approximation

3.5   One Step Approach

Short Shell. The approximation for the short shell (l = 4 m) is not difficult. The deformed shape of it is presented in figure 4.

The figures 5, 6, 7, 8 and 9 show diagrams of some physical internal forces for the considered case. Here $\mathcal{N}_{11}(x)\,\cos(y)$ is a meridian stretching force and $\mathcal{N}_{12}(x)\,\sin(y)$ is a meridian shear force, $\mathcal{M}_{11}(x)\,\sin(y)$ is a meridian torsion moment, $\mathcal{M}_{12}(x)\,\cos(y)$ is

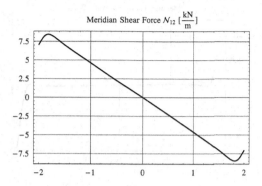

Fig. 6. Short shell approximation

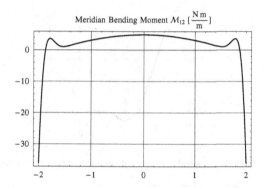

Fig. 7. Short shell approximation

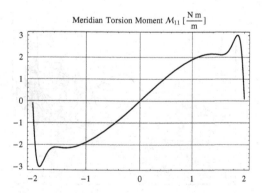

Fig. 8. Short shell approximation

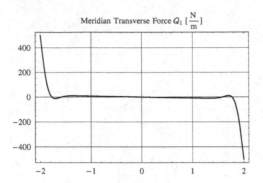

Fig. 9. Short shell approximation

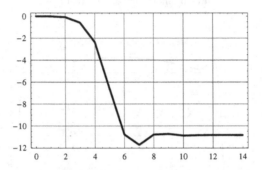

Fig. 10. Short shell approximation: convergence analysis

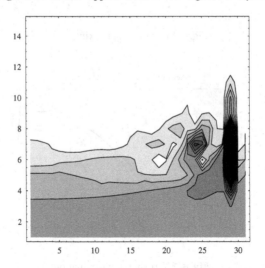

Fig. 11. Short shell: relative error of the convergence

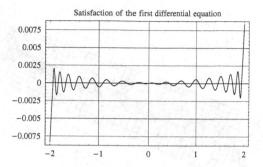

Fig. 12. Short shell approximation

a meridian bending moment, and $Q_1(x) \cos(y)$ is a meridian transverse force. They are defined as follows:

$$N_{11}(x) = -\frac{2\,E\,\mathcal{D}_1'(x)\,h^3}{3\,s_o\,(1-v^2)} + \frac{2\,E\,\mathcal{W}_1'(x)\,h}{1-v^2} + \frac{2\,n\,E\,v\,\mathcal{W}_2(x)\,h}{s_o\,(1-v^2)} - \frac{2\,E\,v\,\mathcal{W}_3(x)\,h}{s_o\,(1-v^2)}, \quad (38)$$

$$N_{12}(x) = -\frac{E\,\mathcal{D}_2'(x)\,h^3}{3\,s_o\,(1+v)} - \frac{E\,n\,\mathcal{W}_1(x)\,h}{s_o\,(1+v)} + \frac{E\,\mathcal{W}_2'(x)\,h}{1+v}, \quad (39)$$

$$M_{11}(x) = \frac{n\,E\,\mathcal{D}_1(x)\,h^3}{3\,s_o\,(1+v)} - \frac{E\,\mathcal{D}_2'(x)\,h^3}{3\,(1+v)} + \frac{E\,\mathcal{W}_2''(x)\,h^3}{3\,s_o\,(1+v)}, \quad (40)$$

$$M_{12}(x) = \frac{2\,E\,\mathcal{D}_1'(x)\,h^3}{3\,(1-v^2)} + \frac{2\,n\,E\,v\,\mathcal{D}_2(x)\,h^3}{3\,s_o\,(1-v^2)} - \frac{2\,E\,\mathcal{W}_1'(x)\,h^3}{3\,s_o\,(1-v^2)}, \quad (41)$$

$$Q_1(x) = \frac{5\,h\,E\,\mathcal{D}_1(x)}{6\,(1+v)} + \frac{5\,h\,E\,\mathcal{W}_3'(x)}{6\,(1+v)}. \quad (42)$$

The figures 10 and 11 present analysis of the approximation convergence. The figure 10 shows convergence of the function value $M_{12}\,(-1.933\,\text{m})$ with respect to the value of **PolyDegree**. The figure 11 demonstrates function value M_{12} in 31 point in the interval $x \in \langle 0, \frac{1}{2} \rangle$; variable on the vertical axis is **PolyDegree**, lighter colour represent smaller relative error with respect to the approximation for the highest value of **PolyDegree** used in computations. It can be seen that the approximation close to exact solution is reached when value of the variable **PolyDegree** is equal to 14. It means that the degrees of the approximating polynomials attain value 28 for even functions and 29 for odd ones.

The obtained functions can be substituted to the approximated equations to observe the error. The figure 12 presents (dis)satisfaction of the first differential equation for the considered case.

Long Shell. If the approximation of the problem of a long cylindrical shell had been done with a full set of boundary conditions, then difficulties would have been experienced with convergence and stability. This is a high-order differential operator and

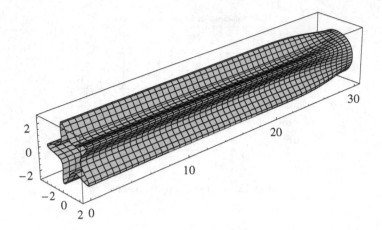

Fig. 13. Long shell: deformation from the plane of symmetry $x = 0$ to the fixed end $x = 30$ (exaggeration 500 times)

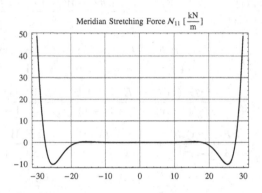

Fig. 14. Long shell, one step approximation approach

therefore the problem is ill-conditioned. The degrees of the approximating polynomials have to be quite large number and the working precision of the computations have to be equal to 512. Despite of this the loss of the precision has been over 300 digits.

The results of this approximation are presented in figures 14, 15, 16, 17 and 18.

The figures 19 and 20 show analysis of the approximation convergence. The figure 19 shows convergence of the function value $M_{12}(-29\,\text{m})$ with respect to the value of **PolyDegree**. The figure 20 demonstrates function value M_{12} in 31 point in the interval $x \in \langle 0, \frac{l}{2} \rangle$; variable on the vertical axis is **PolyDegree**, lighter colour represent smaller relative error with respect to the approximation for the highest value of **PolyDegree** used in computations. It can be seen that the approximation close to exact solution is reached when value of the variable **PolyDegree** is equal to 46. It means that the degrees of the approximating polynomials attain value 92 for even functions and 93 for odd ones. The computational process becomes very long because the system has a lot integrals to be computed symbolically.

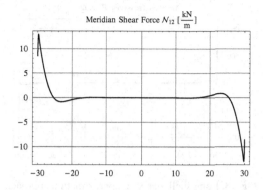

Fig. 15. Long shell, one step approximation approach

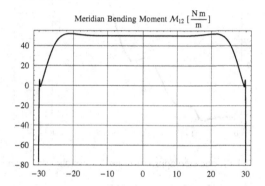

Fig. 16. Long shell, one step approximation approach

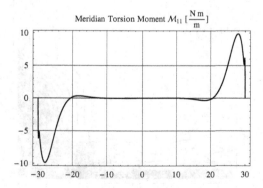

Fig. 17. Long shell, one step approximation approach

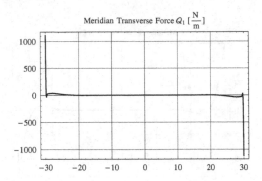

Fig. 18. Long shell, one step approximation approach

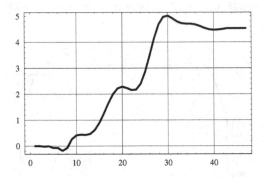

Fig. 19. Long shell, one step approximation approach: convergence analysis

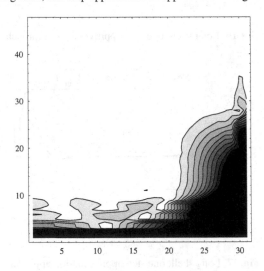

Fig. 20. Long shell, one step approximation approach: relative error of the convergence

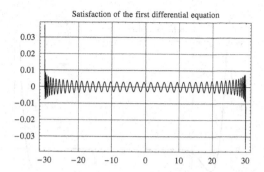

Fig. 21. Long shell, one step approximation approach

The figure 21 show (dis)satisfaction of the first differential equation for the considered case.

3.6 Two Step Approach

One of the most important features of the RLS method is the scale factors by which better or worse satisfaction a selected equation or boundary condition can be enforced. The idea of the base solution presented below has been developed by experimenting with scale factors. When decreasing a scale factor in some boundary conditions a zero has been put in instead of a very small number and it has turned out to still produce a very good approximation.

Base Solution. It has been found that it is possible to approximate the system (28) with the RLS method, taking into account only the essential boundary conditions (29) and neglecting the boundary–layer ones (30). As it has been already mentioned they are simply multiplied by zero, for example:

```
In[26]:= BoundaryCondition[5] :=
             BoundaryCondition[5] = 𝒟₁[ax] e 0 ;
```

It is an unexpected situation but it can be interpreted physically - the problem is statically indeterminate if all boundary conditions on fixed edge are satisfied. Some the boundary conditions which are not essential for the overall stability can be released. It has been found that it is enough to apply polynomials much lower degree than for one step approach to obtain quite good results. The results of this approximation are shown in figures 22, 23, 24, 25 and 26.

The figures 27 and 28 show analysis of the approximation convergence. The figure 27 shows convergence of the function value $M_{12}(-29\,\text{m})$ with respect to the value of **PolyDegree**. The figure 28 demonstrates function value M_{12} in 31 point in the interval $x \in \langle 0, \frac{l}{2} \rangle$; variable on the vertical axis is **PolyDegree**, lighter colour represent smaller relative error with respect to the approximation for the highest value of

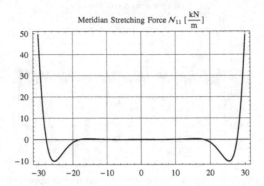

Fig. 22. Long cylindrical shell, "base" approximation

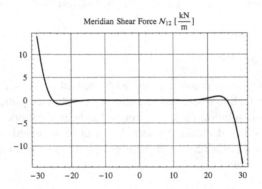

Fig. 23. Long shell, "base" approximation

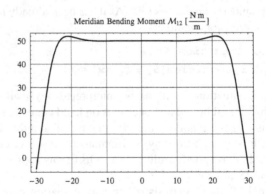

Fig. 24. Long shell, "base" approximation

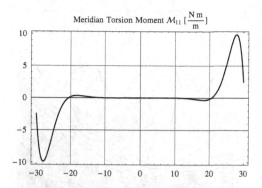

Fig. 25. Long shell, "base" approximation

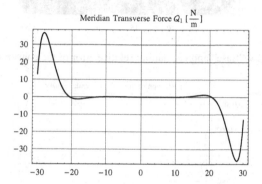

Fig. 26. Long shell, "base" approximation

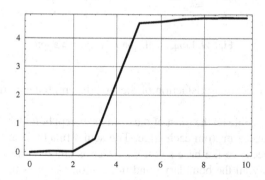

Fig. 27. "Base" approximation: convergence analysis

PolyDegree used in computations. It can be seen that the approximation close to exact solution is reached when value of the variable **PolyDegree** is equal to 10. It means that the degree of the approximating polynomial attains value 20 for even functions and 21 for odd ones.

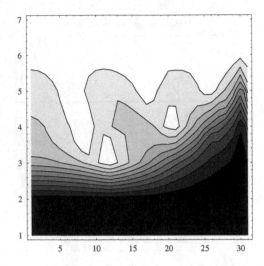

Fig. 28. "Base" approximation: relative error of the convergence

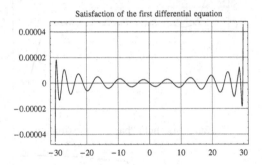

Fig. 29. Long shell, "base" approximation

The figure 29 shows (dis)satisfaction of the first differential equation for the considered case.

The solution is feasible on the most of the domain except boundary layers which sizes are limited to about 1.2 m from each edge. The actual functions are highly oscillating but they quickly decay to the base solution, which is smooth. Hence, it is practically impossible to satisfy all the boundary conditions in one step, as the problem becomes ill–conditioned in the Lyapunov sense and is slowly convergent. The base solution is not a membrane because the moments and shear forces are not zero functions. Note that neglecting of some boundary conditions in other methods usually results in a singular system. Here a well–conditioned and non–singular problem is obtained.

Boundary-Layer Refinement. The base solution allows refinement of the approximation at the boundary layers.

We are taking into account limited domain so the parameters describing the position of the boundaries take values:

$In[27]:=$ **ax = -30;**

$In[28]:=$ **bx = -288/10;**

As none symmetry can be expected the approximation polynomials should contain both odd and even functions.

$In[29]:=$ W_1 **[x_] :=** $\displaystyle\sum_{i=0}^{\text{PolyDegree}}$ **c[5 i] MonicChebyshevT** $\left[i, \dfrac{2(x-bx)}{ax-bx}-1\right];$

$In[30]:=$ W_2 **[x_] :=**

$\displaystyle\sum_{i=0}^{\text{PolyDegree}}$ **c[5 i + 1] MonicChebyshevT** $\left[i, \dfrac{2(x-bx)}{ax-bx}-1\right];$

A full set of boundary conditions are now taken into account, for the fixed edge the boundary conditions are expressed in displacements.

$In[31]:=$ **BoundaryCondition[1] :=**

   BoundaryCondition[1] = D_1 **[ax] e;**

$In[32]:=$ **BoundaryCondition[2] :=**

   BoundaryCondition[2] = D_2 **[ax] e;**

$In[33]:=$ **BoundaryCondition[3] :=**

   BoundaryCondition[3] = W_1 **[ax] e;**

$In[34]:=$ **BoundaryCondition[4] :=**

   BoundaryCondition[4] = W_2 **[ax] e;**

$In[35]:=$ **BoundaryCondition[5] :=**

   BoundaryCondition[5] = W_3 **[ax] e;**

The boundary conditions on the "artificial" edge $x = 28.8$ are expressed in forces. It is assumed that the forces are equal to those obtained from the base approximation. To avoid further loss of precision we can change numerical values of base approximation with a **Rationalize** function to "exact" fractions. It is a kind of cheating but ensures symbolic computations of the coefficients of the system of algebraic equations. Moreover the calculations are quicker.

$In[36]:=$ **rd1 = Rationalize** $\left[\text{n11[bx]}, 10^{-\text{Precision[n11[bx]]}}\right];$

$In[37]:=$ **rd2 = Rationalize** $\left[\text{n12[bx]}, 10^{-\text{Precision[n12[bx]]}}\right];$

$In[38]:=$ **rd3 = Rationalize** $\left[\text{m11[bx]}, 10^{-\text{Precision[m11[bx]]}}\right];$

$In[39]:=$ **rd4 = Rationalize** $\left[\text{m12[bx]}, 10^{-\text{Precision[m12[bx]]}}\right];$

$In[40]:=$ **rd5 = Rationalize** $\left[\text{q1[bx]}, 10^{-\text{Precision[q1[bx]]}}\right];$

In[41]:= **BoundaryCondition[6] :=**

  BoundaryCondition[6] = $\dfrac{\mathcal{N}_{11}[\mathbf{bx}] - \mathbf{rd1}}{\mathbf{h}}$;

In[42]:= **BoundaryCondition[7] :=**

  BoundaryCondition[7] = $\dfrac{\mathcal{N}_{12}[\mathbf{bx}] - \mathbf{rd2}}{\mathbf{h}}$;

In[43]:= **BoundaryCondition[8] :=**

  BoundaryCondition[8] = $\dfrac{\mathcal{M}_{11}[\mathbf{bx}] - \mathbf{rd3}}{\mathbf{h}^3}$;

In[44]:= **BoundaryCondition[9] :=**

  BoundaryCondition[9] = $\dfrac{\mathcal{M}_{12}[\mathbf{bx}] - \mathbf{rd4}}{\mathbf{h}^3}$;

In[45]:= **BoundaryCondition[10] :=**

  BoundaryCondition[10] = $\dfrac{\mathcal{Q}_1[\mathbf{bx}] - \mathbf{rd5}}{\mathbf{h}^3}$;

Satisfying all boundary conditions on the fixed edge layer conditions results in the refined functions which diagrams are shown in figures 30, 31, 32, 33, 34, 35, 36 and 37 compared to the base solution. It can be observed that the taking into account additional boundary conditions results in a small decrease of displacements. It can be interpreted physically as a local increase of the shell stiffness connected with taking into consideration more constrains in the boundary conditions. The obtained approximation error is allowable for engineering purposes.

The figures 38 and 39 show analysis of the approximation convergence. The figure 38 demonstrates convergence of the function value $\mathcal{M}_{12}(-29\,m)$ with respect to the value of **PolyDegree**. The figure 39 demonstrates function value $\mathcal{M}_{12}$ in 31 point in the interval $x \in \langle 0, \frac{l}{2} \rangle$; variable on the vertical axis is **PolyDegree**, lighter color represent smaller relative error with respect to the approximation for the highest value of **PolyDegree** used in computations. It can be seen that the approximation close to exact solution is reached when value of the variable **PolyDegree** is equal to 25. It means that the degree of the approximating polynomial attains the same value.

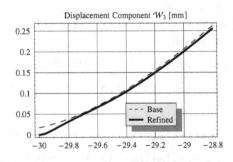

Fig. 30. Boundary layer refinement of the " base" approximation

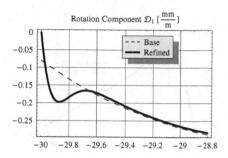

Fig. 31. Boundary layer refinement of the " base" approximation

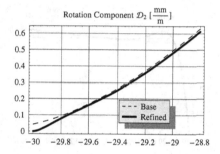

Fig. 32. Boundary layer refinement of the " base" approximation

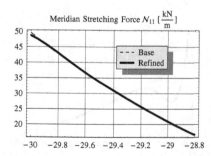

Fig. 33. Boundary layer refinement of the " base" approximation

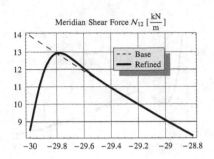

Fig. 34. Boundary layer refinement of the " base" approximation

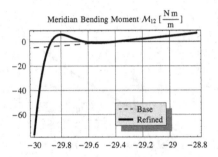

Fig. 35. Boundary layer refinement of the " base" approximation

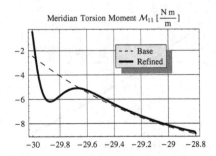

Fig. 36. Boundary layer refinement of the " base" approximation

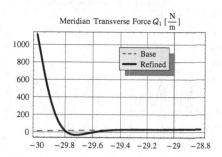

Fig. 37. Boundary layer refinement of the " base" approximation

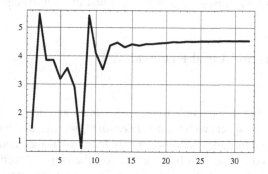

Fig. 38. Boundary layer refinement of the " base" approximation: convergence analysis

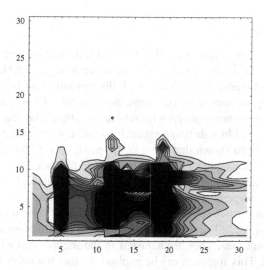

Fig. 39. Boundary layer refinement of the " base" approximation: relative error of the convergence

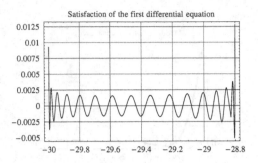

Fig. 40. Boundary layer refinement of the " base" approximation

The figure 40 presents (dis)satisfaction of the first differential equation for the considered case.

4   Conclusions and Final Remarks

4.1   Symbolic Tasks

It has been shown that it is possible to solve the linear and nonlinear tensor symbolic problems in different aspects with the *MathTensor*™ package and formulate equations from the very general formulas to the ones ready for numerical computations. Each task requires different tools. There has been shown how the built in functions of the system and the package can be used efficiently. The tools are intuitive and relatively simple, nevertheless the full responsibility for the results obtained belongs to the user of the system and package.

4.2   Numerical Tasks

The RLS method allows simpler algorithm to be built if there is no need to use functions that have to satisfy boundary conditions. Therefore it is applicable to more general purposes like multidimensional problems with discontinuities of boundary conditions.

The results of approximation are functions, not numbers. Therefore the results can be used directly in further computations without interpolation. The quality of the approximation can be evaluated by substituting functions into approximated equations.

Moreover it has been shown that there is the possibility of the neglecting of some boundary conditions for boundary layer problems and satisfy them locally in the next step. This phenomenon for the problem being considered has a physical interpretation. The mathematical interpretation seems to be an open problem.

According to that the two step approach to the boundary layer problems of theory of shells has been proposed. The first step consists in approximation with negligence of some boundary conditions. They are adjusted in the second step when only boundary layer is considered. This approach can be applied to other boundary layer tasks.

MATHEMATICA as a (fully) integrated environment of symbolic and numerical computation and graphics is a very effective tool for the complex analysis of symbolic (tensor)

calculations and can be applied to in the entire computational and publication process. Its external package **MathTensor**™ is an effective tool of tensor analysis with the theory of shells. However, its long expected upgrade should be better adopted for the current possibilities of *MATHEMATICA* 4.1.

Acknowledgements

The author would like to express gratitude to organisers of the SNSC'2001 conference, especially Professor Franz Winkler and Mrs. Ramona Poechinger, for the kind invitation and nice welcome.

The grant from Wolfram Research, Inc., has supported the author with a free upgrade of the system *MATHEMATICA* 4.1.

Some Notations

There are some symbols used for tensor problems discussed in the section 2.

a — determinant of the first differential form

a_{ij} — the first differential form of the reference surface, metric tensor on the reference surface

b_{ij} — the first differential form of the reference surface, curvature tensor

d — rotation vector

d_i — derivative of rotation vector

d_i — physical component of the rotation vector

d^i — contravariant component of the rotation vector in the curvilinear basis

g — determinant of the metric tensor

g_{ij} — metric tensor

$2h$ — shell thickness

H — mean curvature

K — Gaussian curvature

N^{ij} — tensor of stretching (tensile) forces

M^{ij} — tensor of moments

P^i — vector of loads

Q^i — vector of transverse forces

r — parameterisation vector

r_i — component of the curvilinear basis

w — displacement vector

w_i — derivative of displacement vector

w_i — physical component of the displacement vector

w^i — contravariant component of the displacement vector in the curvilinear basis

w^3 — normal displacement component

$\delta_i{}^j$ — Kronecker delta

$\overset{*}{\gamma}_{ij}$ — strain tensor in 3D space

γ_{ij} — the first strain tensor of the reference surface

ϵ — coefficient of the thermal expansion

ε_{ij} — antisymmetric object under interchange of any indices

ρ_{ij} — the second strain tensor of the reference surface
ϑ_{ij} — the third strain tensor of the reference surface
τ^{ij} — stress tensor
Other notations are explained in the text.

References

[1] Başar Y., Ding Y. (1990): Theory and finite-element formulation for shell structures undergoing finite rotations. In: Voyiadjis G.Z., Karamanlidis D. (eds) Advances in the theory of plates and shells. Elsevier Science Publishers B.V., Amsterdam, 3–26

[2] Bielak S. (1990): Theory of Shells. Studies and Monographs, 30, Opole University of Technology, Opole

[3] Naghdi P.M. (1963): Foundations of Elastic Shell Theory. chapter 1, North-Holland Publishing CO., Amsterdam, 2–90

[4] Parker L., Christensen S.M. (1994): *MathTensor*™: A System for Doing Tensor Analysis by Computer. Addison-Wesley, Reading, MA, USA

[5] Walentyński, R.A. (1995): Statics of Sloped Shells of Revolution, Comparison of Analytic and Numerical Approaches. PhD Thesis, Silesian University of Technology, Gliwice

[6] Walentyński, R.A. (1997): A least squares method for solving initial-boundary value problems. In: Keränen V., Mitic P., Hietamäki A. (eds) Innovations in Mathematics, Proceedings of the Second International Mathematica Symposium. Rovaniemen Ammattikorkeakoulun Julkaisuja, Computational Mechanics Publications, Boston, MA; Southampton, 483–490

[7] Walentyński R.A. (1997): Computer assisted analytical solution of initial-boundary value problems. In: Garstecki A., Rakowski J., (eds) Computer Methods in Mechanics, Proceedings of the 13th PCCMM. Poznań University of Technology, Poznań, 1363–1400

[8] Walentyński R.A. (1998): Refined constitutive shell equations. In: Chróścielewski J. et al., (eds), Shell Structures, Theory and Applications, Proceedings of the 6th SSTA Conference. Technical University of Gdańsk, Gdańsk–Jurata, 275–276

[9] Walentyński R.A. (1999): Geometrical description of shell with computer algebra system. In Proceedings of the 11th International Scientific Conference. Brno, Technical University of Brno, 51–54

[10] Walentyński R.A. (1999): Computer assisted refinement of shell's constitutive equations. In: Łakota W. et al. (eds) Computer Methods in Mechanics, Proceedings of the 14th PCCMM. Rzeszów University of Technology, Rzeszów, 381–382

[11] Walentyński R.A. (1999): Refined constitutive shell equations with *MathTensor*™. In: Keränen V. (ed) Mathematics with Power: Proceedings of the Third International *MATHEMATICA* Symposium'99. Research Institute for Symbolic Computations, Hagenberg-Linz, http://south.rotol.ramk.fi/~keranen/IMS99/

[12] Walentyński R.A. (2001): Refined least squares method for shells boundary value problems. In: Tazawa Y. et al. (eds) Symbolic Computation: New Horizons, Proceedings of the Fourth International Mathematica Symposium, Tokyo Denki University Press, Tokyo, 511–518, electronic extended version on the conference CDROM

[13] Walentyński R.A. (2001): Solving symbolic tasks in theory of shells of shell with computer algebra system. In: Kralik J. (ed) Proceedings of the 1st International Scientific Seminar: New Trends in Statics and Dynamics of Buildings, STU, Bratislava, 183–188

[14] Wolfram S. (1999): The *MATHEMATICA* book. Cambridge University Press and Wolfram Research, Inc., New York, Champaign, fourth edition

[15] Zwillinger D. (1989): Handbook of differential equations. Academic Press Inc., New York, second edition

A Symbolic Procedure
for the Diagonalization of Linear PDEs
in Accelerated Computational Engineering

Alireza R. Baghai-Wadji

Vienna University of Technology, Gusshausstr. 27-29, A-1040, Vienna, Austria
alireza.baghai-wadji@tuwien.ac.at

Abstract. In this contribution we conjecture that systems of linearized Partial Differential Equations, viewed as consistent models for physically realizable systems, are diagonalizable. While this property is interesting for its own sake, our discussion will focus on technicalities and implications for advanced accelerated computing. We will demonstrate that diagonalization with respect to a chosen spatial coordinate systematically "reshuffles" the original PDEs and creates equivalent differential forms. It turns out that diagonalization automatically distinguishes the variables in the interface- and boundary conditions defined on surfaces normal to the diagonalization direction.

1 Introduction

Since early 90's in the past century we have been focusing on the acceleration of computer simulations of Partial Differential Equations and accuracy enhancement of the numerical solutions. We have devised a symbolic procedure for systematically and automatically diagonalizing the linear PDEs commonly encountered in mathematical physics and computational engineering applications.

This presentation provides a detailed account of the ideas behind diagonalization, and briefly discusses its implications for designing robust, accurate and accelerated algorithms for solving Boundary Value Problems in both integral- and differential form. We start with a conjecture stating that *all linearized PDEs, viewed as consistent models for physically realizable systems, diagonalizable.* D-forms are equivalent to the parent PDEs. However, they explicitly distinguish the space coordinate with respect to which the diagonalization has been performed. D-forms allow several useful interpretations which are generally deeply hidden in the originating PDEs. We demonstrate the technical details by converting two systems of PDEs into their equivalent D-forms: (i) the electromagnetic wave equations for vector fields in fully bianisotropic and transversally inhomogeneous media, and (ii) magnetostatic equations. For the diagonalization of several other PDEs we refer to [1], [2]. We propose a symbolic notation, which can easily be automated, and replaces tedious calculations by a *simply-by-inspection* manipulatory procedure. In this respect, the present work is a continuation of the work in [1] and [2]. For reasons to be made clear soon we refer to D-representations as

F. Winkler and U. Langer (Eds.): SNSC 2001, LNCS 2630, pp. 347–360, 2003.

the Huygens' Principle in Differential Form. This interpretation can favorably be used for (i) generating novel stencils in the finite difference method, (ii) constructing functionals in the finite element method, and most importantly, (iii) formulating and regularizing singular surface integrals in the Boundary Element Method applications [1], [2]. D-forms in spectral domain are algebraic eigenvalue equations. Using the associated eigenpairs, Green's functions in question can easily be constructed [1], [2]. Furthermore, D-forms facilitate the investigation of the asymptotic properties of the eigenpairs in the far- and near-field in the spectral domain [1], [2]. In accordance with Heisenberg's uncertainty principle these asymptotic expansions correspond to the near- and far-fields in the spatial domain, respectively. In particular, utilizing the far-field expansions in the spectral domain an easy-to-implement recipe for the regularization of singular surface integrals emerges.

2   Diagonalization and Its Interpretation

Consider a differentiable 3-manifold $\mathbf{M}^3$. Let $\mathcal{A}u = 0$ represent a system of linear homogeneous PDEs in Ω with given conditions on the boundary $\partial\Omega$, with Ω and $\partial\Omega \subset \mathbf{M}^3$. At a point $p \in \Omega$ consider the local coordiante system (ξ, η, ζ). We refer to the ζ−direction as the "normal" direction (N-direction), and the plane normal to ζ as the "tangent" plane (T-plane) at p.

Diagonalization with respect to the N-direction transforms $\mathcal{A}u = 0$ into an equivalent form $\mathcal{L}\Psi = \partial_\zeta \Psi$ (D-form) with distinct properties:

1. $\mathcal{L} = \mathcal{L}(\xi, \eta)$. (For time dependent problems the list of arguments contains the time variable t, or, ω, its dual variable in the Fourier domain.)
2. The components of Ψ are *exclusively* those field variables which enter into the interface- (and boundary) conditions on the T−plane. Without any jump discontinuities we obtain $\Psi(\zeta^+) = \Psi(\zeta^-)$, which we preferably will write as $\Psi^< = \Psi^>$.
3. Let Ψ be known and continuously differentiable to any order required in the ϵ−neighborhood $\mathcal{N}(p, \epsilon)$ ($\subset T$−plane). Construct $\mathcal{L}\Psi$ on $\mathcal{N}(p, \epsilon)$. Then, according to $\mathcal{L}\Psi = \partial_\zeta \Psi$, the rate of change of Ψ in the N-direction is uniquely determined at p.
4. The property that $\mathcal{L}$ merely depends on ξ and η allows further conclusions. Differentiate both sides of $\mathcal{L}\Psi = \partial_\zeta \Psi$ with respect to ζ, and interchange the order of ∂_ζ and $\mathcal{L}$ at the L.H.S. to obtain $\mathcal{L}\partial_\zeta \Psi = \partial_\zeta^{(2)} \Psi$. Substitute for $\partial_\zeta \Psi$ from the D-form and introduce $\mathcal{L}^{(2)} = \mathcal{L}\mathcal{L}$ to get $\mathcal{L}^{(2)}\Psi = \partial_\zeta^{(2)} \Psi$. By an induction argument $\mathcal{L}^{(n)}\Psi = \partial_\zeta^{(n)} \Psi$ for $n \in N$.
5. Therefore, in the neighborhood of p in the N-direction, we have:

$$\Psi(\xi, \eta, \zeta + h) = \sum_n \frac{h^n}{n!} \left\{ \mathcal{L}^{(n)}\Psi \right\}_{(\xi,\eta,\zeta)} \qquad (1)$$

6. Consider a closed surface $\gamma \subset \Omega$. At each point (ξ, η, ζ) of γ proceed as sketched above and construct the field Ψ' at all points $(\xi, \eta, \zeta + h)$, such that the entirety of these points build the surface γ', with $\gamma' \subset \Omega$. This interpretation suggests referring to the diagonalization as the Huygence principle in differential form. These considerations allow the construction of useful and fast algorithms for the BEM [3]-[9], Finite Difference Method [10], and the Finite Element Method applications.

7. Consider the Fourier transform of $\mathcal{L}\Psi = \partial_\zeta \Psi$ with respect to the spatial coordinates in the T-plane:

$$\overline{L}(k_\xi, k_\eta)\overline{\Psi} = \lambda\overline{\Psi}, \tag{2}$$

where the bar indicates variables in the $(k_\xi, k_\eta)-$ spectral domain.

8. In view of (2) consider the eigenvalues $\lambda(k_\xi, k_\eta)$ and the associated eigenvectors $\overline{\Psi}(k_\xi, k_\eta)$ for asymptotically large k_ξ and k_η. According to the Heisenberg uncertainty principle the far field in the spectral domain corresponds to the near field in the spatial domain. Therefore, it can be postulated that the $(k_\xi, k_\eta \to \infty)$ asymptotics of the eigenpairs provide the ingredients for expanding the near fields in the spatial domain. Indeed we have shown [1], [2] that this is valid. Furthermore, the asymptotic eigenpairs can be used to regularize surface integrals, with (weak-, strong-, or hyper-) singular kernels [11]: A major source for numerical difficulties in the BEM applications originate in the inadequate treatment of the kernel singularities in surface integrals. This may lead to inaccurate near field distributions and thus divergent or near-divergent numerical results. Regularization techniques using spectral domain asymptotics allow us, however, to "zoom" into these difficulties and "temper" their behavior to any degree we desire.

9. Consider $\mathcal{A}u = s$ with the source function $s = J\delta(\zeta - \zeta_0)$. Here, $\delta(\cdot)$ stands for the Dirac delta function. Diagonalization applied to this equation automatically generates the interface equations. Thereby, in contrast to the expositions in standard text books, there is no need for introducing a thin pillbox, and the subsequent limiting process.

3   Procedural Details of the Diagonalization

We choose a cartesian coordinate system (x, y, z), and perform the diagonalization with respect to the $z-$coordiante. A few comments and preparatory calculations are in place.

3.1   Domain Decomposition

Adopt the following representation for the field[1] $\Psi(x, y, z)$.

$$\Psi(x, y, z) = \Psi^<(x, y, z)\mathcal{H}(z_0 - z) + \Psi^>(x, y, z)\mathcal{H}(z - z_0), \tag{3}$$

---

[1] The differentiability of the involved functions, to any order required, has been assumed.

$\Psi(.)$ may be scalar or a vector. $\mathcal{H}(\cdot)$ is the Heaviside's step function: $\mathcal{H}(z)$ is 0 for $z < 0$, and 1 for $z > 0$. Obviously we have

$$\partial_a \Psi(x,y,z) = \left\{ \partial_a \Psi^<(x,y,z) \right\} \mathcal{H}(z_0 - z) + \left\{ \partial_a \Psi^>(x,y,z) \right\} \mathcal{H}(z - z_0), \quad (4)$$

with a representing x or y. Furthermore, with $\partial_z \mathcal{H}(z_0 - z) = -\delta(z_0 - z) = -\delta(z - z_0)$ and $\partial_z \mathcal{H}(z - z_0) = \delta(z - z_0)$ we obtain

$$\begin{aligned}
\partial_z \Psi(x,y,z) = &\left\{ \partial_z \Psi^<(x,y,z) \right\} \mathcal{H}(z_0 - z) + \left\{ \partial_z \Psi^>(x,y,z) \right\} \mathcal{H}(z - z_0) \\
&+ \left\{ -\Psi^<(x,y,z) + \Psi^>(x,y,z) \right\} \delta(z - z_0).
\end{aligned} \quad (5)$$

3.2   A Convenient Symbolic Representation of the Curl Operator

In order to facilitate our manipulations and to allow symbolic computations we write the curl operator in the form

$$\nabla \times = \begin{bmatrix} 0 & -\partial_z & \partial_y \\ \partial_z & 0 & -\partial_x \\ -\partial_y & \partial_x & 0 \end{bmatrix} = \partial_x \underbrace{\begin{bmatrix} 0 & 0 & 0 \\ 0 & 0 & -1 \\ 0 & 1 & 0 \end{bmatrix}}_{N_1} + \partial_y \underbrace{\begin{bmatrix} 0 & 0 & 1 \\ 0 & 0 & 0 \\ -1 & 0 & 0 \end{bmatrix}}_{N_2} + \partial_z \underbrace{\begin{bmatrix} 0 & -1 & 0 \\ 1 & 0 & 0 \\ 0 & 0 & 0 \end{bmatrix}}_{N_3}$$

$$= \partial_x N_1 + \partial_y N_2 + \partial_z N_3, \quad (6)$$

The matrices N_i $(i = 1, 2, 3)$ are defined in (6). Note that

$$N_i^T N_i = \begin{bmatrix} 1 - \delta_{1i} & 0 & 0 \\ 0 & 1 - \delta_{2i} & 0 \\ 0 & 0 & 1 - \delta_{3i} \end{bmatrix} ; \quad i = 1, 2, 3. \quad (7)$$

(Here, δ_{mn} stands for the Kronecker symbol.) These properties motivate appropriate decompositions of the 3×3 identity matrix I, which will be used for the diagonalization, e.g. consider the decomposition induced by N_1:

$$I = \begin{bmatrix} 1 & 0 & 0 \\ 0 & 0 & 0 \\ 0 & 0 & 0 \end{bmatrix} + \begin{bmatrix} 0 & 0 & 0 \\ 0 & 1 & 0 \\ 0 & 0 & 1 \end{bmatrix} = \underbrace{\begin{bmatrix} 1 & 0 & 0 \\ 0 & 0 & 0 \\ 0 & 0 & 0 \end{bmatrix} \begin{bmatrix} 1 & 0 & 0 \\ 0 & 0 & 0 \\ 0 & 0 & 0 \end{bmatrix}}_{U_1} + \underbrace{\begin{bmatrix} 0 & 0 & 0 \\ 0 & 0 & 1 \\ 0 & -1 & 0 \end{bmatrix} \begin{bmatrix} 0 & 0 & 0 \\ 0 & 0 & -1 \\ 0 & 1 & 0 \end{bmatrix}}_{N_1}$$

$$= U_1 U_1 + N_1^T N_1 \quad (8)$$

The matrix U_1 has been introduced in (8). Similarly, we obtain (9) by introducing U_2 and U_3 mutatis mutandis[2]:

$$I = U_i U_i + N_i^T N_i; \quad i = 2, 3. \quad (9)$$

---

[2] In [1], [2] we have suggested an alternative decomposition of the identity matrix, which was devised for the diagonalization of PDEs arising in the theory of linear elasticity, and piezoelectricity.

The relationships in (10) hold true (0 stands for the 3×3 null matrix):

$$U_i N_i = N_i U_i = U_i N_i^T = N_i^T U_i = 0; \quad i = 1, 2, 3. \tag{10}$$

It is instructive to examine further properties of N_i and establish a connection between these matrices and the proper rotation matrices in R^3.

Matrices N_i and Proper Rotation Matrices R_i. In view of

$$N_1 N_2 = \begin{bmatrix} 0 & 0 & 0 \\ 1 & 0 & 0 \\ 0 & 0 & 0 \end{bmatrix} \quad \text{and} \quad N_2 N_1 = \begin{bmatrix} 0 & 1 & 0 \\ 0 & 0 & 0 \\ 0 & 0 & 0 \end{bmatrix} \tag{11}$$

we obtain $[N_1, N_2] \overset{\text{Def}}{=} N_1 N_2 - N_2 N_1 = N_3$. Therefore, the constituent matrices N_1, N_2, and N_3 of the curl operator satisfy the defining commutation relations of the Lie algebra **so**(3), [12]:

$$[N_1, N_2] = N_3, \quad [N_2, N_3] = N_1, \quad [N_3, N_1] = N_2. \tag{12}$$

Rotation Group SO(3). Next we establish a relationship between N_i and the proper rotation matrices R_i defined in (13). It is known that every proper rotation in R^3 can be expressed in terms of a succession of three proper rotations about the $x-$, $y-$, and $z-$ axis. The corresponding one-parameter subgroups can be generated by the matrices $R_1(\alpha)$, $R_2(\beta)$, and $R_3(\gamma)$:

$$\begin{bmatrix} 1 & 0 & 0 \\ 0 & \cos\alpha & -\sin\alpha \\ 0 & \sin\alpha & \cos\alpha \end{bmatrix}, \quad \begin{bmatrix} \cos\beta & 0 & \sin\beta \\ 0 & 1 & 0 \\ -\sin\beta & 0 & \cos\beta \end{bmatrix}, \quad \begin{bmatrix} \cos\gamma & -\sin\gamma & 0 \\ \sin\gamma & \cos\gamma & 0 \\ 0 & 0 & 1 \end{bmatrix}. \tag{13}$$

Each of these matrices satisfy the following properties: (i) They are orthonormal, meaning that $R_i^T(\theta) R_i(\theta) = I$. (ii) They are proper (special) in the sense that $\det \{R_i\} = 1$. (iii) The group multiplication is an analytic function of parameter θ. More precisely, $R_i(\theta_1) R_i(\theta_2) = R_i(\theta_1 + \theta_2)$. (iv) The inverse operation is an analytic function of parameter θ: $R_i^{-1}(\theta) = R_i(-\theta) = R_i^T(\theta)$.

These matrices form a one-parameter Lie group and since they are *special* and *orthonormal* in R^3, they are called the special orthogonal group in 3 dimensions and denoted by **SO**(3). The basic vectors of the Lie algebra **so**(3) (the infinitesimal generators) are obtained by differentiating $R_i(\theta)$ with respect to the corresponding angle parameter θ and consecutively setting θ equal to zero. This establishes the relationship we have been looking for [12]:

$$\left. \frac{dR_1(\alpha)}{d\alpha} \right|_{\alpha=0} = N_1, \quad \left. \frac{dR_2(\beta)}{d\beta} \right|_{\beta=0} = N_2, \quad \left. \frac{dR_3(\gamma)}{d\gamma} \right|_{\gamma=0} = N_3. \tag{14}$$

A Convenient Symbolic Representation of the Nabla Operator

$$\boldsymbol{\nabla} = \begin{bmatrix} \partial_x \\ \partial_y \\ \partial_z \end{bmatrix} = \underbrace{\begin{bmatrix} 1 & 0 & 0 \\ 0 & 1 & 0 \\ 0 & 0 & 0 \end{bmatrix}}_{N_{12}} \begin{bmatrix} \partial_x \\ \partial_y \\ 0 \end{bmatrix} + \underbrace{\begin{bmatrix} 0 & 0 & 0 \\ 0 & 0 & 0 \\ 0 & 0 & 1 \end{bmatrix}}_{U_3} \begin{bmatrix} 0 \\ 0 \\ \partial_z \end{bmatrix} \tag{15a}$$

$$= N_{12} \begin{bmatrix} \partial_x \\ \partial_y \\ 0 \end{bmatrix} + U_3 \begin{bmatrix} 0 \\ 0 \\ \partial_z \end{bmatrix} \tag{15b}$$

4   Maxwell's Equations in Differential Form

Consider the Maxwell's equations in differential form with $\boldsymbol{J}$ and ρ being, respectively, the electric current density and the charge density:

$$\boldsymbol{\nabla} \times \boldsymbol{H} - \frac{\partial \boldsymbol{D}}{\partial t} = \boldsymbol{J} \tag{16a}$$

$$\boldsymbol{\nabla} \times \boldsymbol{E} + \frac{\partial \boldsymbol{B}}{\partial t} = 0 \tag{16b}$$

$$\boldsymbol{\nabla} \cdot \boldsymbol{B} = 0 \tag{16c}$$

$$\boldsymbol{\nabla} \cdot \boldsymbol{D} = \rho \tag{16d}$$

Applying the dot product of $\boldsymbol{\nabla}$ to both sides of (16a) and using (16d) we obtain the continuity equation $\partial\rho/\partial t + \boldsymbol{\nabla} \cdot \boldsymbol{J} = 0$. The field variables in (16) are as follows. $\boldsymbol{E}$: electric field; $\boldsymbol{D}$: dielectric displacement; $\boldsymbol{H}$: magnetic field; $\boldsymbol{B}$: magnetic induction.

4.1   Constitutive Equations

General Bianisotropic Media. In fully bianisotropic media we have the relationships in (17) for the field vectors $\boldsymbol{D}$ and $\boldsymbol{B}$. The material parameters $\varepsilon_{ij}, \xi_{ij}, \zeta_{ij}$, and μ_{ij}; $i, j = 1, 2, 3$ constitute 3×3 matrices.

$$\boldsymbol{D} = \underline{\underline{\varepsilon}}\boldsymbol{E} + \underline{\underline{\xi}}, \quad \boldsymbol{B} = \underline{\underline{\zeta}}\boldsymbol{E} + \underline{\underline{\mu}}\boldsymbol{H}. \tag{17}$$

Let $\boldsymbol{x}$ represent $\boldsymbol{E}$ or $\boldsymbol{H}$, and $\underline{\underline{\kappa}}$ either of the matrices in (17). General physical properties, such as passivity and reciprocity, translate into specific constraints imposed on the involved material matrices. In present discussion we merely require the existence of these matrices. This means that for general bianisotropic media, for example, we require that $\underline{\underline{\kappa}}\boldsymbol{x} \neq 0$ for every $\boldsymbol{x} \neq 0$. The D-procedure will result in a set of necessary conditions on the material matrices.

General Anisotropic Media. Fully anisotropic media are characterized by the fact that $\underline{\underline{\xi}} = \underline{\underline{\zeta}} = 0$, and that $\underline{\underline{\varepsilon}}\boldsymbol{x} \neq 0$, and $\underline{\underline{\mu}}\boldsymbol{x} \neq 0$ for $\boldsymbol{x} \neq 0$.

Isotropic Media. Here we have $\underline{\xi} = \underline{\zeta} = 0$, in addition to $\underline{\underline{\varepsilon}}x = \alpha I$ and $\underline{\underline{\mu}}x = \beta I$ with $\alpha, \beta \in R^+$.

4.2   Maxwell's Equations in Static Limit ($\partial/\partial t \equiv 0$)

$$\nabla \times H = J, \quad \nabla \times E = 0, \quad \nabla \cdot B = 0, \quad \nabla \cdot D = \rho. \tag{18}$$

Assuming $D = \underline{\underline{\varepsilon}}E$ and $B = \underline{\underline{\mu}}H$, (18) decouples into (19) and (20).

Electrostatic Equations in Anisotropic Dielectric Media

$$\nabla \times E = 0, \quad \nabla \cdot D = \rho, \quad D = \underline{\underline{\varepsilon}}E. \tag{19}$$

Magnetostatic Equations in Anisotropic Magnetic Media

$$\nabla \times H = J, \quad \nabla \cdot B = 0, \quad B = \underline{\underline{\mu}}H. \tag{20}$$

4.3   Diagonalization of the Electrodynamic Equations

Our objective is the diagonalization of the electrodynamic equations in general bianisotropic, source-free media[3] with respect to a coordiante variable, say, z. We adopt a harmonic time-dependence of the fields according to $e^{-j\omega t}$. Then with $\partial/\partial t \Longleftrightarrow -j\omega$ and introducing $\partial/\partial(j\omega x) = \partial/\partial\tilde{x}$, $\partial/\partial(j\omega y) = \partial/\partial\tilde{y}$, and $\partial/\partial(j\omega z) = \partial/\partial\tilde{z}$, we obtain

$$[\partial_{\tilde{x}}N_1 + \partial_{\tilde{y}}N_2 + \partial_{\tilde{z}}N_3[\,H = -\underline{\underline{\varepsilon}}E - \underline{\zeta}H, \tag{21a}$$

$$[\partial_{\tilde{x}}N_1 + \partial_{\tilde{y}}N_2 + \partial_{\tilde{z}}N_3]\,E = \underline{\xi}E + \underline{\underline{\mu}}H. \tag{21b}$$

Multiply (21) by N_3^T, and U_3 from the LHS, and insert the decomposition $U_3U_3 + N_3N_3^T$ for the identity matrix between the material matrices and the field vectors at the RHS, to arrive at (22).

$$[\partial_{\tilde{x}}N_3^T N_1 + \partial_{\tilde{y}}N_3^T N_2]\,H + \partial_{\tilde{z}}N_3^T N_3 H \tag{22a}$$
$$= -(N_3^T \underline{\underline{\varepsilon}}N_3)(N_3^T E) - (N_3^T \underline{\underline{\varepsilon}}U_3)(U_3 E) - (N_3^T \underline{\zeta}N_3)(N_3^T H) - (N_3^T \underline{\zeta}U_3)(U_3 H)$$

$$[\partial_{\tilde{x}}N_3^T N_1 + \partial_{\tilde{y}}N_3^T N_2]\,E + \partial_{\tilde{z}}N_3^T N_3 E \tag{22b}$$
$$= (N_3^T \underline{\xi}N_3)(N_3^T E) + (N_3^T \underline{\xi}U_3)(U_3 E) + (N_3^T \underline{\underline{\mu}}N_3)(N_3^T H) + (N_3^T \underline{\underline{\mu}}U_3)(U_3 H)$$

$$[\partial_{\tilde{x}}U_3 N_1 + \partial_{\tilde{y}}U_3 N_2]\,H + \partial_{\tilde{z}}U_3 N_3 H \tag{22c}$$
$$= -(U_3\underline{\underline{\varepsilon}}N_3)(N_3^T E) - (U_3\underline{\underline{\varepsilon}}U_3)(U_3 E) - (U_3\underline{\zeta}N_3)(N_3^T H) - (N_3^T \underline{\zeta}U_3)(U_3 H)$$

$$[\partial_{\tilde{x}}U_3 N_1 + \partial_{\tilde{y}}U_3 N_2]\,E + \partial_{\tilde{z}}U_3 N_3 E \tag{22d}$$
$$= (U_3\underline{\xi}N_3)(N_3^T E) + (U_3\underline{\xi}U_3)(U_3 E) + (U_3\underline{\underline{\mu}}N_3)(N_3^T H) + (U_3\underline{\underline{\mu}}U_3)(U_3 H)$$

---

[3] Details regarding the inclusion of sources will be exemplified in Sec. 4.4.

To convey the underlying idea consider (22a) in greater detail:

$$
\begin{bmatrix} 0 & 0 & -\partial_{\tilde{x}} \\ 0 & 0 & -\partial_{\tilde{y}} \\ 0 & 0 & 0 \end{bmatrix} \boldsymbol{H} + \partial_{\tilde{z}} \begin{bmatrix} 1 & 0 & 0 \\ 0 & 1 & 0 \\ 0 & 0 & 0 \end{bmatrix} \boldsymbol{H}
$$

$$
= - \begin{pmatrix} \varepsilon_{22} & -\varepsilon_{21} & 0 \\ -\varepsilon_{12} & \varepsilon_{11} & 0 \\ 0 & 0 & 0 \end{pmatrix} \begin{pmatrix} E_2 \\ -E_1 \\ 0 \end{pmatrix} - \begin{pmatrix} 0 & 0 & \varepsilon_{23} \\ 0 & 0 & -\varepsilon_{13} \\ 0 & 0 & 0 \end{pmatrix} \begin{pmatrix} 0 \\ 0 \\ E_3 \end{pmatrix}
$$

$$
- \begin{pmatrix} \zeta_{22} & -\zeta_{21} & 0 \\ -\zeta_{12} & \zeta_{11} & 0 \\ 0 & 0 & 0 \end{pmatrix} \begin{pmatrix} H_2 \\ -H_1 \\ 0 \end{pmatrix} - \begin{pmatrix} 0 & 0 & \zeta_{23} \\ 0 & 0 & -\zeta_{13} \\ 0 & 0 & 0 \end{pmatrix} \begin{pmatrix} 0 \\ 0 \\ H_3 \end{pmatrix} \tag{23}
$$

Obviously, we have been able to achieve several goals by one stroke: (i) The (trivial) third equation ($0 = 0$) can be discarded. (ii) The "transversal" $(x, y)-$, and the "normal" $z-$dependences have been distinctly separated. (iii) At the LHS the matrices are built such, that the $x-$ and $y-$ derivatives act on H_3 only. At the same time, the $z-$derivative acts on the components H_1 and H_2 only. Equation (22b) leads to similar conclusions *mutatis mutandis*.

Next consider (22c). Written explicitly we have:

$$
\begin{bmatrix} 0 & 0 & 0 \\ 0 & 0 & 0 \\ -\partial_{\tilde{y}} & \partial_{\tilde{x}} & 0 \end{bmatrix} \boldsymbol{H} + \partial_{\tilde{z}} \begin{bmatrix} 0 & 0 & 0 \\ 0 & 0 & 0 \\ 0 & 0 & 0 \end{bmatrix} \boldsymbol{H}
$$

$$
= - \begin{pmatrix} 0 & 0 & 0 \\ 0 & 0 & 0 \\ \varepsilon_{32} & -\varepsilon_{31} & 0 \end{pmatrix} \begin{pmatrix} E_2 \\ -E_1 \\ 0 \end{pmatrix} - \begin{pmatrix} 0 & 0 & 0 \\ 0 & 0 & 0 \\ 0 & 0 & \varepsilon_{33} \end{pmatrix} \begin{pmatrix} 0 \\ 0 \\ E_3 \end{pmatrix}
$$

$$
- \begin{pmatrix} 0 & 0 & 0 \\ 0 & 0 & 0 \\ \zeta_{32} & -\zeta_{31} & 0 \end{pmatrix} \begin{pmatrix} H_2 \\ -H_1 \\ 0 \end{pmatrix} - \begin{pmatrix} 0 & 0 & 0 \\ 0 & 0 & 0 \\ 0 & 0 & \zeta_{33} \end{pmatrix} \begin{pmatrix} 0 \\ 0 \\ H_3 \end{pmatrix} \tag{24}
$$

Several facts are noteworthy: (i) The $z-$ variation has been eliminated. (ii) The (trivial) first two equations ($0 = 0$) can be discarded. (iii) The remaining equation allows us to express the "normal" field components E_3 and H_3 in terms of the "transversal" field components and their derivatives with respect to the "transversal" variables. Equation (22d) leads to similar conclusions *mutatis mutandis*. In summary we obtain

$$
\partial_{\tilde{z}} \begin{pmatrix} E_1 \\ E_2 \\ H_1 \\ H_2 \end{pmatrix} = \begin{pmatrix} \xi_{21} & \xi_{22} & \mu_{21} & \mu_{22} \\ -\xi_{11} & -\xi_{12} & -\mu_{11} & -\mu_{12} \\ -\varepsilon_{21} & -\varepsilon_{22} & -\zeta_{21} & -\zeta_{22} \\ \varepsilon_{11} & \varepsilon_{12} & \zeta_{11} & \zeta_{12} \end{pmatrix} \begin{pmatrix} E_1 \\ E_2 \\ H_1 \\ H_2 \end{pmatrix}
$$

$$
+ \begin{pmatrix} \xi_{23} & \mu_{23} \\ -\xi_{13} & -\mu_{13} \\ -\varepsilon_{23} & -\zeta_{23} \\ \varepsilon_{13} & \zeta_{13} \end{pmatrix} \begin{pmatrix} E_3 \\ H_3 \end{pmatrix} + \begin{bmatrix} \partial_{\tilde{x}} & 0 \\ \partial_{\tilde{y}} & 0 \\ 0 & \partial_{\tilde{x}} \\ 0 & \partial_{\tilde{y}} \end{bmatrix} \begin{pmatrix} E_3 \\ H_3 \end{pmatrix} \tag{25}
$$

$$\begin{pmatrix} E_3 \\ H_3 \end{pmatrix} = \begin{pmatrix} \varepsilon_{33} & \zeta_{33} \\ -\xi_{33} & -\mu_{33} \end{pmatrix}^{-1} \begin{pmatrix} -\varepsilon_{31} & -\varepsilon_{32} & -\zeta_{31} & -\zeta_{32} \\ \xi_{31} & \xi_{32} & \mu_{31} & \mu_{32} \end{pmatrix} \begin{pmatrix} E_1 \\ E_2 \\ H_1 \\ H_2 \end{pmatrix}$$

$$+ \begin{pmatrix} \varepsilon_{33} & \zeta_{33} \\ -\xi_{33} & -\mu_{33} \end{pmatrix}^{-1} \begin{bmatrix} 0 & 0 & \partial_{\tilde{y}} & -\partial_{\tilde{x}} \\ \partial_{\tilde{y}} & -\partial_{\tilde{x}} & 0 & 0 \end{bmatrix} \begin{pmatrix} E_1 \\ E_2 \\ H_1 \\ H_2 \end{pmatrix} \tag{26}$$

Substituting[4] (26) into (25) and introducing $\Psi = (E_1, E_2, H_1, H_2)^T$ lead to

$$\mathcal{L}(\varepsilon_{ij}, \xi_{ij}, \zeta_{ij}, \mu_{ij}, \partial_{\tilde{x}}, \partial_{\tilde{y}})\Psi = \partial_{\tilde{z}}\Psi. \tag{27}$$

4.4 Diagonalization of the Magnetostatic Equations

Our starting point are the equations in (20), [13]. We first assume a source free medium ($\boldsymbol{J} \equiv 0$), and then consider fields which are driven by a source function. Equation (20) implies that $\boldsymbol{B}$ can be expressed as the curl of a vector potential $\boldsymbol{A}$. Under these conditions (20) implies

$$\boldsymbol{\nabla} \times \boldsymbol{H} = 0, \tag{28a}$$

$$\boldsymbol{\nabla} \times \boldsymbol{A} = \underline{\mu}\boldsymbol{H}. \tag{28b}$$

Our goal is to investigate the diagonalization of (28) with respect to the $z-$coordiante. The occurance of the curl operator in both equations suggests a procedure similar to that we developed in the previous section. We insert $N_3 N_3^T + U_3 U_3$ for the indentity matrix I between $\underline{\mu}$ and $\boldsymbol{H}$ at the RHS of (28b), and multiply both equations, first by N_3^T, and then by U_3, from the LHS. The resulting equations collapse in

$$\partial_z \begin{pmatrix} H_1 \\ H_2 \end{pmatrix} = \begin{bmatrix} \partial_x \\ \partial_y \end{bmatrix} H_3, \tag{29a}$$

$$\partial_z \begin{pmatrix} A_1 \\ A_2 \end{pmatrix} = \begin{bmatrix} \partial_x \\ \partial_y \end{bmatrix} A_3 + \begin{pmatrix} \mu_{21} & \mu_{22} \\ -\mu_{11} & -\mu_{12} \end{pmatrix} \begin{pmatrix} H_1 \\ H_2 \end{pmatrix} + \begin{pmatrix} \mu_{23} \\ -\mu_{13} \end{pmatrix} H_3, \tag{29b}$$

$$[-\partial_y \ \partial_x] \begin{pmatrix} H_1 \\ H_2 \end{pmatrix} = 0, \tag{29c}$$

$$[-\frac{1}{\mu_{33}}\partial_y \ \frac{1}{\mu_{33}}\partial_x] \begin{pmatrix} A_1 \\ A_2 \end{pmatrix} - (\frac{\mu_{31}}{\mu_{33}} \ \frac{\mu_{32}}{\mu_{33}}) \begin{pmatrix} H_1 \\ H_2 \end{pmatrix} = H_3. \tag{29d}$$

---

[4] Note that the existence of the above inverse matrix is necessary to perform our manipulations. In particular its determinant should nonzero: $\varepsilon_{33}\mu_{33} - \zeta_{33}\xi_{33} \neq 0$. Note that diagonalization with respect to the $x-$, and $y-$ coordinates results in similar inequalities. Therfeore, we conclude that for physically realizable systems the necessary conditions $\varepsilon_{ii}\mu_{ii} - \zeta_{ii}\xi_{ii} \neq 0$, ($i = 1, 2, 3$) have to be satisfied.

The appearance of the $z-$ derivatives of H_1, H_2, A_1, and A_2 in (29a) and (29b) suggest that these are the only field components which will enter into the D-form (permissible, reduced variables). Therefore, H_3 and A_3 must be eliminated by using (29c) and (29d). However, we realize that this is not possible: While (29d) is an expression for H_3 in terms of the admissible field variables[5], there is no equation available which would allow the elimination of A_3. (Equations (29c) and (29d) do not involve A_3.) *Therefore, according to our conjecture, (28) does not model a physically realizable system.* Alternatively, it can be stated that: *Assuming a general 3×3 matrix $\underline{\underline{\mu}}$ and allowing spatial variation with respect to all the three spatial coordiantes, the equations (28) are not consistent.*

Therefore, the primary objective is now to create conditions under which A_3 disappears from our equations. We can achieve this goal by assuming that there is no variation in one of the transversal directions, and/or specializing the structure of $\underline{\underline{\mu}}$. Thus, assume $\partial/\partial y \equiv 0$. Then (29) reduces to (30).

$$\partial_z H_1 = \partial_x H_3 \tag{30a}$$

$$\partial_z H_2 = 0 \tag{30b}$$

$$\partial_z A_1 = \partial_x A_3 + \mu_{21} H_1 + \mu_{22} H_2 + \mu_{23} H_3 \tag{30c}$$

$$\partial_z A_2 = -\mu_{11} H_1 - \mu_{12} H_2 - \mu_{13} H_3 \tag{30d}$$

$$\partial_x H_2 = 0 \tag{30e}$$

$$H_3 = \frac{1}{\mu_{33}} \partial_x A_2 - \frac{\mu_{31}}{\mu_{33}} H_1 - \frac{\mu_{32}}{\mu_{33}} H_2 \tag{30f}$$

Interpretation

1. Equations (30b) and (30e) imply that H_2 is a trivial solution.
2. In (30d) the undesirable field component A_3 does not appear. In order to eliminate the trivial solution H_2 from (30d) we have to set $\mu_{12} = 0$.
3. Equation (30f) is needed to eliminate H_3. To eliminate the trivial solution H_2 from this equation the condition $\mu_{32} = 0$ is necessary.
4. Setting $\mu_{21} = 0$, H_1 disappears from (30c). However, H_1 enters this equation over H_3(see (30f))! Therfeore, $\mu_{23} = 0$.

In summary it can be stated that under the conditions that

$$\underline{\underline{\mu}} = \begin{pmatrix} \mu_{11} & 0 & \mu_{13} \\ 0 & \mu_{22} & 0 \\ \mu_{31} & 0 & \mu_{33} \end{pmatrix} \qquad \text{and} \qquad \partial_y \equiv 0 \tag{31}$$

(30) decouples into (32) and (33).

$$\partial_z H_1 = \partial_x H_3 \tag{32a}$$

$$\partial_z A_2 = -\mu_{11} H_1 - \mu_{13} H_3 \tag{32b}$$

$$H_3 = \frac{1}{\mu_{33}} \partial_x A_2 - \frac{\mu_{31}}{\mu_{33}} H_1 \tag{32c}$$

---

[5] In view of (29d) $\mu_{33} \neq 0$. Analogeously, diagonalization with respect to $x-$, and $y-$ coordinates require that $\mu_{11} \neq 0$, and $\mu_{22} \neq 0$, respectively.

$$\partial_z H_2 = 0 \tag{33a}$$

$$\partial_z A_1 = \partial_x A_3 + \mu_{22} H_2 \tag{33b}$$

$$\partial_x H_2 = 0 \tag{33c}$$

Focus on the latter system first: (33a) and (33c) imply that $H_2 = \text{const}$ in the entire space. Considering finite energy functions we have $H_2 = 0$. Therefore, from (33b) we obtain $\partial_z A_1 = \partial_x A_3$, which cannot be diagonalized, unless A_1 and A_3 vanish simultaneously. Therefore: $A_1 = A_3 = H_2 = 0$.

Consider (32) next: Substitute (32c) into (32a) and (32b), and rearrange to obtain the desired D-form:

$$\partial_z \begin{pmatrix} A_2 \\ H_1 \end{pmatrix} = \begin{pmatrix} -(\mu_{13}/\mu_{33})\partial_x & -(\mu_{11}\mu_{33} - \mu_{13}\mu_{31})/\mu_{33} \\ \partial_x(1/\mu_{33})\partial_x & -\partial_x(\mu_{31}/\mu_{33}) \end{pmatrix} \begin{pmatrix} A_2 \\ H_1 \end{pmatrix}. \tag{34}$$

It should be pointed out that the coefficients in our PDEs may vary as functions of x, which is the independent variable on the T-plane[6].

Inclusion of Idealized Sources. Consider in (20) a uniform ($J_2 = \text{const}$) electric current line:

$$\boldsymbol{J} = J_2 \boldsymbol{e}_2 \delta(x - x_0)\delta(z - z_0) \tag{35}$$

($\boldsymbol{e}_2$: unit vector in y-direction) Note that since this "source distribution" satisfies the condition $\partial/\partial_y \equiv 0$, the consistency requirement is satisfied.

Repeating the D-procedure in this Section along with our formulation in terms of the Heaviside's function in (5) we obtain

$$\left\{ \begin{pmatrix} -(\mu_{13}/\mu_{33})\partial_x & -(\mu_{11}\mu_{33} - \mu_{13}\mu_{31}/\mu_{33}) \\ \partial_x(1/\mu_{33})\partial_x & -\partial_x(\mu_{31}/\mu_{33}) \end{pmatrix} \begin{pmatrix} A_2 \\ H_1 \end{pmatrix} \right\}^< \mathcal{H}(z_0 - z)$$

$$+ \left\{ \begin{pmatrix} -(\mu_{13}/\mu_{33})\partial_x & -(\mu_{11}\mu_{33} - \mu_{13}\mu_{31}/\mu_{33}) \\ \partial_x(1/\mu_{33})\partial_x & -\partial_x(\mu_{31}/\mu_{33}) \end{pmatrix} \begin{pmatrix} A_2 \\ H_1 \end{pmatrix} \right\}^> \mathcal{H}(z - z_0)$$

$$= \left\{ \partial_z \begin{pmatrix} A_2 \\ H_1 \end{pmatrix} \right\}^< \mathcal{H}(z_0 - z) + \left\{ \partial_z \begin{pmatrix} A_2 \\ H_1 \end{pmatrix} \right\}^> \mathcal{H}(z - z_0) \tag{36}$$

$$+ \left\{ -\begin{pmatrix} A_2 \\ H_1 \end{pmatrix}^< + \begin{pmatrix} A_2 \\ H_1 \end{pmatrix}^> \right\} \delta(z - z_0) + \begin{pmatrix} 0 \\ -J_2\delta(x - x_0) \end{pmatrix} \delta(z - z_0)$$

Note that the definition range of $\mathcal{H}(z_0 - z)$, $\mathcal{H}(z - z_0)$, and $\delta(z - z_0)$ are mutually exclusive and their union spans the entire z-axis. Using this property and individually equating to zero terms which are associated with $\mathcal{H}(z_0 - z)$, $\mathcal{H}(z - z_0)$, and $\delta(z - z_0)$, we obtain the governing equations in $z < z_0$, and $z > z_0$, and the interface condition at $z = z_0$, respectively.

---

[6] A condition for obtaining meaningful results is that $\mu_{11}\mu_{33} - \mu_{13}\mu_{31} \neq 0$. Similarly, diagonalization with respect to the $x-$ and $y-$ directions results in the conditions $\mu_{22}\mu_{33} - \mu_{23}\mu_{32} \neq 0$ and $\mu_{11}\mu_{22} - \mu_{12}\mu_{21} \neq 0$.

4.5   Diagonalization of the Electrostatic Equations

Consider (19). Let $\underline{\underline{\varepsilon}}$ be a general 3×3 matrix. In contrast to Sec. 4.4, our analysis here is three-dimensional ($\partial/\partial y \neq 0$).

Equation (19) implies that $\boldsymbol{E}$ can be written as a gradient of a scalar potential function $\boldsymbol{E} = -\nabla\varphi$. Therefore, in source-free media the governing equations are $\nabla \cdot \boldsymbol{D} = 0$, and $\boldsymbol{D} = -\underline{\underline{\varepsilon}}\nabla\varphi$.

It can be shown [1], [2] that the following is valid[7]

$$\begin{pmatrix} L_{11} & L_{12} \\ L_{21} & L_{22} \end{pmatrix} \begin{pmatrix} \varphi \\ D_3 \end{pmatrix} = \partial_z \begin{pmatrix} \varphi \\ D_3 \end{pmatrix}. \tag{37}$$

Written explicitely L_{ij} ($i, j = 1, 2$) are:

$$L_{11} = -\left(\varepsilon_{13}/\varepsilon_{33} \;\; \varepsilon_{23}/\varepsilon_{33} \right) \begin{bmatrix} \partial_x \\ \partial_y \end{bmatrix} = L_{22}^T \tag{38a}$$

$$L_{12} = -1/\varepsilon_{33} \tag{38b}$$

$$L_{21} = \begin{bmatrix} \partial_x \\ \partial_y \end{bmatrix} \begin{pmatrix} (\varepsilon_{11}\varepsilon_{33} - \varepsilon_{13}^2)/\varepsilon_{33} & (\varepsilon_{12}\varepsilon_{33} - \varepsilon_{13}\varepsilon_{23})/\varepsilon_{33} \\ (\varepsilon_{21}\varepsilon_{33} - \varepsilon_{13}\varepsilon_{23})/\varepsilon_{33} & (\varepsilon_{22}\varepsilon_{33} - \varepsilon_{23}^2)/\varepsilon_{33} \end{pmatrix} \begin{bmatrix} \partial_x \\ \partial_y \end{bmatrix} \tag{38c}$$

5   Conclusion

In this contribution we conjectured that systems of linearized Partial Differential Equations, considered as consistent models for physically realizable systems, are diagonalizable. We examined two problems in computational electromagnetics and showed that the associated PDEs are diagonalizable. We made clear that the diagonalization with respect to a given spatial coordinate is a process for systematically "reshuffling" the original PDEs, and creating equivalent reduced differential forms. Furthermore, we demosntrated that the diagonalization procedure automatically selects the variables which enter into the interface- and boundary eqautions related to surfaces normal to the diagonalization direction.

It should be pointed out that using our regularization technique moments of Green's functions associated with a given Boundary Value Problems can be written in coordinate-free, frequency-, and material independent forms, and therefore, these moments can be regarded as *universal functions* for the underlying class of problems [3]. The universal functions are generally astonishingly smooth; they can be precalculated, stored, and thus retrieved as often as required (data recycling). This capability allows us to separate the scientific computing efforts from the pre- and postprocessing steps in simulations. These considerations have been instrumental in the development of the Fast-MoM, which is an accelerated

---

[7] Notice that we are requiring that $\varepsilon_{33} \neq 0$. The diagonalization with respect to the $x-$ and $y-$ coordinates would require that $\varepsilon_{11} \neq 0$ and $\varepsilon_{22} \neq 0$, respectively. Therefore, necessary conditions for physical realizability of the present problem are $\varepsilon_{ii} \neq 0$, with $i = 1, 2, 3$.

form of the conventional Method of Moments [4]-[7]. Using precalculated universal functions the computation times for the calculation of *impedance matrices* in the BEM applications reduce to the times required for retrieving data from the chosen storage medium. However, in spite of this advancement, a major drawback in the BEM still remains to be removed: the impedance matrices are dense. Several techniques have been suggested in literature for obtaining sparse matrices, each with its own limitation. We have suggested a procedure which consists of constructing problem-specific orthogonal sequences of basis functions derived from the involved Green's functions [18], [19], and [20]. The idea is to expand the unknowns in our problems in terms of basis functions which embed in their structures intrinsic features of the underlying PDEs: Using Meyer's orthogonalization technique [14]-[17], and Green's functions, we have constructed functions which are orthonormal to their integer-translates [18], [19], and [20]. In the case of Laplace operator, we have been able to prove that the resulting functions support a multiresolution analysis, and leads to *Green's-functions-based scaling functions and wavelets*.

Our present research includes the investigation of alternative *localization* techniques for generating sparse impedance matrices in the MoM applications, e.g. coherent states and Wannier fucntions. Furthermore, we are investigating the *existence* question of the diagonalized forms: Why is it possible to diagonalize a certain systems of linearized PDEs in the first place? And, viewed from a physical prospective, what are the necessary and sufficient conditions for their formulation.

References

1. Baghai-Wadji, A.R. (2001): Theory and Applications of Green's Functions. In: Ruppel, C.C.W., Fjeldly, T.A. (eds) Advances in Surface Acoustic Wave Technology, Systems and Applications Vol. 2. World Scientific
2. Baghai-Wadji, A.R. (1994): A Unified Approach for Construction of Green's Functions. Habilitation Thesis, Vienna University of Technology, Vienna
3. Baghai-Wadji, A.R. (1997): Fast-MoM: A Method-of-Moments Formulation for Fast Computations. ACES, Vol. 12, No. 2, 75–80
4. Harrington, R.F. (1968): Field Computation by Moment Methods. Macmillan, New York
5. M. M. Ney, "Method of moments as applied to electromagnetic problems," *IEEE Trans. Microwave Theory Tech.*, vol. MTT-33, pp. 972-980, 1985.
6. Muskhelishvili, N.I. (1953): Singular Integral Equations. Holland
7. Brebbia, C.A., Telles, J.C.F., Wroble, L.C. (1984): Boundary Element Techniques. Springer Verlag
8. A. R. Baghai-Wadji A.R., Penunuri, D. (1995): Coordinate-Free, Frequency-Independent Universal Functions for BAW Analysis in SAW Devices. Proc. IEEE-SU, 287–290
9. Ramberger, S., Baghai-Wadji, A.R., (2002): Exact Eigenvector- and Eigenvalue Derivatives with Applications to Asymptotic Green's Functions. Proc. ACES, Monterey

10. Varis, K., Baghai–Wadji, A.R. (2001): Hyrid Planewave/Finite-Difference Transfer Method for Solving Photonic Crystals in Finite Thickness Slabs. Proc. EDMO, Vienna, 161–166
11. Kanwal, R.P. (1983): Generalized Functions: Academic Press
12. Adams, B.G. (1994): Algebraic Approach to Simple Quantum Systems. Springer Verlag
13. Manuzuri-Shalmani, M.T. (1996): Spatial Domain Analysis of Planar Structures in Anisotropic Media. PhD Dissertation, Vienna University of Technology, Vienna
14. Y. Meyer, Y. (1992): Wavelets and Operators. Cambridge University Press
15. Daubechies, I. (1992): Ten Lectures on Wavelets. SIAM
16. G.G. Walter, G.G., (1994): Wavelets and Other Orthogonal Systems with Applications. CRC Press
17. Hernandez, E., Weiss, G. (1996): A First Course on Wavelets. CRC Press
18. Baghai-Wadji, A.R., Walter, G.G. (2000): Green's Function-Based Wavelets. Proc. IEEE-SU, Puerto Rico
19. Baghai-Wadji, A.R., Walter, G.G. (2000): Green's Function-Based Wavelets: Selected Properties. Proc. IEEE-SU, Puerto Rico
20. Baghai-Wadji, A.R., Walter, G.G. (2002): Green's Function Induced Wavelets- and Wavelet-like Orthogonal Systems for EM Applications. Proc. ACES, Monterey

Generation of the Quasi-solitons in the Lasers: Computer Algebra Approach to an Analysis

Vladimir L. Kalashnikov

Institut für Photonik, TU Wien, Gusshausstr. 27/387, A-1040 Vienna, Austria
vladimir.kalashnikov@tuwien.ac.at

1 Introduction

The recent progress in the generation of the extremely short laser pulses has allowed to reach the pulse durations of few femtoseconds and the peak intensities higher than 10^{15} W/cm^2 (for review see [1]). Their applications cover the range from the technology and the medicine to the sophisticated spectroscopical researches and the nuclear fusion. The femtosecond lasers possess the rich nonlinear properties, which dramatically complicate the field dynamics. There are two main approaches to the theoretical analysis of the femtosecond pulse dynamics. In fact, these approaches reproduce the mainstreams of the nonlinear physics in general.

The former is based on the numerical simulations of the field dynamics and the field-matter interaction. Even a comparatively simple model of this type needs the computations taking about of one week (1.2 GHz Athlon) for the simulation of the real lasing taking less than 10 μs. The main obstacle in the direct numerical simulations is not only the limits of the computer performance, but also the large set of the parameters governing the laser dynamics. This excludes the direct enumeration for the analysis of the system's global properties. Moreover, the obtained results, as a rule, are very complicated for the clear physical comprehension.

On the other hand, the use of the analytical and semi-analytical methods can give us the Ariadna's thread in the labyrinth of the nonlinear phenomena. The basis of the analytical approach to the research of the femtosecond laser dynamics is the so-called soliton model [2]. In the framework of this approach, the nonlinear evolution is approximated by the distributed model, which, as a rule, leads to the analog of the Landau-Ginzburg's type equation [3]. The stable soliton-like solutions of this equation are considered as the ultrashort pulses (see, for example, [2,4]). This approach reduces the problem to the algebraic one and allows the computer algebra realization enhancing the physical comprehension of the underlying phenomena and the result's reproducibility.

However, the main obstacle here is the lack of the exact solutions describing a non-steady-state evolution. Moreover, a lot of the important problems lies at the border of the basic approximations validity that demands the additional analysis involving the numerical simulations.

Here we present the general approach to the analysis of the ultrashort pulse dynamics in the solid-state lasers, which is based on the analytical and semi-

F. Winkler and U. Langer (Eds.): SNSC 2001, LNCS 2630, pp. 361–374, 2003.

analytical consideration and allows the computer algebra realization. The comparisons with the experiment and the numerical results demonstrate the validity of the models under consideration and give the clear and comparatively simple understanding of the underlaying phenomena.

2   Dimension Reduction

As it was mentioned, the analytical approach, as a rule, is based on the soliton model of the ultrashort pulse generation. Here we shall demonstrate that the dimension reduction allows to obtain the physically important results on the basis of the relatively simple maps.

Let us consider the four-dimensional model taking into account the evolution of the pulse peak intensity, the wave front radius of curvature, the laser beam diameter and the intracavity net-gain [5].

During the past years a great deal of attention has been paid to the method of the ultrashort pulse generation called as the Kerr-lens mode locking. The basic idea here is the saturation of the diffraction loss by the high-intensive field due to its self-focusing in the laser active medium [2]. The propagation of the field $a(z, r, t)$ (z is the longitudinal coordinate, r is the radial coordinate for the cylindrically symmetric laser field, t is the local time) through nonlinear medium can be described on the basis of the nonlinear parabolic equation [6]:

$$\left(\frac{\partial}{\partial z} + i\frac{\Delta_\perp}{2k} + i\beta|a|^2 \right) a = 0, \tag{1}$$

where the laser beam has the Gaussian radial profile:

$$a(r) = E_0 \rho^{-1/2} exp\left[\frac{-k_0 r^2}{2\rho} - \frac{ikr^2}{2R} \right]. \tag{2}$$

R is the radius of the wave front curvature, $\rho = \pi w^2/\lambda$ is related to the laser beam diameter w, $k = 2\pi/\lambda$, λ is the wavelength, $\Delta_\perp = (1/r^2)(\partial/\partial r)[r^2(\partial/\partial r)]$, $\beta = k\, n_2/n$ is the coefficient of the Kerr-type nonlinearity, n_2 and n are the nonlinear and linear refraction coefficients, respectively. We shall suppose that $|a|^2$ has the dimension of the field intensity I.

From Eqs. (1, 2) the maps for R and ρ can be obtained:

$$\rho_{z+1} = \frac{(C_1 x + C_2)^2 + C_3}{C_1}, \tag{3}$$

$$R_{z+1} = \frac{\rho_{z+1}}{C_1 x + C_2} \tag{4}$$

where z is the field's round-trip number (i.e. we use the normalization of the propagation distance to the laser cavity length), x is the active crystal length, $C_1 = \rho_z/R_z^2 + C_3/\rho_z$, $C_2 = \rho_z/R_z$ and $C_3 = 1 - 2\beta\rho_z I_z$.

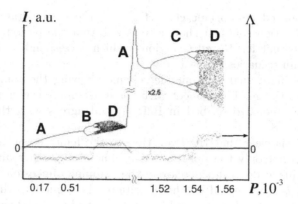

Fig. 1. The ultrashort pulse intensity I and the Lyapunov's parameter Λ versus the dimensionless pump P. $T_{cav}=10$ ns, $t_p=1$ ps.

Then the maps for the field intensity and the dimensionless gain in the active medium are:

$$I_{z+1} = I_z \exp \left(g_z - l - kr^2/D' \right), \tag{5}$$

$$g_{z+1} = g_z + (g_m - g_z) P - \frac{2g_z I_z t_p}{E_g} - \frac{g_z T_{cav}}{T_r}, \tag{6}$$

where g and l are the gain and loss coefficients, respectively, g_m is the maximal gain corresponding to the full population inversion of the quasi-two level active medium, T_{cav} is the cavity period, T_r is the gain relaxation time, the E_g parameter has a sense of the gain saturation energy. The saturable diffraction loss is described by the last term in Eq. (5) (D' is the diameter of the intracavity aperture). The key control parameter is $P=\sigma_a I_p T_{cav}/(h\nu_a)$ (here σ_a is the pump absorption cross-section, I_p is the pump intensity, ν_a is the frequency of the pumping field), which is the dimensionless pump energy stored during T_{cav}. It is very conventional to normalize the field intensity to the nonlinear parameter β. Then the third term in Eq. (6), which contains the definition of the soliton-like ultrashort pulse energy $2I_z t_p$ (t_p is the pulse width), has to be rewritten as $2\tau I_z$ ($\tau=t_p/(\beta E_g)$). Let us confine ourself to the picosecond pulse generation. In this case the pulse duration is defined by the net-group-delay dispersion and the geometrical configuration of the laser cavity, so that the pulse duration is approximately constant and is not mapped.

The result of the simulations on the basis of this simplest model is shown in Fig. 1 for the parameters corresponding to the picosecond Ti:sapphire laser. One can see, that the variation of the P parameter causes the dramatic changes in the field intensity behavior (left axes) and in the corresponding Lyapunov's parameter Λ (right axes). There exist the regions of the stable operation (A), which change into the regions with the oscillatory and chaotical behavior (B, C and D). So, the region of the pump providing the ultrashort pulse generation has the pronounced heterogenic character with the steady-state, chaotic and period multiplication zones. It should be noted, that the period-multiplication behavior is different for B and C regions, which correspond to Ruelle-Takens and Fiegenbaum scenarios, respectively.

It is surprisingly, that such simple scheme allowing the computer algebra realization is in the good agreement with the exact numerical simulation based on the detailed model described in Ref. [7] and agrees with the experiment reported in Ref. [8].

However, it is clear that the transition to the femtosecond generation demands to map not only the pulse intensity, but also another ultrashort pulse parameters. This leads to the increase of the mapping dimension. The most appropriate way is to be based on the distributed laser model, which takes into consideration the saturable gain and loss action, the linear and spectral dissipation, the phase nonlinearity and the group-delay dispersion. As it was mentioned above, this leads to the generalized Landau-Ginzburg's equation:

$$\frac{\partial a(z,t)}{\partial z} = \left[g - l + \left(t_f^2 + id\right)\frac{\partial^2}{\partial t^2} - i\beta \left|a(z,t)\right|^2 + \hat{L}(|a|^2)\right] a(z,t), \qquad (7)$$

where we excluded the explicit dependence on r by means of the assumption that the spatial evolution contributes only to the nonlinear operator $\hat{L}(|a|^2)$ causing the change of the field loss on the aperture due to self-focusing. The last process is governed by the field intensity that validates our approximation. Moreover, we introduced the action of the spectral filter with the inverse bandwidth t_f and the group-delay dispersion with the corresponding coefficient d.

For the pulse-like solutions the boundary conditions correspond to the vanishing at $t \to \pm\infty$ field. However, in the case of the numerical simulations we have to put obviously the periodic boundary condition through the identification of $t = -T_{cav}/2$ with $t = T_{cav}/2$.

The form of $\hat{L}$ can be different. In the weak-nonlinear limit we have $\hat{L}=\sigma |a|^2$ (cubic nonlinear Landau-Ginzburg's equation). The more realistic approximation is $\hat{L}= \sigma |a|^2 - \mu |a|^4$ (quintic nonlinear Landau-Ginzburg's equation). Here the key parameter is $\sigma=10^{-10} \div 10^{-12}$ cm^2/W, which has a sense of the inverse intensity of the diffraction loss saturation and can be controlled by the cavity alignment [9]. The saturation of the nonlinearity due to the higher-order nonlinear term with the μ parameter prevents from the collapse of the soliton-like solution of Eq. (7) [4].

The cubic Landau-Ginzburg's equation has a soliton-like solution in the form $a(t)=a_0 \cosh\left((t - \delta)/t_p\right)^{-1-i\psi} \exp(i\phi)$, where a_0, t_p, δ, ψ and ϕ are the ultrashort pulse amplitude, its width, the time delay, the chirp and the phase delay, respectively. However, the general non-steady-state pulse-like solution is not known even for this simplest form of $\hat{L}$. In order to analyze the field evolution in the general case and for the more complicated nonlinearities we have to perform the numerical simulations.

Nevertheless there is the possibility for the dimension reduction, which is based on the so-called aberrationless approximation [10,11]: let us suppose that the pulse shape conserves approximately, but the pulse parameters depend on z. Then the backward substitution into Eq. (7) results in the set of the first-order ordinary differential equations for the pulse parameters evolution. Such system allows the comparatively simple consideration in the framework of the computer algebra system.

Let us consider the cubic Landau-Ginzburg's equation, but with the saturation of the nonlinearity to avoid the collapse. The simplest choice is $\hat{L}=\sigma_0(1 - \sigma_0 a_0(z)^2/2)\,|a(z,t)|^2$, where the weak-nonlinear coefficient σ_0 is saturated by the evolutive pulse peak intensity $a_0(z)^2$. Then from Eq. (7) we have the four-dimensional mapping for the ultrashort pulse parameters:

$$t_p(z+1) = t_p(z)+$$
$$\frac{4 - 7d\psi(z) - 3\psi(z)^2 \left(\phi\psi(z) - 2\sigma a_0(z)^2 - \psi(z)\beta a_0(z)^2\right) t_p(z)^2}{2t_p(z)^2},$$

$$\psi(z+1) = \left(1 - 2\sigma a_0(z)^2\right)\psi(z) + \left(\phi - \beta a_0(z)^2\right)\psi(z)^2+$$
$$\phi - 3\beta a_0(z)^2 - \frac{5d + 3\psi(z) + 5d\psi(z)^2 + 3\psi(z)^2}{t_p(z)^2}, \qquad (8)$$

$$a_0(z+1) = a_0(z) \left[1 + \frac{d\psi(z) - 1 + \left(g(z) - l - \sigma a_0(z)^2\right) t_p(z)^2}{t_p(z)^2}\right],$$

$$g(z+1) = g(z) \exp\left(-2\,\tau\, a_0(z)^2 t_p(z)^2 - \frac{T_{cav}}{T_r} - P\right)+$$
$$\frac{g_m P}{P + T_{cav}/T_r} \left[1 - \exp\left(-\frac{T_{cav}}{T_r} - P\right)\right],$$

where $\sigma = \sigma_0(1 - \sigma_0 a_0(z)^2/2)$, $\delta = 0$, $\phi = \beta a_0(z)^2 + (d + \psi(z))/t_p(z)^2$ and $\tau = t_f/(\beta E_g)$ (note the replacement of t_p by t_f). As the initial conditions we can take the pulse parameters corresponding to the quasi-soliton solution of the cubic Landau-Ginzburg's equation.

The stable stationary points of this system correspond to the steady-state generation of the ultrashort pulse. But there exists also the unsteady-state operation, which has regular as well as irregular character. The corresponding regions of the group-delay dispersion are shown in Fig. 2, which depicts the dependance of the pulse duration t_p and its chirp ψ on the d coefficient for the steady-state regime. The A region corresponds to the chaotic behavior of the pulse parameters; B shows the region, where there is no the quasi-soliton solution (divergent mapping); in the C region there are the periodical oscillations of the pulse parameters (most pronounced for a_0^2).

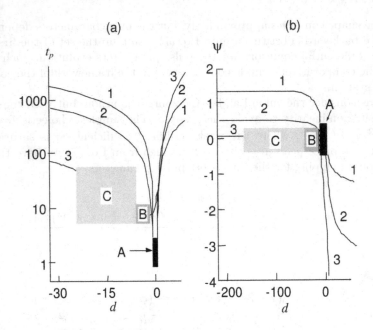

Fig. 2. The steady-state pulse duration (a) and the chirp (b) versus the d coefficient for the time scale and the intensity normalized to t_f and β, respectively. $\sigma_0=10$ (curve 1), 1 (curve 2), 0.1 (curve 3). $l=0.05$, $g_m=0.5$, $P=4\times10^{-4}$, $x=5$ mm, $\lambda=800$ nm, $T_{cav}=10$ ns, $T_r=3$ μs.

The appearance of the regular as well as irregular oscillations, as a result of the $\mid d \mid$ decrease, was observed experimentally [8,11] and their basic characteristics are in the good agreement with the theoretically predicted on the basis of our simple mapping procedure. Unfortunately, the attempt to describe the B region on the basis of the model under consideration fails due to the strong deviation from the soliton model, i. e. due to the invalidity of the abberation-less approximation. The numerical simulations based on Eq. (7) with $\hat{L}=-\gamma/(1+\sigma \mid a\mid^2)$ (the arbitrary strong amplitude nonlinearity) demonstrate the multiple pulse operation in this region (Fig. 3). The main features of the transition to the multiple pulse generation are: 1) there exists the σ parameter producing the minimal extension of the A and B regions; 2) the decrease or the increase of σ relatively to this optimal value expands the region of the multiple pulse operation.

The similar behavior is typical also for the femtosecond lasers with the so-called slow saturable absorber, i. e. the absorber with the energy-dependent nonlinearity: $\hat{L}=-\gamma/(1+\Sigma \int_{-\infty}^{t} \mid a(z,t')\mid^2 dt')$, where $\Sigma=1/E_a$, E_a is the loss saturation energy and we do not take into account the longitudinal relaxation in the absorber (the corresponding relaxation time is much larger than t_p) [12].

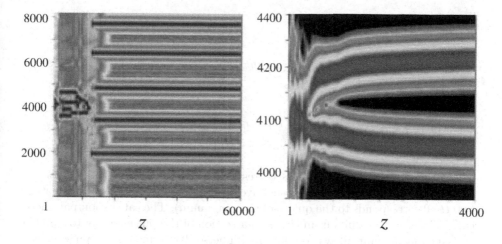

Fig. 3. Contour-plots of the field intensity logarithm below (a) and above (b) optimal σ. Vertical axes is the number of the simulation lattice points (their total number is 2^{13} that corresponds to the 102 ps time window).

In the next section we shall demonstrate that, in spite of the absence of the analytical solution for the description of the multiple pulses, the computer algebra approach allows the comprehension of their sources.

3 Perturbation Analysis

Even beyond the rigorous limits of its validity the linear stability analysis can shed light upon the nature of the nonlinear system instability. We can not exactly describe the solution resulting from the perturbation growth, but can comprehend the character of the processes causing the pulse transformation. Here we try to fill up the gaps in the above presented analysis. We shall consider the destabilizing phenomena in the vicinity of the stability boundaries. As the basic model the ultrashort pulse laser with a slow saturable absorber will be considered.

The simplifying idealization intended for the description of the mode locking region heterogeneity (Fig. 1) and for the understanding of the destabilization scenarios difference (Fig. 3) is to neglect the imaginary terms in Eq. (7). This simplification is valid in the picosecond region of the ultrashort pulse duration and allows to comprehend the basic sources of the phenomena under consideration.

Let the pulse duration be much less than the longitudinal relaxation time in the saturable absorber (for the review of the physical model under consideration see [13]) and much larger than the transversal relaxation time (i. e. we consider the incoherent regime of the light-matter interaction). Then in the weak-nonlinear limit (i. e. when the pulse energy $\int\limits_{-\infty}^{t} |a(t')|^2\, dt' \ll E_a, E_g$) the master Eq. (7) results in:

$$\frac{\partial a(z,t)}{\partial z} = g_0 \left[1 - \varXi \int_0^t a(z,t')^2 dt' + \left(\varXi \int_0^t a(z,t')^2 dt' \right)^2 \right] a(z,t) - \tag{9}$$

$$\gamma_0 \left[1 - \int_0^t a(z,t')^2 dt' + \left(\int_0^t a(z,t')^2 dt' \right)^2 \right] a(z,t) + t_f^2 \frac{\partial^2 a(z,t)}{\partial t^2} - la(z,t),$$

where the first and second terms correspond to the gain and loss saturation, respectively; g_0 and γ_0 are the respective saturable gain and loss coefficients at $t=0$ ($t=0$ corresponds to the quasi-soliton maximum). The saturation parameter is $\varXi=E_a/E_g$ that results from the normalization of the pulse energy to E_a. The weak-nonlinear limit allows to omit the higher-order terms on the pulse energy in Eq. (9). This equation has the soliton-like solution [14]: $a(t)=a_0/\cosh((t - \delta)/t_p)$, where $t_p^2 = 1/(g_0 - l - \gamma_0)$, $a_0^2 = \sqrt{2}\,(g_0 - l - \gamma_0)/\sqrt{\gamma_0 - \varXi^2 g_0}$, $\delta = -\sqrt{2}\,(\varXi g_0 - \gamma_0)/\sqrt{\gamma_0 - \varXi^2 g_0}$.

Let us consider the stability of this solution in the framework of the linear stability analysis. The seeding perturbation $\xi(t)\exp(\varsigma z)$ obeys after linearization of Eq. (9) the following equation:

$$\frac{\partial^2 \xi(t)}{\partial t^2} + \delta \frac{\partial \xi(t)}{\partial t} +$$

$$\xi(t)\,(\gamma_0 - \varXi g_0) \int_0^t a(t')^2 dt' + \xi(t)\,(\varXi^2 g_0 - \gamma_0) \left[\int_0^t a(t')^2 dt' \right]^2 + \tag{10}$$

$$a(t)\,(\gamma_0 - \varXi g_0) \int_0^t a(t')\xi(t')dt' + (g_0 - \gamma_0 - l - \varsigma) = 0.$$

The simplest way is to analyze Eq. (10) in the long- and short-wavelength limits that allows either to neglect the second-order derivative in the comparison with the first-order one or to exclude the integral over the perturbation by virtue of the Riemann-Lebesgue's theorem. In the first case there is no the perturbation damped at $t = \pm\infty$. In the second case we have the solution for the perturbation expressed through the hypergeometric functions, which is too awkward to be written here (see Maple 6 worksheet at Waterloo Maple Application Center [15]). The basic picture is close to that in the quantum mechanics: the loss and gain saturations by the soliton-like pulse form the "potential well", which allows the existence of both the discrete and the continuous spectrum of the perturbations. It should be noted, that our approximation is valid for large "quantum numbers", that is in the vicinity of the continuous spectrum. If the allowed "levels" lie below the critical level with eigenvalue ("energy") $g_0 - \gamma_0$, then $\varsigma > 0$ and there are the asymptotically growing perturbations, i.e. the pulse is unstable in

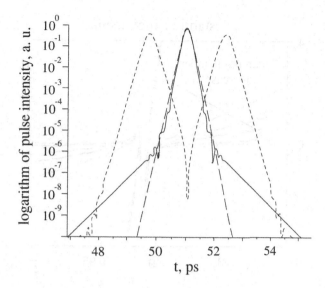

Fig. 4. Logarithm of the numerical profile of the pulse intensity (solid). Dashed - unperturbed soliton profile, Dotted - double pulses generation as a result of the perturbation growth.

the framework of linear analysis. The undamped perturbation belonging to the discrete spectrum is visible in the numerical simulations as the slowly decaying wings outside the soliton wings (compare solid curve with dashed one in Fig. 4). The amplitude of this perturbation is proportional to the pulse energy.

The increase of the pulse energy due to the $\varXi$ decrease, i. e. the decrease of E_a, results in the increase of the "bounded" (i. e. belonging to the discrete spectrum) perturbation. This causes the pulse splitting and the multiple pulse generation (dotted curve in Fig. 4 and the right-hand side picture in Fig. 3).

The contribution of the continuous spectrum is quite different. Its appearance results from the increase of the spectral loss for the shortest pulses or from the nonlinear pulse spectrum broadening (the latter takes a place in the presence of the imaginary terms in the master equation). In this case the pulse energy decreases and the amplified background noise appears. The satellite pulses are formed by this noise amplification. This causes the multiple pulse generation (the left-hand side picture in Fig. 3) or the chaotic behavior.

Our relatively simple analytical model explains also the heterogeneous character of the mode locking region (see Fig. 5). The increment of the perturbation growth, which is negative if the soliton is stable, can oscillate around zero. An example of such an oscillation in the form of the "kink" is shown in Fig. 5. As a result, there exist the regions of the ultrashort pulse stability, which alternate with the regions of the pulse instability.

Hence, we may completely analyze the main stability properties of the quasi-solitons in the mode-locked lasers, which allows the computer algebra realization.

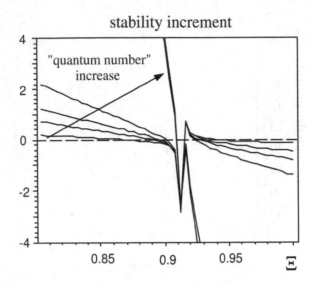

Fig. 5. Stability increment ς vs $\varXi$ for the growing "quantum number" (i. e. the growing order of the perturbation).

4   Coherent Effects

The above presented analytical and semi-analytical models do not take into consideration the coherent nature of the field-matter interaction, which is important for the ultrashort pulse duration shorter than 50 fs. The usual approach here is numerical simulation (see, for example, [16]). Here we evolve the analytical model of the quasi-soliton generation in the lasers with a coherent semiconductor absorber. Our approach allows the direct computer algebra realization [17].

In the presence of the coherent interaction with the absorber the $\hat{L}$ operator in Eq. (7) has to be replaced by the expression

$$\frac{2\pi N z_a \omega \mu}{c} u - \frac{2\pi N z_a \mu}{c}\frac{dv}{dt},$$

where $N = \gamma E_a/(\hbar \omega z_a) \sim 10^{18}$ cm^{-3} is the carrier density in the semiconductor absorber, ω is the field carrier frequency, z_a is the absorber thickness (~ 10 nm), μ is the dipole momentum ($\sim 0.28 \times e$ coulomb$\times$nanometer for GaAs/AlAs absorber, e is the elementary charge). For the semiconductor absorber $E_a \simeq 10 \div 100$ μJ/cm^2. u and v are the slowly varying envelopes of the polarization quadrature components, which obey the Bloch equations (w is the population difference) [18]:

$$\frac{du}{dt} = (\Delta - \frac{d\theta}{dt})v + qaw, \qquad \frac{dv}{dt} = -(\Delta - \frac{d\theta}{dt})u, \qquad \frac{dw}{dt} = -qau, \qquad (11)$$

where Δ is the mismatch between the optical resonance and the pulse carrier frequency, θ is the instant field phase, $q = \mu/\hbar$. We suppose that the pulse duration is shorter than both longitudinal and transverse relaxation times in the absorber. For the latter we use $t_{coh}=50$ fs.

Now let us introduce in $\hat{L}$ also the fast absorber term $\sigma|a|^2$ in order to take into account the self-focusing in the active medium and then to make some trick for the sake of the analytical consideration of the soliton-like solution of Eqs. (7, 11). The transition to the field area $\alpha = \int\limits_{-\infty}^{t} a(t')dt'$ yields:

$$\left[(g - l + i\phi)\frac{d}{dt} + \delta\frac{d^2}{dt^2} + (1 + id)\frac{d^3}{dt^3} + \frac{\sigma - i\beta}{\eta^2}\left(\frac{d\psi(t)}{dt}\right)^2\frac{d}{dt}\right]\alpha(t) - \tag{12}$$

$$\frac{\gamma}{t_{coh}}\sin(\alpha(t)) = 0.$$

Here we introduced the η parameter, which is governed by the ratio between the size of the generation mode in the active medium and in the semiconductor absorber or by the reflectivity of the upper surface of semiconductor saturable absorber. Also we normalized the time to t_f and the field to qt_f. Then the intensity and time units for the Ti:sapphire laser are 0.2 TW/cm^2 and 2.5 fs, respectively.

The trick $\alpha(t) - x$, $d\alpha(t)/dt = y(x)$ with $d=0$ and $\beta=0$ transforms this third-order equation in the second-order one:

$$\left[\left(\frac{d^2y}{dx^2}\right)y + \left(\frac{dy}{dx}\right)^2 + \delta\frac{dy}{dx} + (g - l)\right]y - \frac{\gamma}{t_{coh}}\sin(x) = 0, \tag{13}$$

which has not the soliton-like solution but allows the approximate solution in the form $y(x) = a_0\sin(x/2) + \sum\limits_{m\geq 1} a_m\sin(\frac{(m+1)\pi}{2})$. In the lowest order this solution corresponds to the generation of the 2π-area pulse (the so-called self-induced transparency pulse) and gives for the pulse amplitude, the time delay and the duration $a_0 = 2\sqrt{2(g - l)}$, $\delta = \gamma/(2(g - l)t_{coh})$ and $t_p = 2/a_0$, respectively. The positivity of $g - l$ imposes the initial absorber loss constraint $\gamma > g - l$. Only in this case there is no background amplification (i. e. there exists no continuous spectrum for the ultrashort pulse perturbations). This condition is obviously disadvantageous because the γ growth increases the lasing threshold. The presence of the phase factors $(d, \beta \neq 0)$ does not improve the situation, although in the chirp-free limit this allows to reduce Eq. (12) to the first-order equation:

$$\left[\delta\frac{dy(x)}{dx} + \frac{\beta}{\eta^2 d}y(x)^2 + (\alpha - \gamma - \frac{\phi}{d})\right]y(x) - \frac{\gamma}{t_{coh}}\sin x = 0, \tag{14}$$

which has the same approximated solution, but with the slightly transformed parameters: $\phi = 3\beta a_0^2/(4\eta^2) + d(g - l)$, $\delta = 4\gamma/(t_{coh}a_0^2)$, $a_0 = 2\sqrt{3(g - l)}$, $t_p = 2/a_0$.

The fundamental improvement is caused by $\sigma \neq 0$. In this case there exists the soliton-like solution with *sech*-shaped profile and

$$a_0 = \frac{2}{t_p}, \; t_p = \frac{1}{\sqrt{l-g}}, \; \delta = \frac{\gamma}{t_{coh}(l-g)}, \; \sigma = \eta^2/2. \tag{15}$$

The confinement on the value of σ results from the coherent absorber induced selection of only 2π-pulses, although the pulse amplitude and the duration are defined by the fast saturable absorber, which is produced by the Kerr-lens mode locking. The main advantage is the negative value of $g - l - \gamma$ for the arbitrary γ. This produces the stability against background amplification automatically and allows to reduce the pulse duration down to sub-10 fs for the moderate pumping. Moreover, the pulse parameters mapping obtained on the basis of the abberationless approximation

$$\frac{da_0}{dz} = 2\frac{(g-l)\eta^2 t_p^2 - \eta^2 + 4\sigma}{\eta^2 t_p^3}, \; \frac{dt_p}{dz} = 4\frac{\eta^2 - 2\sigma}{a_0\eta^2 t_p^2},$$

$$\theta = \frac{2(d + 4\frac{\beta}{\eta^2})}{a_0 t_p^3}, \delta = 2\frac{\gamma t_p}{t_{coh} a_0}. \tag{16}$$

demonstrates the pulse automodulational stability.

Additionally the presented model describes the *sech*-shaped chirp-free pulses with π-area and the chirped pulses with variable area. The respective substitutions are

$$u(t) = u_0/\cosh(\frac{t}{t_p}), \; v(t) = v_0/\cosh(\frac{t}{t_p}),$$

$$w(t) = \tanh(\frac{t}{t_p}), \; \frac{d\phi(t)}{dt} = \frac{\psi}{t_p}\tanh(\frac{t}{t_p}),$$

where $a_0 = \sqrt{1 + \psi^2}/t_p, u_0 = -1/\sqrt{1 + \psi^2}, v_0 = \psi/\sqrt{1 + \psi^2}$.

As our analysis demonstrates, in the contrast with the propagation in the solely semiconductor layer, the π-pulses can be stable in the laser due to the fast saturable absorber action.

Hence, the pure analytical extension of the soliton model allows to describe the coherent effects. As a result, the computer algebra simulation replaces the cumbersome numerical analysis and allows the very clear physical interpretation.

5   Conclusion

We presented the self-consistent approach to the analysis of the ultrashort pulse generation in the mode-locked lasers, which is based on the dimension reduction and the analysis of the soliton-like states of the dynamical equations describing the action of the main lasing factor in the incoherent as well as the coherent limits. The numerical simulations in the finite dimensional model in the combination with the stability analysis of the soliton-like solutions of the master dynamical equations allow the computer algebra realization including the on-line calculators. This can essentially supplement the full dimensional numerical analysis and partially replaces their, especially, in the problem of the global optimization.

The approach under consideration is realized as the Maple 6 worksheet [15,19] and takes less than 15 min of calculation a 900 MHz Athlon processor.

For the sake of completeness, let us note some problems, which are still unsolved. The methods under consideration take into account only single pulse regimes or, as a direct but nontrivial generalization, the complexes of the *unbounded* multiple pulses. However, there exist the reach structures of the strongly interacting pulses, which can be considered only numerically. We suppose, that the situation can be improved by the means of the extension of the single soliton-like solution. There are the first attempts in this direction, which are based on the Painlevé analysis [20]. In order to be applicable the cumbersome algebraic calculations taking place in this approach require the computer algebra realization likewise the direct soliton methods (as an example of the computer algebra algorithmization, the Hirota's method is considered in [19]). We plan the further development of the computer algebra approach in this direction.

Acknowledgements

This work was supported by the Austrian National Science Fund, Project M-611. The author is grateful to Dr. E. Sorokin and Dr. I.T. Sorokina for the helpful discussions about the nature of the multiple pulse complexes.

References

1. Brabec, T., Krausz, F. (2000): Intense few-cycle laser fields: frontiers of nonlinear optics. Rev. Mod. Physics, **72**, 545–591
2. Haus, H.A., Fujimoto, J.G., Ippen, E.P. (1992): Analytic theory of additive pulse and Kerr lens mode locking. IEEE J. Quantum Electron., **28**, 2086–2096
3. Aranson, I.S., Kramer, L. (2002): The world of the complex Ginzburg-Landau equation. Rev. Modern Physics, **74**, 99–143
4. Akhmediev, N.N., Afanasjev, V.V., Soto-Crespo, J.M. (1996): Singularities and special soliton solutions of the cubic-quintic complex Ginzburg-Landau equation. Phys. Rev. E, **53**, 1190–1201
5. Kalashnikov, V.L., Poloyko, I.G., Mikhailov, V.P., von der Linde,D. (1997): Regular, quasi-periodic and chaotic behaviour in cw solid-state Kerr-lens mode-locked lasers. J. Opt. Soc. Am. B, **14**, 2691–2695
6. Akhmanov, S.A., Vysloukh, V.A., and Chirkin, A.S. (1992): Optics of femtosecond laser pulses. Springer, New York
7. Kalashnikov, V.L., Kalosha, V.P., Mikhailov, V.P., Poloyko, I.G. (1995): Self-mode locking of four-mirror cavity solid-state lasers by Kerr self-focusing. J. Opt. Soc. Am. B, **12**, 462–467
8. Xing, Q., Chai, L., Zhang, W., Wang, Ch.-yue (1999): Regular, period-doubling, quasi-periodic, and chaotic behavior in a self-mode-locked Ti:sapphire laser. Optics Commun. **162**, 71–74

9. Kalosha, V.P., Müller, M., Herrmann, J., and Gatz, S. (1998): Spatiotemporal model of femtosecond pulse generation in Kerr-lens mode-locked solid-state lasers. J. Opt. Soc. Am. B, **15**, 535–550

10. Sergeev, A.M., Vanin, E.V., Wise, F.W. (1997): Stability of passively modelocked lasers with fast saturable absorbers. Optics Commun. **140**, 61–64

11. Jasapara, J., Kalashnikov, V.L., Krimer, D.O., Poloyko, I.G., Lenzner, M., Rudolph, W. (2000): Automodulations in cw Kerr-lens modelocked solid-state lasers. J. Opt. Soc. Am. B, **17**, 319–326

12. Kalashnikov, V.L., Krimer, D.O., Poloyko, I.G. (2000): Soliton generation and picosecond collapse in solid-state lasers with semiconductor saturable absorber. J. Opt. Soc. Am. B, **17**, 519–523

13. Kärtner, F.X., Yung, I.D., Keller, U. (1996): Soliton mode-locking with saturable absorbers. IEEE J. Sel. Top. Quantum Electron., **2**, 540–556

14. Haus, H.A. (1975): Theory of mode locking with a slow saturable absorber. IEEE J. Quantum Electron., **11**, 736–746

15. Maple 6 computer algebra realization can be found on http://www.mapleapps. com/categories/science/physics/worksheets/Ultrashort PulseLaser.mws.
(Maple V version: http://www.mapleapps.com/categories/mathematics/ desolving/worksheets/soliton.mws)

16. Kalosha, V.P., Müller, M., Herrmann, J. (1999): Theory of solid-state laser mode locking by coherent semiconductor quantum-well absorbers. J. Opt. Soc. Am. B, **16**, 323–338

17. Kalashnikov, V.L. (2001): Theory of sub-10 fs generation in Kerr-lens mode-locked solid-state lasers with a coherent semiconductor absorber. Optics Commun., **192**, 323–331

18. Allen, L., Eberly, J.H. (1975): Optical resonance and two-level atoms. Wiley, New York

19. Kalashnikov V.L. (2002): Mathematical ultrashort-pulse laser physics. arXiv:physics/0009056

20. Marcq, P., Chaté, H., Conte, R. (1994): Exact solitons of the one-dimensional quintic complex Ginzburg-Landau equation. Physica D, **73**, 305–317

Nonlinear Periodic Waves in Shallow Water

Alexander Shermenev

Wave Research Center, Russian Academy of Sciences,
38, Vavilova Street, Moscow, 117942, Russia,
sher@orc.ru

Abstract. Two classical types of periodic wave motion in polar coordinates has been studied using a computer algebra system. In the case of polar coordinates, the usual perturbation techniques for the nonlinear shallow water equation leads to overdetermined systems of linear algebraic equations for unknown coefficients. The compatibility of the systems is the key point of the investigation. The found coefficients allow to construct solutions to the shallow water equation which are periodic in time. The accuracy of the solutions is the same as of the shallow water equation. Expanding the potential and surface elevation in Fourier series, we express explicitly the coefficients of the first two harmonics as polynomials of Bessel functions. One may speculate that the obtained expressions are the first two terms of an expanded exact three-dimensional solution to the surface wave equations, which describe the axisymmetrical and the simplest unaxisymmetrical surface waves in shallow water.

1 Introduction

Two-dimensional nonlinear shallow water equation

$$f_{xx} + f_{yy} - f_{tt} + \alpha\varepsilon\left(-f_x f_{xt} - f_y f_{yt}\right) + \beta\varepsilon\left(-f_{xx}f_t - f_{yy}f_t\right) = 0 \qquad (1)$$

was introduced by Airy to describe the propagation of surface gravity waves. This is the leading approximation for very long waves of finite amplitude. The same equation describes two-dimensional gas dynamics in an isentropic gas flow.

The aim of this paper is to present a series of solutions to this equation in polar coordinates (θ, r). They are periodic in time, and their accuracy is the same as of equation (1).

In his classic book "Hydrodynamics", §§ 191-195, Lamb [4] considers at least two special cases of long **linear** waves in polar coordinates (θ, r) propagating over a horizontal bottom. The first case is axisymmetric waves caused by a periodic source of energy (3). The second case is the simplest unaxisymmetric waves (4) in a circular basin. The solution we have found gives, in particular, nonlinear corrections to these well known linear solutions.

Different versions of the shallow water equations are related mainly to a choice of basic variables (see [5]). We follow the book of Mei [6] and use the potential at bottom and surface elevation as basic variables.

F. Winkler and U. Langer (Eds.): SNSC 2001, LNCS 2630, pp. 375–386, 2003.

There is a small parameter associated with the shallow water equations: the ratio of amplitude to depth, ε (measure of nonlinearity). As in the classic Airy equations, we retain only the quadratic terms (of orders $O(\varepsilon)$).

The potential at the bottom is expanded in Fourier series in time:

$$f(r,\theta,t) = U(r,\theta) + S^1(r,\theta)\sin(\omega t) + C^1(r,\theta)\cos(\omega t)$$
$$+S^2(r,\theta)\sin(2\omega t) + C^2(r,\theta)\cos(2\omega t) + ...$$
$$+S^m(r,\theta)\sin(m\omega t) + C^m(r,\theta)\cos(m\omega t) + ... \tag{2}$$

The main result of this work consists in the explicit expressions (35) - (42) for functions U, S^1, C^1, S^2 and C^2 up to orders ε, which are homogenous polynomials in Bessel functions $Z_0(\omega r)$ and $Z_1(\omega r)$ and trigonometric functions of angular variable θ. Their coefficients are polynomials in r^{-1} and r. A similar expansion for the surface elevation can be derived. These expressions give a periodic solution to the classic shallow water equation, within the same accuracy as the equations are derived. Therefore, the result can be interpreted as a nonlinear periodic solution of surface wave equations (8) - (11) calculated up to orders ε. The calculations of coefficients and the check of results were performed using the Computer Algebra System "MATHEMATICA".

The two linear solutions used in the book of Lamb [4] are following:

$$f(r,\theta,t) = J_0(\omega r)\sin(\omega t) - Y_0(\omega r)\cos(\omega t) \tag{3}$$
$$f(r,\theta,t) = J_1(\omega r)\cos\theta\sin(\omega t). \tag{4}$$

They describe the axisymmetric wave motion with periodic source located in the center of polar coordinates system and the simplest case of regular unsymmetrical wave motion in a circular basin.

When trying to find the functions $U(r,\theta)$, $S^1(r,\theta)$, $C^1(r,\theta)$, $S^2(r,\theta)$, and $C^2(r,\theta)$, we are forced to solve a Bessel-type non-homogenous differential equation of the second order:

$$Z''(r) + \frac{1}{r}Z'(r) + \left(A + \frac{B}{r^2}\right)Z \tag{5}$$
$$= q_{00}J_0(\omega r)Y_0(\omega r) + q_{01}J_0(\omega r)Y_1(\omega r) + q_{10}J_1(\omega r)Y_0(\omega r) + q_{11}J_1(\omega r)Y_1(\omega r),$$

where q_{ij} are polynomials of r, r^{-1}.

We seek a solution $Z(r)$ in the form

$$Q_{00}J_0(\omega r)Y_0(\omega r) + Q_{01}J_0(\omega r)Y_1(\omega r) + Q_{10}J_1(\omega r)Y_0(\omega r) + Q_{11}J_1(\omega r)Y_1(\omega r) \tag{6}$$

where Q_{ij} are polynomials of r with unknown coefficients. These coefficients are calculated as solutions of overdetermined systems of linear algebraic equations (as in [7] and [8], where the same approach was used for describing the long periodic waves on a slope). In the cases considered, the compatibility of these overdetermined systems appears to be a matter of luck, and we see no reason why the luck should continue. Nevertheless, we take the liberty of making an "experimental" conjecture that the terms derived are only the lowest terms of a three dimensional exact solution to the surface wave equations (8) - (11) over a horizontal bottom. The exact three-dimensional solutions are not known at the moment, but we present an "intermediate" object (solutions of orders ε) which, we hope, describes more accurately the behavior of the long periodic waves as compared with the Lamb linear solution.

2   Basic Equations

The aim of this section is to remind briefly the derivation of shallow water equation in the form suitable for us and to fix notations. We follow mainly to book of Mei [6]. Non-dimensional coordinates are used as follows:

$$x = \frac{x'}{l'_0}, \; y = \frac{y'}{l'_0}, \; z = \frac{z'}{h'_0}, \; t = \frac{g^{\frac{1}{2}} h'^{\frac{1}{2}}_0}{l'_0} t', \; \eta = \frac{\eta'}{a'_0}, \; \varphi = \frac{h'_0}{a'_0 l'_0 g^{\frac{1}{2}} h'^{\frac{1}{2}}_0} \varphi', \; h = \frac{h'}{h'_0} \quad (7)$$

where the prime denotes physical variables, and a'_0, l'_0, and h'_0 denote the characteristic wave amplitude, depth, and wavelength, respectively; η' is the surface elevation; g is the acceleration of gravity; x and y are the horizontal coordinate; z is the vertical coordinate; and t is the time. The scaled governing equation and the boundary conditions for the irrotational wave problem read

$$\varphi_{xx} + \varphi_{yy} + \mu^{-2} \varphi_{zz} = 0, \quad -1 < z < \varepsilon \eta(x, y, t) \quad (8)$$

$$\eta_t + \varepsilon \varphi_x \eta_x + \varepsilon \varphi_y \eta_y - \mu^{-2} \varphi_z = 0, \quad z = \varepsilon \eta(x, y, t) \;, \quad (9)$$

$$\varphi_t + \eta + \frac{1}{2} \varepsilon \left(\varphi_x^2 + \varphi_y^2 \right) + \varepsilon \mu^{-2} \varphi_z^2 = 0, \quad z = \varepsilon \eta(x, y, t) \;, \quad (10)$$

$$\varphi_z = 0, \quad z = -1 \quad (11)$$

where ε and μ are the measures of nonlinearity and frequency dispersion defined by

$$\varepsilon = a'_0 / h'_0, \quad \mu = h'_0 / l'_0. \quad (12)$$

The potential $\varphi(x, y, z, t)$ is expanded in powers of vertical coordinate

$$\varphi(x, y, z, t) = \sum_{m=0}^{\infty} (z + 1)^m F_m(x, y, t) \quad (13)$$

We use ∇ to denote the horizontal gradient $(\partial/\partial x, \partial/\partial y)$. Substituting (13) into (8) and equating to zero the coefficients of each power of $z + 1$, we have

$$F_{m+2} = \frac{-\mu^2 \nabla^2 F_m}{(m + 2)(m + 1)} \quad (14)$$

The boundary condition at the bottom (11) gives

$$F_1 = 0 \quad (15)$$

Denoting $f(x, y, t) \equiv F_0(x, y, t)$ and expanding all expressions in powers of μ, we obtain the first terms of φ

$$\varphi = f - \frac{1}{2!}\mu^2(z+1)^2\nabla^2 f + O\left(\mu^4\right) \tag{16}$$

Expression (16) satisfies (8) and (11). Substitution of (16) into (9) and (10) gives the Boussinesq-type equations for two functions: potential at bottom $f(x, y, t)$ and surface elevation $\eta(x, y, t)$:

$$\eta_t + \varepsilon\nabla\eta \cdot \nabla f + (1 + \varepsilon\eta)\nabla^2 f = 0 \tag{17}$$

$$\eta + f_t + \frac{1}{2}\varepsilon(\nabla f)^2 = 0 \tag{18}$$

The last two equations are equivalent to equations (1.16) and (1.17) of chapter 11 of [4].

Expressing the surface elevation $\eta(x, y, t)$ in terms of $f(x, y, t)$ and its derivatives, we have the single equation of type (1) for the function $f(x, y, t)$.

This equation can be rewritten in polar coordinates (r, θ) as follows:

$$\frac{1}{r^2}f_{\theta\theta} + \frac{1}{r}f_r + f_{rr} - f_{tt} + \alpha\varepsilon\left(-\frac{2}{r^2}f_\theta f_{\theta t} - 2f_r f_{rt}\right) \tag{19}$$

$$+\beta\varepsilon\left(-\frac{1}{r^2}f_t f_{\theta\theta} - \frac{1}{r}f_t f_r - f_t f_{rr}\right) = 0$$

3   Periodic Problem

We suppose that the solution is periodic in time and can be expanded in Fourier series in an area excluding a neighborhood of axis of symmetry and infinity. Making some additional scaling, we can assume that frequency ω is equal to 1 without loss of generality.

$$f(r, \theta, t) = S_{00}^1(r, \theta))\sin(t) + C_{00}^1(r, \theta)\cos(t) \tag{20}$$
$$+S_{10}^2(r, \theta)\varepsilon\ \sin(2t) + C_{10}^2(r, \theta)\varepsilon\ \cos(2t) + ...$$

In the zero-order, we have the following linear problems for $S_{00}^1(r, \theta)$ and $C_{00}^1(r, \theta)$

$$S_{00}^1 + \frac{1}{r^2}S_{00\theta\theta}^1 + \frac{1}{r}S_{00r}^1 + S_{00rr}^1 = 0 \tag{21}$$

$$C_{00}^1 + \frac{1}{r^2}C_{00\theta\theta}^1 + \frac{1}{r}C_{00r}^1 + C_{00rr}^1 = 0 \tag{22}$$

Their general solution expressed in polar coordinates can be presented as a series:

$$\alpha_0^J J_0(r) + \alpha_0^Y Y_0(r)$$
$$+ \left(\alpha_{1S}^J J_1(r) + \alpha_{1S}^Y Y_1(r)\right)\sin\theta + \left(\alpha_{1C}^J J_1(r) + \alpha_{1C}^Y Y_1(r)\right)\cos\theta + ...$$
$$+ \left(\alpha_{nS}^J J_n(r) + \alpha_{nS}^Y Y_n(r)\right)\sin n\theta + \left(\alpha_{nC}^J J_n(r) + \alpha_{nC}^Y Y_n(r)\right)\cos n\theta + ... \tag{23}$$

where α_{nC}^J and α_{nC}^Y are arbitrary constant coefficients providing the convergency. We concentrate our attention on the first two terms of this series (zero and the first harmonics in θ)

Let us denote by $S = S(r)$ and $C = C(r)$ two arbitrary solutions of Bessel equation

$$rZ_{rr} + Z_r + rZ = 0, \qquad (24)$$

Functions $S(r)$, $C(r)$ and their derivatives $S'(r)$, $C'(r)$ can be expressed in terms of Bessel functions as follows:

$$S(r) = a_{11}J_0(r) + a_{12}Y_0(r), \qquad (25)$$
$$S'(r) = -a_{11}J_1(r) - a_{12}Y_1(r) \qquad (26)$$
$$C(r) = a_{21}J_0(r) + a_{22}Y_0(r), \qquad (27)$$
$$C'(r) = -a_{21}J_1(r) - a_{22}Y_1(r) \qquad (28)$$

Then the first two terms of (23) can be written in the form:

$$S_{00}^1(r,\theta) = A_S S - (A_{SSS}S' + A_{SSC}C')\sin\theta - (A_{SCS}S' + A_{SCC}C')\cos\theta \qquad (29)$$
$$C_{00}^1(r,\theta) = A_C C - (A_{CSS}S' + A_{CSC}C')\sin\theta - (A_{CCS}S' + A_{CCC}C')\cos\theta \qquad (30)$$

where A_X and A_{XYZ} are arbitrary constant coefficients (We can assume $A_S = 1$ and $A_C = 1$ almost without loss of generality).

Lamb [4] considers two particular cases of (29), (30):

(i)

$$S_{00}^1(r,\theta) = \varepsilon J_0(r) \qquad (31)$$
$$C_{00}^1(r,\theta) = -\varepsilon Y_0(r) \qquad (32)$$

($A_X = 0$, $A_{XYZ} = 0$, $a_{12} = -\varepsilon$, $a_{21} = \varepsilon$, $a_{11} = a_{22} = 0$). The solution describes axisymmetric wave motion with periodic source located in the center of polar coordinates system.

(ii)

$$S_{00}^1(r,\theta) = 0 \qquad (33)$$
$$C_{00}^1(r,\theta) = -\varepsilon J_1(r)\cos\theta \qquad (34)$$

($A_S = 0$, $A_C = 0$, all A_{XYZ} are equal to zero except for $A_{CCC} = -\varepsilon$). The solution describes the simplest case of regular unsymmetrical wave motion in a circular basin.

The first solution is used in §§ 191-195 of book of Lamb [4] for describing axisymmetric wave motion with periodic source located in the center of polar coordinates system. The second one is used there for describing the simplest case of regular unsymmetrical wave motion in a circular basin. The aim of subsequent consideration is to give next-order corrections to these linear solutions.

4 Calculation of $S_{10}^2(r,\theta)$

$S_{10}^2(r)$ is presented as a sum of four components:

$$S_{10}^2(r) = -\frac{1}{2}\alpha S_{grad2} + \frac{1}{2}\alpha C_{grad2} + \frac{1}{2}\beta S_{mod2} - \frac{1}{2}\beta C_{mod2} \tag{35}$$

where

$$\begin{aligned}
S_{grad2} = {} & \frac{1}{2}rSS' + (A_{SCC}^2 + A_{SSC}^2)\left(\frac{1}{4}rC^2 + \frac{1}{4}rCC' - \frac{1}{4}C'^2\right) \tag{36}\\[4pt]
& + (A_{SCS}^2 + A_{SSS}^2)\left(\frac{1}{4}rS^2 + \frac{1}{4}rSS' - \frac{1}{4}S'^2\right) \\[4pt]
& + (A_{SCC}A_{SCS} + A_{SSC}A_{SSS})\left(\frac{1}{2}CS + \frac{1}{4}C'S + \frac{1}{4}CS' - \frac{1}{2}C'S'\right) \\[4pt]
& + \left(\begin{array}{c} A_{SSC}\left(-\frac{1}{4}rCS + \frac{3}{8}C'S + \frac{1}{8}CS' + \frac{1}{4}rC'S'\right) \\ +A_{SSS}(-\frac{1}{4}rS^2 + \frac{1}{2}SS' + \frac{1}{4}rS'^2) \end{array}\right)\sin\theta \\[4pt]
& + \left(\begin{array}{c} A_{SCC}\left(-\frac{1}{4}rCS + \frac{3}{8}C'S + \frac{1}{8}CS' + \frac{1}{4}rC'S'\right) \\ +A_{SCS}(-\frac{1}{4}rS^2 + \frac{1}{2}SS' + \frac{1}{4}rS'^2) \end{array}\right)\cos\theta \\[4pt]
& + \left(\begin{array}{c} \frac{1}{4}A_{SSC}A_{SCC}CC' + \frac{1}{4}A_{SCS}A_{SSS}SS' \\ +\frac{1}{4}(A_{SCS}A_{SSC} + A_{SCC}A_{SSC})(CS' + C'S) \end{array}\right)\sin 2\theta \\[4pt]
& + \left(\begin{array}{c} \frac{1}{8}(A_{SCC}^2 - A_{SSS}^2)rCC' + \frac{1}{8}(A_{SCS}^2 - A_{SSS}^2)SS' \\ +\frac{1}{4}(A_{SCC}A_{SCS} - A_{SSC}A_{SSS})r(CS' + C'S) \end{array}\right)\cos 2\theta
\end{aligned}$$

$$\begin{aligned}
C_{grad2} = {} & \frac{1}{2}rCC' + (A_{CCC}^2 + A_{CSC}^2)\left(\frac{1}{4}rC^2 + \frac{1}{4}rCC' - \frac{1}{4}C'^2\right) \tag{37}\\[4pt]
& + (A_{CCS}^2 + A_{CSS}^2)\left(\frac{1}{4}rS^2 + \frac{1}{4}rSS' - \frac{1}{4}S'^2\right) \\[4pt]
& + (A_{CCC}A_{CCS} + A_{CSC}A_{CSS})\left(\frac{1}{2}CS + \frac{1}{4}C'S + \frac{1}{4}CS' - \frac{1}{2}C'S'\right) \\[4pt]
& + \left(\begin{array}{c} 2A_{CSC}\left(-\frac{1}{4}rC^2 + \frac{1}{2}rCC' + \frac{1}{4}C'^2\right) \\ +2A_{SSS}(-\frac{1}{4}rCS + \frac{1}{8}C'S + \frac{3}{8}CS' + \frac{1}{4}rC'S') \end{array}\right)\sin\theta \\[4pt]
& + \left(\begin{array}{c} 2A_{CCC}\left(-\frac{1}{4}rC^2 + \frac{1}{2}rCC' + \frac{1}{4}C'^2\right) \\ +2A_{CCS}(-\frac{1}{4}rCS + \frac{1}{8}C'S + \frac{3}{8}CS' + \frac{1}{4}rC'S') \end{array}\right)\cos\theta \\[4pt]
& + \left(\begin{array}{c} \frac{1}{4}A_{CCC}A_{CSC}CC' + \frac{1}{4}A_{CCS}A_{CSS}SS' \\ +\frac{1}{4}(A_{CCS}A_{CSC} + A_{CCC}A_{CSS})(CS' + C'S) \end{array}\right)\sin 2\theta \\[4pt]
& + \left(\begin{array}{c} \frac{1}{8}(A_{CCC}^2 - A_{CSC}^2)rCC' + \frac{1}{8}(A_{CCS}^2 - A_{CSS}^2)SS' \\ +\frac{1}{4}(A_{CCC}A_{CCS} - A_{CSC}A_{CSS})r(CS' + C'S) \end{array}\right)\cos 2\theta
\end{aligned}$$

$$S_{mod2} = \left(\begin{array}{c} \frac{1}{2}rSS' + \frac{1}{2}(A_{SCC}^2 + A_{SSC}^2)\left(-\frac{1}{2}C^2 - \frac{1}{2}rCC'\right) \\ +\frac{1}{2}(A_{SCS}^2 + A_{SSS}^2)\left(-\frac{1}{2}S^2 - \frac{1}{2}rSS'\right) \\ +(A_{CCC}A_{CCS} + A_{CSC}A_{CSS})\left(-\frac{1}{2}CS - \frac{1}{4}rC'S - \frac{1}{4}rCS'\right) \end{array}\right) \tag{38}$$

$$+\left(\begin{array}{c} -2A_{CSS}\left(-\frac{1}{4}rCS+\frac{3}{8}C'S+\frac{1}{8}CS'+\frac{1}{4}rC'S'\right) \\ -2A_{CSC}(-\frac{1}{4}rC^2+\frac{1}{4}rC'^2) \end{array}\right)\sin\theta$$

$$+\left(\begin{array}{c} -2A_{SCC}\left(-\frac{1}{4}rCS-\frac{1}{8}C'S+\frac{1}{8}CS'+\frac{1}{4}rC'S'\right) \\ -2A_{SCS}(-\frac{1}{4}rS^2+\frac{1}{4}rS'^2) \end{array}\right)\cos\theta$$

$$+\left(\begin{array}{c} A_{SCC}A_{SSC}(-\frac{1}{2}rCC'-\frac{1}{2}C'^2) \\ +A_{SCS}A_{SSS}\left(-\frac{1}{2}rSS'-\frac{1}{2}S'^2\right) \\ +(A_{SCS}A_{SSC}+A_{SCC}A_{SSS})\left(-\frac{1}{4}rC'S-\frac{1}{4}rC'S-\frac{1}{2}C'S'\right) \end{array}\right)\sin 2\theta$$

$$+\left(\begin{array}{c} \frac{1}{2}\left(A_{SCC}^2-A_{SSC}^2\right)\left(-\frac{1}{2}rCC'-\frac{1}{2}C'^2\right) \\ +\frac{1}{2}\left(A_{SCS}^2-A_{SSS}^2\right)\left(-\frac{1}{2}rSS'-\frac{1}{2}S'^2\right) \\ +(A_{SCC}A_{SCS}-A_{SSC}A_{SSS})\left(-\frac{1}{4}rC'S-\frac{1}{4}rC'S-\frac{1}{2}C'S'\right) \end{array}\right)\cos 2\theta$$

$$C_{mod2}=\left(\begin{array}{c} \frac{1}{4}rCC'+\frac{1}{2}\left(A_{CCC}^2+A_{CSC}^2\right)\left(-\frac{1}{2}C^2-\frac{1}{2}rCC'\right) \\ +\frac{1}{2}\left(A_{CCS}^2+A_{CSS}^2\right)\left(-\frac{1}{2}S^2-\frac{1}{2}rSS'\right) \\ +(A_{CCC}A_{CCS}+A_{CSC}A_{CSS})\left(-\frac{1}{2}CS-\frac{1}{4}rC'S-\frac{1}{4}rCS'\right) \end{array}\right)\qquad(39)$$

$$+\left(\begin{array}{c} -2A_{CSC}\left(-\frac{1}{4}rCS-\frac{1}{8}C'S+\frac{1}{8}CS'+\frac{1}{4}rC'S'\right) \\ -2A_{CSS}(-\frac{1}{4}rC^2+\frac{1}{4}rC'^2) \end{array}\right)\sin\theta$$

$$+\left(\begin{array}{c} -2A_{CCS}\left(-\frac{1}{4}rCS-\frac{1}{8}C'S+\frac{1}{8}CS'+\frac{1}{4}rC'S'\right) \\ -2A_{CCC}(-\frac{1}{4}rC^2+\frac{1}{4}rC'^2) \end{array}\right)\cos\theta$$

$$+\left(\begin{array}{c} \frac{1}{2}\left(A_{CCC}^2-A_{CSC}^2\right)\left(-\frac{1}{2}rCC'-\frac{1}{2}C'^2\right) \\ +\frac{1}{2}\left(A_{CCS}^2-A_{CSS}^2\right)\left(-\frac{1}{2}rSS'-\frac{1}{2}S'^2\right) \\ +(A_{SCS}A_{SSC}+A_{SCC}A_{SSS})\left(-\frac{1}{4}rC'S-\frac{1}{4}rC'S-\frac{1}{2}C'S'\right) \end{array}\right)\sin 2\theta$$

$$+\left(\begin{array}{c} \frac{1}{2}\left(A_{SCC}^2-A_{SSC}^2\right)\left(-\frac{1}{2}rCC'-\frac{1}{2}C'^2\right) \\ +\frac{1}{2}\left(A_{SCS}^2-A_{SSS}^2\right)\left(-\frac{1}{2}rSS'-\frac{1}{2}S'^2\right) \\ +(A_{SCC}A_{SCS}-A_{SSC}A_{SSS})\left(-\frac{1}{4}rC'S-\frac{1}{4}rC'S-\frac{1}{2}C'S'\right) \end{array}\right)\cos 2\theta$$

5 Calculation of $C_{10}^2(r,\theta)$

$C_{10}^2(r)$ is presented as a sum of two components:

$$C_{10}^2(r)=-\frac{1}{2}\alpha\Pi_{grad}+\frac{1}{2}\beta\Pi_{mod2}\qquad(40)$$

where

$$\Pi_{grad}=\left(\begin{array}{c} -\frac{1}{2}CS-\frac{1}{4}rC'S-\frac{1}{4}rCS' \\ +(A_{CCC}A_{SCC}+A_{CSC}A_{SSC})(\frac{1}{4}C^2+\frac{1}{4}rCC'-\frac{1}{4}C'^2) \\ +(A_{CCS}A_{SCS}+A_{CSS}A_{SSS})(\frac{1}{4}S^2+\frac{1}{4}rSS'-\frac{1}{4}S'^2) \\ +\frac{1}{2}\left(\begin{array}{c}A_{CCS}A_{SCC}+A_{CCC}A_{SCS} \\ +A_{CSS}A_{SSC}+A_{CSC}A_{SSS}\end{array}\right)(\frac{1}{2}CS+\frac{1}{4}rC'S+\frac{1}{4}rCS'-\frac{1}{2}C'S') \end{array}\right)\qquad(41)$$

$$+\left(\begin{array}{c} A_{SSC}\left(-\frac{1}{4}rC^2+\frac{1}{2}CC'+\frac{1}{4}rC'^2\right)+A_{CSS}(-\frac{1}{4}rS^2+\frac{1}{2}SS'+\frac{1}{4}rS'^2) \\ +A_{SSS}(-\frac{1}{4}rCS+\frac{1}{8}C'S+\frac{3}{8}CS'+\frac{1}{4}rC'S') \\ +A_{CSC}(-\frac{1}{4}rCS+\frac{3}{8}C'S+\frac{1}{8}CS'+\frac{1}{4}rC'S') \end{array}\right)\sin\theta$$

$$+\begin{pmatrix} A_{SCC}\left(-\frac{1}{4}rC^2+\frac{1}{2}CC'+\frac{1}{4}rC'^2\right)+A_{CCS}\left(-\frac{1}{4}rS^2+\frac{1}{2}SS'+\frac{1}{4}rS'^2\right) \\ +A_{CCC}\left(-\frac{1}{4}rCS+\frac{3}{8}C'S+\frac{1}{8}CS'+\frac{1}{4}rC'S'\right) \\ +A_{SCS}\left(-\frac{1}{4}rCS+\frac{1}{8}C'S+\frac{3}{8}CS'+\frac{1}{4}rC'S'\right) \end{pmatrix}\cos\theta$$

$$+\begin{pmatrix} (A_{CSC}A_{SCC}+A_{CCC}A_{SSC})\frac{1}{4}rCC' \\ +(A_{CSS}A_{SCS}+A_{CCS}A_{SSS})\frac{1}{4}rSS' \\ +\frac{1}{2}r\begin{pmatrix} A_{CSS}A_{SCC}+A_{CSC}A_{SCS} \\ +A_{CCS}A_{SSC}+A_{CCC}A_{SSS} \end{pmatrix}(CS'+C'S) \end{pmatrix}\sin 2\theta$$

$$+\begin{pmatrix} (A_{CCC}A_{SCC}-A_{CSC}A_{SSC})\frac{1}{4}rCC' \\ +(A_{CCS}A_{SCS}-A_{CSS}A_{SSS})\frac{1}{4}rSS' \\ -\frac{1}{2}r\begin{pmatrix} -A_{CCS}A_{SCC}-A_{CCC}A_{SCS} \\ +A_{CSS}A_{SSC}+A_{CSC}A_{SSS} \end{pmatrix}(CS'+C'S) \end{pmatrix}\cos 2\theta$$

$$\Pi_{mod2}=\begin{pmatrix} \frac{1}{4}r\left(C'S+CS'\right) \\ +\frac{1}{2}\left(A_{CCC}A_{SCC}+A_{CSC}A_{SSC}\right)\left(-\frac{1}{2}C^2-\frac{1}{2}rCC'\right) \\ +\frac{1}{2}\left(A_{CCS}A_{SCS}+A_{CSS}A_{SSS}\right)\left(-\frac{1}{2}S^2-\frac{1}{2}rSS'\right) \\ +\begin{pmatrix} A_{CCS}A_{SCC}+A_{CCC}A_{SCS} \\ +A_{CSS}A_{SSC}+A_{CSC}A_{SSS} \end{pmatrix}\left(-\frac{1}{2}CS-\frac{1}{4}rC'S-\frac{1}{4}rCS'\right) \end{pmatrix}\tag{42}$$

$$+\begin{pmatrix} -A_{CSC}\left(-\frac{1}{4}rCS-\frac{1}{8}C'S+\frac{1}{8}CS'+\frac{1}{4}rC'S'\right) \\ -A_{SSS}\left(-\frac{1}{4}rCS+\frac{1}{8}C'S-\frac{1}{8}CS'+\frac{1}{4}rC'S'\right) \\ -A_{SSC}\left(-\frac{1}{4}rC^2+\frac{1}{4}rC'^2\right)-A_{CSS}\left(-\frac{1}{4}rS^2+\frac{1}{4}rS'^2\right) \end{pmatrix}\sin\theta$$

$$+\begin{pmatrix} -A_{CCC}\left(-\frac{1}{4}rCS-\frac{1}{8}C'S+\frac{1}{8}CS'+\frac{1}{4}rC'S'\right) \\ -A_{SCS}\left(-\frac{1}{4}rCS+\frac{1}{8}C'S-\frac{1}{8}CS'+\frac{1}{4}rC'S'\right) \\ -A_{SCC}\left(-\frac{1}{4}rC^2+\frac{1}{4}rC'^2\right)-A_{CCS}\left(-\frac{1}{4}rS^2+\frac{1}{4}rS'^2\right) \end{pmatrix}\cos\theta$$

$$+\begin{pmatrix} \frac{1}{2}\left(A_{CSC}A_{SCC}+A_{CCC}A_{SSC}\right)\left(-\frac{1}{2}rCC'-\frac{1}{2}C'^2\right) \\ +\frac{1}{2}\left(A_{CSS}A_{SCS}+A_{CCS}A_{SSS}\right)\left(-\frac{1}{2}rSS'-\frac{1}{2}S'^2\right) \\ +\frac{1}{2}\begin{pmatrix} A_{CSS}A_{SCC}+A_{CSC}A_{SCS} \\ +A_{CCS}A_{SSC}+A_{CCC}A_{SSS} \end{pmatrix}\left(-\frac{1}{4}rC'S-\frac{1}{4}rC'S'-\frac{1}{2}C'S'\right) \end{pmatrix}\sin 2\theta$$

$$+\begin{pmatrix} \frac{1}{2}\left(A_{CCC}A_{SCC}-A_{CSC}A_{SSC}\right)\left(-\frac{1}{2}rCC'-\frac{1}{2}C'^2\right) \\ +\frac{1}{2}\left(A_{CCS}A_{SCS}+A_{CSS}A_{SSS}\right)\left(-\frac{1}{2}rSS'-\frac{1}{2}S'^2\right) \\ +\frac{1}{2}\begin{pmatrix} A_{CCS}A_{SCC}+A_{CCC}A_{SCS} \\ -A_{CSS}A_{SSC}-A_{CSC}A_{SSS} \end{pmatrix}\left(-\frac{1}{4}rC'S-\frac{1}{4}rC'S'-\frac{1}{2}C'S'\right) \end{pmatrix}\cos 2\theta$$

6   Examples

We give here explicit corrections for Lamb's linear solutions.

Case i.

Assuming that

$$S_{00}^1(r,\theta)=S\left(r\right),\qquad C_{00}^1(r,\theta)=C\left(r\right)\tag{43}$$

We have

$$f(r,t)=S\sin\left(t\right)+C\cos\left(t\right)\tag{44}$$

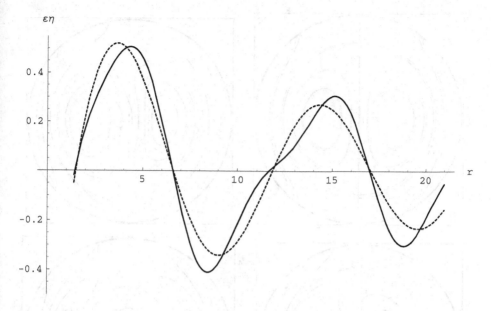

Fig. 1. Axisymmetric progressive waves. Dependence of surface elevation η on radius r. The solid line is (ε)-order solution, and the dashed line is zero-order solution. Face slope is more steep and rear slope is more gentle as compared with zero order Lamb solution. $\varepsilon = 0.1, \quad t = \pi.$

$$+ \varepsilon \left(\frac{1}{2} \left(S^2 - C^2 \right) + \frac{3}{4} r (SS' - CC') \right) \sin(2t)$$

$$+ \varepsilon \left(CS + \frac{3}{4} r (SC' + S'C) \right) \cos(2t).$$

Substituting these expressions into (18), we could derive the following expressions for $\eta(r, \theta, t)$.

$$\eta(r, \theta, t) = \left(-\frac{1}{4} S'^2 - \frac{1}{4} C'^2 \right) \varepsilon + C \sin t - S \cos t \qquad (45)$$

$$+ \left(2CS + \frac{3r}{2} \left(CS' + C'S \right) \right) \varepsilon \sin(2t) \qquad (46)$$

$$+ \left(\left(C^2 - S^2 \right) + \frac{3r}{2} \left(CC' - SS' \right) - \frac{1}{4} \left(C'^2 - S'^2 \right) \right) \varepsilon \cos(2t) \qquad (47)$$

Setting $a_{12} = -1$, $a_{21} = 1$, $a_{11} = a_{22} = 0$, we have $S(r) = -Y_0(r)$ and $C(r) = J_0(r)$. Then the main approximation for $\eta(r, \theta, t)$ is

$$J_0(r) \sin(t) + Y_0(r) \cos(t) \qquad (48)$$

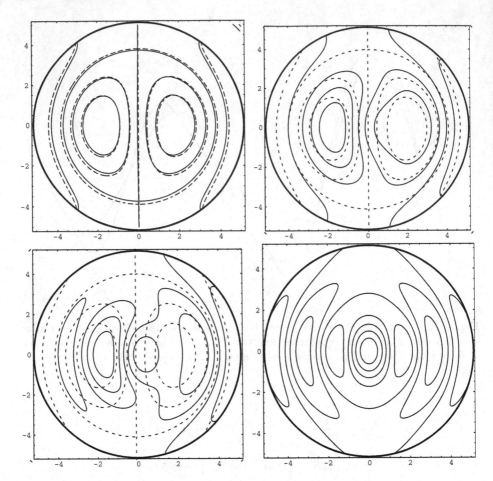

Fig. 2. Contours of unsymmetric waves (case ii) for $t = \frac{3}{4}\pi$, $\frac{5}{6}\pi$, $\frac{11}{12}\pi$, π. The solid lines are (ε)-order solutions, and the dashed lines are zero-order solutions. $\varepsilon = 0.2$.

which is equivalent to

$$\sqrt{\frac{2}{\pi r}} \left(\cos \left(r - \frac{\pi}{4} \right) \sin (t) + \sin \left(r - \frac{\pi}{4} \right) \cos (t) \right) = \sqrt{\frac{1}{2\pi r}} \sin \left(r - \frac{\pi}{4} + t \right) \tag{49}$$

for $r \to +\infty$. So, we can consider this case as progressive wave, whereas the case $a_{11} = 1$, $a_{22} = a_{12} = a_{21} = 0$ is a standing (symmetrical) wave. New solution (solid line) and zero-order Lamb solution (broken line) are shown in Fig. 1.

Case ii.

Assuming that

$$S_{00}^1(r, \theta) = 0, \qquad C_{00}^1(r, \theta) = C'(r) \cos \theta \tag{50}$$

(We suppose in the numerical calculation that if $C = -\varepsilon J_0$)
We have

$$f(r,t) = C' \cos\theta \cos(t) \tag{51}$$
$$+ \varepsilon \left(\frac{3}{8}C^2 + \frac{3}{8}rCC' - \frac{1}{4}C'^2 + \left(\frac{3}{8}rCC' + \frac{1}{8}C'^2 \right) \cos 2\theta \right) \sin(2t).$$

Substituting this expression into (18), we could derive the following expressions for $\eta(r,\theta,t)$.

$$\eta(r,\theta,t) = \left(-\frac{1}{4}\cos^2\theta C^2 - \frac{1}{2r}\cos^2\theta CC' - \frac{1}{4r^2}C'^2 \right)\varepsilon + C'\cos\theta\sin t \tag{52}$$
$$+ \left(\begin{array}{c} \left(-\frac{1}{4}\cos^2\theta - \frac{3}{4}\right)C^2 - \left(\frac{1}{2r}\cos^2\theta + \frac{3}{2}r\cos^2\theta\right)CC' \\ + \left(-\frac{1}{4r^2} + \frac{3}{4} - \frac{1}{2}\cos^2\theta\right)C'^2 \end{array} \right) \varepsilon\cos(2t)$$

The presented nonlinear corrections alter dramatically the topology of contours of surface waves. In particular, the contours vary in time in contrast with Lamb solution and the surface is never flat. Examples of the contours of waves are shown in Fig. 2.

7   Conclusions

Some periodic solutions with the accuracy of ε to the classical surface wave equations (8) - (11) are presented. The intermediate equations are given for illustrating the method of derivation but the expressions (35) - (42) can be proved by substitution into system (8) - (11) (using expression (16) for the potential). Linear versions of these problems when the terms of zero orders in ε only are retained were subject of classical investigation in the book of Lamb [4] §§ 191-195.

We conjecture that these expressions are only the lowest terms of a certain expanded exact three-dimensional solution to surface wave equations (8) - (11). The number of known exact solutions to the surface wave equations is rather small and there are no three-dimensional ones.

The first solution can be used for describing the water dynamics generated by periodic pressure source.

The second one presents the simplest unsymmetrical water waves in circular basin.

The presented formulas can be useful, in particular, for describing wave dynamics in the problems with boundary conditions at the circumferences.

The derived formulas are obtained by the method of unknown coefficients as solutions of some **overdetermined** systems algebraic linear equations. The reason for their solvability remains obscure in the moment.

References

1. Boussinesq, J. (1872): Theorie des ondes et des remous qui se propagent le long d'un canal rectangulaire horisontal, en communiquant au liquide contenu dans ce canal des vitesses sensiblement pareilles de la surface au fond. *J. Math. Pures Appl. 2nd Series* **17**, 55–108
2. Friedrichs, K. O. (1948): On the derivation of the shallow water theory. *Comm. Pure Appl. Math.* **1**, 81–85
3. Grange, J. L. de la (Lagrange) (1788): Mecanique Analitique. v. 2, Paris
4. Lamb (1932): Hydrodynamics. Sixth Ed., Cambridge Univ. Press
5. Madsen, P. A., Schäffer, H. A. (1998): Higher-order Boussinesq-type equations for surface gravity waves: derivation and analysis. *Phil. Trans. R. Soc. Lond.* **A**, **8**, 441–455
6. Mei, C. C. (1983): The Applied Dynamics of Ocean Surface Waves. Wiley
7. Shermenev, A., Shermeneva, M. (2000): Long periodic waves on an even beach. *Physical Review* **E**, No. 5, 6000–6002
8. Shermenev, A. (2001): Nonlinear periodic waves on a beach. *Geophysical and Astrophysical Fluid Dynamics*, No. 1–2, 1–14

Author Index

Lecture Notes in Computer Science

For information about Vols. 1–2654
please contact your bookseller or Springer-Verlag